FOURTH EDITION

Student Solutions Manual for COLLEGE MATHEMATICS

S. T. TAN
Stonehill College

Brooks/Cole Publishing Company
ITP® *An International Thomson Publishing Company*

Pacific Grove • Albany • Belmont • Bonn • Boston • Cincinnati • Detroit • Johannesburg • London
Madrid • Melbourne • Mexico City • New York • Paris • Singapore • Tokyo • Toronto • Washington

Sponsoring Editor: *Melissa Duge Henderson*
Editorial Assistants: *Shelley Gesicki, Joanne Von Zastrow*
Marketing Manager: *Caroline Croley*
Marketing Assistant: *Debra Johnston*
Production: *Dorothy Bell*
Cover Design: *Lisa Henry*
Cover Illustration: *Judith L. Harkness*
Printing and Binding: *Webcom Limited*

For more information, contact:

BROOKS/COLE PUBLISHING COMPANY
511 Forest Lodge Road
Pacific Grove, CA 93950
USA

International Thomson Publishing Europe
Berkshire House 168-173
High Holborn
London WC1V 7AA
England

Thomas Nelson Australia
102 Dodds Street
South Melbourne, 3205
Victoria, Australia

Nelson Canada
1120 Birchmount Road
Scarborough, Ontario
Canada M1K 5G4

International Thomson Editores
Seneca 53
Col. Polanco
11560 México, D. F., México

International Thomson Publishing GmbH
Königswinterer Strasse 418
53227 Bonn
Germany

International Thomson Publishing Asia
60 Albert Street
#15-01 Albert Complex
Singapore 189969

International Thomson Publishing Japan
Hirakawacho Kyowa Building, 3F
2-2-1 Hirakawacho
Chiyoda-ku, Tokyo 102
Japan

Printed in Canada

10 9 8 7 6 5 4 3 2 1

ISBN 0-534-36122-6

COLLEGE MATHEMATICS

CONTENTS

CHAPTER 6 SETS AND COUNTING

CHAPTER 7 PROBABILITY

CHAPTER 8 PROBABILITY DISTRIBUTIONS AND STATISTICS

CHAPTER 9 PRECALCULUS REVIEW

CHAPTER 10 FUNCTIONS, LIMITS, AND THE DERIVATIVE

CHAPTER 11 DIFFERENTIATION

CHAPTER 12 APPLICATIONS OF THE DERIVATIVE

CHAPTER 13 EXPONENTIAL AND LOGARITHMIC FUNCTIONS

CHAPTER 14 INTEGRATION

CHAPTER 15 ADDITIONAL TOPICS IN INTEGRATION

CHAPTER 16 CALCULUS OF SEVERAL VARIABLES

CHAPTER 1

EXERCISES 1.1, page 8

1. The coordinates of A are (3,3) and it is located in Quadrant I.

3. The coordinates of C are (2,-2) and it is located in Quadrant IV.

5. The coordinates of E are (-4,-6) and it is located in Quadrant III.

7. A

9. E, F, and G.

11. F

For Exercises 13-19, refer to the following figure.

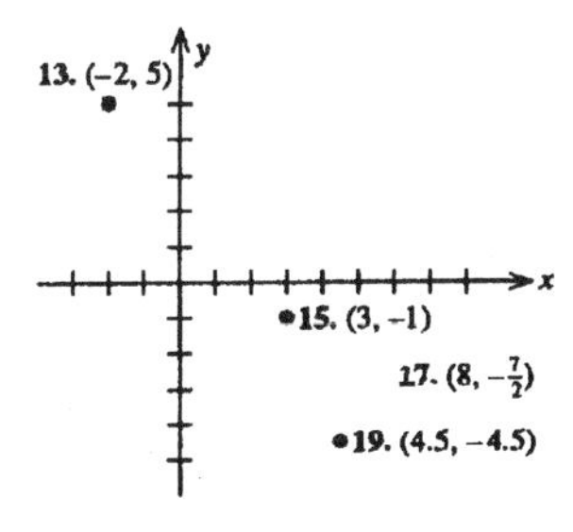

21. Using the distance formula, we find that $\sqrt{(4-1)^2+(7-3)^2}=\sqrt{3^2+4^2}=\sqrt{25}=5$.

23. Using the distance formula, we find that
$$\sqrt{(4-(-1))^2+(9-3)^2}=\sqrt{5^2+6^2}=\sqrt{25+36}=\sqrt{61}.$$

25. The coordinates of the points have the form $(x,-6)$. Since the points are 10 units away from the origin, we have
$$(x-0)^2+(-6-0)^2=10^2$$
$$x^2=64,$$
or $x=\pm 8$. Therefore, the required points are $(-8,-6)$ and $(8,-6)$.

27. The points are shown in the diagram that follows.

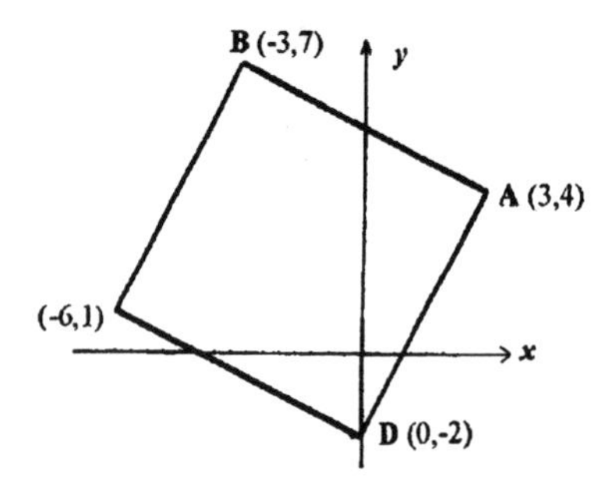

To show that the four sides are equal, we compute the following:

$$d(A,B)=\sqrt{(-3-3)^2+(7-4)^2}=\sqrt{(-6)^2+3^2}=\sqrt{45}$$

$$d(B,C)=\sqrt{[(-6-(-3)]^2+(1-7)^2}=\sqrt{(-3)^2+(-6)^2}=\sqrt{45}$$

$$d(C,D)=\sqrt{[0-(-6)]^2+[(-2)-1]^2}=\sqrt{(6)^2+(-3)^2}=\sqrt{45}$$

$$d(A,D)=\sqrt{(0-3)^2+(-2-4)^2}=\sqrt{(3)^2+(-6)^2}=\sqrt{45}.$$

Next, to show that ΔABC is a right triangle, we show that it satisfies the Pythagorean Theorem. Thus,

$$d(A,C)=\sqrt{(-6-3)^2+(1-4)^2}=\sqrt{(-9)^2+(-3)^2}=\sqrt{90}=3\sqrt{10}$$

and $[d(A,B)]^2+[d(B,C)]^2=90=[d(A,C)]^2$. Similarly, $d(B,D)=\sqrt{90}=3\sqrt{10}$, so ΔBAD is a right triangle as well. It follows that $\angle B$ and $\angle D$ are right angles, and we conclude that $ADCB$ is a square.

29. Referring to the diagram on page 9 of the text, we see that the distance from A to B is given by $d(A,B)=\sqrt{400^2+300^2}=\sqrt{250{,}000}=500$. The distance from B to C is given by

$$d(B,C)=\sqrt{(-800-400)^2+(800-300)^2}=\sqrt{(-1200)^2+(500)^2}$$
$$=\sqrt{1{,}690{,}000}=1300.$$

The distance from C to D is given by

$$d(C,D)=\sqrt{[-800-(-800)]^2+(800-0)^2}=\sqrt{0+800^2}=800.$$

The distance from D to A is given by

$$d(D,A)=\sqrt{[0-(-800)]^2+(0-0)}=\sqrt{640000}=800.$$

Therefore, the total distance covered on the tour, is

$$d(A,B)+d(B,C)+d(C,D)+d(D,A)=500+1300+800+800$$
$$=3400, \quad \text{or 3400 miles.}$$

31. Referring to the following diagram,

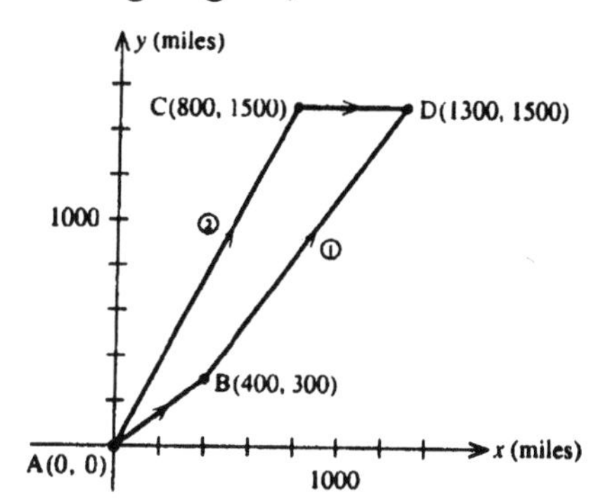

we see that the distance he would cover if he took Route (1) is given by

$$d(A,B)+d(B,D)=\sqrt{400^2+300^2}+\sqrt{(1300-400)^2+(1500-300)^2}$$
$$=\sqrt{250{,}000}+\sqrt{2{,}250{,}000}=500+1500=2000,$$

or 2000 miles. On the other hand, the distance he would cover if he took Route (2) is given by

$$d(A,C)+d(C,D)=\sqrt{800^2+1500^2}+\sqrt{(1300-800)^2}$$
$$=\sqrt{2{,}890{,}000}+\sqrt{250{,}000}=1700+500=2200,$$

or 2200 miles. Comparing these results, we see that he should take Route (1).

33. Calculations to determine VHF requirements:

$$d=\sqrt{25^2+35^2}=\sqrt{625+1225}=\sqrt{1850}\approx 43.01.$$

Models B through D satisfy this requirement.

Calculations to determine UHF requirements:

$$d=\sqrt{20^2+32^2}=\sqrt{400+1024}=\sqrt{1424}=37.74$$

Models C through D satisfy this requirement. Therefore, Model C will allow him to receive both channels at the least cost.

35. a. Let $P(x,y)$ be any point in the plane. Draw a line through P parallel to the y-axis and a line through P parallel to the x-axis (see figure).

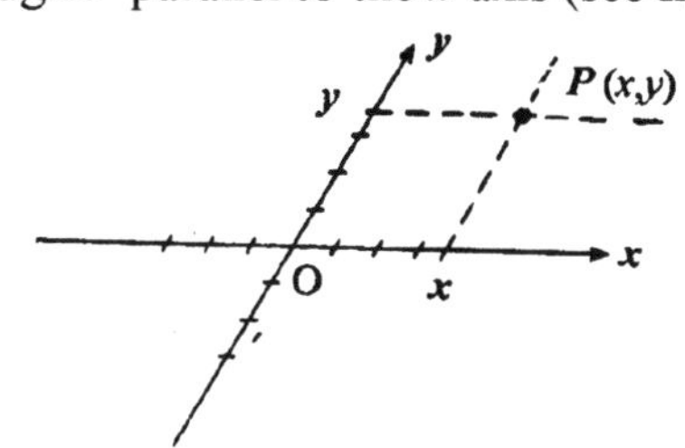

The x-coordinate of P is the number corresponding to the point on the x-axis at which the line through P hits the x-axis. Similarly y is the number that corresponds to the point on the y-axis at which the line parallel to the x-axis crosses the y-axis. To show the converse, reverse the process.
b. You can use the Pythagorean Theorem in the Cartesian Coordinate System . This simplifies the computations greatly.

EXERCISES 1.2, page 22

1. Referring to the figure shown in the text, we see that $m = \dfrac{2-0}{0-(-4)} = \dfrac{1}{2}$.

3. This is a vertical line, and hence its slope is undefined.

5. $m = \dfrac{y_2 - y_1}{x_2 - x_1} = \dfrac{8-3}{5-4} = 5.$

7. $m = \dfrac{y_2 - y_1}{x_2 - x_1} = \dfrac{8-3}{4-(-2)} = \dfrac{5}{6}.$

9. $m = \dfrac{y_2 - y_1}{x_2 - x_1} = \dfrac{d-b}{c-a}.$

11. Since the equation is in the slope-intercept form, we read off the slope $m = 4$.
    a. If x increases by 1 unit, then y increases by 4 units.
    b. If x decreases by 2 units, y decreases by $4(-2) = -8$ units.

13. The slope of the line through A and B is $\dfrac{-10-(-2)}{-3-1} = \dfrac{-8}{-4} = 2$.
    The slope of the line through C and D is $\dfrac{1-5}{-1-1} = \dfrac{-4}{-2} = 2$.
    Since the slopes of these two lines are equal, the lines are parallel.

15. The slope of the line through A and B is $\dfrac{2-5}{4-(-2)} = -\dfrac{3}{6} = -\dfrac{1}{2}$.
    The slope of the line through C and D is $\dfrac{6-(-2)}{3-(-1)} = \dfrac{8}{4} = 2$.
    Since the slopes of these two lines are the negative reciprocals of each other, the lines are perpendicular.

17. The slope of the line through the point $(1, a)$ and $(4,-2)$ is $m_1 = \dfrac{-2-a}{4-1}$ and the slope of the line through $(2,8)$ and $(-7, a+4)$ is $m_2 = \dfrac{a+4-8}{-7-2}$. Since these two lines are parallel, m_1 is equal to m_2. Therefore,

$$\frac{-2-a}{3} = \frac{a-4}{-9}$$
$$-9(-2-a) = 3(a-4)$$
$$18 + 9a = 3a - 12$$
$$6a = -30 \qquad \text{and } a = -5$$

19. An equation of a horizontal line is of the form $y = b$. In this case $b = -3$, so $y = -3$ is an equation of the line.

21. e 23. a 25. f

27. We use the point-slope form of an equation of a line with the point $(3,-4)$ and slope $m = 2$. Thus $y - y_1 = m(x - x_1)$,
and
$$y - (-4) = 2(x-3)$$
$$y + 4 = 2x - 6$$
$$y = 2x - 10.$$

29. Since the slope $m = 0$, we know that the line is a horizontal line of the form $y = b$. Since the line passes through $(-3,2)$, we see that $b = 2$, and an equation of the line is $y = 2$.

31. We first compute the slope of the line joining the points $(2,4)$ and $(3,7)$. Thus,
$$m = \frac{7-4}{3-2} = 3.$$
Using the point-slope form of an equation of a line with the point $(2,4)$ and slope $m = 3$, we find
$$y - 4 = 3(x-2)$$
$$y = 3x - 2.$$

33. We first compute the slope of the line joining the points $(1,2)$ and $(-3,-2)$. Thus,
$$m = \frac{-2-2}{-3-1} = \frac{-4}{-4} = 1.$$

Using the point-slope form of an equation of a line with the point (1,2) and slope $m = 1$, we find

$$y - 2 = x - 1$$
$$y = x + 1.$$

35. We use the slope-intercept form of an equation of a line: $y = mx + b$. Since $m = 3$, and $b = 4$, the equation is $y = 3x + 4$.

37. We use the slope-intercept form of an equation of a line: $y = mx + b$. Since $m = 0$, and $b = 5$, the equation is $y = 5$.

39. We first write the given equation in the slope-intercept form:

$$x - 2y = 0$$
$$-2y = -x$$
$$y = \tfrac{1}{2}x \quad .$$

From this equation, we see that $m = 1/2$ and $b = 0$.

41. We write the equation in slope-intercept form:

$$2x - 3y - 9 = 0$$
$$-3y = -2x + 9$$
$$y = \tfrac{2}{3}x - 3.$$

From this equation, we see that $m = 2/3$ and $b = -3$.

43. We write the equation in slope-intercept form:

$$2x + 4y = 14$$
$$4y = -2x + 14$$
$$y = -\tfrac{2}{4}x + \tfrac{14}{4}$$
$$= -\tfrac{1}{2}x + \tfrac{7}{2}.$$

From this equation, we see that $m = -1/2$ and $b = 7/2$.

45. We first write the equation $2x - 4y - 8 = 0$ in slope- intercept form:

$$2x - 4y - 8 = 0$$
$$4y = 2x - 8$$
$$y = \tfrac{1}{2}x - 2$$

Now the required line is parallel to this line, and hence has the same slope. Using the point-slope equation of a line with $m = 1/2$ and the point $(-2,2)$, we have

$$y - 2 = \tfrac{1}{2}[x - (-2)]$$
$$y = \tfrac{1}{2}x + 3.$$

47. A line parallel to the x-axis has slope 0 and is of the form $y = b$. Since the line is 6 units below the axis, it passes through (0,–6) and its equation is $y = -6$.

49. We use the point-slope form of an equation of a line to obtain
$$y - b = 0(x - a) \quad \text{or} \quad y = b.$$

51. Since the required line is parallel to the line joining (–3,2) and (6,8), it has slope
$$m = \frac{8-2}{6-(-3)} = \frac{6}{9} = \frac{2}{3}.$$
We also know that the required line passes through (–5,–4). Using the point-slope form of an equation of a line, we find
$$y - (-4) = \frac{2}{3}(x - (-5))$$
or $$y = \tfrac{2}{3}x + \tfrac{10}{3} - 4$$
that is $$y = \tfrac{2}{3}x - \tfrac{2}{3}.$$

53. Since the point (–3,5) lies on the line $kx + 3y + 9 = 0$, it satisfies the equation. Substituting $x = -3$ and $y = 5$ into the equation gives
$$-3k + 15 + 9 = 0$$
or $$k = 8.$$

55. $3x - 2y + 6 = 0$

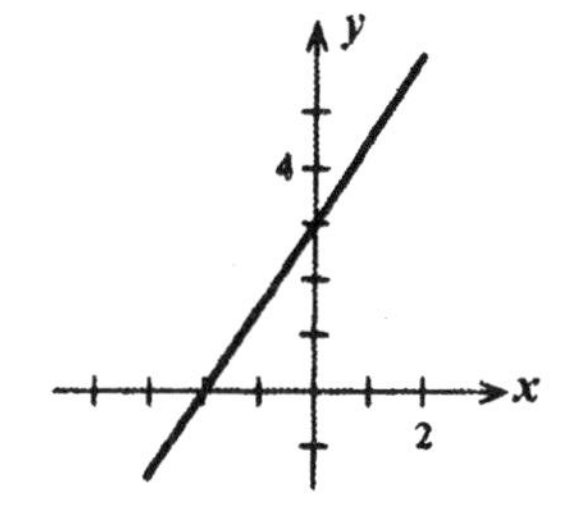

57. $x + 2y - 4 = 0$

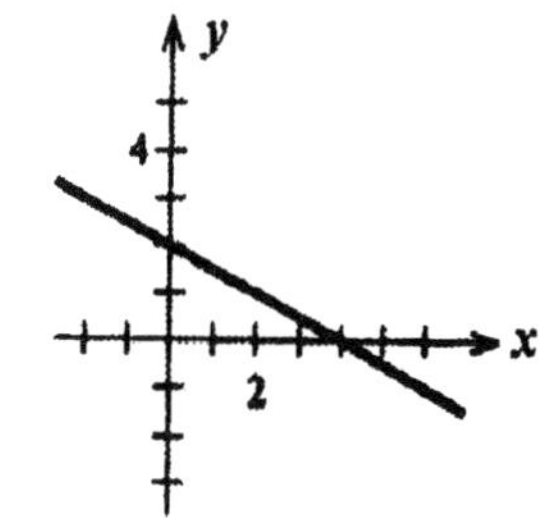

59. $y + 5 = 0$

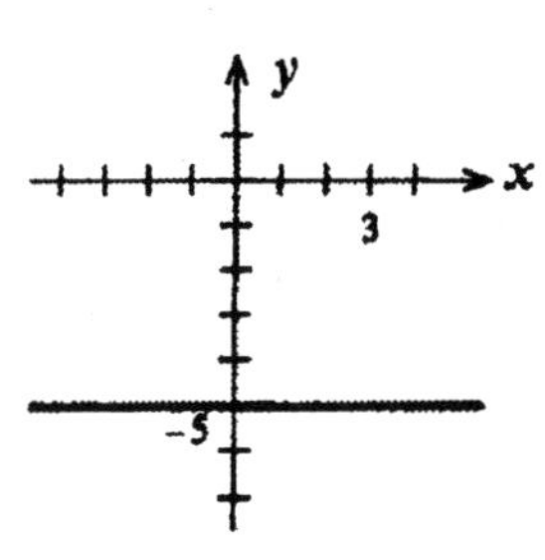

61. Since the line passes through the points $(a, 0)$ and $(0, b)$, its slope is $m = \dfrac{b-0}{0-a} = -\dfrac{b}{a}$.

Then, using the point-slope form of an equation of a line with the point $(a, 0)$ we have

$$y - 0 = -\tfrac{b}{a}(x - a)$$
$$y = -\tfrac{b}{a}x + b$$

which may be written in the form

$$\tfrac{b}{a}x + y = b.$$

Multiplying this last equation by $1/b$, we have

$$\frac{x}{a} + \frac{y}{b} = 1.$$

63. Using the equation $\dfrac{x}{a} + \dfrac{y}{b} = 1$ with $a = -2$ and $b = -4$, we have $-\dfrac{x}{2} - \dfrac{y}{4} = 1$.

Then

$$-4x - 2y = 8$$
$$2y = -8 - 4x$$
$$y = -2x - 4.$$

65. Using the equation $\dfrac{x}{a} + \dfrac{y}{b} = 1$ with $a = 4$ and $b = -1/2$, we have

$$\frac{x}{4} + \frac{y}{-\frac{1}{2}} = 1$$
$$-\tfrac{1}{4}x + 2y = -1$$
$$2y = \tfrac{1}{4}x - 1$$
$$y = \tfrac{1}{8}x - \tfrac{1}{2}.$$

67. The slope of the line passing through A and B is $m = \dfrac{7-1}{1-(-2)} = -\dfrac{6}{3} = 2$,

and the slope of the line passing through B and C is $m = \dfrac{13-7}{4-1} = \dfrac{6}{3} = 2$.

Since the slopes are equal, the points lie on the same line.

69. a. $y = 0.55x$

b. Solving the equation $1100 = 0.55x$ for x, we have $x = \dfrac{1100}{0.55} = 2000$.

71. a. – b.

c. Using the points (0,200) and (100,250), we see that the slope of the required line is $m = \dfrac{250-200}{100} = \dfrac{1}{2}$. Therefore, the required equation is

$$y - 200 = \tfrac{1}{2}x \quad \text{or } y = \tfrac{1}{2}x + 200.$$

d. The approximate cost for producing 54 units of the commodity is

$$\tfrac{1}{2}(54) + 200, \quad \text{or } \$227.$$

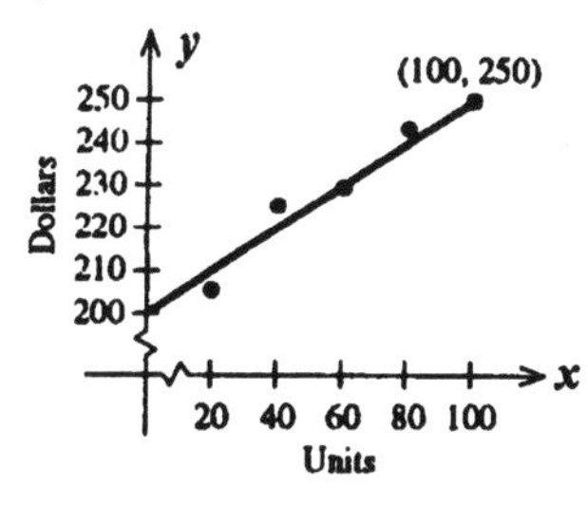

$$y = \tfrac{11}{3}(65) - 112 \ = 126.33, \quad \text{or } 126\tfrac{1}{3} \text{ pounds.}$$

73. a. – b.

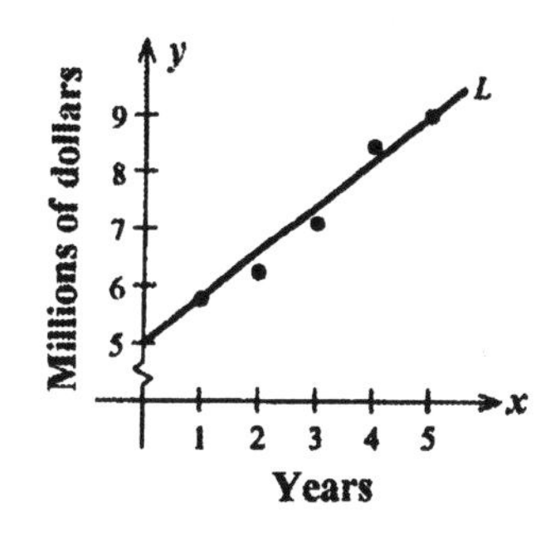

c. The slope of L is $m = \frac{9.0 - 5.8}{5 - 1} = \frac{3.2}{4} = 0.8$. Using the point-slope form of an equation of a line, we have $y - 5.8 = 0.8(x - 1) = 0.8x - 0.8$, or $y = 0.8x + 5$.

d. Using the equation of part (c) with $x = 9$, we have

$y = 0.8(9) + 5 = 12.2$, or \$12.2 million.

75. a. We obtain a family of parallel lines each having slope m.

b. We obtain a family of straight lines all of which pass through the point $(0,b)$.

77. The slope of L_1 is $m_1 = \frac{b - 0}{1 - 0} = b$. The slope of L_2 is $m_2 = \frac{c - 0}{1 - 0} = c$.

Applying the Pythagorean theorem to ΔOAC and ΔOCB gives

$$(OA)^2 = 1^2 + b^2 \quad \text{and} \quad (OB)^2 = 1^2 + c^2.$$

Adding these equations and applying the Pythagorean theorem to ΔOBA gives

$$(AB)^2 = (OA)^2 + (OB)^2 = 1^2 + b^2 + 1^2 + c^2 = 2 + b^2 + c^2.$$

Also $(AB)^2 = (b - c)^2$.

Therefore,
$$(b - c)^2 = 2 + b^2 + c^2$$
$$b^2 - 2bc + c^2 = 2 + b^2 + c^2$$
$$-2bc = 2, \; 1 = -bc.$$

Next, $m_1 m_2 = b \cdot c = bc = -1$, as was to be shown.

USING TECHNOLOGY EXERCISES 1.2, page 29

1.

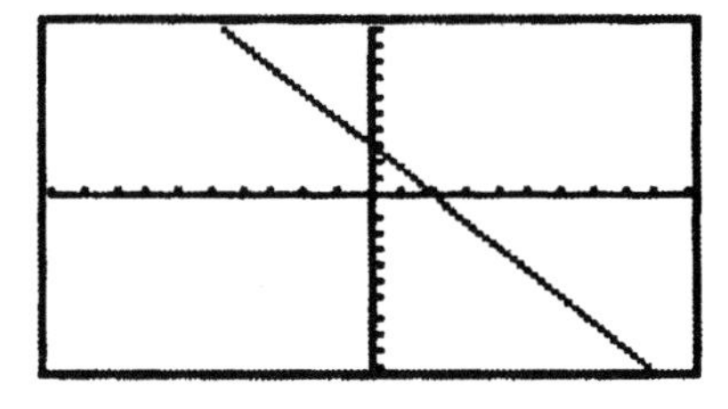

3.

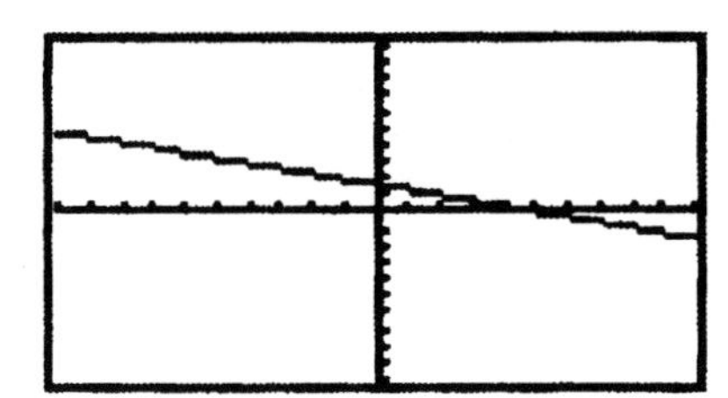

5.

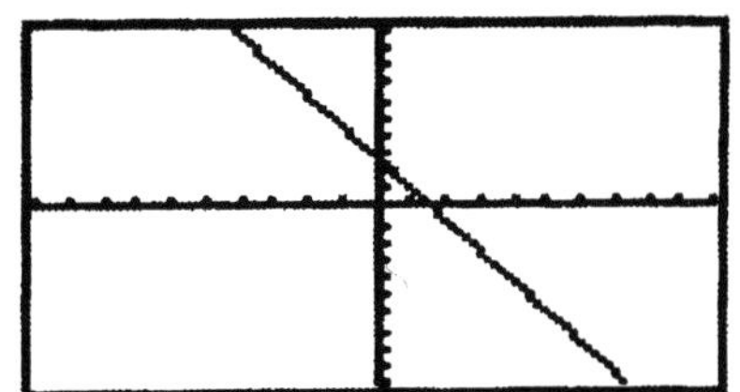
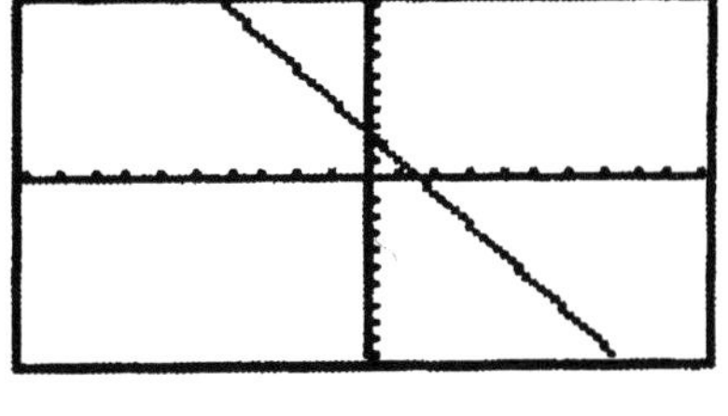

7. a.

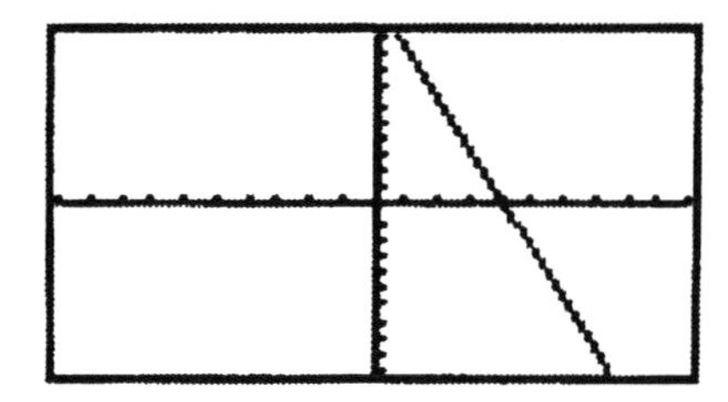

b.

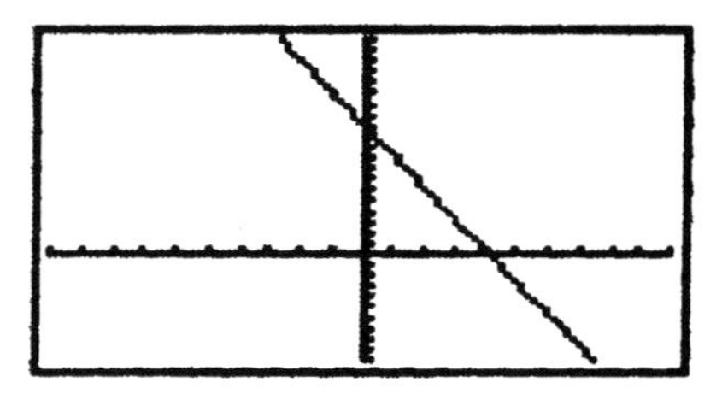

9. a.

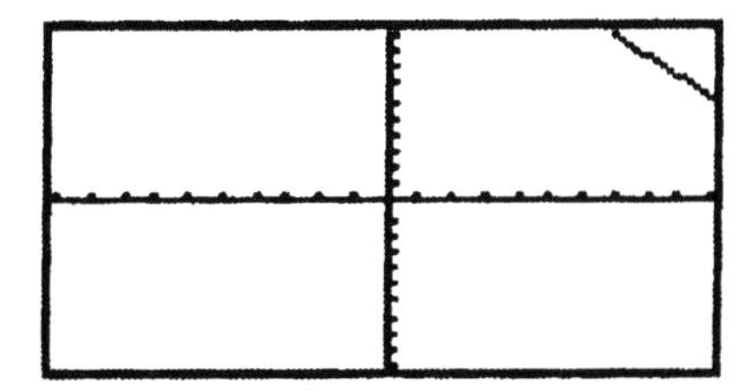

b.

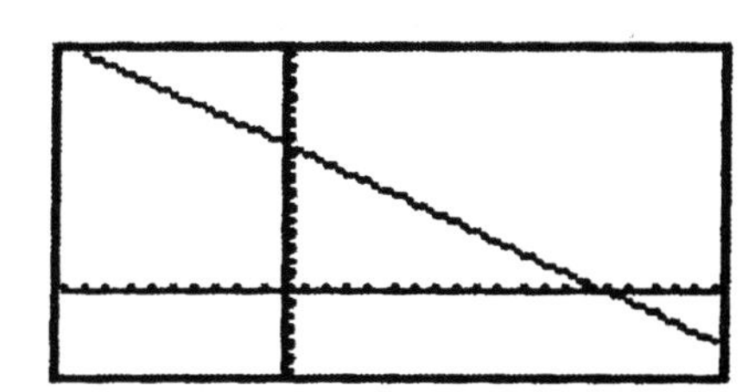

11.

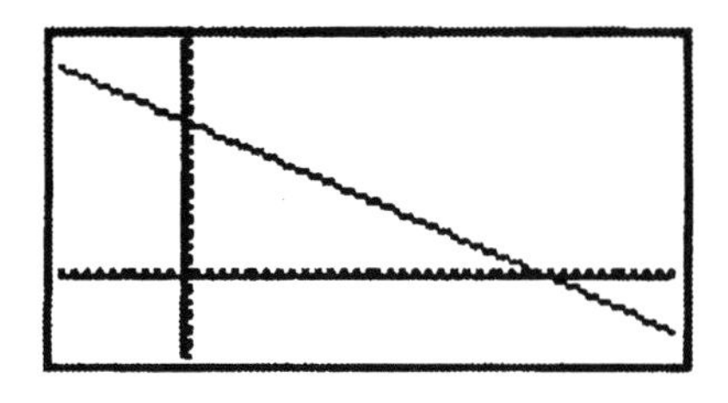

13.

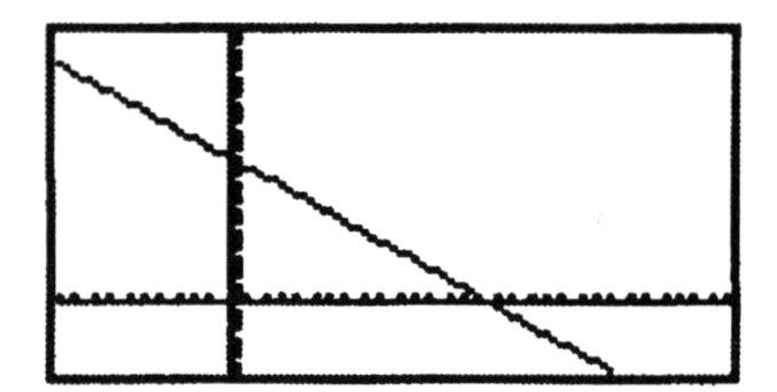

15.

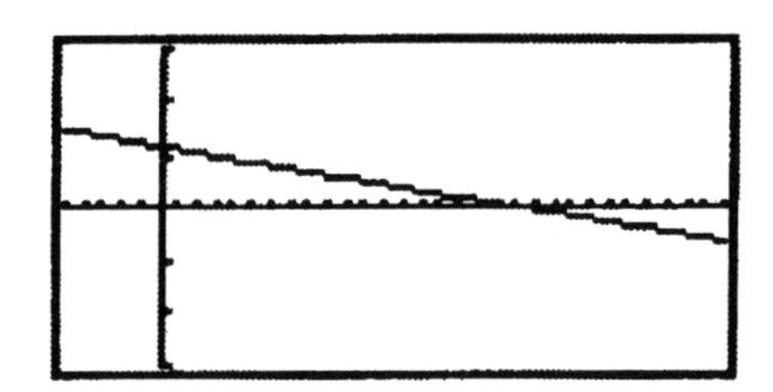

17.

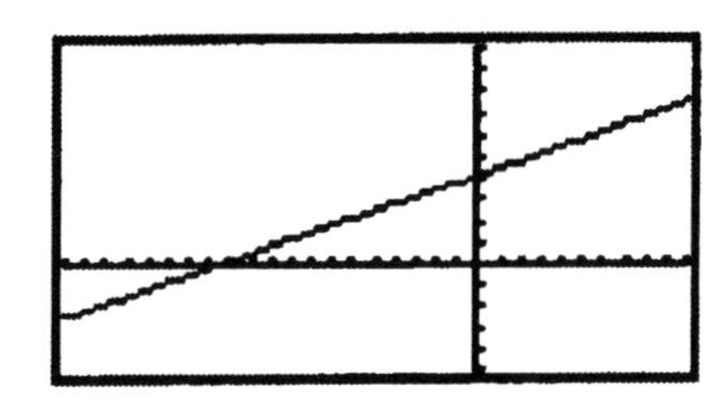

EXERCISES 1.3, page 38

1. Yes. Solving for y in terms of x, we find $3y = -2x + 6$, or $y = -\frac{2}{3}x + 2$.

3. Yes. Solving for y in terms of x, we find $2y = x + 4$, or $y = \frac{1}{2}x + 2$.

5. Yes. Solving for y in terms of x, we have $4y = 2x + 9$, or $y = \frac{1}{2}x + \frac{9}{4}$.

7. y is not a linear function of x because of the quadratic term $2x^2$.

9. y is not a linear function of x because of the term $-3y^2$.

11. a. $C(x) = 8x + 40{,}000$, where x is the number of units produced.
    b. $R(x) = 12x$, where x is the number of units sold.
    c. $P(x) = R(x) - C(x) = 12x - (8x + 40{,}000) = 4x - 40{,}000$.
    d. $P(8{,}000) = 4(8{,}000) - 40{,}000 = -8{,}000$, or a loss of \$8,000.
    $P(12{,}000) = 4(12{,}000) - 40{,}000 = 8{,}000$ or a profit of \$8,000.

13. $f(0) = 2$ gives $m(0) + b = 2$, or $b = 2$. So, $f(x) = mx + 2$. Next, $f(3) = -1$ gives $m(3) + 2 = -1$, or $m = -1$.

15. Let V be the book value of the office building after 1995. Since $V = 1{,}000{,}000$ when $t = 0$, the line passes through (0, 1,000,000). Similarly, when $t = 50$, $V = 0$, so the line passes through (50, 0). Then the slope of the line is given by
$$m = \frac{0 - 1{,}000{,}000}{50 - 0} = -20{,}000.$$
Using the point-slope form of the equation of a line with the point (0, 1,000,000), we have $V - 1{,}000{,}000 = -20{,}000(t - 0)$,

or $\quad V = -20{,}000t + 1{,}000{,}000.$

In 2000, $t = 5$ and $V = -20{,}000(5) + 1{,}000{,}000 = 900{,}000$, or \$900,000.

In 2005, $t = 10$ and $V = -20{,}000(10) + 1{,}000{,}000 = 800{,}000$, or \$800,000.

17. The consumption function is given by $C(x) = 0.75x + 6$. When $x = 0$, we have $C(0) = 0.75(0) + 6 = 6$, or \$6 billion dollars.
    When $x = 50$, $\quad C(50) = 0.75(50) + 6 = 43.5$, or \$43.5 billion dollars.

    When $x = 100$, $\quad C(100) = 0.75(100) + 6 = 81$, or \$81 billion dollars.

19. a. $y = 1.053x$, where x is the monthly benefit before adjustment, and y is the adjusted monthly benefit.
    b. His adjusted monthly benefit will be $(1.053)(620) = 652.86$, or \$652.86.

21. Let the number of tapes produced and sold be x. Then

$$C(x) = 12{,}100 + 0.60x$$
$$R(x) = 1.15x$$

and

$$P(x) = R(x) - C(x) = 1.15x - (12{,}100 + 0.60x)$$
$$= 0.55x - 12{,}100.$$

23. Let the value of the minicomputer after t years be V. When $t = 0$, $V = 60{,}000$ and when $t = 4$, $V = 12{,}000$.

a. Since $\quad m = \dfrac{12{,}000 - 60{,}000}{4} = -\dfrac{48{,}000}{4} = -12{,}000$

the rate of depreciation $(-m)$ is \$12,000.

b. Using the point-slope form of the equation of a line with the point (4, 12,000), we have $\quad V - 12{,}000 = -12{,}000(t - 4)$

or $\quad V = -12{,}000t + 60{,}000.$

c.

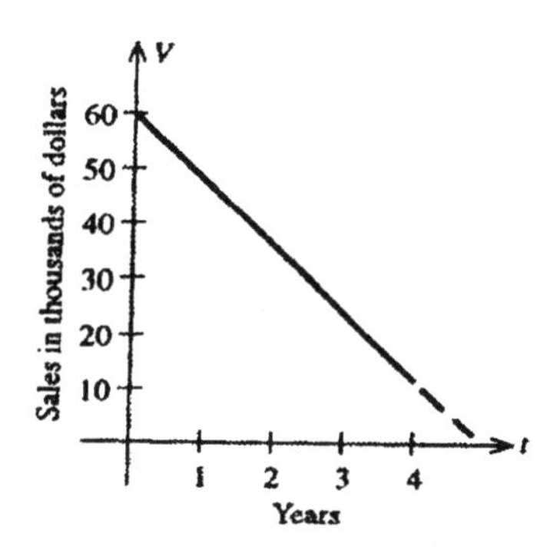

d. When $t = 3$, $V = -12{,}000(3) + 60{,}000 = 24{,}000$, or \$24,000.

25. The formula given in Exercise 24 is $V = C - \frac{C-S}{N}t$.

Then, when C = 1,000,000, N = 50, and S = 0, we have

$$V = 1{,}000{,}000 - \frac{1{,}000{,}000 - 0}{50}t \quad \text{or} \quad V = 1{,}000{,}000 - 20{,}000t.$$

In 2000, t = 5 and V = 1,000,000 – 20,000(5) = 900,000, or \$900,000.
In 2005, t = 10 and V = 1,000,000 – 20,000(10) = 800,000 or \$800,000.

27. a. $D(S) = \frac{Sa}{1.7}$. If we think of D as having the form $D(S) = mS + b$, then

$m = \frac{a}{1.7}$, $b = 0$, and D is a linear function of S.

b. $D(0.4) = \frac{500(0.4)}{1.7} = 117.647$, or approximately 117.65 mg.

29. a. Since the relationship is linear, we can write $F = mC + b$, where m and b are constants. Using the condition $C = 0$ when $F = 32$, we have $32 = b$, and so $F = mC + 32$. Next, using the condition $C = 100$ when $F = 212$, we have

$$212 = 100m + 32 \quad \text{or} \quad m = \tfrac{9}{5}.$$

Therefore, $F = \frac{9}{5}C + 32$.

b. From (a), we see $F = \frac{9}{5}C + 32$. Next, when $C = 20$,

$$F = \tfrac{9}{5}(20) + 32 = 68$$

and so the temperature equivalent to 20°C is 68°F.

c. Solving for C in terms of F, we find $\frac{9}{5}C = F - 32$, or $C = \frac{5}{9}F - \frac{160}{9}$.
When $F = 70$, $C = \frac{5}{9}(70) - \frac{160}{9} = \frac{190}{9}$, or approximately 21.1°C.

31. The slope of L_2 is greater than that of L_1. This tells us that if the manufacturer lowers the unit price for each model clock radio by the same amount, the quantity demanded of model B radios will be greater than that of the model A radios.

33. a. Setting $x = 0$, gives $3p = 18$, or $p = 6$.
Next, setting $p = 0$, gives $2x = 18$, or $x = 9$.

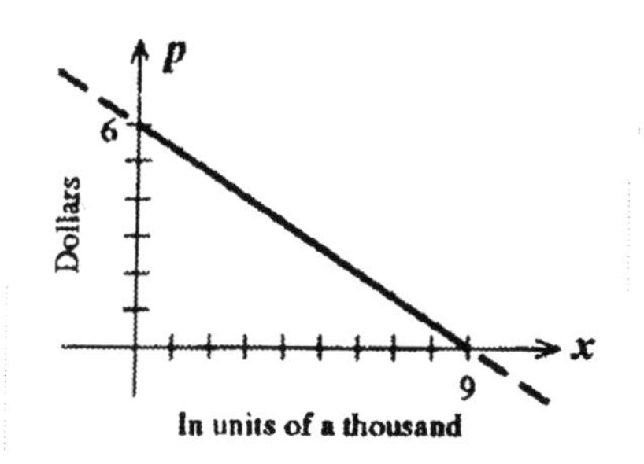

b. When $p = 4$, $2x + 3(4) - 18 = 0$

$$2x = 18 - 12 = 6$$

and $x = 3$. Therefore, the quantity demanded when $p = 4$ is 3000. (Remember x is given in units of a thousand.)

35. a. When $x = 0$, $p = 60$ and when $p = 0$, $-3x = -60$, or $x = 20$.

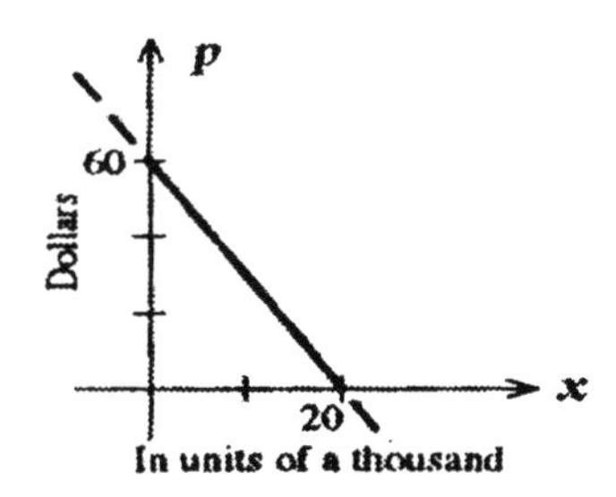

b. When $p = 30$, $30 = -3x + 60$

$3x = 30$ and $x = 10$.

Therefore, the quantity demanded when $p = 30$ is 10,000 units.

37. When $x = 1000$, $p = 55$, and when $x = 600$, $p = 85$. Therefore, the graph of the linear demand equation is the straight line passing through the points (1000, 55) and (600, 85). The slope of the line is

$$\frac{85-55}{600-1000} = -\frac{3}{40}.$$

Using the point (1000, 55) and the slope just found, we find that the required equation is $p - 55 = -\frac{3}{40}(x - 1000)$

$$p = -\frac{3}{40}x + 130$$

When $x = 0$, $p = 130$ which means that there will be no demand above \$130. When $p = 0$, $x = 1733$, which means that 1733 units is the maximum quantity demanded.

39. Since the demand equation is linear, we know that the line passes through the points (1000,9) and (6000,4). Therefore, the slope of the line is given by

$$m = \frac{4-9}{6000-1000} = -\frac{5}{5000} = -0.001.$$

Since the equation of the line has the form $p = ax + b$,

$$9 = -0.001(1000) + b \quad \text{or} \quad b = 10.$$

Therefore, the equation of the line is

$$p = -0.001x + 10.$$

If $p = 7.50$, $7.50 = -0.001x + 10$

$$0.001x = 2.50$$

or $$x = 2500.$$

So, the quantity demanded when the unit price is \$7.50 is 2500 units.

41. a. Setting $x = 0$, we obtain

$$3(0) - 4p + 24 = 0$$
$$-4p = -24$$

or $$p = 6.$$

Setting $p = 0$, we obtain

$$3x - 4(0) + 24 = 0$$
$$3x = -8$$

or $$x = -8/3.$$

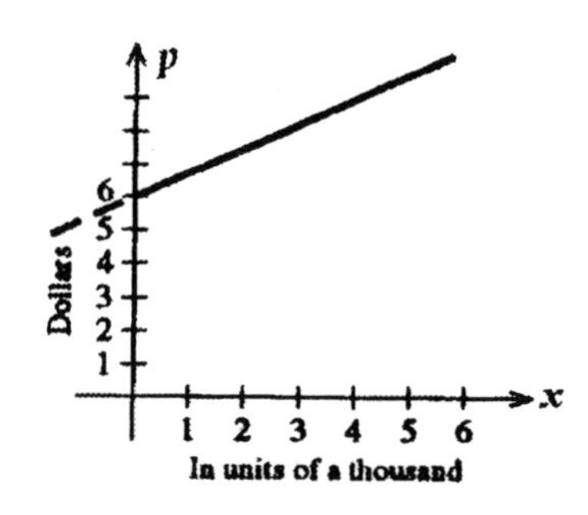

b. When $p = 8$,

$$3x - 4(8) + 24 = 0$$
$$3x = 32 - 24 = 8$$
$$x = 8/3.$$

Therefore, 2667 units of the commodity would be supplied at a unit price of \$8. (Here again x is measured in units of thousands.)

43. a. When $x = 0$, $p = 10$, and when $p = 0$, $x = -5$.

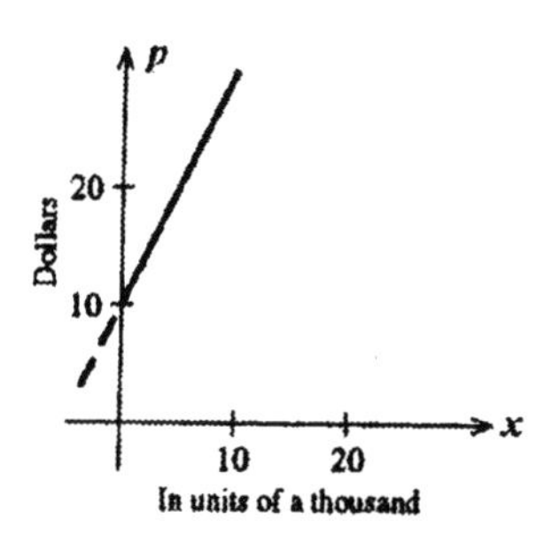

b. $p = 2x + 10$, $14 = 2x + 10$, $2x = 4$, and $x = 2$. Therefore, when $p = 14$, the supplier will make 2000 units of the commodity available.

45. When $x = 10{,}000$, $p = 45$ and when $x = 20{,}000$, $p = 50$. Therefore, the slope of the line passing $(10{,}000, 45)$ and $(20{,}000, 50)$ is

$$m = \frac{50 - 45}{20{,}000 - 10{,}000} = \frac{5}{10{,}000} = 0.0005$$

Using the point- slope form of an equation of a line with the point $(10{,}000, 45)$, we have

$$p - 45 = 0.0005(x - 10{,}000)$$
$$p = 0.0005x - 5 + 45$$

or

$$p = 0.0005x + 40.$$

If $p = 70$, $\quad 70 = 0.0005x + 40$

$$0.0005x = 30 \quad \text{or} \quad x = \frac{30}{0.0005} = 60{,}000 .$$

USING TECHNOLOGY EXERCISES 1.3, page 43

1. 2.2875
3. 2.880952381
5. 7.2851648352
7. 2.4680851064

EXERCISES 1.4, page 52

1. We solve the system $y = 3x + 4$
$$y = -2x + 14.$$
Substituting the first equation into the second yields
$$3x + 4 = -2x + 14$$
$$5x = 10,$$
and $x = 2$. Substituting this value of x into the first equation yields
$$y = 3(2) + 4,$$
or $y = 10$. Thus, the point of intersection is (2,10).

3. We solve the system $2x - 3y = 6$
$$3x + 6y = 16.$$
Solving the first equation for y, we obtain
$$3y = 2x - 6$$
$$y = \tfrac{2}{3}x - 2 \ .$$
Substituting this value of y into the second equation, we obtain
$$3x + 6(\tfrac{2}{3}x - 2) = 16$$
$$3x + 4x - 12 = 16$$
$$7x = 28$$
and $x = 4.$
Then $y = \tfrac{2}{3}(4) - 2 = \tfrac{2}{3}.$
Therefore, the point of intersection is $(4, \tfrac{2}{3})$.

5. We solve the system $\begin{cases} y = -\tfrac{1}{4}x - 5 \\ 2x - \tfrac{3}{2}y = 1 \end{cases}$. Substituting the value of y given in the first equation into the second equation, we obtain
$$2x - \tfrac{3}{2}(\tfrac{1}{4}x - 5) = 1$$
$$2x - \tfrac{3}{8}x + \tfrac{15}{2} = 2$$
$$16x - 3x + 60 = 8$$
$$13x = -52,$$
or $x = -4$. Substituting this value of x in the first equation, we have
$$y = \tfrac{1}{4}(-4) - 5 = -1 - 5,$$
or $y = -6$. Therefore, the point of intersection is (–4,–6).

7. We solve the equation $R(x) = C(x)$, or $15x = 5x + 10{,}000$, obtaining $10x = 10{,}000$, or $x = 1000$. Substituting this value of x into the equation $R(x) = 15x$, we find $R(1000) = 15{,}000$. Therefore, the breakeven point is (1000, 15,000).

9. We solve the equation $R(x) = C(x)$, or $0.4x = 0.2x + 120$, obtaining $0.2x = 120$, or $x = 600$. Substituting this value of x into the equation $R(x) = 0.4x$, we find $R(600) = 240$. Therefore, the breakeven point is (600, 240).

11. a.

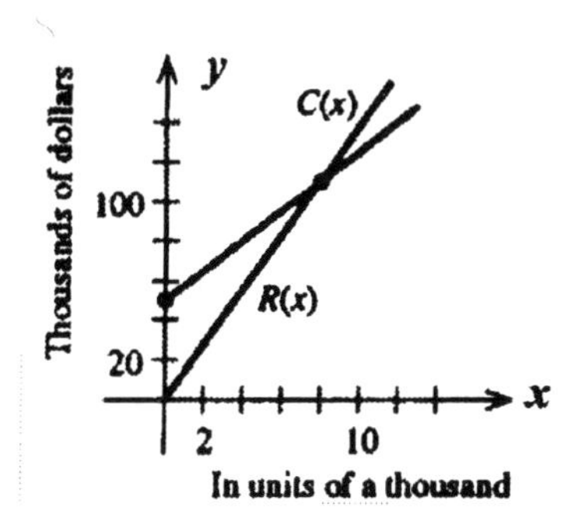

b. We solve the equation $R(x) = C(x)$ or $14x = 8x + 48{,}000$, obtaining $6x = 48{,}000$ or $x = 8000$. Substituting this value of x into the equation $R(x) = 14x$, we find $R(8000) = 14(8000) = 112{,}000$. Therefore, the breakeven point is (8000, 112,000).

c.

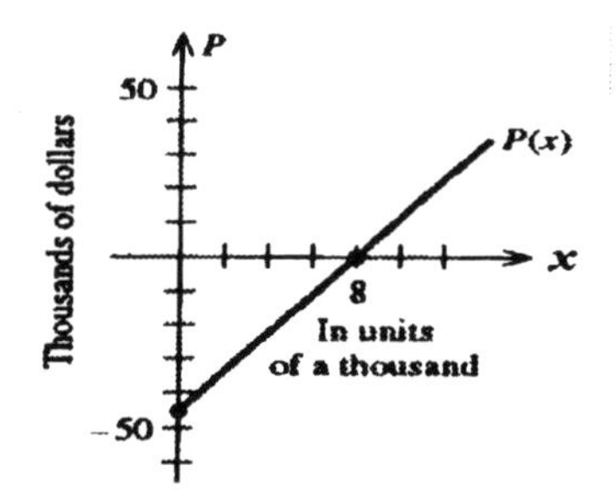

d. $P(x) = R(x) - C(x) = 14x - 8x - 48{,}000 = 6x - 48{,}000$.
The graph of the profit function crosses the x-axis when $P(x) = 0$, or $6x = 48{,}000$ and $x = 8000$. This means that the revenue is equal to the cost when 8000 units are produced and consequently the company breaks even at this point.

13. Let x denote the number of units sold. Then, the revenue function R is given by

$$R(x) = 9x.$$

Since the variable cost is 40 percent of the selling price and the monthly fixed costs are \$50,000, the cost function C is given by

$$C(x) = 0.4(9x) + 50{,}000$$
$$= 3.6x + 50{,}000.$$

To find the breakeven point, we set $R(x) = C(x)$, obtaining

$$9x = 3.6x + 50{,}000$$
$$5.4x = 50{,}000$$
$$x \approx 9259, \text{ or } 9259 \text{ units.}$$

Substituting this value of x into the equation $R(x) = 9x$ gives

$$R(9259) = 9(9259) = 83{,}331.$$

Thus, for a breakeven operation, the firm should manufacture 9259 bicycle pumps resulting in a breakeven revenue of $83,331.

15. a. The cost function associated with using machine *I* is given by

$$C_1(x) = 18{,}000 + 15x.$$

The cost function associated with using machine II is given by

$$C_2(x) = 15{,}000 + 20x.$$

b.

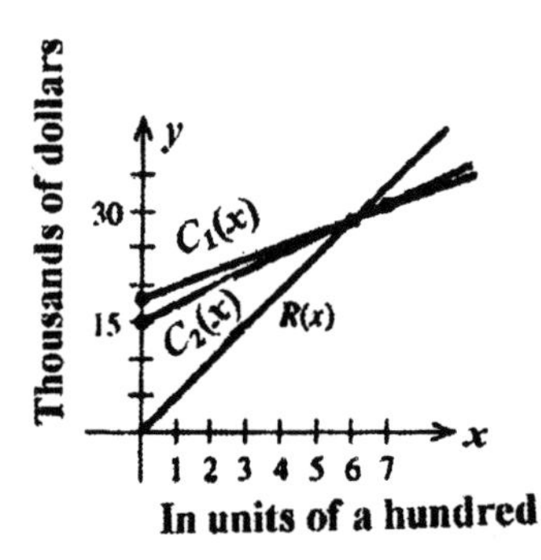

c. Comparing the cost of producing 450 units on each machine, we find

$$C_1(450) = 18{,}000 + 15(450)$$
$$= 24{,}750 \quad \text{or } \$24{,}750 \text{ on machine } I,$$

and

$$C_2(450) = 15{,}000 + 20(450)$$
$$= 24{,}000 \text{ or } \$24{,}000 \text{ on machine } II.$$

Therefore, machine *II* should be used in this case.

Next, comparing the costs of producing 550 units on each machine, we find

$$C_1(550) = 18{,}000 + 15(550)$$
$$= 26{,}250 \text{ or } \$26{,}250 \text{ on machine } I,$$

and

$$C_2(550) = 15{,}000 + 20(550)$$
$$= 26{,}000$$

on machine *II*. Therefore, machine *II* should be used in this instance. Once again, we compare the cost of producing 650 units on each machine and find that

$$C_1(650) = 18{,}000 + 15(650)$$
$$= 27{,}750, \quad \text{or } \$27{,}750 \text{ on machine } I \text{ and}$$
$$C_2(650) = 15{,}000 + 20(650)$$
$$= 28{,}000,$$

or \$28,000 on machine *II*. Therefore, machine *I* should be used in this case.

d. We use the equation $P(x) = R(x) - C(x)$ and find

$$P(450) = 50(450) - 24{,}000 = -1500,$$

or a loss of \$1500 when machine *I* is used to produce 450 units. Similarly,

$$P(550) = 50(550) - 26{,}000 = 1500,$$

or a profit of \$1500 when machine *II* is used to produce 550 units.

Finally, $P(650) = 50(650) - 27{,}750 = 4750$,

or a profit of \$4750 when machine *I* is used to produce 650 units.

17. We solve the system

$$\begin{aligned} 4x + 3p &= 59 \\ 5x - 6p &= -14. \end{aligned}$$

Solving the first equation for p, we find $p = -\frac{4}{3}x + \frac{59}{3}$.

Substituting this value of p into the second equation, we have

$$\begin{aligned} 5x - 6(-\tfrac{4}{3}x + \tfrac{59}{3}) &= -14 \\ 5x + 8x - 118 &= -14 \\ 13x &= 104 \\ x &= 8. \end{aligned}$$

Substituting this value of x into the equation

$$p = -\tfrac{4}{3}x + \tfrac{59}{3}$$

we have $$p = -\tfrac{4}{3}(8) + \tfrac{59}{3} = \tfrac{27}{3} = 9$$

Thus, the equilibrium quantity is 8000 units and the equilibrium price is \$9.

19. We solve the system $p = -2x + 22$

$$p = 3x + 12 \;.$$

Substituting the first equation into the second, we find

$$\begin{aligned} -2x + 22 &= 3x + 12 \\ 5x &= 10 \end{aligned}$$

and $$x = 2.$$

Substituting this value of x into the first equation, we obtain

$$p = -2(2) + 22 = 18.$$

Thus, the equilibrium quantity is 2000 units and the equilibrium price is \$18.

21. Let x denote the number of VCR's produced per week, and p denote the price of each VCR.

a. The slope of the demand curve is given by $\dfrac{\Delta p}{\Delta x} = -\dfrac{20}{250} = -\dfrac{2}{25}$.

Using the point-slope form of the equation of a line with the point (3000, 485), we have $$p - 485 = -\tfrac{2}{25}(x - 3000)$$

$$p = -\tfrac{2}{25}x + 240 + 485$$

or $$p = -0.08x + 725.$$

b. From the given information, we know that the graph of the supply equation passes through the points (0, 300) and (2500, 525). Therefore, the slope of the supply curve is $$m = \frac{525-300}{2500-0} = \frac{225}{2500} = 0.09.$$

Using the point-slope form of the equation of a line with the point (0, 300), we find that $$p - 300 = 0.09x$$
$$p = 0.09x + 300.$$

c. Equating the supply and demand equations, we have
$$-0.08x + 725 = 0.09x + 300$$
$$0.17x = 425$$
or $$x = 2500.$$
Then $$p = -0.08(2500) + 725 = 525.$$
We conclude that the equilibrium quantity is 2500 and the equilibrium price is \$525.

23. We solve the system
$$3x + p = 1500$$
$$2x - 3p = -1200.$$
Solving the first equation for p, we obtain
$$p = 1500 - 3x.$$
Substituting this value of p into the second equation, we obtain
$$2x - 3(1500 - 3x) = -1200$$
$$11x = 3300$$
or $$x = 300.$$
Next, $$p = 1500 - 3(300) = 600.$$
Thus, the equilibrium quantity is 300 and the equilibrium price is \$600.

25. a. We solve the system of equations $p = cx + d$ and $p = ax + b$. Substituting the first into the second gives
$$cx + d = ax + b$$
$$(c - a)x = b - d$$
or $$x = \frac{b-d}{c-a}.$$
Since $a < 0$ and $c > 0$, $c - a \neq 0$ and x is well-defined. Substituting this value of x into the second equation, we obtain
$$p = a\left(\frac{b-d}{c-a}\right) + b = \frac{ab - ad + bc - ab}{c-a} = \frac{bc-ad}{c-a}.$$

Therefore, the equilibrium quantity is $\frac{b-d}{c-a}$ and the equilibrium price is $\frac{bc-ad}{c-a}$.

b. If c is increased, the denominator in the expression for x increases and so x gets smaller. At the same time, the first term in the first equation for p decreases and so p gets larger. This analysis shows that if the unit price for producing the product is increased then the equilibrium quantity decreases while the equilibrium price increases.

c. If b is decreased, the numerator of the expression for x decreases while the denominator stays the same. Therefore x decreases. The expression for p also shows that p decreases. This analysis shows that if the (theoretical) upper bound for the unit price of a commodity is lowered, then both the equilibrium quantity and the equilibrium price drop.

27. Solving the two equations simultaneously to find the point(s) of intersection of L_1 and L_2, we obtain

$$m_1x + b_1 = m_2x + b_2$$
$$(m_1 - m_2)x = b_2 - b_1 \qquad (1)$$

a. If $m_1 = m_2$ and $b_2 \neq b_1$, then there is no solution for (1) and in this case L_1 and L_2 do not intersect.

b. If $m_1 \neq m_2$, then Equation (1) can be solved (uniquely) for x and this shows that L_1 and L_2 intersect at precisely one point.

c. If $m_1 = m_2$ and $b_1 = b_2$, then (1) is satisfied for all values of x and this shows that L_1 and L_2 intersect at infinitely many points.

USING TECHNOLOGY EXERCISES 1.4, page 55

1. (0.6, 6.2) 3. (3.8261, 0.1304) 5. (386.9091, 145.3939)

EXERCISES 1.5, page 61

1. a. We first summarize the data:

x	y	x^2	xy
1	4	1	4
2	6	4	12
3	8	9	24
4	11	16	44
10	29	30	84

The normal equations are $4b + 10m = 29$
$10b + 30m = 84.$
Solving this system of equations, we obtain $m = 2.3$ and $b = 1.5$. So an equation is $y = 2.3x + 1.5$.

b. The scatter diagram and the least squares line for this data follow:

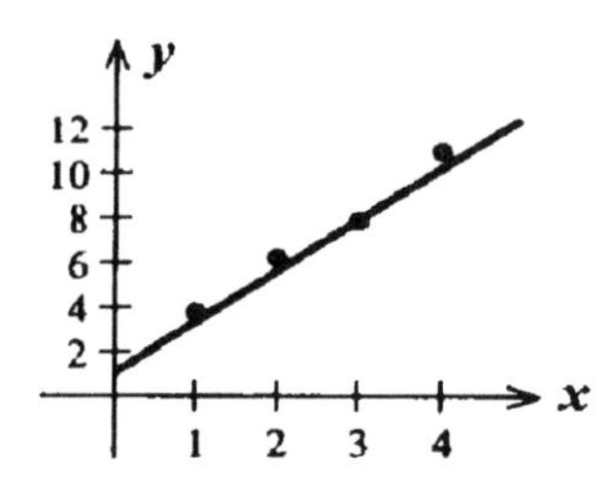

3. a. We first summarize the data:

x	y	x^2	xy
1	4.5	1	4.5
2	5	4	10
3	3	9	9
4	2	16	8
4	3.5	16	14
6	1	36	6
20	19	82	51.5

The normal equations are $6b + 20m = 19$

$20b + 82m = 51.5.$
The solutions are $m \approx -0.7717$ and $b \approx 5.7391$ and so a required equation is $y = -0.772x + 5.739$.
b. The scatter diagram and the least-squares line for these data follow.

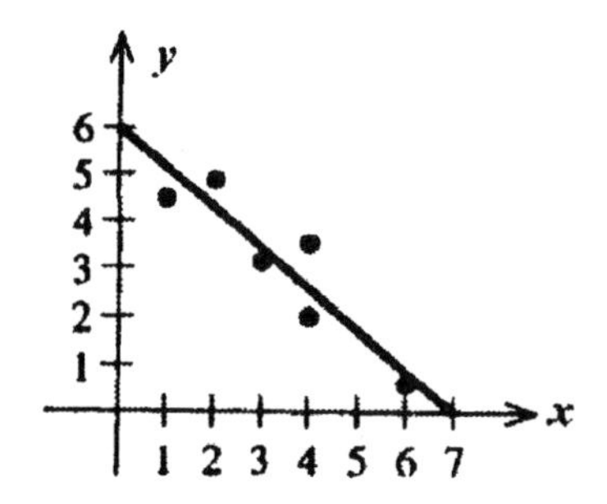

5. a. We first summarize the data:

x	y	x^2	xy
1	3	1	3
2	5	4	10
3	5	9	15
4	7	16	28
5	8	25	40
15	28	55	96

The normal equations are

$$55m + 15b = 96$$
$$15m + 5b = 28.$$

Solving, we find $m = 1.2$ and $b = 2$, so that the required equation is $y = 1.2x + 2$.

b. The scatter diagram and the least-squares line for the given data follow.

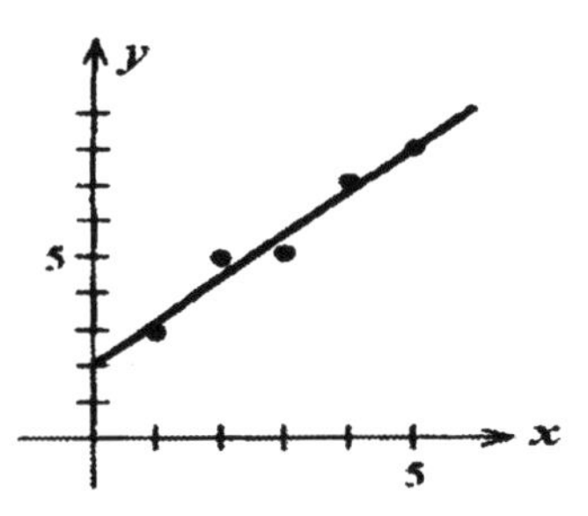

7. a. We first summarize the data:

x	y	x^2	xy
4	0.5	16	2
4.5	0.6	20.25	2.7
5	0.8	25	4
5.5	0.9	30.25	4.95
6	1.2	36	7.2
25	4	127.5	20.85

The normal equations are

$$5b + 25m = 4$$
$$25b + 127.5m = 20.85.$$

The solutions are $m = 0.34$ and $b = -0.9$, and so a required equation is $y = 0.34x - 0.9$.

b. The scatter diagram and the least-squares line for these data follow.

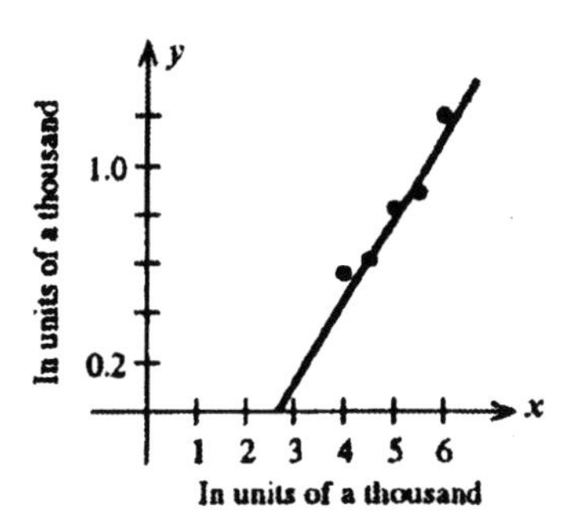

c. If $x = 6.4$, then $y = 0.34(6.4) - 0.9 = 1.276$ and so 1276 completed applications might be expected.

9. a. We first summarize the data:

x	y	x^2	xy
1	436	1	436
2	438	4	876
3	428	9	1284
4	430	16	1720
5	426	25	2130
15	2158	55	6446

The normal equations are

$$5b + 15m = 2158$$
$$15b + 55m = 6446.$$

Solving this system, we find $m = -2.8$ and $b = 440$.
Thus, the equation of the least-squares line is $y = -2.8x + 440$.

b. The scatter diagram and the least-squares line for this data are shown in the figure that follows.

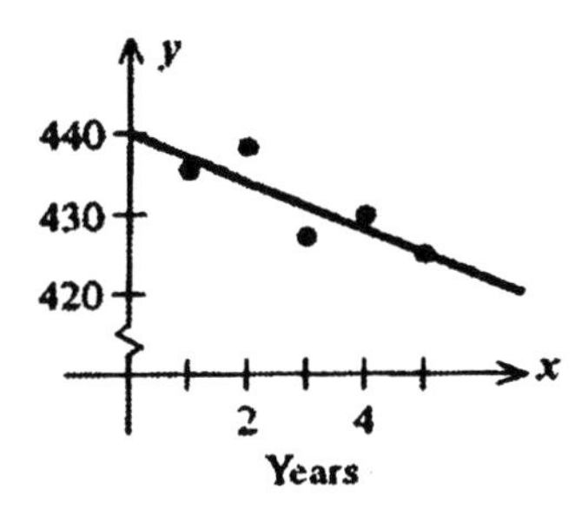

c. Two years from now, the average SAT verbal score in that area will be $y = -2.8(7) + 440 = 420.4$.

11. a. We first summarize the data:

x	y	x^2	xy
0	168	0	0
10	213	100	2130
20	297	400	5940
30	374	900	11220
40	427	1600	17080
51	467	2601	23817
151	1946	5601	60187

The normal equations are

$$6b + 151m = 1946$$
$$151b + 5601m = 60187.$$

The solutions are $m = 6.2264$ and $b = 167.636$ and so a required equation is $y = 6.226x + 167.6$.

b. In 2000, $x = 60$, $y = 6.226(90) + 167.6 \approx 541$. Hence, the expected size of the average farm will be 541 acres.

13. a. We first summarize the data:

x	y	x^2	xy
1	20	1	20
2	24	4	48
3	26	9	78
4	28	16	112
5	32	25	160
15	130	55	418

The normal equations are $5b + 15m = 130$

$$15b + 55m = 418.$$

The solutions are $m = 2.8$ and $b = 17.6$, and so an equation of the line is

$$y = 2.8x + 17.6.$$

b. When $x = 8$, $y = 2.8(8) + 17.6 = 40$. Hence, the state subsidy is expected to be \$40 million for the eighth year.

15. a. We first summarize the data:

x	y	x^2	xy
1	16.7	1	16.7
3	26	9	78
5	33.3	25	166.5
7	48.3	49	338.1
9	57	81	513
11	65.8	121	723.8
13	74.2	169	964.6
15	83.3	225	1249.5
64	404.6	680	4050.2

The normal equations are $8b + 64m = 404.6$

$$64b + 680m = 4050.2.$$

The solutions are $m = 4.8417$ and $b = 11.8417$ and so a required equation is $y = 4.842x + 11.842$.

b. In 1993, $x = 19$, and so $y = 4.842(19) + 11.842 = 103.83$. Hence the estimated number of cans produced in 1993 is 103.8 billion.

USING TECHNOLOGY EXERCISES 1.5, page 65

1. $y = 2.3596x + 3.8639$

3. $y = -1.1948x + 3.5525$

5. a. $y = 13.321x + 72.57$ b. 192 million tons

CHAPTER 1 REVIEW EXERCISES, page 68

1. The distance is $d = \sqrt{(6-2)^2 + (4-1)^2} = \sqrt{4^2 + 3^2} = \sqrt{25} = 5$.

3. The distance is
$$d = \sqrt{[1-(-2)]^2 + [-7-(-3)]^2} = \sqrt{3^2 + (-4)^2} = \sqrt{9+16} = \sqrt{25} = 5.$$

5. An equation is $x = -2$.

7. The slope of L is $m = \dfrac{\frac{7}{2} - 4}{3-(-2)} = -\dfrac{1}{10}$ and an equation of L is
$$y - 4 = -\tfrac{1}{10}[x-(-2)] = -\tfrac{1}{10}x - \tfrac{1}{5},$$
or
$$y = -\tfrac{1}{10}x + \tfrac{19}{5}$$
The general form of this equation is $x + 10y - 38 = 0$.

9. Writing the given equation in the form $y = \frac{5}{2}x - 3$, we see that the slope of the given line is 5/2. So a required equation is
$$y - 4 = \tfrac{5}{2}(x+2) \quad \text{or} \quad y = \tfrac{5}{2}x + 9$$
The general form of this equation is $5x - 2y + 18 = 0$.

11. Using the slope-intercept form of the equation of a line, we have $y = -\frac{1}{2}x - 3$.

13. Rewriting the given equation in the slope-intercept form, we have $4y = -3x + 8$
or
$$y = -\tfrac{3}{4}x + 2$$
and conclude that the slope of the required line is $-3/4$. Using the point-slope form of the equation of a line with the point (2,3) and slope $-3/4$, we obtain
$$y - 3 = -\tfrac{3}{4}(x-2)$$
$$y = -\tfrac{3}{4}x + \tfrac{6}{4} + 3$$
$$= -\tfrac{3}{4}x + \tfrac{9}{2}.$$
The general form of this equation is $3x + 4y - 18 = 0$.

15. Rewriting the given equation in the slope-intercept form $y = \frac{2}{3}x - 8$, we see that the slope of the line with this equation is 2/3. The slope of the required line is – 3/2. Using the point-slope form of the equation of a line with the point (– 2,– 4) and slope – 3/2, we have

$$y - (-4) = -\tfrac{3}{2}[x - (-2)]$$

or

$$y = -\tfrac{3}{2}x - 7 .$$

The general form of this equation is $3x + 2y + 14 = 0$.

17. Setting $x = 0$, gives $5y = 15$, or $y = 3$. Setting $y = 0$, gives $-2x = 15$, or $x = -15/2$. The graph of the equation $-2x + 5y = 15$ follows.

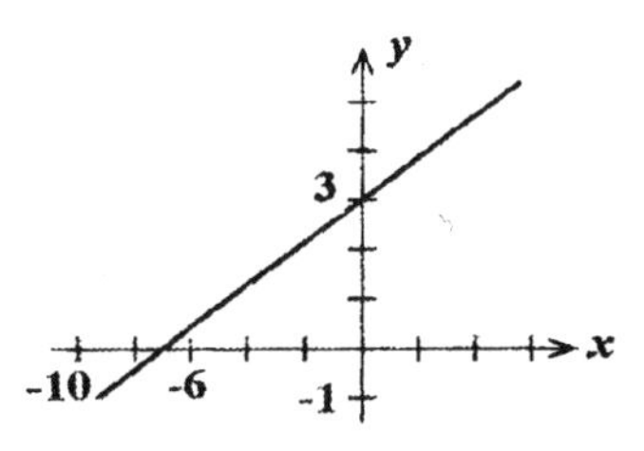

19. Let x denote the time in years. Since the function is linear, we know that it has the form $f(x) = mx + b$.

a. The slope of the line passing through (0, 2.4) and (5, 7.4) is

$$m = \frac{7.4 - 2.4}{5} = 1.$$

Since the line passes through (0, 2.4), we know that the y-intercept is 2.4. Therefore, the required function is $f(x) = x + 2.4$.

b. In 1992 ($x = 3$), the sales were $f(3) = 3 + 2.4 = 5.4$, or \$5.4 million dollars.

21. a. $D(w) = \dfrac{a}{150}w$. The given equation can be expressed in the form $y = mx + b$,

where $m = \dfrac{a}{150}$ and $b = 0$.

b. If $a = 500$ and $w = 35$, $D(35) = \frac{500}{150}(35) = 116\frac{2}{3}$, or approximately 117 mg.

23. Let V denote the value of the machine after t years.

a. The rate of depreciation is

$$-\frac{\Delta V}{\Delta t} = \frac{300{,}000 - 30{,}000}{12} = \frac{270{,}000}{12} = 22{,}500, \quad \text{or } \$22{,}500/\text{year}.$$

b. Using the point-slope form of the equation of a line with the point (0, 300,000) and $m = -22{,}500$, we have

$$V - 300{,}000 = -22{,}500(t - 0)$$
$$V = -22{,}500t + 300{,}000.$$

25. The slope of the demand curve is $\dfrac{\Delta p}{\Delta x} = -\dfrac{10}{200} = -0.05$.

Using the point-slope form of the equation of a line with the point (0, 200), we have

$p - 200 = -0.05(x)$, or $\quad p = -0.05x + 200.$

The graph of the demand equation follows.

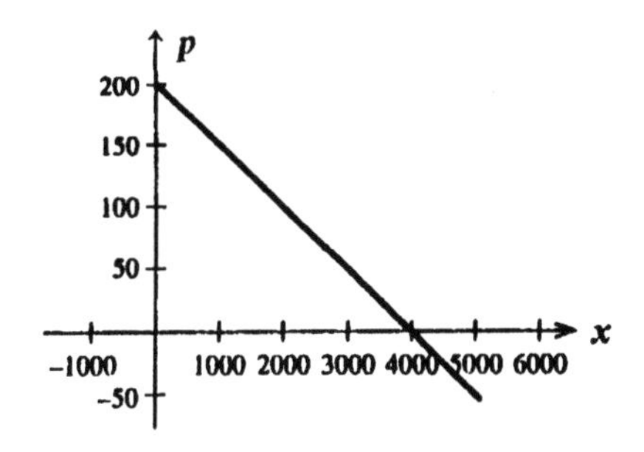

27. We solve the system $\quad 3x + 4y = -6$
$$2x + 5y = -11.$$

Solving the first equation for x, we have $3x = -4y - 6$

and $\quad x = -\frac{4}{3}y - 2$.

Substituting this value of x into the second equation yields

$$2(-\tfrac{4}{3}y - 2) + 5y = -11$$
$$-\tfrac{8}{3}y - 4 + 5y = -11$$
$$\tfrac{7}{3}y = -7, \quad \text{or} \quad y = -3.$$

Then $\quad x = -\frac{4}{3}(-3) - 2 = 4 - 2 = 2.$

Therefore, the point of intersection is (2,– 3).

29. Setting $C(x) = R(x)$, we have $12x + 20{,}000 = 20x$
$$8x = 20{,}000$$

or $\quad x = 2500.$

Next, $\quad R(2500) = 20(2500) = 50{,}000,$

and we conclude that the breakeven point is (2500, 50,000).

31. a. The slope of the line is $m = \dfrac{1 - 0.5}{4 - 2} = 0.25$.

Using the point-slope form of an equation of a line, we have

$$y - 1 = 0.25(x - 4)$$
$$y = 0.25x$$

b. $y = 0.25(6.4) = 1.6$, or 1600 applications.

CHAPTER 2

EXERCISES 2.1, page 78

1. Solving the first equation for x, we find $x = 3y - 1$. Substituting this value of x into the second equation yields
$$4(3y-1)+3y=11$$
$$12y-4+3y=11$$
or
$$y=1.$$
Substituting this value of y into the first equation gives
$$x=3(1)-1=2$$
Therefore, the unique solution of the system is (2,1).

3. Solving the first equation for x, we have $x = 7 - 4y$. Substituting this value of x into the second equation, we have
$$\tfrac{1}{2}(7-4y)+2y=5$$
$$7-4y+4y=10$$
$$7=10.$$
Clearly, this is impossible and we conclude that the system of equations has no solution.

5. Solving the first equation for x, we obtain $x = 7 - 2y$.
Substituting this value of x into the second equation, we have
$$2(7-2y)-y=4$$
$$14-4y-y=4$$
$$-5y=-10$$
and
$$y=2.$$
Then
$$x = 7 - 2(2) = 7 - 4 = 3.$$
We conclude that the solution to the system is (3,2).

7. Solving the first equation for x, we have
$$2x=5y+10$$
and
$$x=\tfrac{5}{2}y+5.$$
Substituting this value of x into the second equation, we have
$$6(\tfrac{5}{2}y+5)-15y=30$$
$$15y+30-15y=30$$
or
$$0=0.$$

This result tells us that the second equation is equivalent to the first. Thus, any ordered pair of numbers (x, y) satisfying the equation

$$2x - 5y = 10 \qquad (\text{or } 6x - 15y = 30)$$

is a solution to the system. In particular, by assigning the value t to x, where t is any real number, we find that

$$y = -2 + \tfrac{2}{5}t$$

so the ordered pair, $(t, \tfrac{2}{5}t - 2)$ is a solution to the system, and we conclude that the system has infinitely many solutions.

9. Solving the first equation for x, we obtain

$$\begin{aligned} 4x - 5y &= 14 \\ 4x &= 14 + 5y \\ x &= \tfrac{14}{4} + \tfrac{5}{4}y = \tfrac{7}{2} + \tfrac{5}{4}y. \end{aligned}$$

Substituting this value of x into the second equation gives

$$\begin{aligned} 2(\tfrac{7}{2} + \tfrac{5}{4}y) + 3y &= -4 \\ 7 + \tfrac{5}{2}y + 3y &= -4 \\ \tfrac{11}{2}y &= -11 \end{aligned}$$

or $\qquad y = -2.$

Then, $\qquad x = \tfrac{7}{2} + \tfrac{5}{4}(-2) = 1.$

We conclude that the ordered pair $(1, -2)$ satisfies the given system of equations.

11. Solving the first equation for x, we obtain

$$\begin{aligned} 2x &= 3y + 6 \\ x &= \tfrac{3}{2}y + 3 \end{aligned}$$

Substituting this value of x into the second equation gives

$$\begin{aligned} 6(\tfrac{3}{2}y + 3) - 9y &= 12 \\ 9y + 18 - 9y &= 12 \\ 18 &= 12. \end{aligned}$$

which is impossible. We conclude that the system of equations has no solution.

13. Solving the first equation for y, we obtain $y = 2x - 3$. Substituting this value of y into the second equation yields

$$\begin{aligned} 4x + k(2x - 3) &= 4, \\ 4x + 2xk - 3k &= 4 \\ 2x(2 + k) &= 4 + 3k \\ x &= \frac{4 + 3k}{2(2 + k)}. \end{aligned}$$

Since x is not defined when the denominator of this last expression is zero, we conclude that the system has no solution when $k = -2$.

15. Let x and y denote the number of acres of corn and wheat planted, respectively. Then $x + y = 500$. Since the cost of cultivating corn is \$42/acre and that of wheat \$30/acre and Mr. Johnson has \$18,600 available for cultivation, we have $42x + 30y = 18600$. Thus, the solution is found by solving the system of equations

$$\begin{aligned} x + \quad y &= \quad 500 \\ 42x + 30y &= 18{,}600 \end{aligned}.$$

17. Let x denote the number of pounds of the \$2.50/lb coffee and y denote the number of pounds of the \$3/lb coffee. Then

$$x + y = 100.$$

Since the blended coffee sells for \$2.80/lb, we know that the blended mixture is worth $(2.80)(100) = \$280$. Therefore,

$$2.50x + 3y = 280.$$

Thus, the solution is found by solving the system of equations

$$\begin{aligned} x + \quad y &= 100 \\ 2.50x + 3y &= 280 \end{aligned}.$$

19. Let x denote the number of children who rode the bus during the morning shift and y denote the number of adults who rode the bus during the morning shift. Then $x + y = 1000$. Since the total fare collected was \$650, we have $0.25x + 0.75y = 650$. Thus, the solution to the problem can be found by solving the system of equations

$$\begin{aligned} x + \qquad y &= 1000 \\ 0.25x + 0.75y &= \quad 650. \end{aligned}$$

21. Let x = the amount of money invested at 6 percent in a savings account
y = the amount of money invested at 8 percent in mutual funds
and z = the amount of money invested at 12 percent in money market certificates.
Since the total interest was \$21,600, we have

$$0.6x + 0.8y + 0.12z = 21{,}600.$$

Also, since the amount of Mr. Carrington's investment in money market certificates is twice the amount of the investment in the savings account, we have

$$z = 2x.$$

Finally, the interest earned from his investment in mutual funds was equal to the interest earned on his money market certificates, so

$$0.08y = 0.12z.$$

Thus, the solution to the problem can be found by solving the system of equations

$$\begin{aligned} 0.06x + 0.08y + 0.12z &= 21{,}600 \\ 2x \qquad\quad - z &= 0 \\ 0.08y - 0.12z &= 0. \end{aligned}$$

23. Let x, y, and z denote the number of compact, intermediate, and full-size cars, respectively, to be purchased. The cost incurred in buying the specified number of cars is $8000x + 12000y + 16000z$. Since the budget is \$1 million, we have the system

$$\begin{aligned} 8000x + 12{,}000y + 16{,}000z &= 1{,}000{,}000 \\ x - 2y \qquad\qquad &= 0 \\ x + y + z &= 100. \end{aligned}$$

25. Let x = the number of ounces of Food I used in the meal
y = the number of ounces of Food II used in the meal
and z = the number of ounces of Food III used in the meal.
Since 100 percent of the daily requirement of proteins, carbohydrates, and iron is to be met by this meal, we have the following system of linear equations:

$$\begin{aligned} 10x + 6y + 8z &= 100 \\ 10x + 12y + 6z &= 100 \\ 5x + 4y + 12z &= 100. \end{aligned}$$

EXERCISES 2.2, page 93

1. $\left[\begin{array}{rr|r} 2 & -3 & 7 \\ 3 & 1 & 4 \end{array}\right]$

3. $\left[\begin{array}{rrr|r} 0 & -1 & 2 & 6 \\ 2 & 2 & -8 & 7 \\ 0 & 3 & 4 & 0 \end{array}\right]$

5. $$\begin{aligned} 3x + 2y &= -4 \\ x - y &= 5 \end{aligned}$$

7. $$\begin{aligned} x + 3y + 2z &= 4 \\ 2x \qquad\qquad &= 5 \\ 3x - 3y + 2z &= 6 \end{aligned}$$

9. Yes. Conditions 1-4 are satisfied (see page 85 of the text).

11. No. Condition 3 is violated. The first nonzero entry in the second row does not lie to the right of the first nonzero entry 1 in row 1.

13. Yes. Conditions 1-4 are satisfied.

15. No. Condition 2 and consequently condition 4 are not satisfied. The first nonzero entry in the last row is not a 1 and the column containing that entry does not have zeros elsewhere.

17. No. Condition 1 is violated. The first row consists entirely of zeros and it lies above row 2.

19. $$\left[\begin{array}{cc|c} 2 & 4 & 8 \\ 3 & 1 & 2 \end{array}\right] \xrightarrow{\frac{1}{2}R_1} \left[\begin{array}{cc|c} 1 & 2 & 4 \\ 3 & 1 & 2 \end{array}\right] \xrightarrow{R_2-3R_1} \left[\begin{array}{cc|c} 1 & 2 & 4 \\ 0 & -5 & -10 \end{array}\right]$$

21. $$\left[\begin{array}{cc|c} -1 & 2 & 3 \\ 6 & 4 & 2 \end{array}\right] \xrightarrow{-R_1} \left[\begin{array}{cc|c} 1 & -2 & -3 \\ 6 & 4 & 2 \end{array}\right] \xrightarrow{R_2-6R_1} \left[\begin{array}{cc|c} 1 & -2 & -3 \\ 0 & 16 & 20 \end{array}\right]$$

23. $$\left[\begin{array}{ccc|c} 2 & 4 & 6 & 12 \\ 2 & 3 & 1 & 5 \\ 3 & -1 & 2 & 4 \end{array}\right] \xrightarrow{\frac{1}{2}R_1} \left[\begin{array}{ccc|c} 1 & 2 & 3 & 6 \\ 2 & 3 & 1 & 5 \\ 3 & -1 & 2 & 4 \end{array}\right] \xrightarrow[R_3-3R_1]{R_2-2R_1} \left[\begin{array}{ccc|c} 1 & 2 & 3 & 6 \\ 0 & -1 & -5 & -7 \\ 0 & -7 & -7 & -14 \end{array}\right]$$

25. $$\left[\begin{array}{ccc|c} 0 & 1 & 3 & 4 \\ 2 & 4 & 1 & 3 \\ 5 & 6 & 2 & -4 \end{array}\right] \xrightarrow[R_3-2R_2]{R_1-3R_2} \left[\begin{array}{ccc|c} -6 & -11 & 0 & -5 \\ 2 & 4 & 1 & 3 \\ 1 & -2 & 0 & -10 \end{array}\right]$$

27. $$\left[\begin{array}{cc|c} 3 & 9 & 6 \\ 2 & 1 & 4 \end{array}\right] \xrightarrow{\frac{1}{3}R_1} \left[\begin{array}{cc|c} 1 & 3 & 2 \\ 2 & 1 & 4 \end{array}\right] \xrightarrow{R_2-2R_1} \left[\begin{array}{cc|c} 1 & 3 & 2 \\ 0 & -5 & 0 \end{array}\right] \xrightarrow{-\frac{1}{5}R_2}$$

$$\left[\begin{array}{cc|c} 1 & 3 & 2 \\ 0 & 1 & 0 \end{array}\right] \xrightarrow{R_1-3R_2} \left[\begin{array}{cc|c} 1 & 0 & 2 \\ 0 & 1 & 0 \end{array}\right]$$

29. $$\left[\begin{array}{ccc|c} 1 & 3 & 1 & 3 \\ 3 & 8 & 3 & 7 \\ 2 & -3 & 1 & -10 \end{array}\right] \xrightarrow[R_3-2R_1]{R_2-3R_1} \left[\begin{array}{ccc|c} 1 & 3 & 1 & 3 \\ 0 & -1 & 0 & -2 \\ 0 & -9 & -1 & -16 \end{array}\right] \xrightarrow{-R_2} \left[\begin{array}{ccc|c} 1 & 3 & 1 & 3 \\ 0 & 1 & 0 & 2 \\ 0 & -9 & -1 & -16 \end{array}\right]$$

$$\xrightarrow[R_3+9R_2]{R_1-3R_2}\left[\begin{array}{ccc|c}1 & 0 & 1 & -3\\0 & 1 & 0 & 2\\0 & 0 & -1 & 2\end{array}\right]\xrightarrow[-R_3]{R_1+R_3}\left[\begin{array}{ccc|c}1 & 0 & 0 & -1\\0 & 1 & 0 & 2\\0 & 0 & 1 & -2\end{array}\right]$$

31. The augmented matrix is equivalent to the system of linear equations

$$\begin{aligned}3x+9y&=6\\2x+\ \ y&=4\end{aligned}$$

The ordered pair (2,0) is the solution to the system.

33. The augmented matrix is equivalent to the system of linear equations

$$\begin{aligned}x+3y+\ z&=\ \ 3\\3x+8y+3z&=\ \ 7\\2x-3y+\ z&=-10\end{aligned}$$

Reading off the solution from the last augmented matrix, which is in row-reduced form,

$$\left[\begin{array}{ccc|c}1 & 0 & 0 & -1\\0 & 1 & 0 & 2\\0 & 0 & 1 & -2\end{array}\right]$$

we have $x=-1$, $y=2$, and $z=-2$.

35. Using the Gauss-Jordan method, we have

$$\left[\begin{array}{cc|c}1 & -2 & 8\\3 & 4 & 4\end{array}\right]\xrightarrow{R_2-3R_1}\left[\begin{array}{cc|c}1 & -2 & 8\\0 & 10 & -20\end{array}\right]\xrightarrow{\frac{1}{10}R_2}\left[\begin{array}{cc|c}1 & -2 & 8\\0 & 1 & -2\end{array}\right]\xrightarrow{R_1+2R_2}\left[\begin{array}{cc|c}1 & 0 & 4\\0 & 1 & -2\end{array}\right]$$

The solution is (4,–2).

37. Using the Gauss-Jordan method, we have

$$\left[\begin{array}{cc|c}2 & -3 & -8\\4 & 1 & -2\end{array}\right]\xrightarrow{\frac{1}{2}R_1}\left[\begin{array}{cc|c}1 & -\frac{3}{2} & -4\\4 & 1 & -2\end{array}\right]\xrightarrow{R_2-4R_1}\left[\begin{array}{cc|c}1 & -\frac{3}{2} & -4\\0 & 7 & 14\end{array}\right]\xrightarrow{\frac{1}{7}R_2}$$

$$\left[\begin{array}{cc|c}1 & -\frac{3}{2} & -4\\0 & 1 & 2\end{array}\right]\xrightarrow{R_1+\frac{3}{2}R_2}\left[\begin{array}{cc|c}1 & 0 & -1\\0 & 1 & 2\end{array}\right].$$ The solution is (–1,2).

39. Using the Gauss-Jordan method, we have

$$\left[\begin{array}{ccc|c} 1 & 1 & 1 & 0 \\ 2 & -1 & 1 & 1 \\ 1 & 1 & -2 & 2 \end{array}\right] \xrightarrow[R_3-R_1]{R_2-2R_1} \left[\begin{array}{ccc|c} 1 & 1 & 1 & 0 \\ 0 & -3 & -1 & 1 \\ 0 & 0 & -3 & 2 \end{array}\right] \xrightarrow{-\frac{1}{3}R_2} \left[\begin{array}{ccc|c} 1 & 1 & 1 & 0 \\ 0 & 1 & \frac{1}{3} & -\frac{1}{3} \\ 0 & 0 & -3 & 2 \end{array}\right] \xrightarrow{R_1-R_2}$$

$$\left[\begin{array}{ccc|c} 1 & 0 & \frac{2}{3} & \frac{1}{3} \\ 0 & 1 & \frac{1}{3} & -\frac{1}{3} \\ 0 & 0 & -3 & 2 \end{array}\right] \xrightarrow{-\frac{1}{3}R_3} \left[\begin{array}{ccc|c} 1 & 0 & \frac{2}{3} & \frac{1}{3} \\ 0 & 1 & \frac{1}{3} & -\frac{1}{3} \\ 0 & 0 & 1 & -\frac{2}{3} \end{array}\right] \xrightarrow[R_2-\frac{1}{3}R_3]{R_1-\frac{2}{3}R_3} \left[\begin{array}{ccc|c} 1 & 0 & 0 & \frac{7}{9} \\ 0 & 1 & 0 & -\frac{1}{9} \\ 0 & 0 & 1 & -\frac{2}{3} \end{array}\right].$$

The solution is $(\frac{7}{9},-\frac{1}{9},-\frac{2}{3})$.

41. $$\left[\begin{array}{ccc|c} 2 & 2 & 1 & 9 \\ 1 & 0 & 1 & 4 \\ 0 & 4 & -3 & 17 \end{array}\right] \xrightarrow{R_1 \leftrightarrow R_2} \left[\begin{array}{ccc|c} 1 & 0 & 1 & 4 \\ 2 & 2 & 1 & 9 \\ 0 & 4 & -3 & 17 \end{array}\right] \xrightarrow{R_2-2R_1} \left[\begin{array}{ccc|c} 1 & 0 & 1 & 4 \\ 0 & 2 & -1 & 1 \\ 0 & 4 & -3 & 17 \end{array}\right] \xrightarrow{\frac{1}{2}R_2}$$

$$\left[\begin{array}{ccc|c} 1 & 0 & 1 & 4 \\ 0 & 1 & -\frac{1}{2} & \frac{1}{2} \\ 0 & 4 & -3 & 17 \end{array}\right] \xrightarrow{R_3-4R_2} \left[\begin{array}{ccc|c} 1 & 0 & 1 & 4 \\ 0 & 1 & -\frac{1}{2} & \frac{1}{2} \\ 0 & 0 & -1 & 15 \end{array}\right] \xrightarrow{-R_3} \left[\begin{array}{ccc|c} 1 & 0 & 1 & 4 \\ 0 & 1 & -\frac{1}{2} & \frac{1}{2} \\ 0 & 0 & 1 & -15 \end{array}\right] \xrightarrow[R_2+\frac{1}{2}R_3]{R_1-R_3}$$

$$\left[\begin{array}{ccc|c} 1 & 0 & 0 & 19 \\ 0 & 1 & 0 & -7 \\ 0 & 0 & 1 & -15 \end{array}\right].$$

The solution is (19,–7,–15).

43. $$\left[\begin{array}{ccc|c} 0 & -1 & 1 & 2 \\ 4 & -3 & 2 & 16 \\ 3 & 2 & 1 & 11 \end{array}\right] \xrightarrow{R_1 \leftrightarrow R_2} \left[\begin{array}{ccc|c} 4 & -3 & 2 & 16 \\ 0 & -1 & 1 & 2 \\ 3 & 2 & 1 & 11 \end{array}\right] \xrightarrow{R_1-R_3} \left[\begin{array}{ccc|c} 1 & -5 & 1 & 5 \\ 0 & -1 & 1 & 2 \\ 3 & 2 & 1 & 11 \end{array}\right]$$

$$\xrightarrow[R_3-3R_1]{-R_2} \left[\begin{array}{ccc|c} 1 & -5 & 1 & 5 \\ 0 & 1 & -1 & -2 \\ 0 & 17 & -2 & -4 \end{array}\right] \xrightarrow[R_3-17R_2]{R_1+5R_2} \left[\begin{array}{ccc|c} 1 & 0 & -4 & -5 \\ 0 & 1 & -1 & -2 \\ 0 & 0 & 15 & 30 \end{array}\right] \xrightarrow{\frac{1}{15}R_3}$$

$$\left[\begin{array}{ccc|c} 1 & 0 & -4 & -5 \\ 0 & 1 & -1 & -2 \\ 0 & 0 & 1 & 2 \end{array}\right] \xrightarrow[R_2+R_3]{R_1+4R_3} \left[\begin{array}{ccc|c} 1 & 0 & 0 & 3 \\ 0 & 1 & 0 & 0 \\ 0 & 0 & 1 & 2 \end{array}\right].$$

The solution is (3,0,2).

45. Using the Gauss-Jordan method, we have

$$\left[\begin{array}{ccc|c} 1 & -2 & 1 & 6 \\ 2 & 1 & -3 & -3 \\ 1 & -3 & 3 & 10 \end{array}\right] \xrightarrow[R_3-R_1]{R_2-2R_1} \left[\begin{array}{ccc|c} 1 & -2 & 1 & 6 \\ 0 & 5 & -5 & -15 \\ 0 & -1 & 2 & 4 \end{array}\right] \xrightarrow{\frac{1}{5}R_2} \left[\begin{array}{ccc|c} 1 & -2 & 1 & 6 \\ 0 & 1 & -1 & -3 \\ 0 & -1 & 2 & 4 \end{array}\right]$$

$$\xrightarrow[R_3+R_2]{R_1+2R_2} \left[\begin{array}{ccc|c} 1 & 0 & -1 & 0 \\ 0 & 1 & -1 & -3 \\ 0 & 0 & 1 & 1 \end{array}\right] \xrightarrow[R_2+R_3]{R_1+R_3} \left[\begin{array}{ccc|c} 1 & 0 & 0 & 1 \\ 0 & 1 & 0 & -2 \\ 0 & 0 & 1 & 1 \end{array}\right].$$

Therefore, the solution is (1,–2,1).

47. Using the Gauss-Jordan method, we have

$$\left[\begin{array}{ccc|c} 2 & 0 & 3 & -1 \\ 3 & -2 & 1 & 9 \\ 1 & 1 & 4 & 4 \end{array}\right] \xrightarrow{R_1 \leftrightarrow R_3} \left[\begin{array}{ccc|c} 1 & 1 & 4 & 4 \\ 3 & -2 & 1 & 9 \\ 2 & 0 & 3 & -1 \end{array}\right] \xrightarrow[R_3-2R_1]{R_2-3R_1} \left[\begin{array}{ccc|c} 1 & 1 & 4 & 4 \\ 0 & -5 & -11 & -3 \\ 0 & -2 & -5 & -9 \end{array}\right]$$

$$\xrightarrow{-\frac{1}{5}R_2} \left[\begin{array}{ccc|c} 1 & 1 & 4 & 4 \\ 0 & 1 & \frac{11}{5} & \frac{3}{5} \\ 0 & -2 & -5 & -9 \end{array}\right] \xrightarrow[R_3+2R_2]{R_1-R_2} \left[\begin{array}{ccc|c} 1 & 0 & \frac{9}{5} & \frac{17}{5} \\ 0 & 1 & \frac{11}{5} & \frac{3}{5} \\ 0 & 0 & -\frac{3}{5} & -\frac{39}{5} \end{array}\right] \xrightarrow{-\frac{5}{3}R_3}$$

$$\left[\begin{array}{ccc|c} 1 & 0 & \frac{9}{5} & \frac{17}{5} \\ 0 & 1 & \frac{11}{5} & \frac{3}{5} \\ 0 & 0 & 1 & 13 \end{array}\right] \xrightarrow[R_2-\frac{11}{5}R_3]{R_1-\frac{9}{5}R_3} \left[\begin{array}{ccc|c} 1 & 0 & 0 & -20 \\ 0 & 1 & 0 & -28 \\ 0 & 0 & 1 & 13 \end{array}\right].$$

Therefore, the solution is (–20, –28, 13).

49. Using the Gauss-Jordan method, we have

$$\left[\begin{array}{ccc|c} 1 & -1 & 3 & 14 \\ 1 & 1 & 1 & 6 \\ -2 & -1 & 1 & -4 \end{array}\right] \xrightarrow[R_3+2R_1]{R_2-R_1} \left[\begin{array}{ccc|c} 1 & -1 & 3 & 14 \\ 0 & 2 & -2 & -8 \\ 0 & -3 & 7 & 24 \end{array}\right] \xrightarrow{\frac{1}{2}R_2} \left[\begin{array}{ccc|c} 1 & -1 & 3 & 14 \\ 0 & 1 & -1 & -4 \\ 0 & -3 & 7 & 24 \end{array}\right]$$

$$\xrightarrow[R_3+3R_2]{R_1+R_2} \left[\begin{array}{ccc|c} 1 & 0 & 2 & 10 \\ 0 & 1 & -1 & -4 \\ 0 & 0 & 4 & 12 \end{array}\right] \xrightarrow{\frac{1}{4}R_3} \left[\begin{array}{ccc|c} 1 & 0 & 2 & 10 \\ 0 & 1 & -1 & -4 \\ 0 & 0 & 4 & 3 \end{array}\right] \xrightarrow[R_2+R_3]{R_1-2R_3} \left[\begin{array}{ccc|c} 1 & 0 & 0 & 4 \\ 0 & 1 & 0 & -1 \\ 0 & 0 & 1 & 3 \end{array}\right]$$

Therefore, the solution is (4, –1, 3).

51. We wish to solve the system of equations

$$\begin{aligned} x + y &= 500 \qquad (x = \text{the number of acres of corn planted}) \\ 42x + 30y &= 18{,}600 \qquad (y = \text{the number of acres of wheat planted}) \end{aligned}$$

Using the Gauss-Jordan method, we find

$$\left[\begin{array}{cc|c} 1 & 1 & 500 \\ 42 & 30 & 18600 \end{array}\right] \xrightarrow{R_2-42R_1} \left[\begin{array}{cc|c} 1 & 1 & 500 \\ 0 & -12 & -2400 \end{array}\right] \xrightarrow{-\frac{1}{12}R_2} \left[\begin{array}{cc|c} 1 & 1 & 500 \\ 0 & 1 & 200 \end{array}\right]$$

$$\xrightarrow{R_1-R_2} \left[\begin{array}{cc|c} 1 & 0 & 300 \\ 0 & 1 & 200 \end{array}\right]$$

The solution to this system of equations is $x = 300$ and $y = 200$. We conclude that Mr. Johnson should plant 300 acres of corn and 200 acres of wheat.

53. Let x denote the number of pounds of the \$2.50/lb coffee and y denote the number of pounds of the \$3.00/lb coffee. Then we are required to solve the system

$$\begin{aligned} x + y &= 100 \\ 2.50x + 3.00y &= 280 \end{aligned}$$

Using the Gauss-Jordan method of elimination, we have

$$\left[\begin{array}{cc|c} 1 & 1 & 100 \\ 2.5 & 3 & 280 \end{array}\right] \xrightarrow{R_2-2.5R_1} \left[\begin{array}{cc|c} 1 & 1 & 100 \\ 0 & 0.5 & 30 \end{array}\right] \xrightarrow{2R_2} \left[\begin{array}{cc|c} 1 & 1 & 100 \\ 0 & 1 & 60 \end{array}\right]$$

$$\xrightarrow{R_1 - R_2} \left[\begin{array}{cc|c} 1 & 0 & 40 \\ 0 & 1 & 60 \end{array}\right].$$

Therefore, 40 pounds of the \$2.50/lb coffee and 60 pounds of the \$3.00/lb coffee should be used in the 100 lb mixture.

55. Let x and y denote the number of children and adults who rode the bus during the morning shift, respectively. Then the solution to the problem can be found by solving the system of equations

$$\begin{aligned} x + \quad y &= 1000 \\ 0.25x + 0.75y &= 650 \end{aligned}$$

Using the Gauss-Jordan elimination method, we have

$$\left[\begin{array}{cc|c} 1 & 1 & 1000 \\ 0.25 & 0.75 & 650 \end{array}\right] \xrightarrow{R_2 - 0.25R_1} \left[\begin{array}{cc|c} 1 & 1 & 1000 \\ 0 & 0.5 & 400 \end{array}\right] \xrightarrow{2R_2} \left[\begin{array}{cc|c} 1 & 1 & 1000 \\ 0 & 1 & 800 \end{array}\right]$$

$$\xrightarrow{R_1 - R_2} \left[\begin{array}{cc|c} 1 & 0 & 200 \\ 0 & 1 & 800 \end{array}\right]$$

We conclude that 800 adults and 200 children rode the bus during the morning shift.

57. Let x, y, and z, denote the amount of money he should invest in a savings account, in mutual funds, and in money market certificates, respectively. Then, we are required to solve the system

$$\begin{aligned} 0.06x + 0.08y + 0.12z &= 21{,}600 \\ 2x \qquad\quad - \quad z &= 0 \\ 0.08y - 0.12z &= 0 \end{aligned}$$

Using the Gauss-Jordan method, we find

$$\left[\begin{array}{ccc|c} 0.06 & 0.08 & 0.12 & 21{,}600 \\ 2 & 0 & -1 & 0 \\ 0 & 0.08 & -0.12 & 0 \end{array}\right] \xrightarrow[\frac{1}{0.08}R_3]{\frac{1}{0.06}R_1} \left[\begin{array}{ccc|c} 1 & \frac{4}{3} & 2 & 360{,}000 \\ 2 & 0 & -1 & 0 \\ 0 & 1 & -\frac{3}{2} & 0 \end{array}\right] \xrightarrow{R_2 - 2R_1}$$

$$\left[\begin{array}{ccc|c} 1 & \frac{4}{3} & 2 & 360{,}000 \\ 0 & -\frac{8}{3} & -5 & -720{,}000 \\ 0 & 1 & -\frac{3}{2} & 0 \end{array}\right] \xrightarrow{-\frac{3}{8}R_2} \left[\begin{array}{ccc|c} 1 & \frac{4}{3} & 2 & 360{,}000 \\ 0 & 1 & \frac{15}{8} & 270{,}000 \\ 0 & 1 & -\frac{3}{2} & 0 \end{array}\right]$$

$$\xrightarrow[R_3-R_2]{R_1-\frac{4}{3}R_2} \left[\begin{array}{ccc|c} 1 & 0 & -\frac{1}{2} & 0 \\ 0 & 1 & \frac{15}{8} & 270{,}000 \\ 0 & 0 & -\frac{27}{8} & -270{,}000 \end{array}\right] \xrightarrow{-\frac{8}{27}R_3} \left[\begin{array}{ccc|c} 1 & 0 & -\frac{1}{2} & 0 \\ 0 & 1 & \frac{15}{8} & 270{,}000 \\ 0 & 0 & 1 & 80{,}000 \end{array}\right]$$

$$\xrightarrow[R_2-\frac{15}{8}R_3]{R_1+\frac{1}{2}R_3} \left[\begin{array}{ccc|c} 1 & 0 & 0 & 40{,}000 \\ 0 & 1 & 0 & 120{,}000 \\ 0 & 0 & 1 & 80{,}000 \end{array}\right]$$

Therefore, Mr. Carrington should invest \$40,000 in a savings account, \$120,000 in mutual funds, and \$80,000 in money-market certificates.

59. Let x, y, and z denote the number of compact, intermediate, and full-size cars, respectively, to be purchased. Then the problem can be solved by solving the system

$$\begin{aligned} 8000x + 12{,}000y + 16{,}000z &= 1{,}000{,}000 \\ x - 2y \phantom{+16{,}000z} &= 0 \\ x + y + z &= 100 \end{aligned}$$

Using the Gauss-Jordan method, we have

$$\left[\begin{array}{ccc|c} 8000 & 12{,}000 & 16{,}000 & 1{,}000{,}000 \\ 1 & -2 & 0 & 0 \\ 1 & 1 & 1 & 100 \end{array}\right] \xrightarrow{R_1 \leftrightarrow R_3} \left[\begin{array}{ccc|c} 1 & 1 & 1 & 100 \\ 1 & -2 & 0 & 0 \\ 8000 & 12000 & 16000 & 1{,}000{,}000 \end{array}\right]$$

$$\xrightarrow[R_3-8000R_1]{R_2-R_1} \left[\begin{array}{ccc|c} 1 & 1 & 1 & 100 \\ 1 & -3 & -1 & -100 \\ 0 & 4000 & 8000 & 200{,}000 \end{array}\right] \xrightarrow{-\frac{1}{3}R_2} \left[\begin{array}{ccc|c} 1 & 1 & 1 & 100 \\ 0 & 1 & \frac{1}{3} & \frac{100}{3} \\ 0 & 4000 & 8000 & 200{,}000 \end{array}\right]$$

$$\xrightarrow[R_3-4000R_2]{R_1-R_2} \left[\begin{array}{ccc|c} 1 & 0 & \frac{2}{3} & \frac{200}{3} \\ 0 & 1 & \frac{1}{3} & \frac{100}{3} \\ 0 & 0 & \frac{200{,}000}{3} & \frac{200{,}000}{3} \end{array}\right] \xrightarrow{\frac{3}{20{,}000}R_3} \left[\begin{array}{ccc|c} 1 & 0 & \frac{2}{3} & \frac{200}{3} \\ 0 & 1 & \frac{1}{3} & \frac{100}{3} \\ 0 & 0 & 1 & 10 \end{array}\right] \xrightarrow[R_2-\frac{1}{3}R_3]{R_1-\frac{2}{3}R_3} \left[\begin{array}{ccc|c} 1 & 0 & 0 & 60 \\ 0 & 1 & 0 & 30 \\ 0 & 0 & 1 & 10 \end{array}\right].$$

We conclude that 60 compact cars, 30 intermediate-size cars, and 10 full-size cars will be purchased.

61. Let x, y, and z, represent the number of ounces of Food *I*, Food *II*, and Food *III* used in the meal, respectively. Then the problem reduces to solving the following system of linear equations:

$$\begin{aligned} 10x + 6y + 8z &= 100 \\ 10x + 12y + 6z &= 100 \\ 5x + 4y + 12z &= 100. \end{aligned}$$

Using the Gauss-Jordan method, we obtain

$$\left[\begin{array}{ccc|c} 10 & 6 & 8 & 100 \\ 10 & 12 & 6 & 100 \\ 5 & 4 & 12 & 100 \end{array}\right] \xrightarrow{\frac{1}{10}R_1} \left[\begin{array}{ccc|c} 1 & \frac{3}{5} & \frac{4}{5} & 10 \\ 10 & 12 & 6 & 100 \\ 5 & 4 & 12 & 100 \end{array}\right] \xrightarrow[R_3-5R_1]{R_2-10R_1}$$

$$\left[\begin{array}{ccc|c} 1 & \frac{3}{5} & \frac{4}{5} & 10 \\ 0 & 6 & -2 & 0 \\ 0 & 1 & 8 & 50 \end{array}\right] \xrightarrow{\frac{1}{6}R_2} \left[\begin{array}{ccc|c} 1 & \frac{3}{5} & \frac{4}{5} & 10 \\ 0 & 1 & -\frac{1}{3} & 0 \\ 0 & 1 & 8 & 50 \end{array}\right] \xrightarrow[R_3-R_2]{R_1-\frac{3}{5}R_2}$$

$$\left[\begin{array}{ccc|c} 1 & 0 & 1 & 10 \\ 0 & 1 & -\frac{1}{3} & 0 \\ 0 & 0 & \frac{25}{3} & 50 \end{array}\right] \xrightarrow{\frac{3}{25}R_3} \left[\begin{array}{ccc|c} 1 & 0 & 1 & 10 \\ 0 & 1 & -\frac{1}{3} & 0 \\ 0 & 0 & 1 & 6 \end{array}\right] \xrightarrow[R_2+\frac{1}{3}R_3]{R_1-R_3}$$

$$\left[\begin{array}{ccc|c} 1 & 0 & 0 & 4 \\ 0 & 1 & 0 & 2 \\ 0 & 0 & 1 & 6 \end{array}\right].$$

We conclude that 4 oz of Food *I*, 2 oz of Food *II*, and 6 oz of Food *III* should be used to prepare the meal.

63. Let x = the number of front orchestra seats sold
y = the number of rear orchestra seats sold
and z = the number of front balcony seats sold for this performance.

Then, we are required to solve the system

$$\begin{aligned} x + y + z &= 1{,}000 \\ 80x + 60y + 50z &= 62{,}800 \\ x + y - 2z &= 400. \end{aligned}$$

Using the Gauss-Jordan method, we find

$$\left[\begin{array}{ccc|c} 1 & 1 & 1 & 1{,}000 \\ 80 & 60 & 50 & 62{,}800 \\ 1 & 1 & -2 & 400 \end{array}\right] \xrightarrow[R_3-R_1]{R_2-80R_1} \left[\begin{array}{ccc|c} 1 & 1 & 1 & 1{,}000 \\ 0 & -20 & -30 & -17{,}200 \\ 0 & 0 & -3 & -600 \end{array}\right] \xrightarrow[-\frac{1}{3}R_3]{-\frac{1}{20}R_2}$$

$$\left[\begin{array}{ccc|c} 1 & 1 & 1 & 1{,}000 \\ 0 & 1 & \frac{3}{2} & 860 \\ 0 & 0 & 1 & 200 \end{array}\right] \xrightarrow{R_1-R_2} \left[\begin{array}{ccc|c} 1 & 0 & -\frac{1}{2} & 140 \\ 0 & 1 & \frac{3}{2} & 860 \\ 0 & 0 & 1 & 200 \end{array}\right] \xrightarrow[R_2-\frac{3}{2}R_3]{R_1+\frac{1}{2}R_3}$$

$$\left[\begin{array}{ccc|c} 1 & 0 & 0 & 240 \\ 0 & 1 & 0 & 560 \\ 0 & 0 & 1 & 200 \end{array}\right].$$

We conclude that tickets for 240 front orchestra seats, 560 rear orchestra seats, and 200 front balcony seats were sold.

USING TECHNOLOGY EXERCISES 2.2, page 98

1. (3,1,–1,2)   3. (5,4,–3,–4)   5. (1,–1,2,0,3)

EXERCISES 2.3, page 108

1. a. The system has one solution.   b. The solution is (3, –1, 2).

3. a. The system has one solution.   b. The solution is (2, 4).

5. a. The system has infinitely many solutions.
   b. Letting $x_3 = t$, we see that the solutions are given by $(4 - t, -2, t)$, where t is a parameter.

7. a. The system has no solution. The last row contains all zeros to the left of the vertical line and a nonzero number (1) to its right.

9. a. The system has infinitely many solutions.
   b. Letting $x_4 = t$, we see that the solutions are given by $(2, -1, 2 - t, t)$, where t is a parameter.

11. a. The system has infinitely many solutions.
    b. Letting $x_3 = s$ and $x_4 = t$, the solutions are given by $(2 - 3s,\ 1 + s,\ s,\ t)$, where s and t are parameters.

13. Using the Gauss-Jordan method, we have

$$\left[\begin{array}{cc|c} 2 & -1 & 3 \\ 1 & 2 & 4 \\ 2 & 3 & 7 \end{array}\right] \xrightarrow{R_1 \leftrightarrow R_2} \left[\begin{array}{cc|c} 1 & 2 & 4 \\ 2 & -1 & 3 \\ 2 & 3 & 7 \end{array}\right] \xrightarrow[R_3 - 2R_1]{R_2 - 2R_1} \left[\begin{array}{cc|c} 1 & 2 & 4 \\ 0 & -5 & -5 \\ 0 & -1 & -1 \end{array}\right] \xrightarrow{-\frac{1}{5}R_2}$$

$$\left[\begin{array}{cc|c} 1 & 2 & 4 \\ 0 & 1 & 1 \\ 0 & -1 & -1 \end{array}\right] \xrightarrow[R_3 + R_2]{R_1 - 2R_2} \left[\begin{array}{cc|c} 1 & 0 & 2 \\ 0 & 1 & 1 \\ 0 & 0 & 0 \end{array}\right].$$

The solution is $(2, 1)$.

15. Using the Gauss-Jordan method, we have

$$\left[\begin{array}{cc|c} 3 & -2 & -3 \\ 2 & 1 & 3 \\ 1 & -2 & -5 \end{array}\right] \xrightarrow{R_1 \leftrightarrow R_3} \left[\begin{array}{cc|c} 1 & -2 & -5 \\ 2 & 1 & 3 \\ 3 & -2 & -3 \end{array}\right] \xrightarrow[R_3 - 3R_1]{R_2 - 2R_1} \left[\begin{array}{cc|c} 1 & -2 & -5 \\ 0 & 5 & 13 \\ 0 & 4 & 12 \end{array}\right] \xrightarrow{\frac{1}{5}R_2}$$

$$\left[\begin{array}{cc|c} 1 & -2 & -5 \\ 0 & 1 & \frac{13}{5} \\ 0 & 4 & 12 \end{array}\right] \xrightarrow[R_3 - 4R_2]{R_1 + 2R_2} \left[\begin{array}{cc|c} 1 & 0 & \frac{1}{5} \\ 0 & 1 & \frac{13}{5} \\ 0 & 0 & \frac{8}{5} \end{array}\right].$$

Since the last row implies the $0 = 8/5$, we conclude that the system of equations is inconsistent and has no solution.

17.
$$\left[\begin{array}{cc|c} 3 & -2 & 5 \\ -1 & 3 & -4 \\ 2 & -4 & 6 \end{array}\right] \xrightarrow{R_1 \leftrightarrow R_2} \left[\begin{array}{cc|c} -1 & 3 & -4 \\ 3 & -2 & 5 \\ 2 & -4 & 6 \end{array}\right] \xrightarrow{-R_1} \left[\begin{array}{cc|c} 1 & -3 & 4 \\ 3 & -2 & 5 \\ 2 & -4 & 6 \end{array}\right] \xrightarrow[R_3 - 2R_1]{R_2 - 3R_1}$$

$$\left[\begin{array}{cc|c} 1 & -3 & 4 \\ 0 & 7 & -7 \\ 0 & 2 & -2 \end{array}\right] \xrightarrow{\frac{1}{7}R_2} \left[\begin{array}{cc|c} 1 & -3 & 4 \\ 0 & 1 & -1 \\ 0 & 2 & -2 \end{array}\right] \xrightarrow[R_3 - 2R_2]{R_1 + 3R_2} \left[\begin{array}{cc|c} 1 & 0 & 1 \\ 0 & 1 & -1 \\ 0 & 0 & 0 \end{array}\right].$$

We conclude that the solution is $(1,-1)$.

19. $$\left[\begin{array}{cc|c} 1 & -2 & 2 \\ 7 & -14 & 14 \\ 3 & -6 & 6 \end{array}\right] \xrightarrow[R_3-3R_1]{R_2-7R_1} \left[\begin{array}{cc|c} 1 & -2 & 2 \\ 0 & 0 & 0 \\ 0 & 0 & 0 \end{array}\right].$$

We conclude that the infinitely many solutions are given by $(2t+2, t)$, where t is a parameter.

21. $$\left[\begin{array}{cc|c} 3 & 2 & 4 \\ -\frac{3}{2} & -1 & -2 \\ 6 & 4 & 8 \end{array}\right] \xrightarrow{\frac{1}{3}R_1} \left[\begin{array}{cc|c} 1 & \frac{2}{3} & \frac{4}{3} \\ -\frac{3}{2} & -1 & -2 \\ 6 & 4 & 8 \end{array}\right] \xrightarrow[R_3-6R_1]{R_2+\frac{3}{2}R_1} \left[\begin{array}{cc|c} 1 & \frac{2}{3} & \frac{4}{3} \\ 0 & 0 & 0 \\ 0 & 0 & 0 \end{array}\right].$$

We conclude that the infinitely many solutions are given by $(\frac{4}{3}-\frac{2}{3}t,\ t)$, where t is a parameter.

23. $$\left[\begin{array}{ccc|c} 2 & -1 & 1 & -4 \\ 3 & -\frac{3}{2} & \frac{3}{2} & -6 \\ -6 & 3 & -3 & 12 \end{array}\right] \xrightarrow{\frac{1}{2}R_1} \left[\begin{array}{ccc|c} 1 & -\frac{1}{2} & \frac{1}{2} & -2 \\ 3 & -\frac{3}{2} & \frac{3}{2} & -6 \\ -6 & 3 & -3 & 12 \end{array}\right] \xrightarrow[R_3+6R_1]{R_2-3R_1} \left[\begin{array}{ccc|c} 1 & -\frac{1}{2} & \frac{1}{2} & -2 \\ 0 & 0 & 0 & 0 \\ 0 & 0 & 0 & 0 \end{array}\right].$$

We conclude that the infinitely many solutions are given by $(-2+\frac{1}{2}s-\frac{1}{2}t,\ s\ ,t)$ where s and t are parameters.

25. $$\left[\begin{array}{ccc|c} 1 & -2 & 3 & 4 \\ 2 & 3 & -1 & 2 \\ 1 & 2 & -3 & -6 \end{array}\right] \xrightarrow[R_3-R_1]{R_2-2R_1} \left[\begin{array}{ccc|c} 1 & -2 & 3 & 4 \\ 0 & 7 & -7 & -6 \\ 0 & 4 & -6 & -10 \end{array}\right] \xrightarrow{\frac{1}{7}R_2} \left[\begin{array}{ccc|c} 1 & -2 & 3 & 4 \\ 0 & 1 & -1 & -\frac{6}{7} \\ 0 & 4 & -6 & -10 \end{array}\right]$$

$$\xrightarrow[R_3-4R_2]{R_1+2R_2} \left[\begin{array}{ccc|c} 1 & 0 & 1 & \frac{16}{7} \\ 0 & 1 & -1 & -\frac{6}{7} \\ 0 & 0 & -2 & -\frac{46}{7} \end{array}\right] \xrightarrow{-\frac{1}{2}R_3} \left[\begin{array}{ccc|c} 1 & 0 & 1 & \frac{16}{7} \\ 0 & 1 & -1 & -\frac{6}{7} \\ 0 & 0 & 1 & \frac{23}{7} \end{array}\right] \xrightarrow[R_2+R_3]{R_1-R_3}$$

$$\left[\begin{array}{ccc|c} 1 & 0 & 0 & -1 \\ 0 & 1 & 0 & \frac{17}{7} \\ 0 & 0 & 1 & \frac{23}{7} \end{array}\right].$$

We conclude that the solution is $(-1, \frac{17}{7}, \frac{23}{7})$.

27. $\left[\begin{array}{ccc|c} 4 & 1 & -1 & 4 \\ 8 & 2 & -2 & 8 \end{array}\right] \xrightarrow{\frac{1}{4}R_1} \left[\begin{array}{ccc|c} 1 & \frac{1}{4} & \frac{1}{4} & 1 \\ 8 & 2 & -2 & 8 \end{array}\right] \xrightarrow{R_2-8R_1} \left[\begin{array}{ccc|c} 1 & \frac{1}{4} & -\frac{1}{4} & 1 \\ 0 & 0 & 0 & 0 \end{array}\right]$

We conclude that the infinitely many solutions are given by $(1-\frac{1}{4}s+\frac{1}{4}t,\ s,\ t)$, where s and t are parameters.

29. $\left[\begin{array}{ccc|c} 2 & 1 & -3 & 1 \\ 1 & -1 & 2 & 1 \\ 5 & -2 & 3 & 6 \end{array}\right] \xrightarrow{R_1 \leftrightarrow R_2} \left[\begin{array}{ccc|c} 1 & -1 & 2 & 1 \\ 2 & 1 & -3 & 1 \\ 5 & -2 & 3 & 6 \end{array}\right] \xrightarrow[R_3-5R_1]{R_2-2R_1} \left[\begin{array}{ccc|c} 1 & -1 & 2 & 1 \\ 0 & 3 & -7 & -1 \\ 0 & 3 & -7 & 1 \end{array}\right] \xrightarrow{\frac{1}{3}R_2}$

$\left[\begin{array}{ccc|c} 1 & -1 & 2 & 1 \\ 0 & 1 & -\frac{7}{3} & -\frac{1}{3} \\ 0 & 3 & -7 & 1 \end{array}\right] \xrightarrow[R_3-3R_2]{R_1+R_2} \left[\begin{array}{ccc|c} 1 & 0 & -\frac{1}{3} & \frac{2}{3} \\ 0 & 1 & -\frac{7}{3} & -\frac{1}{3} \\ 0 & 0 & 0 & 2 \end{array}\right].$

This last row implies that $0 = 2$, which is impossible. We conclude that the system of equations is inconsistent and has no solution.

31. $\left[\begin{array}{ccc|c} 1 & 2 & -1 & -4 \\ 2 & 1 & 1 & 7 \\ 1 & 3 & 2 & 7 \\ 1 & -3 & 1 & 9 \end{array}\right] \xrightarrow[\substack{R_3-R_1 \\ R_4-R_1}]{R_2-2R_1} \left[\begin{array}{ccc|c} 1 & 2 & -1 & -4 \\ 0 & -3 & 3 & 15 \\ 0 & 1 & 3 & 11 \\ 0 & -5 & 2 & 13 \end{array}\right] \xrightarrow{-\frac{1}{3}R_2} \left[\begin{array}{ccc|c} 1 & 2 & -1 & -4 \\ 0 & 1 & -1 & -5 \\ 0 & 1 & 3 & 11 \\ 0 & -5 & 2 & 13 \end{array}\right]$

$\xrightarrow[\substack{R_3-R_2 \\ R_4+5R_2}]{R_1-2R_2} \left[\begin{array}{ccc|c} 1 & 0 & 1 & 6 \\ 0 & 1 & -1 & -5 \\ 0 & 0 & 4 & 16 \\ 0 & 0 & -3 & -12 \end{array}\right] \xrightarrow{\frac{1}{4}R_3} \left[\begin{array}{ccc|c} 1 & 0 & 1 & 6 \\ 0 & 1 & -1 & -5 \\ 0 & 0 & 1 & 4 \\ 0 & 0 & -3 & -12 \end{array}\right] \xrightarrow[\substack{R_2+R_3 \\ R_4+3R_3}]{R_1+\frac{1}{3}R_3} \left[\begin{array}{ccc|c} 1 & 0 & 0 & 2 \\ 0 & 1 & 0 & -1 \\ 0 & 0 & 1 & 4 \\ 0 & 0 & 0 & 0 \end{array}\right].$

We conclude that the solution of the system is (2,–1,4).

33. Let x, y, and z represent the number of compact, mid-sized, and full-size cars, respectively, to be purchased. Then the problem can be solved by solving the system

$$\begin{aligned} x + \quad y + \quad z &= 60 \\ 10000x + 16000y + 22000z &= 840000 \end{aligned}.$$

Using the Gauss-Jordan method, we have

$$\left[\begin{array}{ccc|c} 1 & 1 & 1 & 60 \\ 10000 & 16000 & 22000 & 840000 \end{array}\right] \xrightarrow{R_2-10{,}000R_1} \left[\begin{array}{ccc|c} 1 & 1 & 1 & 60 \\ 0 & 6000 & 12000 & 240000 \end{array}\right]$$

$$\xrightarrow{\frac{1}{6000}R_2} \left[\begin{array}{ccc|c} 1 & 1 & 1 & 60 \\ 0 & 1 & 2 & 40 \end{array}\right] \xrightarrow{R_1-R_2} \left[\begin{array}{ccc|c} 1 & 0 & -1 & 20 \\ 0 & 1 & 2 & 40 \end{array}\right]$$

and we conclude that the solution is $(20 + z, 40 - 2z, z)$. Letting $z = 5$, we see that one possible solution is (25,30,5); that is Hartman should buy 25 compact, 30 mid-sized cars, and 5 full-sized cars. Letting $z = 10$, we see that another possible solution is (30,20,10); that is, 30 compact cars, 20 mid-sized cars, and 10 full-sized cars.

35. Let x, y, and z denote the number of ounces of Food I, Food II, and Food III, respectively, that the dietician includes in the meal. Then the problem can be solved by solving the system

$$\begin{aligned} 400x + 1200y + 800z &= 8800 \\ 110x + 570y + 340z &= 2160 \\ 90x + 30y + 60z &= 1020. \end{aligned}$$

Using the Gauss-Jordan method, we have

$$\left[\begin{array}{ccc|c} 400 & 1200 & 800 & 8800 \\ 110 & 570 & 340 & 2160 \\ 90 & 30 & 60 & 1020 \end{array}\right] \xrightarrow{\frac{1}{400}R_1} \left[\begin{array}{ccc|c} 1 & 3 & 2 & 22 \\ 110 & 570 & 340 & 2160 \\ 90 & 30 & 60 & 1020 \end{array}\right] \xrightarrow[R_3-90R_1]{R_2-110R_1}$$

$$\left[\begin{array}{ccc|c} 1 & 3 & 2 & 22 \\ 0 & 240 & 120 & -260 \\ 0 & -240 & -120 & -960 \end{array}\right] \xrightarrow{\frac{1}{240}R_2} \left[\begin{array}{ccc|c} 1 & 3 & 2 & 22 \\ 0 & 1 & \frac{1}{2} & -\frac{13}{12} \\ 0 & -240 & -120 & -960 \end{array}\right] \xrightarrow[R_3+240R_2]{R_1-3R_2}$$

$$\left[\begin{array}{ccc|c} 1 & 0 & \frac{1}{2} & \frac{225}{12} \\ 0 & 1 & \frac{1}{2} & -\frac{13}{12} \\ 0 & 0 & 0 & -1220 \end{array}\right].$$

This last row implies that $0 = -1220$, which is impossible. We conclude that the system of equations is inconsistent and has no solution--that is, the dietician cannot prepare a meal from these foods and meet the given requirements.

37. a. $$\begin{array}{rcl} x_1 - x_2 & & = 200 \\ x_1 \qquad\qquad - x_5 & & = 100 \\ -x_2 + x_3 \qquad\qquad + x_6 & & = 600 \\ -x_3 + x_4 & & = 200 \\ x_4 - x_5 + x_6 & & = 700. \end{array}$$

b.

$$\left[\begin{array}{rrrrrr|r} 1 & -1 & 0 & 0 & 0 & 0 & 200 \\ 1 & 0 & 0 & 0 & -1 & 0 & 100 \\ 0 & -1 & 1 & 0 & 0 & 1 & 600 \\ 0 & 0 & -1 & 1 & 0 & 0 & 200 \\ 0 & 0 & 0 & 1 & -1 & 1 & 700 \end{array}\right] \xrightarrow{R_2-R_1} \left[\begin{array}{rrrrrr|r} 1 & -1 & 0 & 0 & 0 & 0 & 200 \\ 0 & 1 & 0 & 0 & -1 & 0 & -100 \\ 0 & -1 & 1 & 0 & 0 & 1 & 600 \\ 0 & 0 & -1 & 1 & 0 & 0 & 200 \\ 0 & 0 & 0 & 1 & -1 & 1 & 700 \end{array}\right] \xrightarrow[R_1+R_2]{R_3+R_2}$$

$$\left[\begin{array}{rrrrrr|r} 1 & 0 & 0 & 0 & -1 & 0 & 100 \\ 0 & 1 & 0 & 0 & -1 & 0 & -100 \\ 0 & 0 & 1 & 0 & -1 & 1 & 500 \\ 0 & 0 & -1 & 1 & 0 & 0 & 200 \\ 0 & 0 & 0 & 1 & -1 & 1 & 700 \end{array}\right] \xrightarrow{R_4+R_3} \left[\begin{array}{rrrrrr|r} 1 & 0 & 0 & 0 & -1 & 0 & 100 \\ 0 & 1 & 0 & 0 & -1 & 0 & -100 \\ 0 & 0 & 1 & 0 & -1 & 1 & 500 \\ 0 & 0 & 0 & 1 & -1 & 1 & 700 \\ 0 & 0 & 0 & 1 & -1 & 1 & 700 \end{array}\right] \xrightarrow{R_5-R_4}$$

$$\left[\begin{array}{rrrrrr|r} 1 & 0 & 0 & 0 & -1 & 0 & 100 \\ 0 & 1 & 0 & 0 & 1 & 0 & -100 \\ 0 & 0 & 1 & 0 & -1 & 1 & 500 \\ 0 & 0 & 0 & 1 & -1 & 1 & 700 \\ 0 & 0 & 0 & 0 & 0 & 0 & 0 \end{array}\right].$$

We conclude that the solution is

$(s+100,\ s-100,\ s-t+500,\ s-t+700,\ s,\ t)$.

Taking $s = 150$ and $t = 50$, we see that one possible traffic pattern is

(250, 50, 600, 800, 150, 50).

Similarly, taking $s = 200$, and $t = 100$, we see that another possible traffic pattern is

(300, 100, 600, 800, 200, 100).

c. Taking $t = 0$ and $s = 200$, we see that another possible traffic pattern is

(300, 100, 700, 900, 200, 0).

39. We solve the given system by using the Gauss-Jordan method. We have

$$\left[\begin{array}{cc|c} 2 & 3 & 2 \\ 1 & 4 & 6 \\ 5 & k & 2 \end{array}\right] \xrightarrow{R_1 \leftrightarrow R_2} \left[\begin{array}{cc|c} 1 & 4 & 6 \\ 2 & 3 & 2 \\ 5 & k & 2 \end{array}\right] \xrightarrow[R_3-4R_1]{R_2-2R_1} \left[\begin{array}{cc|c} 1 & 4 & 6 \\ 0 & -5 & -10 \\ 0 & k-20 & -28 \end{array}\right] \xrightarrow{-\frac{1}{5}R_2}$$

$$\left[\begin{array}{cc|c} 1 & 4 & 6 \\ 0 & 1 & 2 \\ 0 & k-20 & -28 \end{array}\right] \xrightarrow[R_3+aR_2]{R_1-4R_2} \left[\begin{array}{cc|c} 1 & 0 & -2 \\ 0 & 1 & 2 \\ 0 & k+a-20 & -28+2a \end{array}\right]$$

From the last matrix, we see that the system has a solution if and only if $x = -2$, $y = 2$, and

$$-28 + 2a = 0, \text{ or } a = 14$$

and $$k + a - 20 = k - 6 = 0, \text{ or } k = 6.$$

(All the entries in the last row of the matrix must be equal to zero.)

USING TECHNOLOGY EXERCISES 2.3, page 111

1. $(1+t, 2+t, t)$; t, a parameter   3. $(-\frac{17}{7}+\frac{6}{7}t, 3-t, -\frac{18}{7}+\frac{1}{7}t, t)$ 5. No solution

EXERCISES 2.4, page 120

1. The size of A is 4×4; the size of B is 4×3; the size of C is 1×5, and the size of is 4×1.

3. These are entries of the matrix B. The entry b_{13} refers to the entry in the first row and third column and is equal to 2. Similarly, $b_{31} = 3$, and $b_{43} = 8$.

5. The column matrix is the matrix D. The transpose of the matrix D is

$$D^T = [1 \;\; 3 \;\; -2 \;\; 0].$$

7. A is of size 3×2; B is of size 3×2; C and D are of size 3×3.

9. $$A + B = \begin{bmatrix} -1 & 2 \\ 3 & -2 \\ 4 & 0 \end{bmatrix} + \begin{bmatrix} 2 & 4 \\ 3 & 1 \\ -2 & 2 \end{bmatrix} = \begin{bmatrix} 1 & 6 \\ 6 & -1 \\ 2 & 2 \end{bmatrix}.$$

11. $\begin{bmatrix} 3 & -1 & 0 \\ 2 & -2 & 3 \\ 4 & 6 & 2 \end{bmatrix} - \begin{bmatrix} 2 & -2 & 4 \\ 3 & 6 & 2 \\ -2 & 3 & 1 \end{bmatrix} = \begin{bmatrix} 1 & 1 & -4 \\ -1 & -8 & 1 \\ 6 & 3 & 1 \end{bmatrix}.$

13. $\begin{bmatrix} 6 & 3 & 8 \\ 4 & 5 & 6 \end{bmatrix} - \begin{bmatrix} 3 & -2 & -1 \\ 0 & -5 & -7 \end{bmatrix} = \begin{bmatrix} 3 & 5 & 9 \\ 4 & 10 & 13 \end{bmatrix}.$

15. $\begin{bmatrix} 1 & 4 & -5 \\ 3 & -8 & 6 \end{bmatrix} + \begin{bmatrix} 4 & 0 & -2 \\ 3 & 6 & 5 \end{bmatrix} - \begin{bmatrix} 2 & 8 & 9 \\ -11 & 2 & -5 \end{bmatrix} = \begin{bmatrix} 3 & -4 & -16 \\ 17 & -4 & 16 \end{bmatrix}.$

17. $\begin{bmatrix} 1.2 & 4.5 & -4.2 \\ 8.2 & 6.3 & -3.2 \end{bmatrix} - \begin{bmatrix} 3.1 & 1.5 & -3.6 \\ 2.2 & -3.3 & -4.4 \end{bmatrix} = \begin{bmatrix} -1.9 & 3.0 & -0.6 \\ 6.0 & 9.6 & 1.2 \end{bmatrix}.$

19. $\frac{1}{2}\begin{bmatrix} 1 & 0 & 0 & -4 \\ 3 & 0 & -1 & 6 \\ -2 & 1 & -4 & 2 \end{bmatrix} + \frac{4}{3}\begin{bmatrix} 3 & 0 & -1 & 4 \\ -2 & 1 & -6 & 2 \\ 8 & 2 & 0 & -2 \end{bmatrix} - \frac{1}{3}\begin{bmatrix} 3 & -9 & -1 & 0 \\ 6 & 2 & 0 & -6 \\ 0 & 1 & -3 & 1 \end{bmatrix}$

$$= \begin{bmatrix} \frac{7}{2} & 3 & -1 & \frac{10}{3} \\ -\frac{19}{6} & \frac{2}{3} & -\frac{17}{2} & \frac{23}{2} \\ \frac{29}{3} & \frac{17}{6} & -1 & -2 \end{bmatrix}.$$

21. $\begin{bmatrix} 2x-2 & 3 & 2 \\ 2 & 4 & y-2 \\ 2z & -3 & 2 \end{bmatrix} = \begin{bmatrix} 3 & u & 2 \\ 2 & 4 & 5 \\ 4 & -3 & 2 \end{bmatrix}.$

Now, by the definition of equality of matrices,

$u = 3$

$2x - 2 = 3$ and $2x = 5$, or $x = 5/2$,

$y - 2 = 5$, and $y = 7$,

$2z = 4$, and $z = 2$.

23. $\begin{bmatrix} 1 & x \\ 2y & -3 \end{bmatrix} - 4\begin{bmatrix} 2 & -2 \\ 0 & 3 \end{bmatrix} = \begin{bmatrix} 3z & 10 \\ 4 & -u \end{bmatrix}$; $\begin{bmatrix} -7 & x+8 \\ 2y & -15 \end{bmatrix} = \begin{bmatrix} 3z & 10 \\ 4 & -u \end{bmatrix}$.

Now, by the definition of equality of matrices,

$$-u = -15, \text{ so } u = 15$$

$$x+8 = 10, \quad \text{so } x = 2$$

$$2y = 4, \quad \text{so } y = 2$$

$$3z = -7, \text{ so } z = -7/3.$$

25. To verify the Commutative Law for matrix addition, let us show that $A+B = B+A$.

Now, $A+B = \begin{bmatrix} 2 & -4 & 3 \\ 4 & 2 & 1 \end{bmatrix} + \begin{bmatrix} 4 & -3 & 2 \\ 1 & 0 & 4 \end{bmatrix} = \begin{bmatrix} 6 & -7 & 5 \\ 5 & 2 & 5 \end{bmatrix}$

$$= \begin{bmatrix} 4 & -3 & 2 \\ 1 & 0 & 4 \end{bmatrix} + \begin{bmatrix} 2 & -4 & 3 \\ 4 & 2 & 1 \end{bmatrix} = B + A.$$

27. $(3+5)A = 8A = 8\begin{bmatrix} 3 & 1 \\ 2 & 4 \\ -4 & 0 \end{bmatrix} = \begin{bmatrix} 24 & 8 \\ 16 & 32 \\ -32 & 0 \end{bmatrix} = 3\begin{bmatrix} 3 & 1 \\ 2 & 4 \\ -4 & 0 \end{bmatrix} + 5\begin{bmatrix} 3 & 1 \\ 2 & 4 \\ -4 & 0 \end{bmatrix}$

$=3A+5A$.

29. $4(A+B) = 4\left[\begin{bmatrix} 3 & 1 \\ 2 & 4 \\ -4 & 0 \end{bmatrix} + \begin{bmatrix} 1 & 2 \\ -1 & 0 \\ 3 & 2 \end{bmatrix}\right] = 4\begin{bmatrix} 4 & 3 \\ 1 & 4 \\ -1 & 2 \end{bmatrix} = \begin{bmatrix} 16 & 12 \\ 4 & 16 \\ -4 & 8 \end{bmatrix}$

$$4A + 4B = 4\begin{bmatrix} 3 & 1 \\ 2 & 4 \\ -4 & 0 \end{bmatrix} + 4\begin{bmatrix} 1 & 2 \\ -1 & 0 \\ 3 & 2 \end{bmatrix} = \begin{bmatrix} 16 & 12 \\ 4 & 16 \\ -4 & 8 \end{bmatrix}.$$

31. $\begin{bmatrix} 3 & 2 & -1 & 5 \end{bmatrix}^T = \begin{bmatrix} 3 \\ 2 \\ -1 \\ 5 \end{bmatrix}$.

33. $\begin{bmatrix} 1 & -1 & 2 \\ 3 & 4 & 2 \\ 0 & 1 & 0 \end{bmatrix}^T = \begin{bmatrix} 1 & 3 & 0 \\ -1 & 4 & 1 \\ 2 & 2 & 0 \end{bmatrix}$.

$$
35. \quad \begin{array}{r} \\ Mr.\ Cross \\ Mr.\ Jones \\ Mr.\ Smith \end{array}
\begin{array}{c} \begin{array}{cccc} 1 & 2 & 3 & 4 \end{array} \\ \begin{bmatrix} 220 & 215 & 210 & 205 \\ 220 & 210 & 200 & 195 \\ 215 & 205 & 195 & 190 \end{bmatrix} \end{array}
$$

37. a. $D = A + B - C$

$$
= \begin{bmatrix} 2820 & 1470 & 1120 \\ 1030 & 520 & 480 \\ 1170 & 540 & 460 \end{bmatrix} + \begin{bmatrix} 260 & 120 & 110 \\ 140 & 60 & 50 \\ 120 & 70 & 50 \end{bmatrix} - \begin{bmatrix} 120 & 80 & 80 \\ 70 & 30 & 40 \\ 60 & 20 & 40 \end{bmatrix}
$$

$$
= \begin{bmatrix} 2960 & 1510 & 1150 \\ 1100 & 550 & 490 \\ 1230 & 590 & 470 \end{bmatrix}.
$$

b. $E = 1.1D = 1.1\begin{bmatrix} 2960 & 1510 & 1150 \\ 1100 & 550 & 490 \\ 1230 & 590 & 470 \end{bmatrix} = \begin{bmatrix} 3256 & 1661 & 1265 \\ 1210 & 605 & 539 \\ 1353 & 649 & 517 \end{bmatrix}.$

USING TECHNOLOGY EXERCISES 2.4, page 123

1. $\begin{bmatrix} 15 & 38.75 & -67.5 & 33.75 \\ 51.25 & 40 & 52.5 & -38.75 \\ 21.25 & 35 & -65 & 105 \end{bmatrix}$ 3. $\begin{bmatrix} -5 & 6.3 & -6.8 & 3.9 \\ 1 & 0.5 & 5.4 & -4.8 \\ 0.5 & 4.2 & -3.5 & 5.6 \end{bmatrix}$

5. $\begin{bmatrix} 16.44 & -3.65 & -3.66 & 0.63 \\ 12.77 & 10.64 & 2.58 & 0.05 \\ 5.09 & 0.28 & -10.84 & 17.64 \end{bmatrix}$ 7. $\begin{bmatrix} 7.4 & 7.2 & 2.9 \\ -0.1 & 5.9 & 1.4 \\ -4 & 3 & -6.9 \\ 1.5 & -1.4 & 11.2 \end{bmatrix}$

EXERCISES 2.5, page 133

1. $\underset{=}{(2 \times \underset{\uparrow}{3})(\underset{\uparrow}{3} \times 5)}$ so AB has order 2×5.

   $(3 \times \underset{\uparrow}{5})(\underset{\uparrow}{2} \times 3)$, with $5 \neq 2$, so BA is not defined.

3. $(1 \times \underset{\uparrow}{7})(\underset{\uparrow}{7} \times 1)$, with $7 = 7$, so AB has order 1×1.

   $(7 \times \underset{\uparrow}{1})(\underset{\uparrow}{1} \times 7)$, with $1 = 1$, so AB has order 7×7.

5. If AB and BA are defined then $n = s$ and $m = t$.

7. $$\begin{bmatrix} 1 & 2 \\ 3 & 0 \end{bmatrix}\begin{bmatrix} 1 \\ -1 \end{bmatrix} = \begin{bmatrix} -1 \\ 3 \end{bmatrix}$$

9. $$\begin{bmatrix} 3 & 1 & 2 \\ -1 & 2 & 4 \end{bmatrix}\begin{bmatrix} 4 \\ 1 \\ -2 \end{bmatrix} = \begin{bmatrix} 9 \\ -10 \end{bmatrix}$$

11. $$\begin{bmatrix} -1 & 2 \\ 3 & 1 \end{bmatrix}\begin{bmatrix} 2 & 4 \\ 3 & 1 \end{bmatrix} = \begin{bmatrix} 4 & -2 \\ 9 & 13 \end{bmatrix}$$

13. $$\begin{bmatrix} 2 & 1 & 2 \\ 3 & 2 & 4 \end{bmatrix}\begin{bmatrix} -1 & 2 \\ 4 & 3 \\ 0 & 1 \end{bmatrix} = \begin{bmatrix} 2 & 9 \\ 5 & 16 \end{bmatrix}$$

15. $$\begin{bmatrix} 0.1 & 0.9 \\ 0.2 & 0.8 \end{bmatrix}\begin{bmatrix} 1.2 & 0.4 \\ 0.5 & 2.1 \end{bmatrix} = 2\begin{bmatrix} 0.1(1.2)+0.9(0.5) & 0.1(0.4)+0.9(2.1) \\ 0.2(1.2)+0.8(0.5) & 0.2(0.4)+0.8(2.1) \end{bmatrix}$$

$$= \begin{bmatrix} 0.57 & 1.93 \\ 0.64 & 1.76 \end{bmatrix}.$$

17. $$\begin{bmatrix} 6 & -3 & 0 \\ -2 & 1 & -8 \\ 4 & -4 & 9 \end{bmatrix}\begin{bmatrix} 1 & 0 & 0 \\ 0 & 1 & 0 \\ 0 & 0 & 1 \end{bmatrix} = \begin{bmatrix} 6 & -3 & 0 \\ -2 & 1 & -8 \\ 4 & -4 & 9 \end{bmatrix}.$$

19. $$\begin{bmatrix} 3 & 0 & -2 & 1 \\ 1 & 2 & 0 & -1 \end{bmatrix} \begin{bmatrix} 2 & 1 & -1 \\ -1 & 2 & 0 \\ 0 & 0 & 1 \\ -1 & -2 & 2 \end{bmatrix} = \begin{bmatrix} 5 & 1 & -3 \\ 1 & 7 & -3 \end{bmatrix}.$$

21. $$4\begin{bmatrix} 1 & -2 & 0 \\ 2 & -1 & 1 \\ 3 & 0 & -1 \end{bmatrix} \begin{bmatrix} 1 & 3 & 1 \\ 1 & 4 & 0 \\ 0 & 1 & -2 \end{bmatrix} = \begin{bmatrix} -4 & -20 & 4 \\ 4 & 12 & 0 \\ 12 & 32 & 20 \end{bmatrix}$$

23. $$\begin{bmatrix} 1 & 0 \\ 0 & 1 \end{bmatrix} \begin{bmatrix} 4 & -3 & 2 \\ 7 & 1 & -5 \end{bmatrix} \begin{bmatrix} 1 & 0 & 0 \\ 0 & 1 & 0 \\ 0 & 0 & 1 \end{bmatrix} = \begin{bmatrix} 1 & 0 \\ 0 & 1 \end{bmatrix} \begin{bmatrix} 4 & -3 & 2 \\ 7 & 1 & -5 \end{bmatrix} = \begin{bmatrix} 4 & -3 & 2 \\ 7 & 1 & -5 \end{bmatrix}.$$

25. To verify the associative law for matrix multiplication, we will show that $(AB)C = A(BC)$.

$$AB = \begin{bmatrix} 1 & 0 & -2 \\ 1 & -3 & 2 \\ -2 & 1 & 1 \end{bmatrix} \begin{bmatrix} 3 & 1 & 0 \\ 2 & 2 & 0 \\ 1 & -3 & -1 \end{bmatrix} = \begin{bmatrix} 1 & 7 & 2 \\ -1 & -11 & -2 \\ -3 & -3 & -1 \end{bmatrix}$$

$$(AB)C = \begin{bmatrix} 1 & 7 & 2 \\ -1 & -11 & -2 \\ -3 & -3 & -1 \end{bmatrix} \begin{bmatrix} 2 & 1 & 0 \\ 1 & -1 & 2 \\ 3 & -2 & 1 \end{bmatrix} = \begin{bmatrix} 15 & -12 & 16 \\ -19 & 16 & -24 \\ -12 & 8 & -7 \end{bmatrix}$$

$$BC = \begin{bmatrix} 3 & 1 & 0 \\ 2 & 2 & 0 \\ 1 & -3 & -1 \end{bmatrix} \begin{bmatrix} 2 & -1 & 0 \\ 1 & -1 & 2 \\ 3 & -2 & 1 \end{bmatrix} = \begin{bmatrix} 7 & -4 & 2 \\ 6 & -4 & 4 \\ -4 & 4 & -7 \end{bmatrix}$$

$$A(BC) = \begin{bmatrix} 1 & 0 & -2 \\ 1 & -3 & 2 \\ -1 & 1 & 1 \end{bmatrix} \begin{bmatrix} 7 & -4 & 2 \\ 6 & -4 & 4 \\ -4 & 4 & -7 \end{bmatrix} = \begin{bmatrix} 15 & -12 & 16 \\ -19 & 16 & -24 \\ -12 & 8 & -7 \end{bmatrix}.$$

27. $$AB = \begin{bmatrix} 1 & 2 \\ 3 & 4 \end{bmatrix} \begin{bmatrix} 2 & 1 \\ 4 & 3 \end{bmatrix} = \begin{bmatrix} 10 & 7 \\ 22 & 15 \end{bmatrix}$$

$$BA = \begin{bmatrix} 2 & 1 \\ 4 & 3 \end{bmatrix}\begin{bmatrix} 1 & 2 \\ 3 & 4 \end{bmatrix} = \begin{bmatrix} 5 & 8 \\ 13 & 20 \end{bmatrix}$$

Therefore, $AB \neq BA$ and matrix multiplication is not commutative.

29. $$AB = \begin{bmatrix} 3 & 0 \\ 8 & 0 \end{bmatrix}\begin{bmatrix} 0 & 0 \\ 4 & 5 \end{bmatrix} = \begin{bmatrix} 0 & 0 \\ 0 & 0 \end{bmatrix}$$

$AB = 0$, but neither A nor B is the zero matrix. Therefore, $AB = 0$, does not imply that A or B is the zero matrix.

31. $$\begin{bmatrix} a & b \\ c & d \end{bmatrix}\begin{bmatrix} 1 & 0 \\ -1 & 3 \end{bmatrix} = \begin{bmatrix} a-b & 3b \\ c-d & 3d \end{bmatrix} = \begin{bmatrix} -1 & -3 \\ 3 & 6 \end{bmatrix}$$

Then $3b = -3$, and $b = -1$

$3d = 6$, and $d = 2$

$a - b = -1$, and $a = b - 1 = -2$.

$c - d = 3$, and $c = d + 3 = 5$

Therefore, $A = \begin{bmatrix} -2 & -1 \\ 5 & 2 \end{bmatrix}$.

33. a.

$$A^T = \begin{bmatrix} 2 & 5 \\ 4 & -6 \end{bmatrix} \text{ and } (A^T)^T = \begin{bmatrix} 2 & 4 \\ 5 & -6 \end{bmatrix} = A$$

b. $$(A+B)^T = \begin{bmatrix} 6 & 12 \\ -2 & -3 \end{bmatrix}^T = \begin{bmatrix} 6 & -2 \\ 12 & -3 \end{bmatrix}$$

$$A^T + B^T = \begin{bmatrix} 2 & 5 \\ 4 & -6 \end{bmatrix} + \begin{bmatrix} 4 & -7 \\ 8 & 3 \end{bmatrix} = \begin{bmatrix} 6 & -2 \\ 12 & -3 \end{bmatrix}$$

c. $$AB = \begin{bmatrix} 2 & 4 \\ 5 & -6 \end{bmatrix}\begin{bmatrix} 4 & 8 \\ -7 & 3 \end{bmatrix} = \begin{bmatrix} -20 & 28 \\ 62 & 22 \end{bmatrix}$$

so $(AB)^T = \begin{bmatrix} -20 & 62 \\ 28 & 22 \end{bmatrix}$.

$$B^T A^T = \begin{bmatrix} 4 & -7 \\ 8 & 3 \end{bmatrix} \begin{bmatrix} 2 & 5 \\ 4 & -6 \end{bmatrix} \begin{bmatrix} -20 & 62 \\ 28 & 22 \end{bmatrix} = (AB)^T$$

35. The given system of linear equations can be represented by the matrix equation $AX = B$, where

$$A = \begin{bmatrix} 2 & -3 \\ 3 & -4 \end{bmatrix}, \quad X = \begin{bmatrix} x \\ y \end{bmatrix}, \text{ and } B = \begin{bmatrix} 7 \\ 8 \end{bmatrix}.$$

37. The given system of linear equations can be represented by the matrix equation $AX = B$, where

$$A = \begin{bmatrix} 2 & -3 & 4 \\ 0 & 2 & -3 \\ 1 & -1 & 2 \end{bmatrix}, \quad X = \begin{bmatrix} x \\ y \\ z \end{bmatrix}, \quad B = \begin{bmatrix} 6 \\ 7 \\ 4 \end{bmatrix}.$$

39. The given system of linear equations can be represented by the matrix equation $AX = B$, where

$$A = \begin{bmatrix} -1 & 1 & 1 \\ 2 & -1 & -1 \\ -3 & 2 & 4 \end{bmatrix}, \quad X = \begin{bmatrix} x_1 \\ x_2 \\ x_3 \end{bmatrix}, \quad B = \begin{bmatrix} 0 \\ 2 \\ 4 \end{bmatrix}.$$

41. a.

$$AB = \begin{bmatrix} 200 & 300 & 100 & 200 \\ 100 & 200 & 400 & 0 \end{bmatrix} \begin{bmatrix} 54 \\ 48 \\ 98 \\ 82 \end{bmatrix} = \begin{bmatrix} 51{,}400 \\ 54{,}200 \end{bmatrix}$$

b. The first entry shows that William's total stock holdings are \$51,400, while Michael's stockholdings are \$54,200.

43. The column vector that represents the profit for each type of house is

$$B = \begin{bmatrix} 20{,}000 \\ 22{,}000 \\ 25{,}000 \\ 30{,}000 \end{bmatrix}.$$

The column vector that gives the total profit for Bond Brothers is

$$AB = \begin{bmatrix} 60 & 80 & 120 & 40 \\ 20 & 30 & 60 & 10 \\ 10 & 15 & 30 & 5 \end{bmatrix} \begin{bmatrix} 20{,}000 \\ 22{,}000 \\ 25{,}000 \\ 30{,}000 \end{bmatrix}$$

$$= \begin{bmatrix} 7{,}160{,}000 \\ 2{,}860{,}000 \\ 1{,}430{,}000 \end{bmatrix}.$$

Therefore, Bond Brothers expects to make \$7,160,000 in New York, \$2,860,000 in Connecticut, and \$1,430,000 in Massachusetts, and the total profit is \$11,450,000.

45. $$AB = \begin{bmatrix} 2700 & 3000 \\ 800 & 700 \\ 500 & 300 \end{bmatrix} \begin{bmatrix} 0.25 & 0.20 & 0.30 & 0.25 \\ 0.30 & 0.35 & 0.25 & 0.10 \end{bmatrix} = \begin{bmatrix} 1575 & 1590 & 1560 & 975 \\ 410 & 405 & 415 & 270 \\ 215 & 205 & 225 & 155 \end{bmatrix}.$$

47. a. $$AC = \begin{bmatrix} 320 & 280 & 460 & 280 \\ 480 & 360 & 580 & 0 \\ 540 & 420 & 200 & 880 \end{bmatrix} \begin{bmatrix} 120 \\ 180 \\ 260 \\ 500 \end{bmatrix} = \begin{bmatrix} 348{,}400 \\ 273{,}200 \\ 632{,}400 \end{bmatrix}.$$

The entries give the total production costs at locations I, II, and III for the month of May as \$348,400, \$273,200, and \$632,400, respectively.

b. $$AD = \begin{bmatrix} 320 & 280 & 460 & 280 \\ 480 & 360 & 580 & 0 \\ 540 & 420 & 200 & 880 \end{bmatrix} \begin{bmatrix} 160 \\ 250 \\ 350 \\ 700 \end{bmatrix} = \begin{bmatrix} 478{,}200 \\ 369{,}800 \\ 877{,}400 \end{bmatrix}.$$

The total revenue realized at locations I, II, and III for the month of May are \$478,200, \$369,800, and \$877,400, respectively.

$$\text{c. } BC = \begin{bmatrix} 210 & 180 & 330 & 180 \\ 400 & 300 & 450 & 40 \\ 420 & 280 & 180 & 740 \end{bmatrix} \begin{bmatrix} 120 \\ 180 \\ 260 \\ 500 \end{bmatrix} = \begin{bmatrix} 233{,}400 \\ 239{,}000 \\ 517{,}600 \end{bmatrix}.$$

The total production costs at locations I, II, and III for the month of June are \$233,400, \$239,000, and \$517,600, respectively.

$$\text{d. } BD = \begin{bmatrix} 210 & 180 & 330 & 180 \\ 400 & 300 & 450 & 40 \\ 420 & 280 & 180 & 740 \end{bmatrix} \begin{bmatrix} 160 \\ 250 \\ 350 \\ 700 \end{bmatrix} = \begin{bmatrix} 320{,}100 \\ 324{,}500 \\ 718{,}200 \end{bmatrix}.$$

The total revenue realized at locations I, II, and III for the month of June are \$320,100, \$324,500, and \$718,200, respectively.

$$\text{e. } (A+B)C = \begin{bmatrix} 530 & 460 & 790 & 460 \\ 880 & 660 & 1030 & 40 \\ 960 & 700 & 380 & 1620 \end{bmatrix} \begin{bmatrix} 120 \\ 180 \\ 500 \\ 700 \end{bmatrix} = \begin{bmatrix} 581{,}800 \\ 512{,}200 \\ 1{,}150{,}000 \end{bmatrix}.$$

The total production costs in May and June are Locations I, II, and III are \$581,800, \$512,200, and \$1,150,000, respectively.

$$\text{f. } (A+B)D = \begin{bmatrix} 530 & 460 & 790 & 460 \\ 880 & 660 & 1030 & 40 \\ 960 & 700 & 380 & 1620 \end{bmatrix} \begin{bmatrix} 160 \\ 250 \\ 350 \\ 700 \end{bmatrix} = \begin{bmatrix} 798{,}300 \\ 694{,}300 \\ 1{,}595{,}600 \end{bmatrix}.$$

The total revenue realized in May and June in Locations I, II, and III are \$798,300, \$694,300, and \$1,595,600, respectively.

$$\text{g. } A(D-C) = \begin{bmatrix} 320 & 280 & 460 & 280 \\ 480 & 360 & 580 & 0 \\ 540 & 420 & 200 & 880 \end{bmatrix} \begin{bmatrix} 40 \\ 70 \\ 90 \\ 200 \end{bmatrix} = \begin{bmatrix} 129{,}800 \\ 96{,}600 \\ 245{,}000 \end{bmatrix}.$$

The profits in Locations I, II, and III in May are $129,800, $96,600, and $245,000, respectively.

h. $B(D-C)=\begin{bmatrix} 86{,}700 \\ 85{,}500 \\ 200{,}600 \end{bmatrix}$

The profits in Locations I, II, and III in June are $86,700, $85,500, and $200,600, respectively.

i. $(A+B)(D-C)=\begin{bmatrix} 216{,}500 \\ 182{,}100 \\ 445{,}600 \end{bmatrix}$.

The profits in Locations I, II, and III in May and June are $216,500, $182,100, $445,600, respectively.

USING TECHNOLOGY EXERCISES 2.5, page 137

1. $\begin{bmatrix} 18.66 & 15.2 & -12 \\ 24.48 & 41.88 & 89.82 \\ 15.39 & 7.16 & -1.25 \end{bmatrix}$

3. $\begin{bmatrix} 20.09 & 20.61 & -1.3 \\ 44.42 & 71.6 & 64.89 \\ 20.97 & 7.17 & -60.65 \end{bmatrix}$

5. $\begin{bmatrix} 32.89 & 13.63 & -57.17 \\ -12.85 & -8.37 & 256.92 \\ 13.48 & 14.29 & 181.64 \end{bmatrix}$

7. $\begin{bmatrix} 18.66 & 24.48 & 15.39 \\ 15.2 & 41.88 & 7.16 \\ -12 & 89.82 & -1.25 \end{bmatrix}$

9. $\begin{bmatrix} 87 & 68 & 110 & 82 \\ 119 & 176 & 221 & 143 \\ 51 & 128 & 142 & 94 \\ 28 & 174 & 174 & 112 \end{bmatrix}$ $\begin{bmatrix} 113 & 117 & 72 & 101 & 90 \\ 72 & 85 & 36 & 72 & 76 \\ 81 & 69 & 76 & 87 & 30 \\ 133 & 157 & 56 & 121 & 146 \\ 154 & 157 & 94 & 127 & 122 \end{bmatrix}$

11. $\begin{bmatrix} 170 & 18.1 & 133.1 & -106.3 & 341.3 \\ 349 & 226.5 & 324.1 & 164 & 506.4 \\ 245.2 & 157.7 & 231.5 & 125.5 & 312.9 \\ 310 & 245.2 & 291 & 274.3 & 354.2 \end{bmatrix}$

EXERCISES 2.6, page 149

1. $\begin{bmatrix} 1 & -3 \\ 1 & -2 \end{bmatrix}\begin{bmatrix} -2 & 3 \\ -1 & 1 \end{bmatrix} = \begin{bmatrix} 1 & 0 \\ 0 & 1 \end{bmatrix}$; $\begin{bmatrix} -2 & 3 \\ -1 & 1 \end{bmatrix}\begin{bmatrix} 1 & -3 \\ 1 & -2 \end{bmatrix} = \begin{bmatrix} 1 & 0 \\ 0 & 1 \end{bmatrix}$

3. $\begin{bmatrix} 3 & 2 & 3 \\ 2 & 2 & 1 \\ 2 & 1 & 1 \end{bmatrix}\begin{bmatrix} -\frac{1}{3} & -\frac{1}{3} & \frac{4}{3} \\ 0 & 1 & -1 \\ \frac{2}{3} & -\frac{1}{3} & -\frac{2}{3} \end{bmatrix} = \begin{bmatrix} 1 & 0 & 0 \\ 0 & 1 & 0 \\ 0 & 0 & 1 \end{bmatrix}$ and

$$\begin{bmatrix} -\frac{1}{3} & -\frac{1}{3} & \frac{4}{3} \\ 0 & 1 & -1 \\ \frac{2}{3} & -\frac{1}{3} & -\frac{2}{3} \end{bmatrix}\begin{bmatrix} 3 & 2 & 3 \\ 2 & 2 & 1 \\ 2 & 1 & 1 \end{bmatrix} = \begin{bmatrix} 1 & 0 & 0 \\ 0 & 1 & 0 \\ 0 & 0 & 1 \end{bmatrix}.$$

5. Using Formula (13), we find

$$A^{-1} = \frac{1}{(2)(3)-(1)(5)}\begin{bmatrix} 3 & -5 \\ -1 & 2 \end{bmatrix} = \begin{bmatrix} 3 & -5 \\ -1 & 2 \end{bmatrix}.$$

7. Since $ad - bc = (3)(2) - (-2)(-3) = 6 - 6 = 0$, the inverse does not exist.

9. $$\left[\begin{array}{rrr|rrr} 2 & -3 & -4 & 1 & 0 & 0 \\ 0 & 0 & -1 & 0 & 1 & 0 \\ 1 & -2 & 1 & 0 & 0 & 1 \end{array}\right] \xrightarrow{R_1 \leftrightarrow R_3} \left[\begin{array}{rrr|rrr} 1 & -2 & 1 & 0 & 0 & 1 \\ 0 & 0 & -1 & 0 & 1 & 0 \\ 2 & -3 & -4 & 1 & 0 & 0 \end{array}\right] \xrightarrow{R_3 - 2R_1}$$

$$\left[\begin{array}{rrr|rrr} 1 & -2 & 1 & 0 & 0 & 1 \\ 0 & 0 & -1 & 0 & 1 & 0 \\ 0 & 1 & -6 & 1 & 0 & -2 \end{array}\right] \xrightarrow{R_2 \leftrightarrow R_3} \left[\begin{array}{rrr|rrr} 1 & -2 & 1 & 0 & 0 & 1 \\ 0 & 1 & -6 & 1 & 0 & -2 \\ 0 & 0 & -1 & 0 & 1 & 0 \end{array}\right] \xrightarrow[-R_3]{R_1 + 2R_2}$$

$$\left[\begin{array}{rrr|rrr} 1 & 0 & -11 & 2 & 0 & -3 \\ 0 & 1 & -6 & 1 & 0 & -2 \\ 0 & 0 & 1 & 0 & -1 & 0 \end{array}\right] \xrightarrow[R_2 + 6R_3]{R_1 + 11R_3} \left[\begin{array}{rrr|rrr} 1 & 0 & 0 & 2 & -11 & -3 \\ 0 & 1 & 0 & 1 & -6 & -2 \\ 0 & 0 & 1 & 0 & -1 & 0 \end{array}\right].$$

Therefore, the required inverse is $\begin{bmatrix} 2 & -11 & -3 \\ 1 & -6 & -2 \\ 0 & -1 & 0 \end{bmatrix}$.

11. $$\left[\begin{array}{rrr|rrr} 4 & 2 & 2 & 1 & 0 & 0 \\ -1 & -3 & 4 & 0 & 1 & 0 \\ 3 & -1 & 6 & 0 & 0 & 1 \end{array}\right] \xrightarrow{R_1-R_3} \left[\begin{array}{rrr|rrr} 1 & 3 & -4 & 1 & 0 & -1 \\ -1 & -3 & 4 & 0 & 1 & 0 \\ 3 & -1 & 6 & 0 & 0 & 1 \end{array}\right]$$

$$\xrightarrow[R_3+5R_1]{R_2+R_1} \left[\begin{array}{rrr|rrr} 1 & 3 & -4 & 1 & 0 & -1 \\ 0 & 0 & 0 & 1 & 1 & -1 \\ 3 & -1 & 6 & 0 & 0 & 1 \end{array}\right]$$

Because there is a row of zeros to the left of the vertical line, we see that the inverse does not exist.

13. $$\left[\begin{array}{rrr|rrr} 1 & 4 & -1 & 1 & 0 & 0 \\ 2 & 3 & -2 & 0 & 1 & 0 \\ -1 & 2 & 3 & 0 & 0 & 1 \end{array}\right] \xrightarrow[R_3+R_1]{R_2-2R_1} \left[\begin{array}{rrr|rrr} 1 & 4 & -1 & 1 & 0 & 0 \\ 0 & -5 & 0 & -2 & 1 & 0 \\ 0 & 6 & 2 & 1 & 0 & 1 \end{array}\right] \xrightarrow{R_2+R_3}$$

$$\left[\begin{array}{rrr|rrr} 1 & 4 & -1 & 1 & 0 & 0 \\ 0 & 1 & 2 & -1 & 1 & 1 \\ 0 & 6 & 2 & 1 & 0 & 1 \end{array}\right] \xrightarrow[R_3-6R_2]{R_1-4R_2} \left[\begin{array}{rrr|rrr} 1 & 0 & -9 & 5 & -4 & -4 \\ 0 & 1 & 2 & -1 & 1 & 1 \\ 0 & 0 & -10 & 7 & -6 & -5 \end{array}\right] \xrightarrow{-\frac{1}{10}R_3}$$

$$\left[\begin{array}{rrr|rrr} 1 & 0 & -9 & 5 & -4 & -4 \\ 0 & 1 & 2 & -1 & 1 & 1 \\ 0 & 0 & 1 & -\frac{7}{10} & \frac{3}{5} & \frac{1}{2} \end{array}\right] \xrightarrow[R_2-2R_3]{R_1+9R_3} \left[\begin{array}{rrr|rrr} 1 & 0 & 0 & -\frac{13}{10} & \frac{7}{5} & \frac{1}{2} \\ 0 & 1 & 0 & \frac{2}{5} & -\frac{1}{5} & 0 \\ 0 & 0 & 1 & -\frac{7}{10} & \frac{3}{5} & \frac{1}{2} \end{array}\right]$$

So $$A^{-1} = \begin{bmatrix} -\frac{13}{10} & \frac{7}{5} & \frac{1}{2} \\ \frac{2}{5} & -\frac{1}{5} & 0 \\ -\frac{7}{10} & \frac{3}{5} & \frac{1}{2} \end{bmatrix}.$$

15. $$\left[\begin{array}{rrrr|rrrr} 1 & 1 & -1 & 1 & 1 & 0 & 0 & 0 \\ 2 & 1 & 1 & 0 & 0 & 1 & 0 & 0 \\ 2 & 1 & 0 & 1 & 0 & 0 & 1 & 0 \\ 2 & -1 & -1 & 3 & 0 & 0 & 0 & 1 \end{array}\right] \xrightarrow[\substack{R_3-2R_1 \\ R_4-2R_1}]{R_2-2R_1} \left[\begin{array}{rrrr|rrrr} 1 & 1 & -1 & 1 & 1 & 0 & 0 & 0 \\ 0 & -1 & 3 & -2 & -2 & 1 & 0 & 0 \\ 0 & -1 & 2 & -1 & -2 & 0 & 1 & 0 \\ 0 & -3 & 1 & 1 & -2 & 0 & 0 & 1 \end{array}\right] \xrightarrow{-R_2}$$

$$\left[\begin{array}{rrrr|rrrr} 1 & 1 & -1 & 1 & 1 & 0 & 0 & 0 \\ 0 & 1 & -3 & 2 & 2 & -1 & 0 & 0 \\ 0 & -1 & 2 & -1 & -2 & 0 & 1 & 0 \\ 0 & -3 & 1 & 1 & -2 & 0 & 0 & 1 \end{array}\right] \xrightarrow[\substack{R_1-R_2 \\ R_3+R_2 \\ R_4+3R_2}]{} \left[\begin{array}{rrrr|rrrr} 1 & 0 & 2 & -1 & -1 & 1 & 0 & 0 \\ 0 & 1 & -3 & 2 & 2 & -1 & 0 & 0 \\ 0 & 0 & -1 & 1 & 0 & -1 & 1 & 0 \\ 0 & 0 & -8 & 7 & 4 & -3 & 0 & 1 \end{array}\right] \xrightarrow{-R_3}$$

$$\left[\begin{array}{rrrr|rrrr} 1 & 0 & 2 & -1 & -1 & 1 & 0 & 0 \\ 0 & 1 & -3 & 2 & 2 & -1 & 0 & 0 \\ 0 & 0 & 1 & -1 & 0 & 1 & -1 & 0 \\ 0 & 0 & -8 & 7 & 4 & -3 & 0 & 1 \end{array}\right] \xrightarrow[\substack{R_1-2R_3 \\ R_2+3R_3 \\ R_4+8R_3}]{} \left[\begin{array}{rrrr|rrrr} 1 & 0 & 0 & 1 & -1 & -1 & 2 & 0 \\ 0 & 1 & 0 & -1 & 2 & 2 & -3 & 0 \\ 0 & 0 & 1 & -1 & 0 & 1 & -1 & 0 \\ 0 & 0 & 0 & -1 & 4 & 5 & -8 & 1 \end{array}\right]$$

$$\xrightarrow[\substack{R_1+R_4 \\ R_2-R_4 \\ R_3-R_4 \\ -R_4}]{} \left[\begin{array}{rrrr|rrrr} 1 & 0 & 0 & 0 & 3 & 4 & -6 & 1 \\ 0 & 1 & 0 & 0 & -2 & -3 & 5 & -1 \\ 0 & 0 & 1 & 0 & -4 & -4 & 7 & -1 \\ 0 & 0 & 0 & 1 & -4 & -5 & 8 & -1 \end{array}\right].$$

So the required inverse is

$$A^{-1} = \begin{bmatrix} 3 & 4 & -6 & 1 \\ -2 & -3 & 5 & -1 \\ -4 & -4 & 7 & -1 \\ -4 & -5 & 8 & -1 \end{bmatrix}.$$

We can verify our result by showing that $A^{-1}A = A$. Thus,

$$\begin{bmatrix} 3 & 4 & -6 & 1 \\ -2 & -3 & 5 & -1 \\ -4 & -4 & 7 & -1 \\ -4 & -5 & 8 & -1 \end{bmatrix} \begin{bmatrix} 1 & 1 & -1 & 1 \\ 2 & 1 & 1 & 0 \\ 2 & 1 & 0 & 1 \\ 2 & -1 & -1 & 3 \end{bmatrix} = \begin{bmatrix} 1 & 0 & 0 & 0 \\ 0 & 1 & 0 & 0 \\ 0 & 0 & 1 & 0 \\ 0 & 0 & 0 & 1 \end{bmatrix}.$$

17. a. $A = \begin{bmatrix} 2 & 5 \\ 1 & 3 \end{bmatrix}$, $X = \begin{bmatrix} x \\ y \end{bmatrix}$, $B = \begin{bmatrix} 3 \\ 2 \end{bmatrix}$;

b. $X = A^{-1}B = \begin{bmatrix} 3 & -5 \\ -1 & 2 \end{bmatrix} \begin{bmatrix} 3 \\ 2 \end{bmatrix} = \begin{bmatrix} -1 \\ 1 \end{bmatrix}$;

19. a. $A=\begin{bmatrix}2 & -3 & -4\\ 0 & 0 & -1\\ 1 & -2 & 1\end{bmatrix}$, $X=\begin{bmatrix}x\\ y\\ z\end{bmatrix}$, $B=\begin{bmatrix}4\\ 3\\ -8\end{bmatrix}$

$$X=A^{-1}B=\begin{bmatrix}2 & -11 & -3\\ 1 & -6 & -2\\ 0 & -1 & 0\end{bmatrix}\begin{bmatrix}4\\ 3\\ -8\end{bmatrix}=\begin{bmatrix}-1\\ 2\\ -3\end{bmatrix}$$

21. a. $A=\begin{bmatrix}1 & 4 & -1\\ 2 & 3 & -2\\ -1 & 2 & 3\end{bmatrix}$, $X=\begin{bmatrix}x\\ y\\ z\end{bmatrix}$, $B=\begin{bmatrix}3\\ 1\\ 7\end{bmatrix}$;

b. $X=A^{-1}B=\begin{bmatrix}-\frac{13}{10} & \frac{7}{5} & \frac{1}{2}\\ \frac{2}{5} & -\frac{1}{5} & 0\\ -\frac{7}{10} & \frac{3}{5} & \frac{1}{2}\end{bmatrix}\begin{bmatrix}3\\ 1\\ 7\end{bmatrix}=\begin{bmatrix}1\\ 1\\ 2\end{bmatrix}$.

23. a. $A=\begin{bmatrix}1 & 1 & -1 & 1\\ 2 & 1 & 1 & 0\\ 2 & 1 & 0 & 1\\ 2 & -1 & -1 & 3\end{bmatrix}$, $X=\begin{bmatrix}x_1\\ x_2\\ x_3\\ x_4\end{bmatrix}$, $B=\begin{bmatrix}6\\ 4\\ 7\\ 9\end{bmatrix}$.

b. $X=A^{-1}B=\begin{bmatrix}3 & 4 & -6 & 1\\ -2 & -3 & 5 & 1\\ -4 & -4 & 7 & -1\\ -4 & -5 & 8 & -1\end{bmatrix}\begin{bmatrix}6\\ 4\\ 7\\ 9\end{bmatrix}=\begin{bmatrix}1\\ 2\\ 0\\ 3\end{bmatrix}$.

25. a. $A=\begin{bmatrix}1 & 2\\ 2 & -1\end{bmatrix}$, $X=\begin{bmatrix}x\\ y\end{bmatrix}$, $B=\begin{bmatrix}b_1\\ b_2\end{bmatrix}$;

b. $X=A^{-1}B=\begin{bmatrix}0.2 & 0.4\\ 0.4 & -0.2\end{bmatrix}\begin{bmatrix}14\\ 5\end{bmatrix}=\begin{bmatrix}4.8\\ 4.6\end{bmatrix}$ and we conclude that $x=4.8$ and $y=4.6$.

$$X = A^{-1}B = \begin{bmatrix} 0.2 & 0.4 \\ 0.4 & -0.2 \end{bmatrix} \begin{bmatrix} 4 \\ -1 \end{bmatrix} = \begin{bmatrix} 0.4 \\ 1.8 \end{bmatrix}$$ and we conclude that $x = 0.4$ and $y = 1.8$.

27. First we find A^{-1}.

$$\left[\begin{array}{ccc|ccc} 1 & 2 & 1 & 1 & 0 & 0 \\ 1 & 1 & 1 & 0 & 1 & 0 \\ 3 & 1 & 1 & 0 & 0 & 1 \end{array}\right] \xrightarrow[R_3-3R_1]{R_2-R_1} \left[\begin{array}{ccc|ccc} 1 & 2 & 1 & 1 & 0 & 0 \\ 0 & -1 & 0 & -1 & 1 & 0 \\ 0 & -5 & -2 & -3 & 0 & 1 \end{array}\right] \xrightarrow{-R_2}$$

$$\left[\begin{array}{ccc|ccc} 1 & 2 & 1 & 1 & 0 & 0 \\ 0 & 1 & 0 & 1 & -1 & 0 \\ 0 & -5 & -2 & -3 & 0 & 1 \end{array}\right] \xrightarrow[R_3+5R_2]{R_1-2R_2} \left[\begin{array}{ccc|ccc} 1 & 0 & 1 & -1 & 2 & 0 \\ 0 & 1 & 0 & 1 & -1 & 0 \\ 0 & 0 & -2 & 2 & -5 & 1 \end{array}\right] \xrightarrow{-\frac{1}{2}R_3}$$

$$\left[\begin{array}{ccc|ccc} 1 & 0 & 1 & -1 & 2 & 0 \\ 0 & 1 & 0 & 1 & -1 & 0 \\ 0 & 0 & 1 & -1 & \frac{5}{2} & -\frac{1}{2} \end{array}\right] \xrightarrow{R_1-R_3} \left[\begin{array}{ccc|ccc} 1 & 0 & 0 & 0 & -\frac{1}{2} & \frac{1}{2} \\ 0 & 1 & 0 & 1 & -1 & 0 \\ 0 & 0 & 1 & -1 & \frac{5}{2} & -\frac{1}{2} \end{array}\right]$$

$$\begin{bmatrix} 1 & 2 & 1 \\ 1 & 1 & 1 \\ 3 & 1 & 1 \end{bmatrix} \begin{bmatrix} x \\ y \\ z \end{bmatrix} = \begin{bmatrix} b_1 \\ b_2 \\ b_3 \end{bmatrix}$$

a. $$\begin{bmatrix} x \\ y \\ z \end{bmatrix} = \begin{bmatrix} 0 & -\frac{1}{2} & \frac{1}{2} \\ 1 & -1 & 0 \\ -1 & \frac{5}{2} & -\frac{1}{2} \end{bmatrix} \begin{bmatrix} 7 \\ 4 \\ 2 \end{bmatrix} = \begin{bmatrix} -1 \\ 3 \\ 2 \end{bmatrix}$$ and we conclude that $x = -1$, $y = 3$, and $z = 2$.

b. $$\begin{bmatrix} x \\ y \\ z \end{bmatrix} = \begin{bmatrix} 0 & -\frac{1}{2} & \frac{1}{2} \\ 1 & -1 & 0 \\ -1 & \frac{5}{2} & -\frac{1}{2} \end{bmatrix} \begin{bmatrix} 5 \\ -3 \\ -1 \end{bmatrix} = \begin{bmatrix} 1 \\ 8 \\ -12 \end{bmatrix}$$

and we conclude that $x = 1$, $y = 8$, and $z = -12$.

29. $$\left[\begin{array}{ccc|ccc} 3 & 2 & -1 & 1 & 0 & 0 \\ 2 & -3 & 1 & 0 & 1 & 0 \\ 1 & -1 & -1 & 0 & 0 & 1 \end{array}\right] \xrightarrow{R_1 \leftrightarrow R_3} \left[\begin{array}{ccc|ccc} 1 & -1 & -1 & 0 & 0 & 1 \\ 2 & -3 & 1 & 0 & 1 & 0 \\ 3 & 2 & -1 & 1 & 0 & 0 \end{array}\right] \xrightarrow[R_3-3R_1]{R_2-2R_1}$$

$$\left[\begin{array}{ccc|ccc} 1 & -1 & -1 & 0 & 0 & 1 \\ 0 & -1 & 3 & 0 & 1 & -2 \\ 0 & 5 & 2 & 1 & 0 & -3 \end{array}\right] \xrightarrow{-R_2} \left[\begin{array}{ccc|ccc} 1 & -1 & -1 & 0 & 0 & 1 \\ 0 & 1 & -3 & 0 & -1 & 2 \\ 0 & 5 & 2 & 1 & 0 & -3 \end{array}\right] \xrightarrow[R_3-5R_2]{R_1+R_2}$$

$$\left[\begin{array}{ccc|ccc} 1 & 0 & -4 & 0 & -1 & 3 \\ 0 & 1 & -3 & 0 & -1 & 2 \\ 0 & 0 & 17 & 1 & 5 & -13 \end{array}\right] \xrightarrow{-\frac{1}{17}R_3} \left[\begin{array}{ccc|ccc} 1 & 0 & -4 & 0 & -1 & 3 \\ 0 & 1 & -3 & 0 & -1 & 2 \\ 0 & 0 & 1 & \frac{1}{17} & \frac{5}{17} & -\frac{13}{17} \end{array}\right]$$

$$\xrightarrow[R_2+3R_3]{R_1+4R_3} \left[\begin{array}{ccc|ccc} 1 & 0 & 0 & \frac{4}{17} & \frac{3}{17} & -\frac{1}{17} \\ 0 & 1 & 0 & \frac{3}{17} & -\frac{2}{17} & -\frac{5}{17} \\ 0 & 0 & 1 & \frac{1}{17} & \frac{5}{17} & -\frac{13}{17} \end{array}\right].$$

Therefore $A^{-1} = \begin{bmatrix} \frac{4}{17} & \frac{3}{17} & -\frac{1}{17} \\ \frac{3}{17} & -\frac{2}{17} & -\frac{5}{17} \\ \frac{1}{17} & \frac{5}{17} & -\frac{13}{17} \end{bmatrix}$.

Next, $\begin{bmatrix} 3 & 2 & -1 \\ 2 & -3 & 1 \\ 1 & -1 & -1 \end{bmatrix}\begin{bmatrix} x \\ y \\ z \end{bmatrix} = \begin{bmatrix} b_1 \\ b_2 \\ b_3 \end{bmatrix}$

a. $\begin{bmatrix} x \\ y \\ z \end{bmatrix} = \begin{bmatrix} \frac{4}{17} & \frac{3}{17} & -\frac{1}{17} \\ \frac{3}{17} & -\frac{2}{17} & -\frac{5}{17} \\ \frac{1}{17} & \frac{5}{17} & -\frac{13}{17} \end{bmatrix}\begin{bmatrix} 2 \\ -2 \\ 4 \end{bmatrix} = \begin{bmatrix} -\frac{2}{17} \\ -\frac{10}{17} \\ -\frac{60}{17} \end{bmatrix}$

We conclude that $x = -2/17$, $y = -10/17$, and $z = -60/17$.

b. $\begin{bmatrix} x \\ y \\ z \end{bmatrix} = \begin{bmatrix} \frac{4}{17} & \frac{3}{17} & -\frac{1}{17} \\ \frac{3}{17} & -\frac{2}{17} & -\frac{5}{17} \\ \frac{1}{17} & \frac{5}{17} & -\frac{13}{17} \end{bmatrix}\begin{bmatrix} 8 \\ -3 \\ 6 \end{bmatrix} = \begin{bmatrix} 1 \\ 0 \\ -5 \end{bmatrix}$. We conclude that $x = 1$, $y = 0$, and $z = -5$.

31. $AX = B_1$ and $AX = B_2$, where

$$A = \begin{bmatrix} 1 & 1 & 1 & 1 \\ 1 & -1 & -1 & 1 \\ 0 & 1 & 2 & 2 \\ 1 & 2 & 1 & -2 \end{bmatrix}, \ X = \begin{bmatrix} x_1 \\ x_2 \\ x_3 \\ x_4 \end{bmatrix}, \ B_1 = \begin{bmatrix} 1 \\ -1 \\ 4 \\ 0 \end{bmatrix} \text{ and } B_2 = \begin{bmatrix} 2 \\ 8 \\ 4 \\ -1 \end{bmatrix}.$$

We first find A^{-1}.

$$\left[\begin{array}{cccc|cccc} 1 & 1 & 1 & 1 & 1 & 0 & 0 & 0 \\ 1 & -1 & -1 & 1 & 0 & 1 & 0 & 0 \\ 0 & 1 & 2 & 2 & 0 & 0 & 1 & 0 \\ 1 & 2 & 1 & -2 & 0 & 0 & 0 & 1 \end{array}\right] \xrightarrow[R_4-R_1]{R_2-R_1} \left[\begin{array}{cccc|cccc} 1 & 1 & 1 & 1 & 1 & 0 & 0 & 0 \\ 0 & -2 & -2 & 0 & -1 & 1 & 0 & 0 \\ 0 & 1 & 2 & 2 & 0 & 0 & 1 & 0 \\ 0 & 1 & 0 & -3 & -1 & 0 & 0 & 1 \end{array}\right]$$

$$\xrightarrow{R_2 \leftrightarrow R_3} \left[\begin{array}{cccc|cccc} 1 & 1 & 1 & 1 & 1 & 0 & 0 & 0 \\ 0 & 1 & 2 & 2 & 0 & 0 & 1 & 0 \\ 0 & -2 & -2 & 0 & -1 & 1 & 0 & 0 \\ 0 & 1 & 0 & -3 & -1 & 0 & 0 & 1 \end{array}\right] \xrightarrow[\substack{R_3+2R_2 \\ R_4-R_2}]{R_1-R_2}$$

$$\left[\begin{array}{cccc|cccc} 1 & 0 & -1 & -1 & 1 & 0 & -1 & 0 \\ 0 & 1 & 2 & 2 & 0 & 0 & 1 & 0 \\ 0 & 0 & 2 & 4 & -1 & 1 & 2 & 0 \\ 0 & 0 & -2 & -5 & -1 & 0 & -1 & 1 \end{array}\right] \xrightarrow{\frac{1}{2}R_3} \left[\begin{array}{cccc|cccc} 1 & 0 & -1 & -1 & 1 & 0 & -1 & 0 \\ 0 & 1 & 2 & 2 & 0 & 0 & 1 & 0 \\ 0 & 0 & 1 & 2 & -\frac{1}{2} & \frac{1}{2} & 1 & 0 \\ 0 & 0 & -2 & -5 & -1 & 0 & -1 & 1 \end{array}\right]$$

$$\xrightarrow[\substack{R_2-2R_3 \\ R_4+2R_3}]{R_1+R_3} \left[\begin{array}{cccc|cccc} 1 & 0 & 0 & 1 & \frac{1}{2} & \frac{1}{2} & 0 & 0 \\ 0 & 1 & 0 & -2 & 1 & -1 & -1 & 0 \\ 0 & 0 & 1 & 2 & -\frac{1}{2} & \frac{1}{2} & 1 & 0 \\ 0 & 0 & 0 & -1 & -2 & 1 & 1 & 1 \end{array}\right] \xrightarrow[\substack{R_3+2R_4 \\ -R_4}]{\substack{R_1+R_4 \\ R_2-2R_4}}$$

$$\left[\begin{array}{cccc|cccc} 1 & 0 & 0 & 0 & -\frac{3}{2} & \frac{3}{2} & 1 & 1 \\ 0 & 1 & 0 & 0 & 5 & -3 & -3 & -2 \\ 0 & 0 & 1 & 0 & -\frac{9}{2} & \frac{5}{2} & 3 & 2 \\ 0 & 0 & 0 & 1 & 2 & -1 & -1 & -1 \end{array}\right]. \quad \text{So } A^{-1} = \begin{bmatrix} -\frac{3}{2} & \frac{3}{2} & 1 & 1 \\ 5 & -3 & -3 & -2 \\ -\frac{9}{2} & \frac{5}{2} & 3 & 2 \\ 2 & -1 & -1 & -1 \end{bmatrix}.$$

a. $$\begin{bmatrix} x_1 \\ x_2 \\ x_3 \\ x_4 \end{bmatrix} = \begin{bmatrix} -\frac{3}{2} & \frac{3}{2} & 1 & 1 \\ 5 & -3 & -3 & -2 \\ -\frac{9}{2} & \frac{5}{2} & 3 & 2 \\ 2 & -1 & -1 & -1 \end{bmatrix} \begin{bmatrix} 1 \\ -1 \\ 4 \\ 0 \end{bmatrix} = \begin{bmatrix} 1 \\ -4 \\ 5 \\ -1 \end{bmatrix}$$

and we conclude that $x_1 = 1$, $x_2 = -4$, $x_3 = 5$, and $x_4 = -1$.

b. $$\begin{bmatrix} x_1 \\ x_2 \\ x_3 \\ x_4 \end{bmatrix} = \begin{bmatrix} -\frac{3}{2} & \frac{3}{2} & 1 & 1 \\ 5 & -3 & -3 & -2 \\ -\frac{9}{2} & \frac{5}{2} & 3 & 2 \\ 2 & -1 & -1 & -1 \end{bmatrix} \begin{bmatrix} 2 \\ 8 \\ 4 \\ -1 \end{bmatrix} = \begin{bmatrix} 12 \\ -24 \\ 21 \\ -7 \end{bmatrix}$$

and we conclude that $x_1 = 12$, $x_2 = -24$, $x_3 = 21$, and $x_4 = -7$.

33. a. Using Formula (13), we find

$$A^{-1} = \frac{1}{(2)(-5)-(-4)(3)} \begin{bmatrix} -5 & -3 \\ 4 & 2 \end{bmatrix} = \begin{bmatrix} -\frac{5}{2} & -\frac{3}{2} \\ 2 & 1 \end{bmatrix}.$$

b. Using Formula (13) once again, we find

$$\left(A^{-1}\right)^{-1} = \frac{1}{(-\frac{5}{2})(1)-2(-\frac{3}{2})} \begin{bmatrix} 1 & \frac{3}{2} \\ -2 & -\frac{5}{2} \end{bmatrix} = \begin{bmatrix} 2 & 3 \\ -4 & -5 \end{bmatrix} = A.$$

35. a. $$ABC = \begin{bmatrix} 2 & -5 \\ 1 & -3 \end{bmatrix} \begin{bmatrix} 4 & 3 \\ 1 & 1 \end{bmatrix} \begin{bmatrix} 2 & 3 \\ -2 & 1 \end{bmatrix}$$

$$= \begin{bmatrix} 2 & -5 \\ 1 & -3 \end{bmatrix} \begin{bmatrix} 2 & 15 \\ 0 & 4 \end{bmatrix} = \begin{bmatrix} 4 & 10 \\ 2 & 3 \end{bmatrix}.$$

Using the formula for finding the inverse of a 2×2 matrix, we find

$$A^{-1} = \begin{bmatrix} 3 & -5 \\ 1 & -2 \end{bmatrix}, \ B^{-1} = \begin{bmatrix} 1 & -3 \\ -1 & 4 \end{bmatrix}, \ C^{-1} = \begin{bmatrix} \frac{1}{8} & -\frac{3}{8} \\ \frac{1}{4} & \frac{1}{4} \end{bmatrix}.$$

b. Using the formula for finding the inverse of a 2×2 matrix, we find

$$(ABC)^{-1} = \begin{bmatrix} -\frac{3}{8} & \frac{5}{4} \\ \frac{1}{4} & -\frac{1}{2} \end{bmatrix}$$

$$C^{-1}B^{-1}A^{-1} = \begin{bmatrix} \frac{1}{8} & -\frac{3}{8} \\ \frac{1}{4} & \frac{1}{4} \end{bmatrix} \begin{bmatrix} 1 & -3 \\ -1 & 4 \end{bmatrix} \begin{bmatrix} 3 & -5 \\ 1 & -2 \end{bmatrix}$$

$$= \begin{bmatrix} \frac{1}{8} & -\frac{3}{8} \\ \frac{1}{4} & \frac{1}{4} \end{bmatrix} \begin{bmatrix} 0 & 1 \\ 1 & -3 \end{bmatrix} = \begin{bmatrix} -\frac{3}{8} & \frac{5}{4} \\ \frac{1}{4} & -\frac{1}{2} \end{bmatrix}.$$

Therefore, $(ABC)^{-1} = C^{-1}B^{-1}A^{-1}$.

37. Let x denote the number of copies of the deluxe edition and y the number of copies of the standard edition demanded per month when the unit prices are p and q dollars, respectively. Then the three systems of linear equations

$$\begin{aligned} 5x + y &= 20000 \\ x + 3y &= 15000 \end{aligned} \qquad \begin{aligned} 5x + y &= 25000 \\ x + 3y &= 15000 \end{aligned} \qquad \begin{aligned} 5x + y &= 25000 \\ x + 3y &= 20000 \end{aligned}$$

give the quantity demanded of each edition at the stated price. These systems may be written in the form $AX = B_1$, $AX = B_2$, and $AX = B_3$, where

$$A = \begin{bmatrix} 5 & 1 \\ 1 & 3 \end{bmatrix}, \quad B_1 = \begin{bmatrix} 20000 \\ 15000 \end{bmatrix}, \quad B_2 = \begin{bmatrix} 25000 \\ 15000 \end{bmatrix}, \quad \text{and} \quad B_3 = \begin{bmatrix} 25000 \\ 20000 \end{bmatrix}$$

Using the formula for finding the inverse of a 2×2 matrix, with $a = 5$, $b = 1$, $c = 1$, $d = 3$, and $D = ad - bc = (5)(3) - (1)(1) = 14$, we find that

$$A^{-1} = \begin{bmatrix} \frac{3}{14} & -\frac{1}{14} \\ -\frac{1}{14} & \frac{5}{14} \end{bmatrix}.$$

a. $\begin{bmatrix} x \\ y \end{bmatrix} = \begin{bmatrix} \frac{3}{14} & -\frac{1}{14} \\ -\frac{1}{14} & \frac{5}{14} \end{bmatrix} \begin{bmatrix} 20,000 \\ 15,000 \end{bmatrix} = \begin{bmatrix} 3,214 \\ 3,929 \end{bmatrix}$

b. $\begin{bmatrix} x \\ y \end{bmatrix} = \begin{bmatrix} \frac{3}{14} & -\frac{1}{14} \\ -\frac{1}{14} & \frac{5}{14} \end{bmatrix} \begin{bmatrix} 25,000 \\ 15,000 \end{bmatrix} = \begin{bmatrix} 4,286 \\ 3,571 \end{bmatrix}$

c. $\begin{bmatrix} x \\ y \end{bmatrix} = \begin{bmatrix} \frac{3}{14} & -\frac{1}{14} \\ -\frac{1}{14} & \frac{5}{14} \end{bmatrix} \begin{bmatrix} 25,000 \\ 20,000 \end{bmatrix} = \begin{bmatrix} 3,929 \\ 5,357 \end{bmatrix}.$

39. Let x, y, and z (in millions of dollars) be the amount awarded to organization I, II, and III, respectively. Then we have

$$\begin{aligned} 0.6x + 0.4y + 0.2z &= 9.2 \qquad (8.2) \\ 0.3x + 0.3y + 0.6z &= 9.6 \qquad (7.2) \\ 0.1x + 0.3y + 0.2z &= 5.2 \qquad (3.6). \end{aligned}$$

The quantities within the brackets are for part (b). We can rewrite the systems as $AX = B_1$, and $AX = B_2$. Put

$$X = \begin{bmatrix} x \\ y \\ z \end{bmatrix}, \quad A = \begin{bmatrix} 6 & 4 & 2 \\ 3 & 3 & 6 \\ 1 & 3 & 2 \end{bmatrix}, \quad B_1 = \begin{bmatrix} 92 \\ 96 \\ 52 \end{bmatrix}, \text{ and } B_2 = \begin{bmatrix} 82 \\ 72 \\ 36 \end{bmatrix}.$$

To find A^{-1}, we use the Gauss-Jordan method:

$$\left[\begin{array}{ccc|ccc} 6 & 4 & 2 & 1 & 0 & 0 \\ 3 & 3 & 6 & 0 & 1 & 0 \\ 1 & 3 & 2 & 0 & 0 & 1 \end{array}\right] \xrightarrow{R_1 \leftrightarrow R_3} \left[\begin{array}{ccc|ccc} 1 & 3 & 2 & 0 & 0 & 1 \\ 3 & 3 & 6 & 0 & 1 & 0 \\ 6 & 4 & 2 & 1 & 0 & 0 \end{array}\right] \xrightarrow[R_3 - 6R_1]{R_2 - 3R_1}$$

$$\left[\begin{array}{ccc|ccc} 1 & 3 & 2 & 0 & 0 & 1 \\ 0 & -6 & 0 & 0 & 1 & -3 \\ 0 & -14 & -10 & 1 & 0 & -6 \end{array}\right] \xrightarrow{-\frac{1}{6}R_2} \left[\begin{array}{ccc|ccc} 1 & 3 & 2 & 0 & 0 & 1 \\ 0 & 1 & 0 & 0 & -\frac{1}{6} & \frac{1}{2} \\ 0 & -14 & -10 & 1 & 0 & -6 \end{array}\right] \xrightarrow[R_3 + 14R_2]{R_1 - 3R_2}$$

$$\left[\begin{array}{ccc|ccc} 1 & 0 & 2 & 0 & \frac{1}{2} & -\frac{1}{2} \\ 0 & 1 & 0 & 0 & -\frac{1}{6} & \frac{1}{2} \\ 0 & 0 & -10 & 1 & -\frac{7}{3} & 1 \end{array}\right] \xrightarrow{-\frac{1}{10}R_3} \left[\begin{array}{ccc|ccc} 1 & 0 & 2 & 1 & \frac{1}{2} & -\frac{1}{2} \\ 0 & 1 & 0 & 0 & -\frac{1}{6} & \frac{1}{2} \\ 0 & 0 & 1 & -\frac{1}{10} & \frac{7}{30} & -\frac{1}{10} \end{array}\right] \xrightarrow{R_1 - 2R_3}$$

$$\left[\begin{array}{ccc|ccc} 1 & 0 & 0 & \frac{1}{5} & \frac{1}{30} & -\frac{3}{10} \\ 0 & 1 & 0 & 0 & -\frac{1}{6} & \frac{1}{2} \\ 0 & 0 & 1 & -\frac{1}{10} & \frac{7}{30} & -\frac{1}{10} \end{array}\right].$$

a. $$X = A^{-1}B_1 = \begin{bmatrix} \frac{1}{5} & \frac{1}{30} & -\frac{3}{10} \\ 0 & -\frac{1}{6} & \frac{1}{2} \\ -\frac{1}{10} & \frac{7}{30} & -\frac{1}{10} \end{bmatrix} \begin{bmatrix} 92 \\ 96 \\ 52 \end{bmatrix} = \begin{bmatrix} 6 \\ 10 \\ 8 \end{bmatrix}$$

that is, $x = 6$, $y = 10$, and $z = 8$, and Organization I will receive \$6 million, Organization II will receive \$10 million, and Organization III will receive \$8 million.

b. $X = A^{-1}B_1 = \begin{bmatrix} \frac{1}{5} & \frac{1}{30} & -\frac{3}{10} \\ 0 & -\frac{1}{6} & \frac{1}{2} \\ -\frac{1}{10} & \frac{7}{30} & -\frac{1}{10} \end{bmatrix}\begin{bmatrix} 82 \\ 72 \\ 36 \end{bmatrix} = \begin{bmatrix} 8 \\ 6 \\ 5 \end{bmatrix}$

that is, $x = 8$, $y = 6$, and $z = 5$, and Organization I will receive \$8 million, Organization II will receive \$6 million, and Organization III will receive \$5 million.

USING TECHNOLOGY EXERCISES 2.6, page 153

1. $\begin{bmatrix} 0.36 & 0.04 & -0.36 \\ 0.06 & 0.05 & 0.20 \\ -0.19 & 0.10 & 0.09 \end{bmatrix}$

3. $\begin{bmatrix} 0.01 & -0.09 & 0.31 & -0.11 \\ -0.25 & 0.58 & -0.15 & -0.02 \\ 0.86 & -0.42 & 0.07 & -0.37 \\ -0.27 & 0.01 & -0.05 & 0.31 \end{bmatrix}$

5. $\begin{bmatrix} 0.30 & 0.85 & -0.10 & -0.77 & -0.11 \\ -0.21 & 0.10 & 0.01 & -0.26 & 0.21 \\ 0.03 & -0.16 & 0.12 & -0.01 & 0.03 \\ -0.14 & -0.46 & 0.13 & 0.71 & -0.05 \\ 0.10 & -0.05 & -0.10 & -0.03 & 0.11 \end{bmatrix}$

EXERCISES 2.7, page 163

1. a. The amount of agricultural products consumed in the production of \$100 million worth of manufactured goods is given by (100)(0.10), or \$10 million.

   b. The amount of manufactured goods required to produce \$200 million of all goods in the economy is given by $200(0.1 + 0.4 + 0.3) = 160$, or \$160 million.

   c. From the input-output matrix, we see that the agricultural sector consumes the greatest amount of agricultural products, namely, 0.4 units, in the production of each unit of goods in that sector. The manufacturing and transportation sectors consume the least, 0.1 units each.

3. Multiplying both sides of the given equation on the left by $(I-A)^{-1}$, we see that

$$X=(I-A)^{-1}D.$$

Now, $(I-A)=\begin{bmatrix}1 & 0\\ 0 & 1\end{bmatrix}-\begin{bmatrix}0.4 & 0.2\\ 0.3 & 0.1\end{bmatrix}=\begin{bmatrix}0.6 & -0.2\\ -0.3 & 0.9\end{bmatrix}.$

Using the formula for finding the inverse of a 2×2 matrix, we find

$$(I-A)^{-1}=\begin{bmatrix}1.875 & 0.417\\ 0.625 & 1.25\end{bmatrix}.$$

Then, $(I-A)^{-1}X=\begin{bmatrix}1.875 & 0.417\\ 0.625 & 1.25\end{bmatrix}\begin{bmatrix}10\\ 12\end{bmatrix}=\begin{bmatrix}23.754\\ 21.25\end{bmatrix}.$

5. We first compute

$$(I-A)=\begin{bmatrix}1 & 0\\ 0 & 1\end{bmatrix}-\begin{bmatrix}0.5 & 0.2\\ 0.2 & 0.5\end{bmatrix}=\begin{bmatrix}0.5 & -0.2\\ -0.2 & 0.5\end{bmatrix}$$

Using the formula for finding the inverse of a 2×2 matrix, we find

$$(I-A)^{-1}=\begin{bmatrix}2.381 & 0.952\\ 0.952 & 2.381\end{bmatrix}.$$

Then $\begin{bmatrix}x\\ y\end{bmatrix}=\begin{bmatrix}2.381 & 0.952\\ 0.952 & 2.381\end{bmatrix}\begin{bmatrix}10\\ 20\end{bmatrix}=\begin{bmatrix}42.85\\ 57.14\end{bmatrix}$

7. We verify

$$(I-A)(I-A)^{-1}=\begin{bmatrix}0.92 & -0.60 & -0.30\\ -0.04 & 0.98 & -0.01\\ -0.02 & 0 & 0.94\end{bmatrix}\begin{bmatrix}1.13 & 0.69 & 0.37\\ 0.05 & 1.05 & 0.03\\ 0.02 & 0.02 & 1.07\end{bmatrix}=\begin{bmatrix}1 & 0 & 0\\ 0 & 1 & 0\\ 0 & 0 & 1\end{bmatrix}$$

9. a. $A=\begin{bmatrix}0.2 & 0.4\\ 0.3 & 0.3\end{bmatrix}$ and

$$(I-A)=\begin{bmatrix}1 & 0\\ 0 & 1\end{bmatrix}-\begin{bmatrix}0.2 & 0.4\\ 0.3 & 0.3\end{bmatrix}=\begin{bmatrix}0.8 & -0.4\\ -0.3 & 0.7\end{bmatrix}.$$

Using the formula for finding the inverse of a 2×2 matrix, we find

$$(I - A)^{-1} = \begin{bmatrix} 1.591 & 0.909 \\ 0.682 & 1.818 \end{bmatrix}.$$

Then $\begin{bmatrix} x \\ y \end{bmatrix} = \begin{bmatrix} 1.591 & 0.909 \\ 0.682 & 1.818 \end{bmatrix} \begin{bmatrix} 120 \\ 140 \end{bmatrix} = \begin{bmatrix} 318.18 \\ 336.36 \end{bmatrix}$

To fullfill consumer demand, \$318.2 million worth of agricultural goods and \$336.4 million worth of manufactured goods should be produced.

b. The net value of goods consumed in the internal process of production is

$$AX = X - D = \begin{bmatrix} 318.18 \\ 336.36 \end{bmatrix} - \begin{bmatrix} 120 \\ 140 \end{bmatrix} = \begin{bmatrix} 198.18 \\ 196.36 \end{bmatrix}.$$

or \$198.2 million of agricultural goods and \$196.4 million worth of manufactured goods.

11. a.

$$(I - A) = \begin{bmatrix} 1 & 0 & 0 \\ 0 & 1 & 0 \\ 0 & 0 & 1 \end{bmatrix} - \begin{bmatrix} 0.4 & 0.1 & 0.1 \\ 0.1 & 0.4 & 0.3 \\ 0.2 & 0.2 & 0.2 \end{bmatrix} = \begin{bmatrix} 0.6 & -0.1 & -0.1 \\ -0.1 & 0.6 & -0.3 \\ -0.2 & -0.2 & 0.8 \end{bmatrix}$$

Using the methods of Section 2.6 we next compute the inverse of $(1 - A)^{-1}$ and use this value to find

$$X = (1 - A)^{-1} D = \begin{bmatrix} 1.875 & 0.446 & 0.402 \\ 0.625 & 2.054 & 0.848 \\ 0.625 & 0.625 & 1.563 \end{bmatrix} \begin{bmatrix} 200 \\ 100 \\ 60 \end{bmatrix} = \begin{bmatrix} 443.7 \\ 381.3 \\ 281.3 \end{bmatrix}.$$

Therefore, to fulfull demand, \$443.7 million worth of agricultural products, \$381.3 million worth of manufactured products, and \$281.3 million worth of transportation sservices should be produced.

b. To meet the gross output, the value of goods and transportation consumed in the internal process of production is

$$AX = X - D = \begin{bmatrix} 443.7 \\ 381.3 \\ 281.3 \end{bmatrix} - \begin{bmatrix} 200 \\ 100 \\ 60 \end{bmatrix} = \begin{bmatrix} 243.7 \\ 281.3 \\ 221.3 \end{bmatrix},$$

or \$243.7 million worth of agricultural products, \$281.3 million worth of manufactured services, and \$221.3 million worth of transportation services.

13. We want to solve the equation $(I - A)X = D$ for X, the total output matrix. First, we compute

$$(I - A) = \begin{bmatrix} 1 & 0 \\ 0 & 1 \end{bmatrix} - \begin{bmatrix} 0.4 & 0.2 \\ 0.3 & 0.5 \end{bmatrix} = \begin{bmatrix} 0.6 & -0.2 \\ -0.3 & 0.5 \end{bmatrix}.$$

Using the formula for finding the inverse of a 2×2 matrix, we find

$$(I - A)^{-1} = \begin{bmatrix} 2.08 & 0.833 \\ 1.25 & 2.4 \end{bmatrix}$$

Therefore,

$$X = (I - A)^{-1}D = \begin{bmatrix} 2.08 & 0.833 \\ 1.25 & 2.4 \end{bmatrix}\begin{bmatrix} 12 \\ 24 \end{bmatrix} = \begin{bmatrix} 45 \\ 75 \end{bmatrix}.$$

We conclude that \$45 million worth of goods of one industry and \$75 million worth of goods of the other industry must be produced.

15. First, we compute

$$I - A = \begin{bmatrix} 1 & 0 & 0 \\ 0 & 1 & 0 \\ 0 & 0 & 1 \end{bmatrix} - \begin{bmatrix} 0.2 & 0.4 & 0.2 \\ 0.5 & 0 & 0.5 \\ 0 & 0.2 & 0 \end{bmatrix} = \begin{bmatrix} 0.8 & -0.4 & -0.2 \\ -0.5 & 1 & -0.5 \\ 0 & -0.2 & 1 \end{bmatrix}.$$

Next, using the Gauss-Jordan method, we find

$$(I - A)^{-1} = \begin{bmatrix} 1.8 & 0.88 & 0.80 \\ 1 & 1.6 & 1 \\ 0.2 & 0.32 & 1.20 \end{bmatrix}$$

Then $\begin{bmatrix} x \\ y \\ z \end{bmatrix} = \begin{bmatrix} 1.8 & 0.88 & 0.80 \\ 1 & 1.6 & 1 \\ 0.2 & 0.32 & 1.20 \end{bmatrix}\begin{bmatrix} 10 \\ 5 \\ 15 \end{bmatrix} = \begin{bmatrix} 34.4 \\ 33 \\ 21.6 \end{bmatrix}$

We conclude that \$34.4 million worth of goods of one industry, \$33 million worth of a second industry, and \$21.6 million worth of a third industry should be produced.

USING TECHNOLOGY EXERCISES 2.7, page 164

1. The final outputs of the first, second, third, and fourth industries are 602.62, 502.30, 572.57, and 523.46 units, respectively.

3. The final outputs of the first, second, third, and fourth industries are 143.06, 132.98, 188.59, and 125.53 units, respectively.

CHAPTER 2, REVIEW EXERCISES, page 169

1. $$\begin{bmatrix} 1 & 2 \\ -1 & 3 \\ 2 & 1 \end{bmatrix} + \begin{bmatrix} 1 & 0 \\ 0 & 1 \\ 1 & 2 \end{bmatrix} = \begin{bmatrix} 2 & 2 \\ -1 & 4 \\ 3 & 3 \end{bmatrix}.$$

3. $$\begin{bmatrix} -3 & 2 & 1 \end{bmatrix} \begin{bmatrix} 2 & 1 \\ -1 & 0 \\ 2 & 1 \end{bmatrix} = \begin{bmatrix} -6 & -2 \end{bmatrix}.$$

5. By the equality of matrices, $x = 2$, $z = 1$, $y = 3$ and $w = 3$.

7. By the equality of matrices,

$$a + 3 = 6, \text{ or } a = 3.$$
$$-1 = e + 2, \text{ or } e = -3; b = 4$$
$$c + 1 = -1, \text{ or } c = -2; d = 2$$
$$e + 2 = -1, \text{ and } e = -3.$$

9. $$2A + 3B = 2\begin{bmatrix} 1 & 3 & 1 \\ -2 & 1 & 3 \\ 4 & 0 & 2 \end{bmatrix} + 3\begin{bmatrix} 2 & 1 & 3 \\ -2 & -1 & -1 \\ 1 & 4 & 2 \end{bmatrix} = \begin{bmatrix} 2 & 6 & 2 \\ -4 & 2 & 6 \\ 8 & 0 & 4 \end{bmatrix} + \begin{bmatrix} 6 & 3 & 9 \\ -6 & -3 & -3 \\ 3 & 12 & 6 \end{bmatrix}$$

$$= \begin{bmatrix} 8 & 9 & 11 \\ -10 & -1 & 3 \\ 11 & 12 & 10 \end{bmatrix}.$$

11. $$3A = \begin{bmatrix} 1 & 3 & 1 \\ -2 & 1 & 3 \\ 4 & 0 & 2 \end{bmatrix} = \begin{bmatrix} 3 & 9 & 3 \\ -6 & 3 & 9 \\ 12 & 0 & 6 \end{bmatrix}$$

and $$2(3A) = 2\begin{bmatrix} 3 & 9 & 3 \\ -6 & 3 & 9 \\ 12 & 0 & 6 \end{bmatrix} = \begin{bmatrix} 6 & 18 & 6 \\ -12 & 6 & 18 \\ 24 & 0 & 12 \end{bmatrix}.$$

13. $$B - C = \begin{bmatrix} 2 & 1 & 3 \\ -2 & -1 & -1 \\ 1 & 4 & 2 \end{bmatrix} - \begin{bmatrix} 3 & -1 & 2 \\ 1 & 6 & 4 \\ 2 & 1 & 3 \end{bmatrix} = \begin{bmatrix} -1 & 2 & 1 \\ -3 & -7 & -5 \\ -1 & 3 & -1 \end{bmatrix}$$

and so $$A(B-C) = \begin{bmatrix} 1 & 3 & 1 \\ -2 & 1 & 3 \\ 4 & 0 & 2 \end{bmatrix}\begin{bmatrix} -1 & 2 & 1 \\ -3 & -7 & -5 \\ -1 & 3 & -1 \end{bmatrix} = \begin{bmatrix} -11 & -16 & -15 \\ -4 & -2 & -10 \\ -6 & 14 & 2 \end{bmatrix}.$$

15. $$BC = \begin{bmatrix} 2 & 1 & 3 \\ -2 & -1 & -1 \\ 1 & 4 & 2 \end{bmatrix}\begin{bmatrix} 3 & -1 & 2 \\ 1 & 6 & 4 \\ 2 & 1 & 3 \end{bmatrix} = \begin{bmatrix} 13 & 7 & 17 \\ -9 & -5 & -11 \\ 11 & 25 & 24 \end{bmatrix}$$

$$ABC = \begin{bmatrix} 1 & 3 & 1 \\ -2 & 1 & 3 \\ 4 & 0 & 2 \end{bmatrix}\begin{bmatrix} 13 & 7 & 17 \\ -9 & -5 & -11 \\ 11 & 25 & 24 \end{bmatrix} = \begin{bmatrix} -3 & 17 & 8 \\ -2 & 56 & 27 \\ 74 & 78 & 116 \end{bmatrix}.$$

17. Using the Gauss-Jordan elimination method, we find

$$\left[\begin{array}{cc|c} 2 & -3 & 5 \\ 3 & 4 & -1 \end{array}\right] \xrightarrow{\frac{1}{2}R_1} \left[\begin{array}{cc|c} 1 & -\frac{3}{2} & \frac{5}{2} \\ 3 & 4 & -1 \end{array}\right] \xrightarrow{R_2-3R_1} \left[\begin{array}{cc|c} 1 & -\frac{3}{2} & \frac{5}{2} \\ 0 & \frac{17}{2} & -\frac{17}{2} \end{array}\right]$$

$$\xrightarrow{\frac{2}{17}R_2} \left[\begin{array}{cc|c} 1 & -\frac{3}{2} & \frac{5}{2} \\ 0 & 1 & -1 \end{array}\right] \xrightarrow{R_1+\frac{3}{2}R_2} \left[\begin{array}{cc|c} 1 & 0 & 1 \\ 0 & 1 & -1 \end{array}\right]$$

We conclude that $x = 1$ and $y = -1$.

19. $\left[\begin{array}{ccc|c} 1 & -1 & 2 & 5 \\ 3 & 2 & 1 & 10 \\ 2 & -3 & -2 & -10 \end{array}\right] \xrightarrow[R_3-2R_1]{R_2-3R_1} \left[\begin{array}{ccc|c} 1 & -1 & 2 & 5 \\ 0 & 5 & -5 & -5 \\ 0 & -1 & -6 & -20 \end{array}\right] \xrightarrow{\frac{1}{5}R_2} \left[\begin{array}{ccc|c} 1 & -1 & 2 & 5 \\ 0 & 1 & -1 & -1 \\ 0 & -1 & -6 & -20 \end{array}\right]$

$$\xrightarrow[R_3+R_2]{R_1+R_2} \left[\begin{array}{ccc|c} 1 & 0 & 1 & 4 \\ 0 & 1 & -1 & -1 \\ 0 & 0 & -7 & -21 \end{array}\right] \xrightarrow{-\frac{1}{7}R_3} \left[\begin{array}{ccc|c} 1 & 0 & 1 & 4 \\ 0 & 1 & -1 & -1 \\ 0 & 0 & 1 & 3 \end{array}\right] \xrightarrow[R_2+R_3]{R_1-R_3}$$

$= \left[\begin{array}{ccc|c} 1 & 0 & 0 & 1 \\ 0 & 1 & 0 & 2 \\ 0 & 0 & 1 & 3 \end{array}\right]$. Therefore, $x = 1$, $y = 2$, and $z = 3$.

21. $\left[\begin{array}{ccc|c} 3 & -2 & 4 & 11 \\ 2 & -4 & 5 & 4 \\ 1 & 2 & -1 & 10 \end{array}\right] \xrightarrow{R_1-R_2} \left[\begin{array}{ccc|c} 1 & 2 & -1 & 7 \\ 2 & -4 & 5 & 4 \\ 1 & 2 & -1 & 10 \end{array}\right] \xrightarrow[R_3-R_1]{R_2-2R_1} \left[\begin{array}{ccc|c} 1 & 2 & -1 & 7 \\ 0 & -8 & 7 & -10 \\ 0 & 0 & 0 & 3 \end{array}\right].$

Since this last row implies that $0 = 3$!, we conclude that the system has no solution.

23. $\left[\begin{array}{ccc|c} 3 & -2 & 1 & 4 \\ 1 & 3 & -4 & -3 \\ 2 & -3 & 5 & 7 \\ 1 & -8 & 9 & 10 \end{array}\right] \xrightarrow{R_2-R_3} \left[\begin{array}{ccc|c} 1 & 1 & -4 & -3 \\ 1 & 3 & -4 & -3 \\ 2 & -3 & 5 & 7 \\ 1 & -8 & 9 & 10 \end{array}\right] \xrightarrow[\substack{R_3-2R_1 \\ R_4-R_1}]{R_2-R_1} \left[\begin{array}{ccc|c} 1 & 1 & -4 & -3 \\ 0 & 2 & 0 & 0 \\ 0 & -5 & 13 & 13 \\ 0 & -9 & 13 & 13 \end{array}\right]$

$$\xrightarrow{\frac{1}{2}R_2} \left[\begin{array}{ccc|c} 1 & 1 & -4 & -3 \\ 0 & 1 & 0 & 0 \\ 0 & -5 & 13 & 13 \\ 0 & -9 & 13 & 13 \end{array}\right] \xrightarrow[\substack{R_3+5R_2 \\ R_4+9R_2}]{R_1-R_2} \left[\begin{array}{ccc|c} 1 & 0 & -4 & -3 \\ 0 & 1 & 0 & 0 \\ 0 & 0 & 13 & 13 \\ 0 & 0 & 13 & 13 \end{array}\right]$$

$$\xrightarrow{\frac{1}{13}R_3} \left[\begin{array}{ccc|c} 1 & 0 & -4 & -3 \\ 0 & 1 & 0 & 0 \\ 0 & 0 & 1 & 1 \\ 0 & 0 & 13 & 13 \end{array}\right] \xrightarrow[R_4-13R_3]{R_1+4R_3} \left[\begin{array}{ccc|c} 1 & 0 & 0 & 1 \\ 0 & 1 & 0 & 0 \\ 0 & 0 & 1 & 1 \\ 0 & 0 & 0 & 0 \end{array}\right].$$

Therefore, $x = 1$, $y = 0$, and $z = 1$.

25. $A^{-1} = \dfrac{1}{(3)(2)-(1)(1)}\begin{bmatrix} 2 & -1 \\ -1 & 3 \end{bmatrix} = \begin{bmatrix} \frac{2}{5} & -\frac{1}{5} \\ -\frac{1}{5} & \frac{3}{5} \end{bmatrix}$.

27. $A^{-1} = \dfrac{1}{(3)(2)-(2)(4)}\begin{bmatrix} 2 & -4 \\ -2 & 3 \end{bmatrix} = \begin{bmatrix} -1 & 2 \\ 1 & -\frac{3}{2} \end{bmatrix}$.

29. $\left[\begin{array}{rrr|rrr} 2 & 3 & 1 & 1 & 0 & 0 \\ 1 & -1 & 2 & 0 & 1 & 0 \\ 1 & 2 & 1 & 0 & 0 & 1 \end{array}\right] \xrightarrow{R_1 - R_2} \left[\begin{array}{rrr|rrr} 1 & 4 & -1 & 1 & -1 & 0 \\ 0 & -1 & 2 & 0 & 1 & 0 \\ 1 & 2 & 1 & 0 & 0 & 1 \end{array}\right] \xrightarrow[R_3 - R_1]{R_2 - R_1}$

$\left[\begin{array}{rrr|rrr} 1 & 4 & -1 & 1 & -1 & 0 \\ 0 & -5 & 3 & -1 & 2 & 0 \\ 0 & -2 & 2 & -1 & 1 & 1 \end{array}\right] \xrightarrow{R_2 - 3R_3} \left[\begin{array}{rrr|rrr} 1 & 4 & -1 & 1 & -1 & 0 \\ 0 & 1 & -3 & 2 & -1 & -3 \\ 0 & -2 & 2 & -1 & 1 & 1 \end{array}\right] \xrightarrow[R_3 + 2R_2]{R_1 - 4R_2}$

$\left[\begin{array}{rrr|rrr} 1 & 0 & 11 & -7 & 3 & 12 \\ 0 & 1 & -3 & 2 & -1 & -3 \\ 0 & 0 & -4 & 3 & -1 & -5 \end{array}\right] \xrightarrow{-\frac{1}{4}R_3} \left[\begin{array}{rrr|rrr} 1 & 0 & 11 & -7 & 3 & 12 \\ 0 & 1 & -3 & 2 & -1 & -3 \\ 0 & 0 & 1 & -\frac{3}{4} & \frac{1}{4} & \frac{5}{4} \end{array}\right] \xrightarrow[R_2 + 3R_3]{R_1 - 11R_3}$

$\left[\begin{array}{rrr|rrr} 1 & 0 & 0 & \frac{5}{4} & \frac{1}{4} & -\frac{7}{4} \\ 0 & 1 & 0 & -\frac{1}{4} & -\frac{1}{4} & \frac{3}{4} \\ 0 & 0 & 1 & -\frac{3}{4} & \frac{1}{4} & \frac{5}{4} \end{array}\right]$. So $A^{-1} = \begin{bmatrix} \frac{5}{4} & \frac{1}{4} & -\frac{7}{4} \\ -\frac{1}{4} & -\frac{1}{4} & \frac{3}{4} \\ -\frac{3}{4} & \frac{1}{4} & \frac{5}{4} \end{bmatrix}$.

31. $\left[\begin{array}{rrr|rrr} 1 & 2 & 4 & 1 & 0 & 0 \\ 3 & 1 & 2 & 0 & 1 & 0 \\ 1 & 0 & -6 & 0 & 0 & 1 \end{array}\right] \xrightarrow[R_3 - R_1]{R_2 - 3R_1} \left[\begin{array}{rrr|rrr} 1 & 2 & 4 & 1 & 0 & 0 \\ 0 & -5 & -10 & -3 & 1 & 0 \\ 0 & -2 & -10 & -1 & 0 & 1 \end{array}\right] \xrightarrow{R_2 - 3R_3}$

$\left[\begin{array}{rrr|rrr} 1 & 2 & 4 & 1 & 0 & 0 \\ 0 & 1 & 20 & 0 & 1 & -3 \\ 0 & -2 & -10 & -1 & 0 & 1 \end{array}\right] \xrightarrow[R_3 + 2R_2]{R_1 - 2R_2} \left[\begin{array}{rrr|rrr} 1 & 0 & -36 & 1 & -2 & 6 \\ 0 & 1 & 20 & 0 & 1 & -3 \\ 0 & 0 & 30 & -1 & 2 & -5 \end{array}\right] \xrightarrow{\frac{1}{30}R_3}$

$$\left[\begin{array}{ccc|ccc} 1 & 0 & -36 & 1 & -2 & 6 \\ 0 & 1 & 20 & 0 & 1 & -3 \\ 0 & 0 & 1 & -\frac{1}{30} & \frac{1}{15} & -\frac{1}{6} \end{array}\right] \xrightarrow[R_2-20R_3]{R_1+36R_3} \left[\begin{array}{ccc|ccc} 1 & 0 & 0 & -\frac{1}{5} & \frac{2}{5} & 0 \\ 0 & 1 & 0 & \frac{2}{3} & -\frac{1}{3} & \frac{1}{3} \\ 0 & 0 & 1 & -\frac{1}{30} & \frac{1}{15} & -\frac{1}{6} \end{array}\right]$$

So $$A^{-1} = \begin{bmatrix} -\frac{1}{5} & \frac{2}{5} & 0 \\ \frac{2}{3} & -\frac{1}{3} & \frac{1}{3} \\ -\frac{1}{30} & \frac{1}{15} & -\frac{1}{6} \end{bmatrix}.$$

33. $(A^{-1}B)^{-1} = B^{-1}(A^{-1})^{-1} = B^{-1}A$. Now

$$B^{-1} = \frac{1}{(3)(2)-4(1)}\begin{bmatrix} 2 & -1 \\ -4 & 3 \end{bmatrix} = \begin{bmatrix} 1 & -\frac{1}{2} \\ -2 & \frac{3}{2} \end{bmatrix}.$$

$$B^{-1}A = \begin{bmatrix} 1 & -\frac{1}{2} \\ -2 & \frac{3}{2} \end{bmatrix}\begin{bmatrix} 1 & 2 \\ -1 & 2 \end{bmatrix} = \begin{bmatrix} \frac{3}{2} & 1 \\ -\frac{7}{2} & -1 \end{bmatrix}.$$

35. $$2A - C = \begin{bmatrix} 2 & 4 \\ -2 & 4 \end{bmatrix} - \begin{bmatrix} 1 & 1 \\ -1 & 2 \end{bmatrix} = \begin{bmatrix} 1 & 3 \\ -1 & 2 \end{bmatrix}.$$

$$(2A - C)^{-1} = \frac{1}{(1)(2)-(-1)(3)}\begin{bmatrix} 2 & -3 \\ 1 & 1 \end{bmatrix} = \begin{bmatrix} \frac{2}{5} & -\frac{3}{5} \\ \frac{1}{5} & \frac{1}{5} \end{bmatrix}.$$

37. $$A = \begin{bmatrix} 2 & 3 \\ 1 & -2 \end{bmatrix}, \quad X = \begin{bmatrix} x \\ y \end{bmatrix}, \quad C = \begin{bmatrix} -8 \\ 3 \end{bmatrix}$$

$$A^{-1} = \frac{1}{(-2)(2)-(1)(3)}\begin{bmatrix} -2 & -3 \\ -1 & 2 \end{bmatrix} = \begin{bmatrix} \frac{2}{7} & \frac{3}{7} \\ \frac{1}{7} & -\frac{2}{7} \end{bmatrix}$$

$$\begin{bmatrix} x \\ y \end{bmatrix} = A^{-1}B = \begin{bmatrix} \frac{2}{7} & \frac{3}{7} \\ \frac{1}{7} & -\frac{2}{7} \end{bmatrix}\begin{bmatrix} -8 \\ 3 \end{bmatrix} = \begin{bmatrix} -1 \\ -2 \end{bmatrix}.$$

39. Put

$$X = \begin{bmatrix} x \\ y \\ z \end{bmatrix}, \quad A = \begin{bmatrix} 1 & -2 & 4 \\ 2 & 3 & -2 \\ 1 & 4 & -6 \end{bmatrix}, \quad C = \begin{bmatrix} 13 \\ 0 \\ -15 \end{bmatrix}.$$

Then $AX = C$ and $X = A^{-1}C$. To find A^{-1},

$$\left[\begin{array}{ccc|ccc} 1 & -2 & 4 & 1 & 0 & 0 \\ 2 & 3 & -2 & 0 & 1 & 0 \\ 1 & 4 & -6 & 0 & 0 & 1 \end{array}\right] \xrightarrow[R_3 - R_1]{R_2 - 2R_1} \left[\begin{array}{ccc|ccc} 1 & -2 & 4 & 1 & 0 & 0 \\ 0 & 7 & -10 & -2 & 1 & 0 \\ 0 & 6 & -10 & -1 & 0 & 1 \end{array}\right] \xrightarrow{R_2 - R_3}$$

$$\left[\begin{array}{ccc|ccc} 1 & -2 & 4 & 1 & 0 & 0 \\ 0 & 1 & 0 & -1 & 1 & -1 \\ 0 & 6 & -10 & -1 & 0 & 1 \end{array}\right] \xrightarrow[R_3 - 6R_2]{R_1 + 2R_2} \left[\begin{array}{ccc|ccc} 1 & 0 & 4 & -1 & 2 & -2 \\ 0 & 1 & 0 & -1 & 1 & -1 \\ 0 & 0 & -10 & 5 & -6 & 7 \end{array}\right] \xrightarrow{-\frac{1}{10}R_3}$$

$$\left[\begin{array}{ccc|ccc} 1 & 0 & 4 & -1 & 2 & -2 \\ 0 & 1 & 0 & -1 & 1 & -1 \\ 0 & 0 & 1 & -\frac{1}{2} & \frac{3}{5} & -\frac{7}{10} \end{array}\right] \xrightarrow{R_1 - 4R_3} \left[\begin{array}{ccc|ccc} 1 & 0 & 0 & 1 & -\frac{2}{5} & \frac{4}{5} \\ 0 & 1 & 0 & -1 & 1 & -1 \\ 0 & 0 & 1 & -\frac{1}{2} & \frac{3}{5} & -\frac{7}{10} \end{array}\right].$$

So $$A^{-1} = \begin{bmatrix} 1 & -\frac{2}{5} & \frac{4}{5} \\ -1 & 1 & -1 \\ -\frac{1}{2} & \frac{3}{5} & -\frac{7}{10} \end{bmatrix}.$$

Therefore, $$X = A^{-1}C = \begin{bmatrix} 1 & -\frac{2}{5} & \frac{4}{5} \\ -1 & 1 & -1 \\ -\frac{1}{2} & \frac{3}{5} & -\frac{7}{10} \end{bmatrix} \begin{bmatrix} 13 \\ 0 \\ -15 \end{bmatrix} = \begin{bmatrix} 1 \\ 2 \\ 4 \end{bmatrix}$$

that is, $x = 1$, $y = 2$, and $z = 4$.

41. $$\begin{bmatrix} x \\ y \\ z \end{bmatrix} = \begin{bmatrix} 600 & 1000 & 800 & 1400 \\ 700 & 800 & 600 & 1200 \\ 1200 & 800 & 1000 & 900 \end{bmatrix} \begin{bmatrix} 1.60 \\ 1.20 \\ 1.50 \\ 1.30 \end{bmatrix} = \begin{bmatrix} 5180 \\ 4540 \\ 5550 \end{bmatrix}.$$

The total revenue is \$5180 at station A, \$4540 at station B, and \$5550 at station C.

43. We wish to solve the system of equations

$$\begin{aligned} 2x + 2y + 3z &= 210 \\ 2x + 3y + 4z &= 270 \\ 3x + 4y + 3z &= 300. \end{aligned}$$

Using the Gauss–Jordan method of elimination, we find

$$\left[\begin{array}{ccc|c} 2 & 2 & 3 & 210 \\ 2 & 3 & 4 & 270 \\ 3 & 4 & 3 & 300 \end{array}\right] \xrightarrow{\frac{1}{2}R_1} \left[\begin{array}{ccc|c} 1 & 1 & \frac{3}{2} & 105 \\ 2 & 3 & 4 & 270 \\ 3 & 4 & 3 & 300 \end{array}\right] \xrightarrow[R_3-3R_1]{R_2-2R_1} \left[\begin{array}{ccc|c} 1 & 1 & \frac{3}{2} & 105 \\ 0 & 1 & 1 & 60 \\ 0 & 1 & -\frac{3}{2} & -15 \end{array}\right]$$

$$\xrightarrow[R_3-R_2]{R_1-R_2} \left[\begin{array}{ccc|c} 1 & 0 & \frac{1}{2} & 45 \\ 0 & 1 & 1 & 60 \\ 0 & 0 & -\frac{5}{2} & -75 \end{array}\right] \xrightarrow{-\frac{2}{5}R_3} \left[\begin{array}{ccc|c} 1 & 0 & \frac{1}{2} & 45 \\ 0 & 1 & 1 & 60 \\ 0 & 0 & 1 & 30 \end{array}\right] \xrightarrow[R_2-R_3]{R_1-\frac{1}{2}R_3} \left[\begin{array}{ccc|c} 1 & 0 & 0 & 30 \\ 0 & 1 & 0 & 30 \\ 0 & 0 & 1 & 30 \end{array}\right]$$

So $x = y = z = 30$. Therefore, Desmond should produce 30 of each type of pendant.

CHAPTER 3

EXERCISES 3.1, page 180

1. $4x - 8 < 0$ implies $x < 2$. The graph of the inequality is at the right.

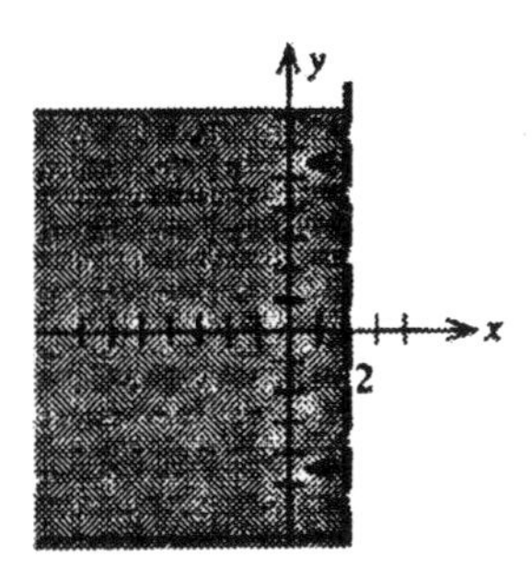

3. $x - y \leq 0$ implies $x \leq y$. The graph of the inequality is at the right.

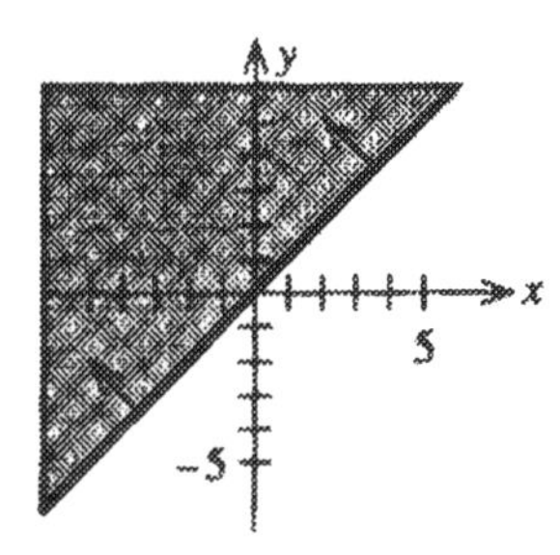

5. The graph of the inequality $x \leq -3$ is at the right.

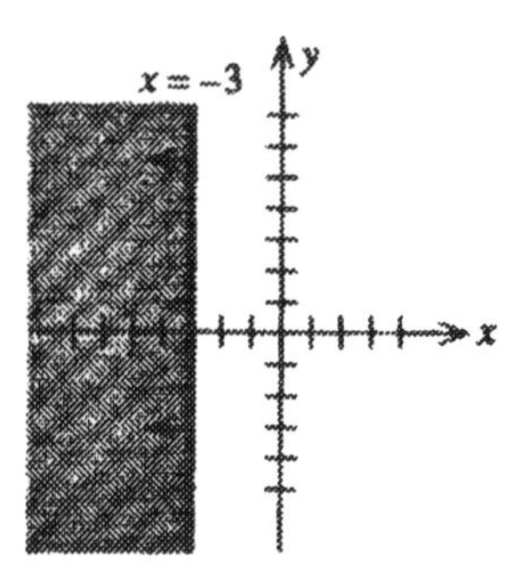

7. We first sketch the straight line with equation $2x + y = 4$. Next, picking the test point (0,0), we have

$$2(0) + (0) = 0 \le 4.$$

We conclude that the half-plane containing the origin is the required half-plane.

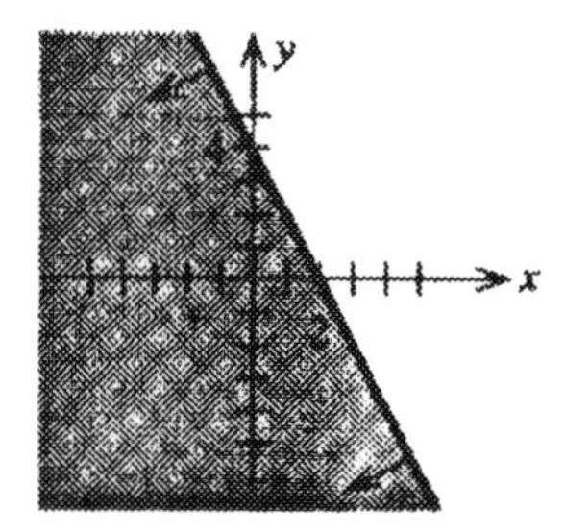

9. We first sketch the graph of the straight line

$$4x - 3y = -24.$$

Next, picking the test point (0,0), we see that

$$4(0) \quad - 3(0) = 0 \not< -24.$$

We conclude that the half-plane not containing the origin is the required half-plane. The graph of this inequality is at the right.

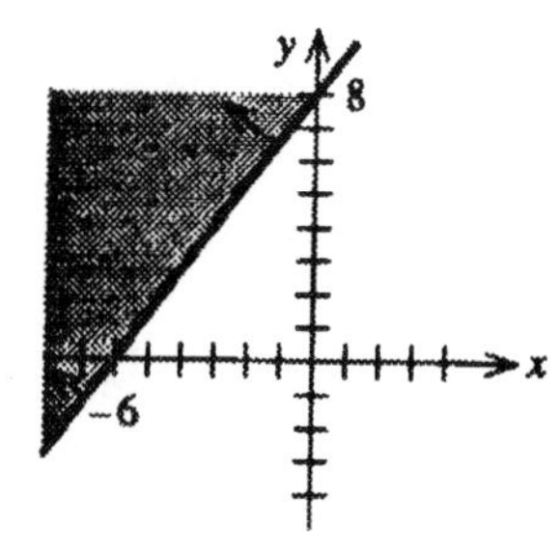

11. The system of linear inequalities that describes the shaded region is

$$x \ge 1,\ x \le 5,\ y \ge 2, \text{ and } y \le 4.$$

We may also combine the first and second inequalities and the third and fourth inequalities and write

$$1 \le x \le 5 \quad \text{and} \quad 2 \le y \le 4.$$

13. The system of linear inequalities that describes the shaded region is

$$2x - y \ge 2,\ 5x + 7y \ge 35, \text{ and } x \le 4.$$

15. The system of linear inequalities that describes the shaded region is

$$7x + 4y \le 140,\ x + 3y \ge 30, \text{ and } x \ - y \ge \ -10.$$

17. The system of linear inequalities that describes the shaded region is

$$x + y \ge 7,\ x \ge 2,\ y \ge 3, \text{ and } \ y \le 7.$$

19. The required solution set is shown at the right. To find the coordinates of A, we solve the system

$$2x+4y=16$$
$$-x+3y=7,$$

giving $A=(2,3)$. Observe that a dotted line is used to show that no point on the line constitutes a solution to the given problem. Observe that this is an unbounded solution set.

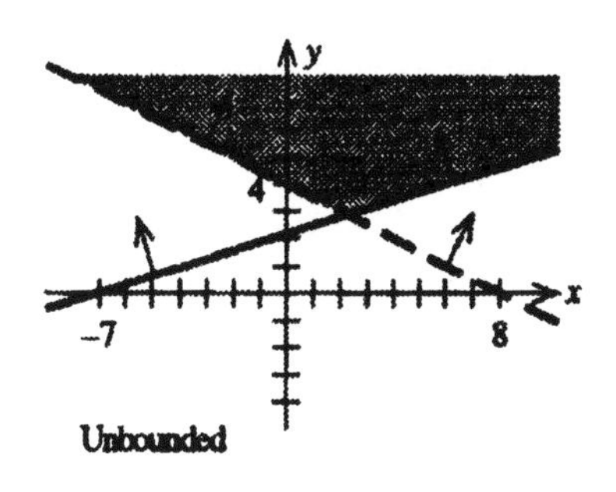

21. The solution set is shown in the figure at the right. Observe that the set is unbounded. To find the coordinates of A, we solve the system

$$x-\ y=0$$
$$2x+3y=10$$

giving $A=(2,2)$. Observe that this is an unbounded solution set.

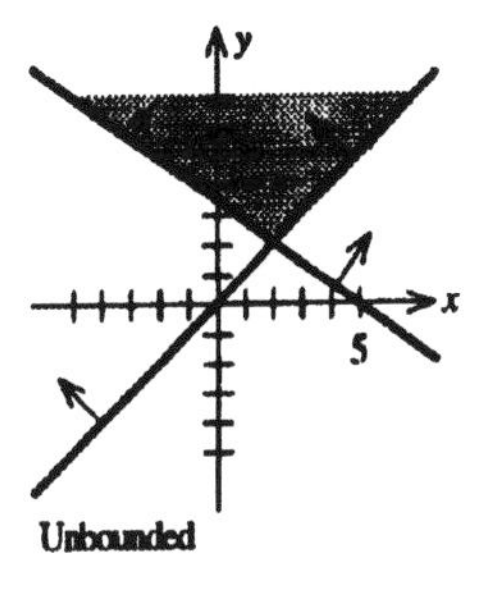

23. The half-planes defined by the two inequalities are shown in the figure at the right. Since the two half-planes have no points in common, we conclude that the given system of inequalities has no solution. (The empty set is a bounded set.)

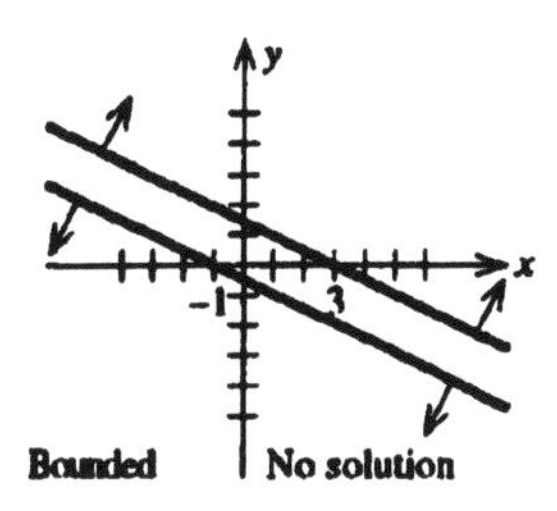

25. The half-planes defined by the three inequalities are shown in the figure at the right. The point A is found by solving the system

$$x+y=6$$
$$x\qquad=3$$

giving $A=(3,3)$. Observe that this is a bounded solution set.

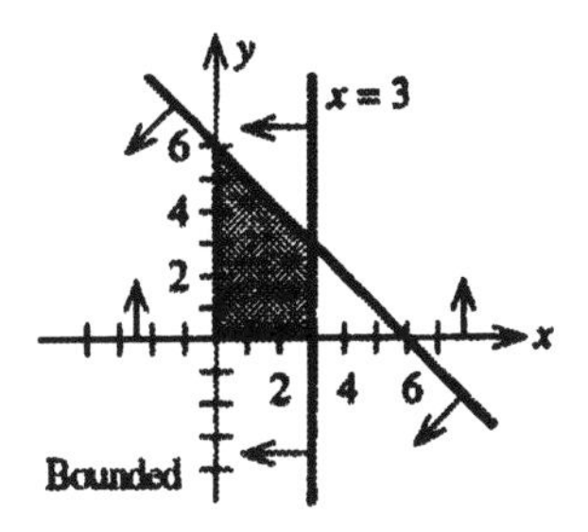

27. The half-planes defined by the given inequalities are shown in the figure at the right. Observe that the two lines described by the equations

$$3x - 6y = 12 \quad \text{and} \quad -x + 2y = 4$$

do not intersect because they are parallel. The solution set is unbounded.

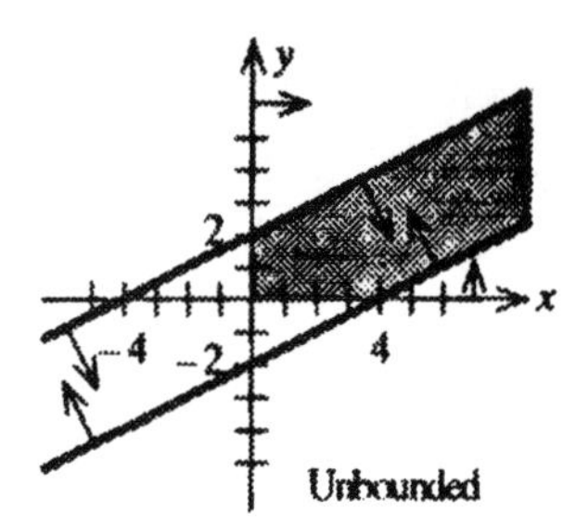

29. The required solution set is shown in the figure at the right. The coordinates of A are found by solving the system

$$3x - 7y = -24$$

$$x + 3y = \quad 8$$

giving $(-1,3)$.
The solution set is unbounded.

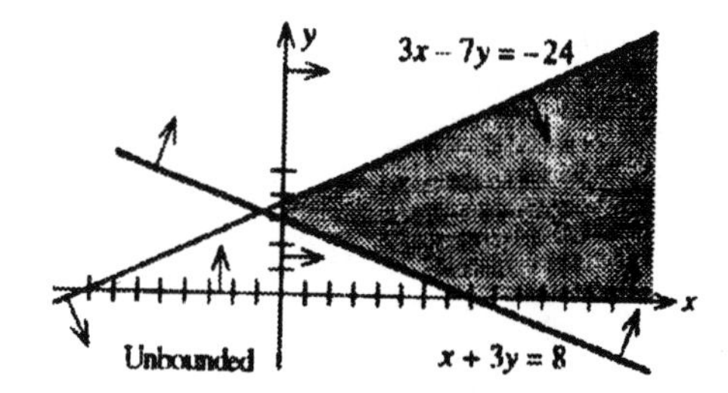

31. The required solution set is shown in the figure that follows.

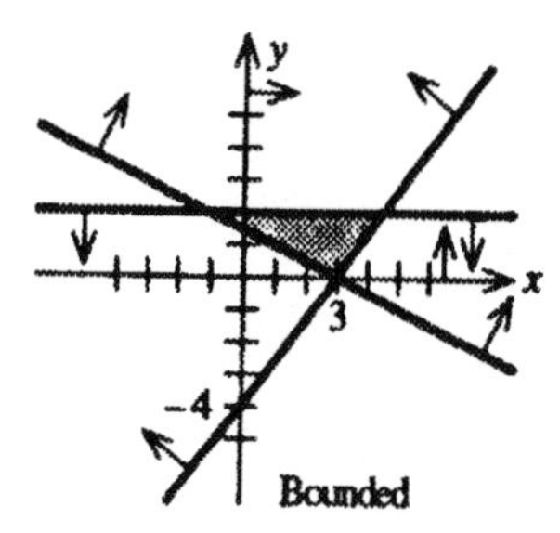

The solution set is bounded.

33. The required solution set is shown in the figure at the right. The solution set has vertices at (0,6), (5,0), (4,0), and (1,3). The solution set is bounded.

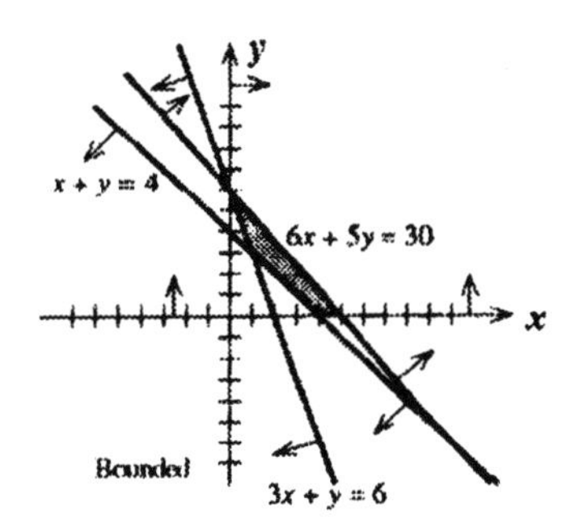

35. The required solution set is shown in the figure at the right. The unbounded solution set has vertices at (2,8), (0,6), (0,3),and (2,2).

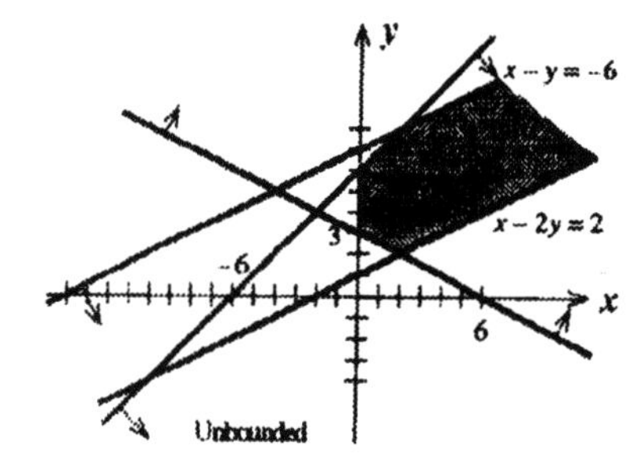

EXERCISES 3.2, page 188

1. We tabulate the given information:

	Product A	Product B	Time available
Machine I	6	9	300
Machine II	5	4	180
Profit per unit ($)	3	4	

Let x and y denote the number of units of Product A and Product B to be produced. Then the required linear programming problem is:

Maximize $P = 3x + 4y$ subject to the constraints

$$6x + 9y \leq 300$$
$$5x + 4y \leq 180$$
$$x \geq 0, y \geq 0$$

3. Let x denote the number of model A hibachis to be produced and y denote the number of model B hibachis to be produced. Since only 1000 pounds of cast iron are available, we must have

$$3x + 4y \leq 1000.$$

The restriction that only 20 hours of labor are available per day implies that

$$6x + 3y \leq 1200. \quad \text{(time in minutes)}$$

Then the profit on the production of these hibachis is given by

$$P = 2x + 1.5y.$$

The additional restriction that at least 150 model A hibachis be produced each day implies that

$$x \geq 150.$$

Summarizing, we have the following linear programming problem:

Maximize $P = 2x + 1.5y$ subject to

$$3x + 4y \leq 1000$$
$$6x + 3y \leq 1200$$
$$x \geq 150, y \geq 0$$

5. Let x and y denote the amount of food A and food B, respectively, used to prepare a meal. Then the requirement that the meal contain a minimum of 400 mg of calcium implies

$$30x + 25y \geq 400.$$

Similarly, the requirements that the meal contain at least 10 mg of iron and 40 mg of vitamin C imply that

$$x + 0.5y \geq 10$$
$$2x + 5y \geq 40.$$

The cholesterol content is given by

$$C = 2x + 5y.$$

Therefore, the linear programming problem is

Minimize $C = 2x + 5y$ subject to

$$30x + 25y \geq 400$$
$$x + 0.5y \geq 10$$
$$2x + 5y \geq 40$$
$$x \geq 0, y \geq 0$$

7. Let x denote the number of picture tubes shipped from location I to city A and let y denote the number of picture tubes shipped from location I to city B. Since the number of picture tubes required by the two factories in city A and city B are 3000 and 4000, respectively, the number of picture tubes shipped from location II to city A and city B, are $(3000 - x)$ and $(4000 - y)$, respectively. These numbers are shown in the following schematic,

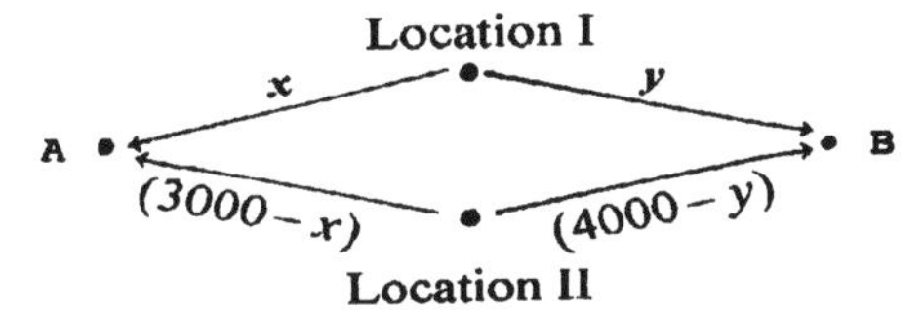

Referring to the schematic and the shipping schedule, we find that the total shipping costs incurred by the company are given by

$$C = 3x + 2y + 4(3000 - x) + 5(4000 - y)$$
$$= 32000 - x - 3y$$

The production constraints on Location I and II lead to the inequalities

$$x + y \leq 6000$$
$$(3000 - x) + (4000 - y) \leq 5000$$

This last inequality simplifies to

$$x + y \geq 2000.$$

The requirements of the two factories lead to the inequalities

$$x \geq 0, y \geq 0, 3000 - x \geq 0, \text{ and } 4000 - y \geq 0.$$

These last two inequalities may be written as $x \leq 3000$ and $y \leq 4000$.
Summarizing, we have the following linear programming problem:

Minimize C = 32,000 - x - $3y$ subject to

$$x + y \leq 6000$$
$$x + y \geq 2000$$
$$x \leq 3000$$
$$y \leq 4000$$
$$x \geq 0, \ y \geq 0$$

9. Let x, y, and z denote the number of units produced of products A, B, and C, respectively. From the given information, we formulate the following linear programming problem:

Maximize $P = 18x + 12y + 15z$ subject to

$$\begin{aligned} 2x + y + 2z &\le 900 \\ 3x + y + 2z &\le 1080 \\ 2x + 2y + z &\le 840 \\ x \ge 0, y \ge 0, z \ge 0 \end{aligned}$$

11. We first tabulate the given information:

Dept.	MODELS A	MODELS B	MODELS C	Time available
Fabrication	$\frac{5}{4}$	$\frac{3}{2}$	$\frac{3}{2}$	310
Assembly	1	1	$\frac{3}{4}$	205
Finishing	1	1	$\frac{1}{2}$	190

Let x, y, and z denote the number of units of model A, model B, and model C to be produced, respectively. Then the required linear programming problem is

Maximize $P = 26x + 28y + 24z$ subject to

$$\begin{aligned} \tfrac{5}{4}x + \tfrac{3}{2}y + \tfrac{3}{2}z &\le 310 \\ x + y + \tfrac{3}{4}z &\le 205 \\ x + y + \tfrac{1}{2}z &\le 190 \\ x \ge 0, y \ge 0, z \ge 0 \end{aligned}$$

13. The shipping costs are tabulated in the following table.

	Warehouse A	Warehouse B	Warehouse C
Plant I	60	60	80
Plant II	80	70	50

Letting x_1 denote the number of pianos shipped from plant I to warehouse A, x_2 the number of pianos shipped from plant I to warehouse B, and so we have

	Warehouse A	B	C	Max. Production
Plant I	x_1	x_2	x_3	300
Plant II	x_4	x_5	x_6	250
Min. Req.	200	150	200	

From the two tables we see that the total monthly shipping cost is given by

$$C = 60x_1 + 60x_2 + 80x_3 + 80x_4 + 70x_5 + 50x_6.$$

Next, the production constraints on plants I and II lead to the inequalities

$$x_1 + x_4 \geq 200$$
$$x_2 + x_5 \geq 150$$
$$x_3 + x_6 \geq 200$$

Summarizing we have the following linear programming problem:

Minimize $C = 60x_1 + 60x_2 + 80x_3 + 80x_4 + 70x_5 + 50x_6$ subject to

$$x_1 + x_2 + x_3 \leq 300$$
$$x_4 + x_5 + x_6 \leq 250$$
$$x_1 + x_4 \geq 200$$
$$x_2 + x_5 \geq 150$$
$$x_3 + x_6 \geq 200$$
$$x_1 \geq 0,\ x_2 \geq 0,\ \ldots,\ x_6 \geq 0$$

15. Let x, y, and z denote the number (in thousands) of bottles of formula I, formula II, and formula III, respectively, produced. Then the profit function to be maximized is

$$P = 180x + 200y + 300z.$$

Next, the limitation on time implies that

$$2.5x + 3y + 4z \leq 70.$$

Similarly, the restrictions on the amount of ingredients available imply that

$$x \leq 9,\ y \leq 12 \text{ and } z \leq 6.$$

Summarizing, we have the following linear programming problem:

Maximize $P = 180x + 200y + 300z$ subject to

$$\begin{aligned} \tfrac{5}{2}x + 3y + 4z &\le 70 \\ x &\le 9 \\ y &\le 12 \\ z &\le 6 \\ x \ge 0, y \ge 0, z &\ge 0. \end{aligned}$$

EXERCISES 3.3, page 201

1. Evaluating the objective function at each of the corner points we obtain the following table.

Vertex	$Z = 2x + 3y$
(1,1)	5
(8,5)	31
(4,9)	35
(2,8)	28

From the table, we conclude that the maximum value of Z is 35 and it occurs at the vertex (4,9). The minimum value of Z is 5 and it occurs at the vertex (1,1).

3. Evaluating the objective function at each of the corner points we obtain the following table.

Vertex	$Z = 3x + 4y$
(0,20)	80
(3,10)	49
(4,6)	36
(9,0)	27

From the graph, we conclude that there is no maximum value since Z is unbounded. The minimum value of Z is 27 and it occurs at the vertex (9,0).

5. Evaluating the objective function at each of the corner points we obtain the following table.

Vertex	$Z = x + 4y$
(0,6)	24
(4,10)	44
(12,8)	44
(15,0)	15

From the table, we conclude that the maximum value of Z is 44 and it occurs at every point on the line segment joining the points (4,10) and (12,8). The minimum value of Z is 15 and it occurs at the vertex (15,0).

7. The problem is to maximize $P = 2x + 3y$ subject to

$$x + y \leq 6$$
$$x \leq 3$$
$$x \geq 0, y \geq 0$$

The feasible set S for the problem is shown in the following figure, and the values of the function P at the vertices of S are summarized in the accompanying table.

Vertex	$P = 2x + 3y$
$A(0,0)$	0
$B(3,0)$	6
$C(3,3)$	15
$D(0,6)$	18

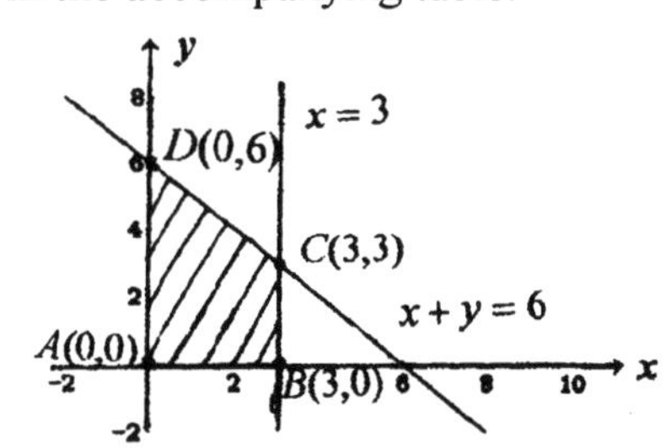

We conclude that P attains a maximum value of 18 when $x = 0$ and $y = 6$.

9. The problem is to minimize $C = 2x + 10y$ subject to

$$5x + 2y \geq 40$$
$$x + 2y \geq 20$$
$$y \geq 3, \ x \geq 0$$

The feasible set S for the problem is shown in the following figure and the values of the function C at the vertices of S are summarized in the accompanying table.

Vertex	$C = 2x + 10y$
$A(0,20)$	200
$B(5,\frac{15}{2})$	85
$C(14,3)$	58

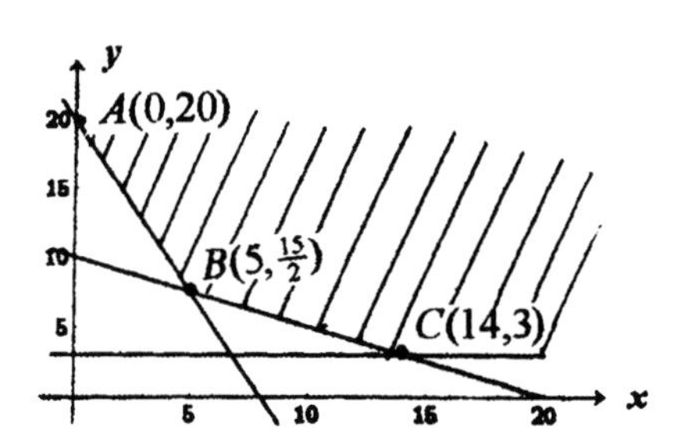

We conclude that C attains a minimum value of 58 when $x = 14$ and $y = 3$.

11. The problem is to minimize $C = 6x + 3y$ subject to

$$\begin{aligned} 4x + y &\geq 40 \\ 2x + y &\geq 30 \\ x + 3y &\geq 30 \\ x \geq 0, y &\geq 0 \end{aligned}$$

The feasible set S is shown in the following figure, and the values of C at each of the vertices of S are shown in the accompanying table.

Vertex	$C = 6x + 3y$
$A(0,40)$	120
$B(5,20)$	90
$C(12,6)$	90
$D(30,0)$	180

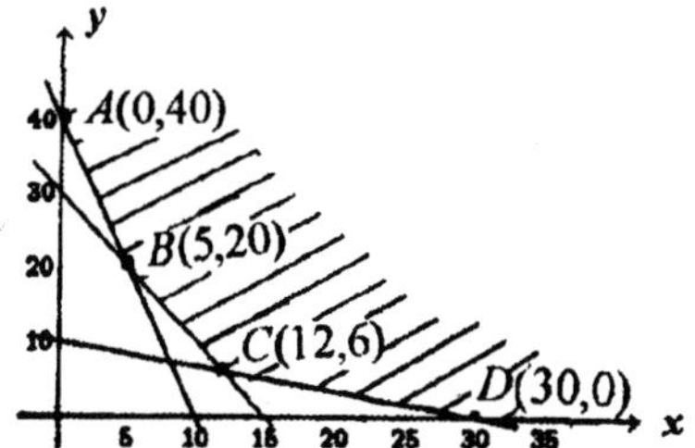

We conclude that C attains a minimum value of 90 at any point (x, y) lying on the line segment joining (5,20) to (12,6).

13. The problem is to minimize $C = 10x + 15y$ subject to

$$\begin{aligned} x + y &\leq 10 \\ 3x + y &\geq 12 \\ -2x + 3y &\geq 3 \\ x \geq 0,\ y &\geq 0 \end{aligned}$$

The feasible set is shown in the following figure, and the values of C at each of the vertices of S are shown in the accompanying table.

Vertex	$C = 10x + 15y$
$A(3,3)$	75
$B(\frac{27}{5}, \frac{23}{5})$	123
$C(1,9)$	145

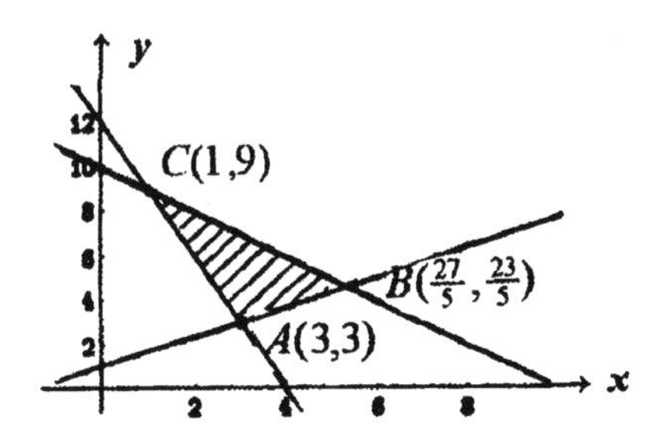

We conclude that C attains a minimum value of 75 when $x = 3$ and $y = 3$.

15. The problem is to maximize $P = 3x + 4y$ subject to

$$x + 2y \leq 50$$
$$5x + 4y \leq 145$$
$$2x + \ \ y \geq 25$$
$$y \geq 5, x \geq 0$$

The feasible set S is shown in the figure that follows, and the values of P at each of the vertices of S are shown in the accompanying table.

Vertex	$C = 3x + 4y$
$A(10,5)$	50
$B(25,5)$	95
$C(15, \frac{35}{2})$	115
$D(0,25)$	100

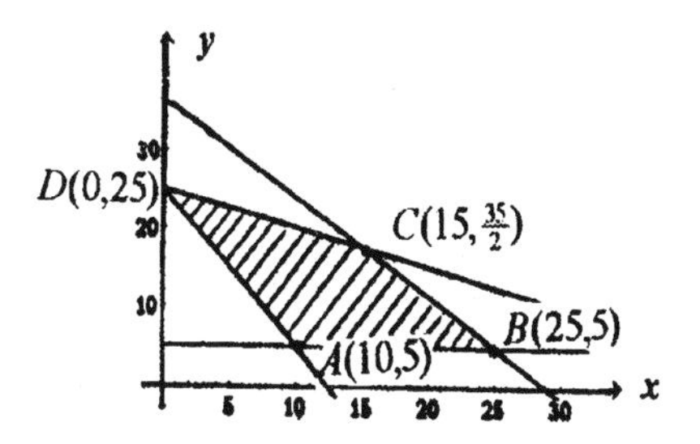

We conclude that P attains a maximum value of 115 when $x = 15$ and $y = 35/2$.

17. The problem is to maximize $P = 2x + 3y$ subject to

$$x + \ \ y \leq \ 48$$
$$x + 3y \geq \ 60$$
$$9x + 5y \leq 320$$
$$x \geq 10, y \geq 0$$

The feasible set S is shown in the figure that follows, and the values of P at each of the vertices of S are shown in the accompanying table.

Vertex	$C = 2x + 3y$
$A(10, \frac{50}{3})$	25
$B(30,10)$	90
$C(20,28)$	124
$D(10,38)$	134

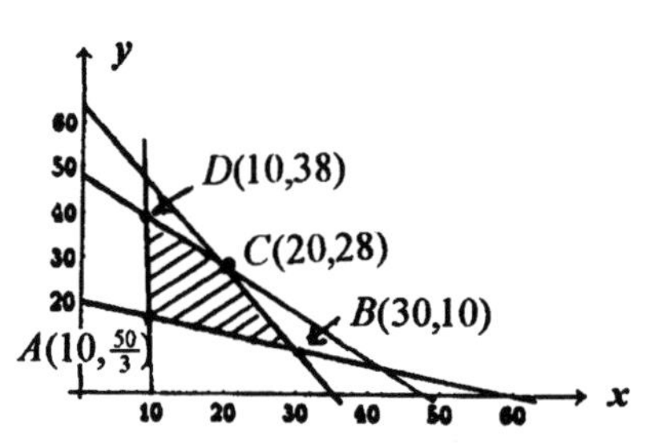

We conclude that P attains a maximum value of 134 when $x = 10$ and $y = 38$.

19. The problem is to find the maximum and minimum value of $P = 10x + 12y$ subject to

$$5x + 2y \geq 63$$
$$x + y \geq 18$$
$$3x + 2y \leq 51$$
$$x \geq 0, y \geq 0$$

The feasible set is shown at the right and the value of P at each of the vertices of S are shown in the accompanying table.

Vertex	$P = 10x + 12y$
$A(9,9)$	198
$B(15,3)$	186
$C(6, \frac{33}{2})$	258

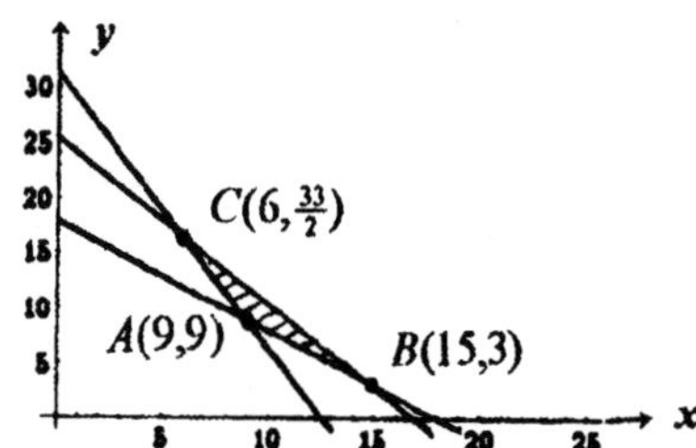

P attains a maximum value of 258 when $x = 6$ and $y = 33/2$. The minimum value of P is 186. It is attained when $x = 15$ and $y = 3$.

21. The problem is to find the maximum and minimum value of $P = 2x + 4y$ subject to

$$x + y \leq 20$$
$$-x + y \leq 10$$
$$x \leq 10$$
$$x + y \geq 5$$
$$y \geq 5, \; x \geq 0$$

The feasible set is shown in the figure that follows, and the value of P at each of the vertices of S are shown in the accompanying table.

Vertex	$P = 2x + 4y$
$A(0,5)$	20
$B(10,5)$	40
$C(10,10)$	60
$D(5,15)$	70
$E(0,10)$	40

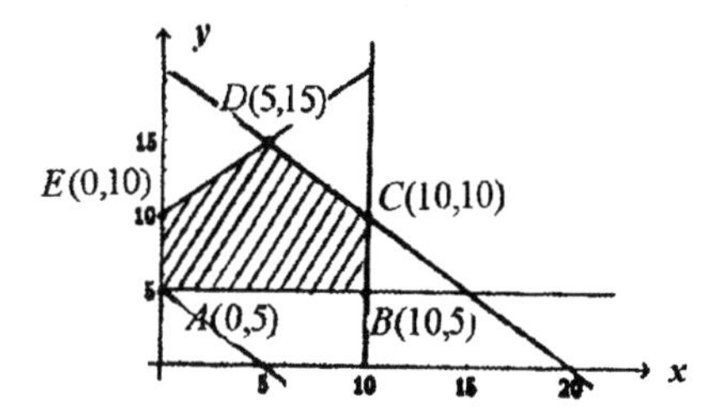

P attains a maximum value of 70 when $x = 5$ and $y = 15$. The minimum value of P is 20. It is attained when $x = 0$ and $y = 5$.

23. Let x and y denote the number of model A and model B fax machines produced in each shift. Then the restriction on manufacturing costs implies

$$200x + 300y \leq 600{,}000,$$

and the limitation on the number produced implies

$$x + y \leq 2{,}500.$$

The total profit is $P = 25x + 40y$. Summarizing, we have the following linear programming problem.

Maximize $P = 25x + 40y$ subject to

$$\begin{aligned} 200x + 300y &\leq 600{,}000 \\ x + \quad y &\leq \quad 2{,}500 \\ x \geq 0,\ y &\geq 0 \end{aligned}$$

The graph of the feasible set S and the associated table of values of P follow.

Vertex	$C = 25x + 40y$
$A(0,0)$	0
$B(2500,0)$	62,500
$C(1500,1000)$	77,500
$D(0,2000)$	80,000

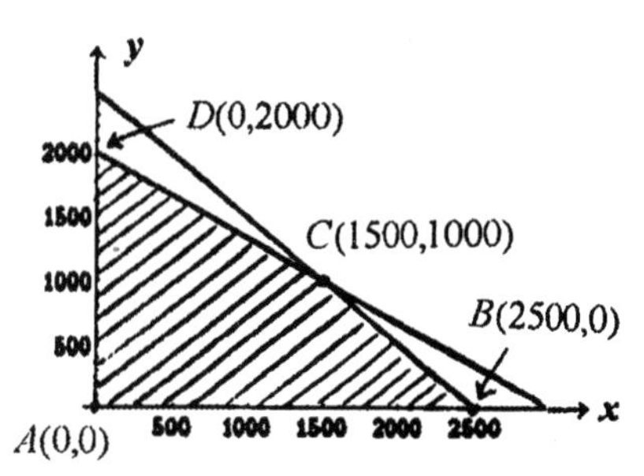

P attains a maximum value of 80,000 when $x = 0$ and $y = 2000$. Thus, by producing 2000 model B fax machines in each shift, the company will realize an optimal profit of \$80,000.

25. Refer to the solution of Exercise 2, Section 3.2, The problem is

Maximize $P = 2x + 1.5y$ subject to

$$3x + 4y \le 1000$$
$$6x + 3y \le 1200$$
$$x \ge 0,\ y \ge 0$$

The graph of the feasible set S and the associated table of values of P follow.

Vertex	$C = 2x + 1.5y$
$A(0,0)$	0
$B(200,0)$	400
$C(120,160)$	480
$D(0,250)$	375

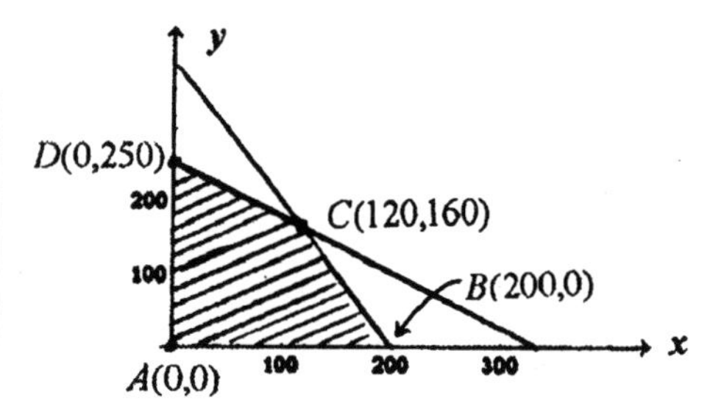

P attains a maximum value of 480 when $x = 120$ and $y = 160$. Thus, by producing 120 model A hibachis and 160 model B hibachis in each shift, the company will realize an optimal profit of \$480.

27. Refer to the solution of Exercise 4, Section 3.2. The linear programming problem is

Maximize $P = 0.1x + 0.12y$ subject to

$$x + y \le 20$$
$$x - 4y \ge 0$$
$$x \ge 0,\ y \ge 0$$

The feasible set S for the problem is shown in the figure at the right, and the value of P at each of the vertices of S is shown in the accompanying table.

Vertex	$C = 0.1x + 0.12y$
$A(0,0)$	0
$B(16,4)$	2.08
$C(20,0)$	2.00

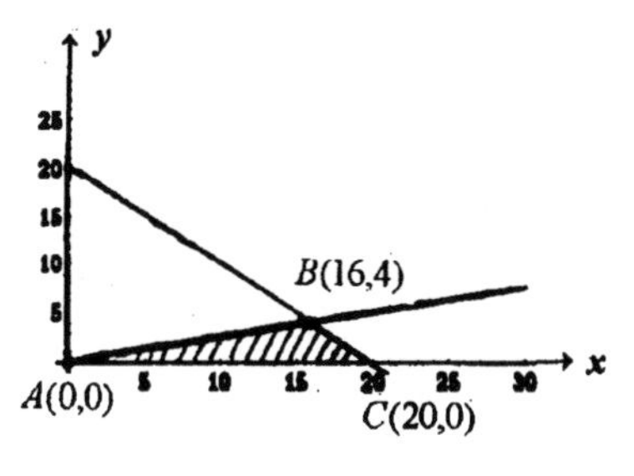

The maximum value of P is attained when $x = 16$ and $y = 4$. Thus, by extending \$16 million in housing loans and \$4 million in automobile loans, the company will realize a return of \$2.08 million on its loans.

29. Refer to Exercise 6, Section 3.2. The problem is

Maximize $P = 150x + 200y$ subject to

$$40x + 60y \leq 7400$$

$$20x + 25y \leq 3300$$

$$x \geq 0,\ y \geq 0$$

The graph of the feasible set S and the associated table of values of P follow.

Vertex	$P = 150x + 200y$
$A(0,0)$	0
$B(165,0)$	24,750
$C(65,80)$	25,750
$D(0,123)$	24,600

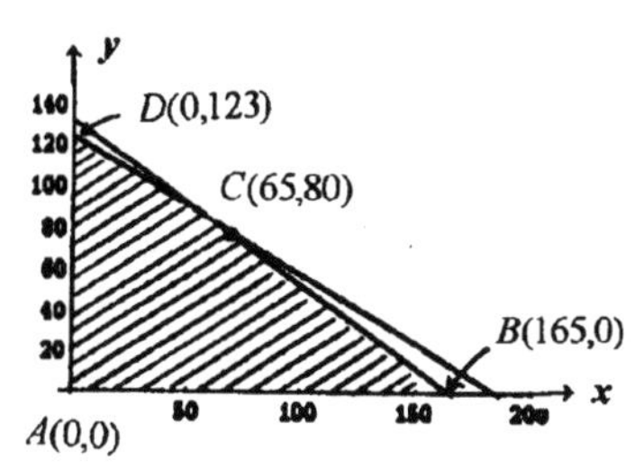

P attains a maximum value of 25,750 when $x = 65$ and $y = 80$. Thus, by producing 65 acres of crop A and 80 acres of crop B, the farmer will realize a maximum profit of \$25,750.

31. Refer to the solution of Exercise 7, Section 3.2.

Minimize $C = 32{,}000 - x - 3y$ subject to

$$x + y \leq 6000$$

$$x + y \geq 2000$$

$$x \leq 3000$$

$$y \leq 4000$$

$$x \geq 0,\ y \geq 0$$

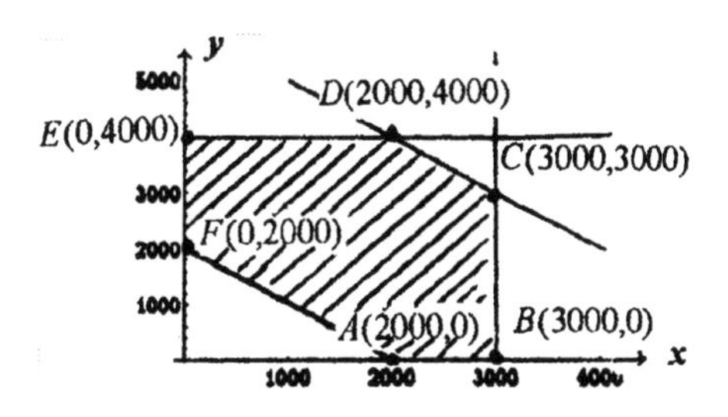

Vertex	$C = 32{,}000 - x - 3y$
$A(2000,0)$	30,000
$B(3000,0)$	29,000
$C(3000,3000)$	20,000
$D(2000,4000)$	18,000
$E(0,4000)$	20,000
$F(0,2000)$	26,000

Since x denotes the number of picture tubes shipped from location I to city A and y denotes the number of picture tubes shipped from location I to city B, we see that the company should ship 2000 tubes from location I to city A and 4000 tubes from location I to city B. Since the number of picture tubes required by the two factories in city A and city B are 3000 and 4000, respectively, the number of picture tubes shipped from location II to city A and city B, are

$$(3000 - x) = 3000 - 2000 = 1000 \text{ and } (4000 - y) = 4000 - 4000 = 0$$

respectively. The minimum shipping cost will then be \$18,000.

33. Let x denote Patricia's investment in growth stocks and y denote the value of her investment in speculative stocks, where both x and y are measured in thousands of dollars. Then the return on her investments is given by $P = 0.15x + 0.25y$. Since her investment may not exceed \$30,000, we have the constraint $x + y \le 30$. The condition that her investment in growth stocks be at least 3 times as much as her investment in speculative stocks translates into the inequality $x \ge 3y$. Thus, we have the following linear programming problem:

Maximize $P = 0.15x + 0.25y$ subject to

$$x + y \le 30$$
$$x - 3y \ge 0$$
$$x \ge 0,\ y \ge 0$$

Vertex	$C = 0.15x + 0.25y$
$A(0,0)$	0
$B(30,0)$	4.5
$C(\frac{45}{2}, \frac{15}{2})$	5.25

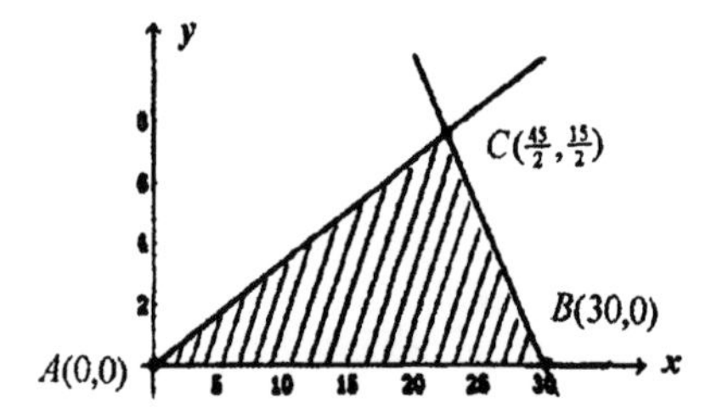

The graph of the feasible set S is shown in the figure and the value of P at each of the vertices of S is shown in the accompanying table. The maximum value of P occurs when $x = 22.5$ and $y = 7.5$. Thus, by investing \$22,500 in growth stocks and \$7,500 in speculative stocks. Patricia will realize a return of \$5250 on her investments.

35. Let x denote the number of urban families and let y denote the number of suburban families interviewed by the company. Then, the amount of money paid to Trendex

will be

$$P = 1500 + 2(x + y) - 1.1x - 1.25y = 1500 + 0.9x + 0.75y.$$

Since a maximum of 1500 families are to be interviewed, we have

$$x + y \leq 1500.$$

Next, the condition that at least 500 urban families are to be interviewed translates into the condition $x \geq 500$. Finally the condition that at least half of the families interviewed must be from the suburban area gives

$$y \geq \tfrac{1}{2}(x + y) \quad \text{or} \quad y - x \geq 0$$

Thus, we are led to the following programming problem:

Maximize $P = 1500 + 0.9x + 0.75y$ subject to

$$x + y \leq 1500$$
$$y - x \geq 0$$
$$x \geq 500, y \geq 0$$

The graph of the feasible set S for this problem follows and the value of P at each of the vertices of S is given in the accompanying table.

Vertex	$P = 1500 + 0.9x + 0.75y$
$A(500,500)$	2325
$B(750,750)$	2735.50
$C(500,1000)$	2700

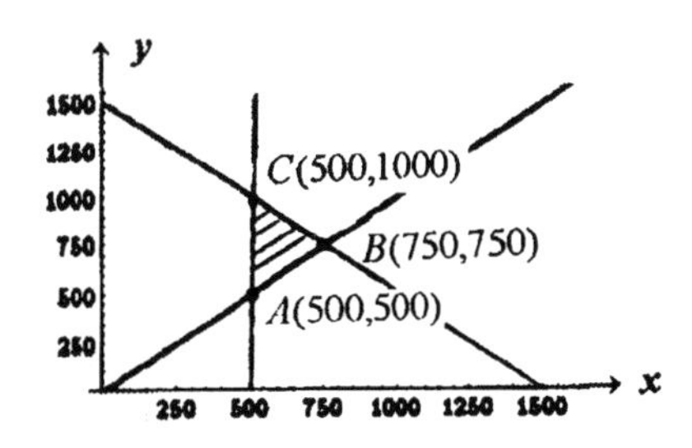

Using the method of corners, we conclude that the profit will be maximized when $x = 750$ and $y = 750$. Thus, a maximum profit of \$2,737.50 will be realized when 750 urban and 750 suburban families are interviewed.

37. a. True. Since $a > 0$, the term ax can be made as large as we please by taking x sufficiently large (because S is unbounded) and therefore P is unbounded as well.

b. True. Maximizing $P = ax + by$ on S is the same as minimizing

$$Q = -P -(ax + by) = -ax - by = AX + By,$$

where $A \geq 0$ and $B \geq 0$. Since $x \geq 0$ and $y \geq 0$, the linear function Q, and therefore P, has at least one optimal solution.

CHAPTER 3 REVIEW EXERCISES, page 208

1. Evaluating Z at each of the corners of the feasible set S, we obtain the following table.

Vertex	$Z = 2x + 3y$
(0,0)	0
(5,0)	10
(3,4)	18
(0,6)	18

We conclude that Z attains a minimum value of 0 when $x = 0$ and $y = 0$, and a maximum value of 18 when x and y lie on the line segment joining (3,4) and (0,6).

3. The graph of the feasible set S is shown at the right.

Vertex	$Z = 3x + 5y$
$A(0,0)$	0
$B(5,0)$	15
$C(3,2)$	19
$D(0,4)$	20

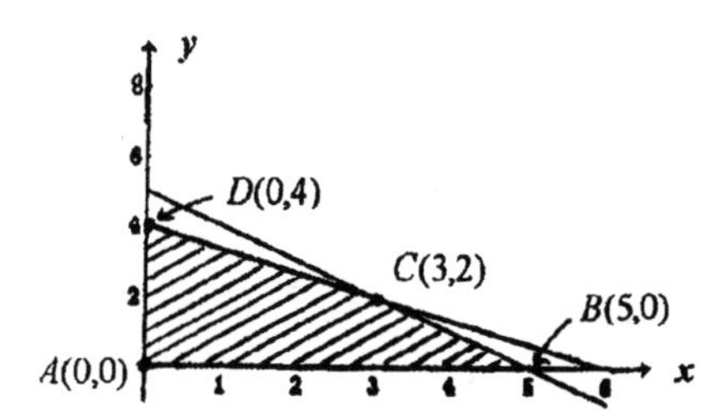

We conclude that the maximum value of P is 20 when $x = 0$ and $y = 4$.

5. The values of the objective function $C = 2x + 5y$ at the corners of the feasible set are given in the following table. The graph of the feasible set S follows.

Vertex	$C = 2x + 5y$
$A(0,0)$	0
$B(4,0)$	8
$C(3,4)$	26
$D(0,5)$	25

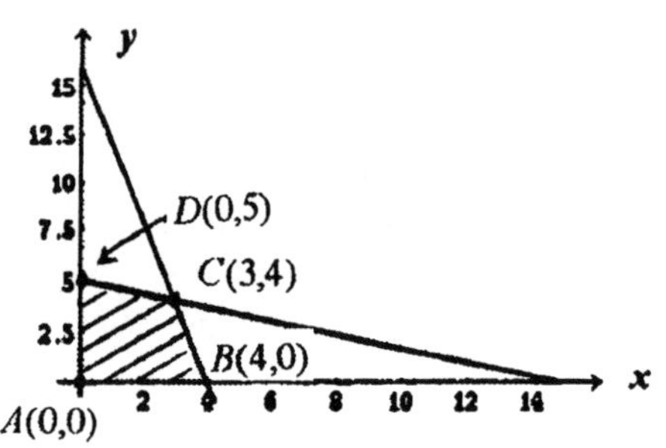

We conclude that the minimum value of C is 0 when $x = 0$ and $y = 0$.

7. The values of the objective function $P = 3x + 2y$ at the vertices of the feasible set are given in the following table. The graph of the feasible set follows.

Vertex	$P = 3x + 2y$
$A(0, \frac{28}{5})$	$\frac{56}{5}$
$B(7,0)$	21
$C(8,0)$	24
$D(3,10)$	29
$E(0,12)$	24

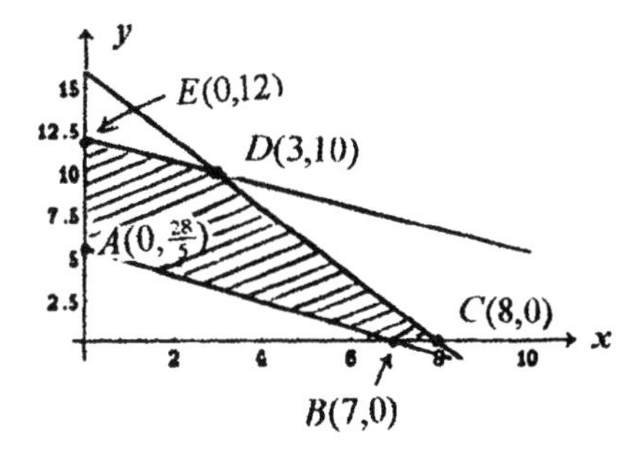

We conclude that P attains a maximum value of 29 when $x = 3$ and $y = 10$.

9. The graph of the feasible set S is shown at the right. The values of the objective function $C = 2x + 7y$ at each of the corner points of the feasible set S are shown in the table that follows.

Vertex	$C = 2x + 7y$
$A(20,0)$	40
$B(10,3)$	41
$C(0,9)$	63

We conclude that C attains a minimum value of 40 when $x = 20$ and $y = 0$.

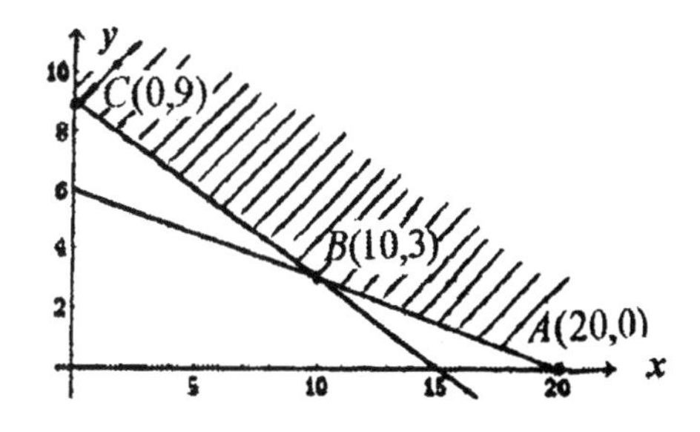

11. The graph of the feasible set S is shown in the following figure. We conclude that Q attains a maximum value of 22 when $x = 22$ and $y = 0$, and a minimum value of $5\frac{1}{2}$ when $x = 3$ and $y = \frac{5}{2}$.

Vertex	$Q = x + y$
$A(2,5)$	7
$B(3, \frac{5}{2})$	$5\frac{1}{2}$
$C(8,0)$	8
$D(22,0)$	22

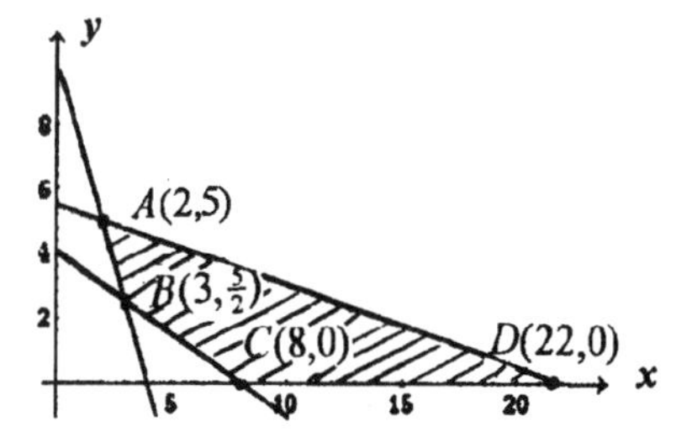

13. Suppose the investor puts x and y thousand dollars into the stocks of company A and company B, respectively. Then the mathematical formulation leads to the linear programming problem:

Maximize $P = 0.14x + 0.20y$ subject to

$$x + \quad y \le 80$$
$$0.01x + 0.04y \le 2$$
$$x \ge 0,\ y \ge 0$$

The feasible set S for this problem is shown in figure at the right and the values at each corner point are given in the accompanying table.

Vertex	$P = 0.14x + 0.20y$
$A(0,0)$	0
$B(80,0)$	11.2
$C(40,40)$	13.6
$D(0,50)$	10

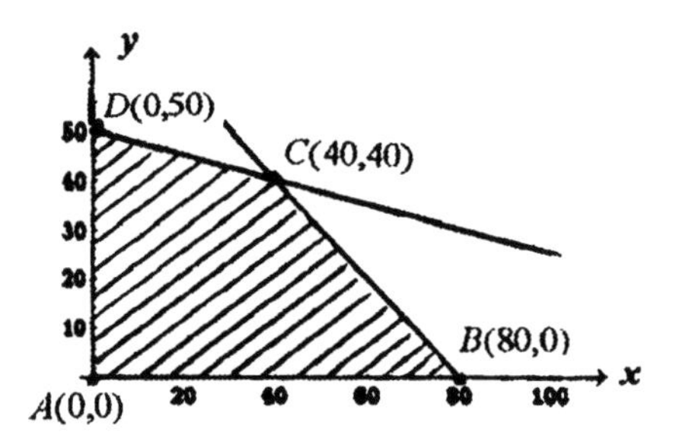

P attains a maximum value of 13.6 when $x = 40$ and $y = 40$. Thus, by investing \$40,000 in the stocks of each company, the investor will achieve a maximum return of \$13,600.

15. Let x denote the number of model A hibachis and y the number of required is

$$3x + 4y \le 1000$$

and the number of minutes of labor used each day is

$$6x + 3y \le 1200.$$

One additional constraint specifies that $y \ge 180$. The daily profit is $P = 2x + 1.5y$. Therefore, we have the following linear programming problem:

Maximize $P = 2x + 1.5y$ subject to

$$3x + 4y \le 1000$$
$$6x + 3y \le 1200$$
$$x \ge 0,\ \ y \ge 180$$

The graph of the feasible set S is shown in the figure that follows.

Vertex	$P = 2x + 1.5y$
$A(0,180)$	270
$B(0,250)$	375
$C(93\frac{1}{3},180)$	$456\frac{2}{3}$

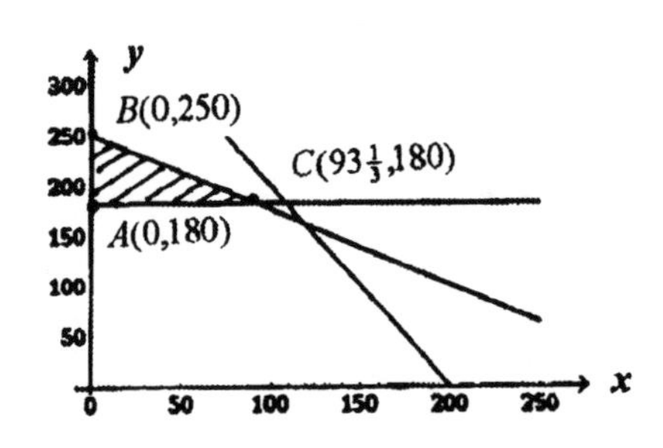

Thus, the optimal profit of \$456 is realized when 93 units of model A hibachis and 180 units of model B hibachis are produced.

CHAPTER 4

EXERCISES 4.1, page 227

1. All entries in the last row of the simplex tableau are nonnegative and an optimal solution has been reached. We find

$$x = 30/7,\ y = 20/7,\ u = 0,\ v = 0, \text{ and } P = 220/7.$$

3. The simplex tableau is not in final form because there is an entry in the last row that is negative. The entry in the first row, second column, is the next pivot element and has a value of 1/2.

5. The simplex tableau is in final form. We find

$$x = 1/3,\ y = 0,\ z = 13/3,\ u = 0,\ v = 6,\ w = 0 \text{ and } P = 17.$$

7. The simplex tableau is not in final form because there are two entries in the last row that are negative. The entry in the third row, second column, is the pivot element and has a value of 1.

9. The simplex tableau is in final form. The solutions are

$$x = 30,\ y = 0,\ z = 0,\ u = 10,\ v = 0, \text{ and } P = 60,$$

and $$x = 0,\ y = 30,\ z = 0,\ u = 10,\ v = 0, \text{ and } P = 60.$$

(There are infinitely many answers).

11.

	x	y	u	v	P	Const	Ratio
p.r. →	1	2	1	0	0	12	$12/2 = 6$
	3	2	0	1	0	24	$24/2 = 12$
	−10	−12	0	0	1	0	
		↑ p.c.					

$\xrightarrow{\frac{1}{2}R_1}$

x	y	u	v	P	Const
$\frac{1}{2}$	1	$\frac{1}{2}$	0	0	6
3	2	0	1	0	24
−10	−12	0	0	1	0

$\xrightarrow[R_3+12R_1]{R_2-2R_1}$

	x	y	u	v	P	Const	Ratio
	$\frac{1}{2}$	1	$\frac{1}{2}$	0	0	6	$6/(1/2)=12$
p.r. →	2	0	−1	1	0	12	$12/2=6$
	−4	0	6	0	1	72	
	↑ p.c.						

$\xrightarrow{\frac{1}{2}R_2}$

x	y	u	v	P	Const
$\frac{1}{2}$	1	$\frac{1}{2}$	0	0	6
1	0	$-\frac{1}{2}$	$\frac{1}{2}$	0	6
−4	0	6	0	1	72

$\xrightarrow[R_3+4R_2]{R_1-\frac{1}{2}R_2}$

x	y	u	v	P	Const
0	1	$\frac{3}{4}$	$-\frac{1}{4}$	0	3
1	0	$-\frac{1}{2}$	$\frac{1}{2}$	0	6
0	0	4	2	1	96

The last tableau is in final form. We find that $x = 6$, $y = 3$, $u = 0$, $v = 0$, and $P = 96$.

13. We obtain the following sequence of tableaus:

	x	y	u	v	w	P	Const.	Ratio
	3	1	1	0	0	0	24	24
	2	1	0	1	0	0	18	18
p.r. →	1	3	0	0	1	0	24	8
	−4	−6	0	0	0	1	0	
		↑ p.c.						

$\xrightarrow{\frac{1}{3}R_3}$

x	y	u	v	w	P	Const.
3	1	1	0	0	0	24
2	1	0	1	0	0	18
$\frac{1}{3}$	1	0	0	$\frac{1}{3}$	0	8
−4	−6	0	0	0	1	0

$\xrightarrow[R_4+6R_3]{R_1-R_3,\ R_2-R_3}$

	x	y	u	v	w	P	Const.	Ratio
$p.r. \to$	$\frac{8}{3}$	0	1	0	$-\frac{1}{3}$	0	16	6
	$\frac{5}{3}$	0	0	1	$-\frac{1}{3}$	0	10	6
	$\frac{1}{3}$	1	0	0	$\frac{1}{3}$	0	8	24
	-2	0	0	0	2	1	48	
	$\uparrow$ $p.c.$							

(Observe that we have a choice here.)

$\xrightarrow{\frac{3}{8}R_1}$

x	y	u	v	w	P	Const.
1	0	$\frac{3}{8}$	0	$-\frac{1}{8}$	0	6
$\frac{5}{3}$	0	0	1	$-\frac{1}{3}$	0	10
$\frac{1}{3}$	1	0	0	$\frac{1}{3}$	0	8
-2	0	0	0	2	1	48

$\xrightarrow[R_4+2R_1]{R_2-\frac{5}{3}R_1 \quad R_3-\frac{1}{3}R_1}$

x	y	u	v	w	P	Const.
1	0	$\frac{3}{8}$	0	$-\frac{1}{8}$	0	6
0	0	$-\frac{5}{8}$	1	$-\frac{1}{8}$	0	0
0	1	$-\frac{1}{8}$	0	$\frac{3}{8}$	0	6
0	0	$\frac{3}{4}$	0	$\frac{7}{4}$	1	60

We deduce that $x = 6$, $y = 6$, $u = 0$, $v = 0$, $w = 0$, and $P = 60$.

15. We obtain the following sequence of tableaus:

	x	y	z	u	v	P	Const.	Ratio
	1	1	1	1	0	0	8	$8/1 = 8$
$p.r. \to$	3	2	4	0	1	0	24	$24/4 = 6$
	-3	-4	-5	0	0	1	0	
			$\uparrow$ $p.c$					

$\xrightarrow{\frac{1}{4}R_2}$

x	y	z	u	v	P	Const.
1	1	1	1	0	0	8
$\frac{3}{4}$	$\frac{1}{2}$	1	0	$\frac{1}{4}$	0	6
−3	−4	−5	0	0	1	0

$\xrightarrow{R_1-R_2,\ R_3+5R_2}$

	x	y	z	u	v	P	Const.
p.r.→	$\frac{1}{4}$	$\frac{1}{2}$	0	1	$-\frac{1}{4}$	0	2
	$\frac{3}{4}$	$\frac{1}{2}$	1	0	$\frac{1}{4}$	0	6
	$\frac{3}{4}$	$-\frac{3}{2}$	0	0	$\frac{5}{4}$	1	30
		↑ p.c.					

Ratio
$2/(1/2) = 4$
$6/(1/2) = 12$

$\xrightarrow{2R_1}$

x	y	z	u	v	P	Const.
$\frac{1}{2}$	1	0	2	$-\frac{1}{2}$	0	4
$\frac{3}{4}$	$\frac{1}{2}$	1	0	$\frac{1}{4}$	0	6
$\frac{3}{4}$	$-\frac{3}{2}$	0	0	$\frac{5}{4}$	1	30

$\xrightarrow{R_2-\frac{1}{2}R_1,\ R_3+\frac{3}{2}R_1}$

x	y	z	u	v	P	Const.
$\frac{1}{2}$	1	0	2	$-\frac{1}{2}$	0	4
$\frac{1}{2}$	0	1	−1	$\frac{1}{2}$	0	4
$\frac{3}{2}$	0	0	3	$\frac{1}{2}$	1	36

This last tableau is in final form. We find that $x = 0$, $y = 4$, $z = 4$, $u = 0$, $v = 0$, and $P = 36$.

17.

	x	y	z	u	v	w	P	Const.
	3	10	5	1	0	0	0	120
p.r.→	5	2	8	0	1	0	0	6
	8	10	3	0	0	1	0	105
	−3	−4	−1	0	0	0	1	0
		↑ p.c.						

Ratio
$120/10 = 12$
$6/2 = 3$
$105/10 = 21/2$

$\xrightarrow{\frac{1}{2}R_2}$

x	y	z	u	v	w	P	Const.
3	10	5	1	0	0	0	120
$\frac{5}{2}$	1	4	0	$\frac{1}{2}$	0	0	3
8	10	3	0	0	1	0	105
−3	−4	−1	0	0	0	1	0

$\xrightarrow{\substack{R_1-10R_2 \\ R_3-10R_2 \\ R_4+4R_2}}$

x	y	z	u	v	w	P	Const.
−22	0	−35	1	−5	0	0	90
$\frac{5}{2}$	1	4	0	$\frac{1}{2}$	0	0	3
−17	0	−37	0	−5	1	0	75
7	0	15	0	2	0	1	12

The last tableau is in final form. We find that $x = 0$, $y = 3$, $z = 0$, $u = 90$, $w = 75$, and $P = 12$.

19. We obtain the following sequence of tableaus:

	x	y	z	u	v	w	P	Const.	Ratio
	1	1	1	1	0	0	0	20	20
p.r. →	2	4	3	0	1	0	0	42	$10\frac{1}{2}$
	2	0	3	0	0	1	0	30	---
	−4	−6	−5	0	0	0	1	0	
		↑ p.c.							

$\xrightarrow{\frac{1}{4}R_2}$

x	y	z	u	v	w	P	Const.
1	1	1	1	0	0	0	20
$\frac{1}{2}$	1	$\frac{3}{4}$	0	$\frac{1}{4}$	0	0	$\frac{21}{2}$
2	0	3	0	0	1	0	30
−4	−6	−5	0	0	0	1	0

$\xrightarrow{\substack{R_1-R_2 \\ R_4+6R_2}}$

	x	y	z	u	v	w	P	Const.	Ratio
	$\frac{1}{2}$	0	$\frac{1}{4}$	1	$-\frac{1}{4}$	0	0	$\frac{19}{2}$	19
	$\frac{1}{2}$	1	$\frac{3}{4}$	0	$\frac{1}{4}$	0	0	$\frac{21}{2}$	21
p.r.→	2	0	3	0	0	1	0	30	15
	−1	0	$-\frac{1}{2}$	0	$\frac{3}{2}$	0	1	63	
	↑ p.c.								

$\xrightarrow{\frac{1}{2}R_3}$

x	y	z	u	v	w	P	Const.
$\frac{1}{2}$	0	$\frac{1}{4}$	1	$-\frac{1}{4}$	0	0	$\frac{19}{2}$
$\frac{1}{2}$	1	$\frac{3}{4}$	0	$\frac{1}{4}$	0	0	$\frac{21}{2}$
1	0	$\frac{3}{2}$	0	0	$\frac{1}{2}$	0	15
−1	0	$-\frac{1}{2}$	0	$\frac{3}{2}$	0	1	63

$\xrightarrow[R_4+R_3]{R_1-\frac{1}{2}R_3,\; R_2-\frac{1}{2}R_3}$

x	y	z	u	v	w	P	Const.
0	0	$-\frac{1}{2}$	1	$-\frac{1}{4}$	$-\frac{1}{4}$	0	2
0	1	0	0	$\frac{1}{4}$	$-\frac{1}{4}$	0	3
1	0	$\frac{3}{2}$	0	0	$\frac{1}{2}$	0	15
0	0	1	0	$\frac{3}{2}$	$\frac{1}{2}$	1	78

So the solution is $x = 15$, $y = 3$, $z = 0$, $u = 2$, $v = 0$, $w = 0$, and $P = 78$.

21. We obtain the following sequence of tableaus:

	x	y	z	u	v	w	P	Const.	Ratio
p.r.→	2	1	1	1	0	0	0	10	$10/2 = 5$
	3	5	1	0	1	0	0	45	$45/3 = 15$
	2	5	1	0	0	1	0	40	$40/2 = 20$
	−12	−10	−5	0	0	0	1	0	
	↑ p.c.								

$\xrightarrow{\frac{1}{2}R_1}$

x	y	z	u	v	w	P	Const.
1	$\frac{1}{2}$	$\frac{1}{2}$	$\frac{1}{2}$	0	0	0	5
3	5	1	0	1	0	0	45
2	5	1	0	0	1	0	40
−12	−10	−5	0	0	0	1	0

$\xrightarrow[R_4+12R_1]{R_2-3R_1,\ R_3-2R_1}$

	x	y	z	u	v	w	P	Const.
	1	$\frac{1}{2}$	$\frac{1}{2}$	$\frac{1}{2}$	0	0	0	5
	0	$\frac{7}{2}$	$-\frac{1}{2}$	$-\frac{3}{2}$	1	0	0	30
p.r. →	0	4	0	−1	0	1	0	30
	0	−4	1	6	0	0	1	60
		↑ p.c						

Ratio

$5/(1/2) = 10$

$30/(7/2) = 60/7$

$30/4 = 15/2$

$\xrightarrow{\frac{1}{4}R_3}$

x	y	z	u	v	w	P	Const.
1	$\frac{1}{2}$	$\frac{1}{2}$	$\frac{1}{2}$	0	0	0	5
0	$\frac{7}{2}$	$-\frac{1}{2}$	$-\frac{3}{2}$	1	0	0	30
0	1	0	$-\frac{1}{4}$	0	$\frac{1}{4}$	0	$\frac{15}{2}$
0	−4	1	6	0	0	1	60

$\xrightarrow[R_4+4R_3]{R_1-\frac{1}{2}R_3,\ R_2-\frac{7}{2}R_3}$

x	y	z	u	v	w	P	Const.
1	0	$\frac{1}{2}$	$\frac{5}{8}$	0	$-\frac{1}{8}$	0	$\frac{5}{4}$
0	0	$-\frac{1}{2}$	$-\frac{5}{8}$	1	$-\frac{7}{8}$	0	$\frac{15}{4}$
0	1	0	$-\frac{1}{4}$	0	$\frac{1}{4}$	0	$\frac{15}{2}$
0	0	1	5	0	1	1	90

This last tableau is in final form, and we conclude that $x = 5/4$, $y = 15/2$, $z = 0$, $u = 0$, $v = 15/4$, $w = 0$, and $P = 90$.

23.

	x	y	z	s	t	u	v	P	Const.	Ratio
p.r. →	1	2	5	1	0	0	0	0	30	$30/5=6$
	2	1	1	0	1	0	0	0	10	10
	3	2	1	0	0	1	0	0	12	12
	1	1	1	0	0	0	1	0	8	8
	−4	−1	−5	0	0	0	0	1	0	
			↑ p.c.							

$\xrightarrow{\frac{1}{5}R_1}$

x	y	z	s	t	u	v	P	Const.
$\frac{1}{5}$	$\frac{2}{5}$	1	$\frac{1}{5}$	0	0	0	0	6
2	1	1	0	1	0	0	0	10
3	2	1	0	0	1	0	0	12
1	1	1	0	0	0	1	0	8
−4	−1	−5	0	0	0	0	1	0

$\xrightarrow[R_5+5R_1]{R_2-R_1,\ R_3-R_1,\ R_4-R_1}$

x	y	z	s	t	u	v	P	Const.
$\frac{1}{5}$	$\frac{2}{5}$	1	$\frac{1}{5}$	0	0	0	0	6
$\frac{9}{5}$	$\frac{3}{5}$	0	$-\frac{1}{5}$	1	0	0	0	4
1	$\frac{4}{7}$	0	$-\frac{1}{14}$	0	$\frac{5}{14}$	0	0	$\frac{15}{7}$
$\frac{4}{5}$	$\frac{3}{5}$	0	$-\frac{1}{5}$	0	0	1	0	2
−3	1	0	1	0	0	0	1	30

$\xrightarrow{\frac{5}{14}R_3}$ (applied to the third row)

$\xrightarrow[R_5+3R_3]{R_1-\frac{1}{5}R_3,\ R_2-\frac{9}{5}R_3,\ R_4-\frac{4}{5}R_3}$

x	y	z	s	t	u	v	P	Const.
0	$\frac{2}{7}$	1	$\frac{3}{14}$	0	$-\frac{1}{14}$	0	0	$\frac{39}{7}$
0	$-\frac{3}{7}$	0	$-\frac{1}{14}$	1	$-\frac{9}{14}$	0	0	$\frac{1}{7}$
1	$\frac{4}{7}$	0	$-\frac{1}{14}$	0	$\frac{5}{14}$	0	0	$\frac{15}{7}$
0	$\frac{1}{7}$	0	$-\frac{1}{7}$	0	$-\frac{2}{7}$	1	0	$\frac{2}{7}$
0	$\frac{19}{7}$	0	$\frac{11}{14}$	0	$\frac{15}{14}$	0	1	$\frac{255}{7}$

The last tableau is in final form. We see that $x = 15/7$, $y = 0$, $z = 39/7$, $s = 0$, $t = 1/7$, $u = 0$, $v = 2/7$, and $P = 255/7$.

25. Pivoting about the x-column in the initial simplex tableau, we have

	x	y	z	u	v	P	Const.	Ratio
	3	3	−2	1	0	0	100	100 / 3
p.r.→	5	5	3	0	1	0	150	150 / 5
	−2	−2	4	0	0	1	0	
	↑ p.c.							

$\xrightarrow{\frac{1}{5}R_2}$

x	y	z	u	v	P	Const.
3	3	−2	1	0	0	100
1	1	$\frac{3}{5}$	0	$\frac{1}{5}$	0	30
−2	−2	4	0	0	1	0

$\xrightarrow[R_3+2R_2]{R_1-3R_2}$

x	y	z	u	v	P	Const.
0	0	$-\frac{19}{5}$	1	$-\frac{3}{5}$	0	10
1	1	$\frac{3}{5}$	0	$\frac{1}{5}$	0	30
0	0	$\frac{26}{5}$	0	$\frac{2}{5}$	1	60

and we see that one optimal solution occurs when $x = 30$, $y = 0$, $z = 0$, and $P = 60$. Similarly, pivoting about the y-column, we obtain another optimal solution: $x = 0$, $y = 30$, $z = 0$, and $P = 60$.

27. Let the number of model A and model B fax machines made each month be x and y, respectively. Then we have the following linear programming problem:

Maximize $P = 25x + 40y$ subject to

$$\begin{aligned} 200x + 300y &\leq 600{,}000 \\ x + \quad y &\leq \quad 2{,}500 \\ x \geq 0,\ y &\geq 0 \end{aligned}$$

Using the simplex method, we obtain the following sequence of tableaus:

x	y	u	v	P	Const	Ratio
200	300	1	0	0	600,000	2000
1	1	0	1	0	2,500	2500
–25	–40	0	0	1	0	

$\xrightarrow{\frac{1}{300}R_1}$

x	y	u	v	P	Const
$\frac{2}{3}$	1	$\frac{1}{300}$	0	0	2000
1	1	0	1	0	2500
–25	–40	0	0	1	0

$\xrightarrow[R_3+40R_1]{R_2-R_1}$

x	y	u	v	P	Const
$\frac{2}{3}$	1	$\frac{1}{300}$	0	0	2000
$\frac{1}{3}$	0	$-\frac{1}{300}$	1	0	500
$\frac{5}{3}$	0	$\frac{2}{15}$	0	1	80,000

We conclude that the maximum monthly profit is $80,000, and this occurs when 0 model *A* and 2000 model *B* fax machines are produced.

29. Suppose the farmer plants x acres of Crop *A* and y acres of crop *B*. Then the problem is

$$\text{Maximize } P = 150x + 200y \text{ subject to}$$
$$40x + 60y \le 7400$$
$$20x + 25y \le 3300$$
$$x \ge 0,\ y \ge 0$$

Using the simplex method, we obtain the following sequence of tableaus:

	x	y	u	v	P	Const	Ratio
$p.r. \to$	40	60	1	0	0	7400	$7400/60 = 370/3$
	20	25	0	1	0	3300	$3300/25 = 132$
	-150	-200	0	0	1	0	
		$\uparrow$ $p.c.$					

$\xrightarrow{\frac{1}{60}R_1}$

x	y	u	v	P	Const
$\frac{2}{3}$	1	$\frac{1}{60}$	0	0	$\frac{370}{3}$
20	25	0	1	0	3300
−150	−200	0	0	1	0

$\xrightarrow[R_3+200R_1]{R_2-25R_1}$

	x	y	u	v	P	Const	Ratio
	$\frac{2}{3}$	1	$\frac{1}{60}$	0	0	$\frac{370}{3}$	185
p.r. →	$\frac{10}{3}$	0	$-\frac{5}{12}$	1	0	$\frac{650}{3}$	65
	$-\frac{50}{3}$	0	$\frac{10}{3}$	0	1	$\frac{74,000}{3}$	
	↑ p.c.						

$\xrightarrow{\frac{3}{10}R_2}$

x	y	u	v	P	Const
$\frac{2}{3}$	1	$\frac{1}{60}$	0	0	$\frac{370}{3}$
1	0	$-\frac{1}{8}$	$\frac{3}{10}$	0	65
$-\frac{50}{3}$	0	$\frac{10}{3}$	0	1	$\frac{74,000}{3}$

$\xrightarrow[R_3+\frac{50}{3}R_2]{R_1-\frac{2}{3}R_2}$

x	y	u	v	P	Const
0	1	$\frac{1}{10}$	$-\frac{1}{5}$	0	80
1	0	$-\frac{1}{8}$	$\frac{3}{10}$	0	65
0	0	$\frac{5}{4}$	5	1	25,750

The last tableau is in final form. We find $x = 65$, $y = 80$, and $P = 25{,}750$. So the maximum profit of $25,750 is realized by planting 65 acres of Crop *A* and 80 acres of Crop *B*.

31. Suppose the Excelsior Company buys x, y, and z minutes of morning, afternoon, and evening commercials, respectively. Then we wish to maximize

$$P = 200{,}000x + 100{,}000y + 600{,}000z \text{ subject to}$$

$$\begin{aligned} 3000x + 1000y + 12{,}000z &\le 102{,}000 \\ z &\le 6 \\ x + \quad y + \quad z &\le 25 \\ x \ge 0, y \ge 0, z &\ge 0 \end{aligned}$$

Using the simplex method, we obtain the following sequence of tableaus.

x	y	z	u	v	w	P	Const.	Ratio
3000	1000	12,000	1	0	0	0	102,000	17 / 2
0	0	1	0	1	0	0	6	6
1	1	1	0	0	1	0	25	25
–200,000	–100,000	–600,000	0	0	0	1	0	

$\xrightarrow{R_1-12{,}000R_2,\; R_3-R_2,\; R_4+600{,}000R_2}$

x	y	z	u	v	w	P	Const.	Ratio
3000	1000	0	1	–12,000	0	0	30,000	10
0	0	1	0	1	0	0	6	--
1	1	0	0	–1	1	0	19	19
–200,000	–100,000	0	0	600,000	0	1	3,600,000	

$\xrightarrow{\frac{1}{3000}R_1}$

x	y	z	u	v	w	P	Const.
1	$\frac{1}{3}$	0	$\frac{1}{3000}$	–4	0	0	10
0	0	1	0	1	0	0	6
1	1	0	0	–1	1	0	19
–200,000	–100,000	0	0	600,000	0	1	3,600,000

$\xrightarrow{R_3-R_1,\; R_4+200{,}000R_1}$

x	y	z	u	v	w	P	Const.
1	$\frac{1}{3}$	0	$\frac{1}{3000}$	–4	0	0	10
0	0	1	0	1	0	0	6
0	$\frac{2}{3}$	0	$-\frac{1}{3000}$	3	1	0	9
0	$-\frac{100{,}000}{3}$	0	$\frac{200}{3}$	–200,000	0	1	5,600,000

$\xrightarrow{\frac{1}{3}R_3}$

x	y	z	u	v	w	P	Const.
1	$\frac{1}{3}$	0	$\frac{1}{3000}$	–4	0	0	10
0	0	1	0	1	0	0	6
0	$\frac{2}{9}$	0	$-\frac{1}{9000}$	1	$\frac{1}{3}$	0	3
0	$-\frac{100{,}000}{3}$	0	$\frac{200}{3}$	–200,000	0	1	5,600,000

$\xrightarrow{R_1+4R_3,\; R_2-R_3,\; R_4+200{,}000R_3}$

x	y	z	u	v	w	P	Const.
1	$\frac{11}{9}$	0	$-\frac{1}{9000}$	0	$\frac{4}{3}$	0	22
0	$-\frac{2}{9}$	1	$\frac{1}{9000}$	0	$-\frac{1}{3}$	0	3
0	$\frac{2}{9}$	0	$-\frac{1}{9000}$	1	$\frac{1}{3}$	0	3
0	$\frac{100,000}{9}$	0	$\frac{400}{9}$	0	$\frac{200,000}{3}$	1	6,200,000

We conclude that $x = 22$, $y = 0$, $z = 3$, $u = 0$, $v = 3$, and $P = 6,200,000$. Therefore, the company should buy 22 minutes of morning and 3 minutes of evening advertising time, thereby maximizing their exposure to 6,200,000 viewers.

33. We first tabulate the given information:

MODELS

Dept.	A	B	C	**Time available**
Fabrication	$\frac{5}{4}$	$\frac{3}{2}$	$\frac{3}{2}$	310
Assembly	1	1	$\frac{3}{4}$	205
Finishing	1	1	$\frac{1}{2}$	190
Profit	26	28	24	

Let x, y, and z denote the number of units of model A, model B, and model C to be produced, respectively. Then the required linear programming problem is

$$\begin{aligned} &\text{Maximize } P = 26x + 28y + 24z \text{ subject to} \\ &\tfrac{5}{4}x + \tfrac{3}{2}y + \tfrac{3}{2}z \le 310 \\ &x + y + \tfrac{3}{4}z \le 205 \\ &x + y + \tfrac{1}{2}z \le 190 \\ &x \ge 0, y \ge 0, z \ge 0 \end{aligned}$$

Using the simplex method, we obtain the following tableaus:

	x	y	z	u	v	w	P	Const.	Ratio
	$\frac{5}{4}$	$\frac{3}{2}$	$\frac{3}{2}$	1	0	0	0	310	248
	1	1	$\frac{3}{4}$	0	1	0	0	205	205
p.r.→	1	1	$\frac{1}{2}$	0	0	1	0	190	190
	−26	−28	−24	0	0	0	1	0	
		↑ p.c.							

$\xrightarrow{R_1-\frac{3}{2}R_3,\ R_2-R_3,\ R_4+28R_3}$

	x	y	z	u	v	w	P	Const.	Ratio
p.r.→	$-\frac{1}{4}$	0	$\frac{3}{4}$	1	0	$-\frac{3}{2}$	0	25	$33\frac{1}{3}$
	0	0	$\frac{1}{4}$	0	1	−1	0	15	60
	1	1	$\frac{1}{2}$	0	0	1	0	190	380
	2	0	−10	0	0	28	1	5320	
			↑ p.c.						

$\xrightarrow{\frac{4}{3}R_1}$

x	y	z	u	v	w	P	Const.
$-\frac{1}{3}$	0	1	$\frac{4}{3}$	0	−2	0	$\frac{100}{3}$
0	0	$\frac{1}{4}$	0	1	−1	0	15
1	1	$\frac{1}{2}$	0	0	1	0	190
2	0	−10	0	0	28	1	5320

$\xrightarrow{R_2-\frac{1}{4}R_1,\ R_3-\frac{1}{2}R_1,\ R_4+10R_1}$

	x	y	z	u	v	w	P	Const.	Ratio
	$-\frac{1}{3}$	0	1	$\frac{4}{3}$	0	−2	0	$\frac{100}{3}$	−−
p.r.→	$\frac{1}{12}$	0	0	$-\frac{1}{3}$	1	$-\frac{1}{2}$	0	$\frac{20}{3}$	80
	$\frac{7}{6}$	1	0	$-\frac{2}{3}$	0	2	0	$\frac{520}{3}$	$148\frac{4}{7}$
	$-\frac{4}{3}$	0	0	$\frac{40}{3}$	0	8	1	$\frac{16960}{3}$	
	↑ p.c.								

$\xrightarrow{12R_2}$

x	y	z	u	v	w	P	Const.
$-\frac{1}{3}$	0	1	$\frac{4}{3}$	0	−2	0	$\frac{100}{3}$
1	0	0	−4	12	−6	0	80
$\frac{7}{6}$	1	0	$-\frac{2}{3}$	0	2	0	$\frac{520}{3}$
$-\frac{4}{3}$	0	0	$\frac{40}{3}$	0	8	1	$\frac{16960}{3}$

$\xrightarrow{R_1+\frac{1}{3}R_2,\ R_3-\frac{7}{6}R_2,\ R_4+\frac{4}{3}R_2}$

x	y	z	u	v	w	P	Const.
0	0	1	0	4	−4	0	60
1	0	0	−4	12	−6	0	80
0	1	0	4	−14	9	0	80
0	0	0	8	16	0	1	5760

The last tableau is in final form. We see that $x = 80$, $y = 80$, $z = 60$, and $P = 5760$. So, by producing 80 units each of Models A and B, and 60 units of Model C, the company stands to make a profit of \$5760.

35. Let x, y, and z denote the number (in thousands) of bottles of formula I, formula II, and formula III, respectively, produced. The resulting linear programming problem is

$$\text{Maximize } P = 180x + 200y + 300z \text{ subject to}$$

$$\begin{aligned} \tfrac{5}{2}x + 3y + 4z &\le 70 \\ x &\le 9 \\ y &\le 12 \\ z &\le 6 \\ x \ge 0,\ y \ge 0,\ z &\ge 0 \end{aligned}$$

Using the simplex method, we have

	x	y	z	s	t	u	v	P	Const.	Ratio
	$\frac{5}{2}$	3	4	1	0	0	0	0	70	$17\frac{1}{2}$
	1	0	0	0	1	0	0	0	9	--
	0	1	0	0	0	1	0	0	12	--
p.r.→	0	0	1	0	0	0	1	0	6	6
	−180	−200	−300	0	0	0	0	1	0	
			↑ p.c.							

$\xrightarrow[R_5+300R_4]{R_1-4R_4}$

	x	y	z	s	t	u	v	P	Const.	Ratio
	$\frac{5}{2}$	3	0	1	0	0	−4	0	46	$15\frac{1}{3}$
	1	0	0	0	1	0	0	0	9	--
p.r. →	0	1	0	0	0	1	0	0	12	12
	0	0	1	0	0	0	1	0	6	--
	−180	−200	0	0	0	0	300	1	1800	
		↑ p.c.								

$\xrightarrow[R_5+200R_3]{R_1-3R_3}$

x	y	z	s	t	u	v	P	Const.	Ratio
$\frac{5}{2}$	0	0	1	0	−3	−4	0	10	4
1	0	0	0	1	0	0	0	9	9
0	1	0	0	0	1	0	0	12	--
0	0	1	0	0	0	1	0	6	--
−180	0	0	0	0	200	300	1	4200	

$\xrightarrow{\frac{2}{5}R_1}$

x	y	z	s	t	u	v	P	Const.
1	0	0	$\frac{2}{5}$	0	$-\frac{6}{5}$	$-\frac{8}{5}$	0	4
1	0	0	0	1	0	0	0	9
0	1	0	0	0	1	0	0	12
0	0	1	0	0	0	1	0	6
−180	0	0	0	0	200	300	1	4200

$\xrightarrow[R_5+180R_1]{R_2-R_1}$

x	y	z	s	t	u	v	P	Const.	Ratio
1	0	0	$\frac{2}{5}$	0	$-\frac{6}{5}$	$-\frac{8}{5}$	0	4	--
1	0	0	$-\frac{2}{5}$	1	$\frac{6}{5}$	$\frac{8}{5}$	0	5	$\frac{25}{6}$
0	1	0	0	0	1	0	0	12	12
0	0	1	0	0	0	1	0	6	--
0	0	0	72	0	−16	12	1	4920	

$\xrightarrow{\frac{5}{6}R_2}$

x	y	z	s	t	u	v	P	Const.
1	0	0	$\frac{2}{5}$	0	$-\frac{6}{5}$	$-\frac{8}{5}$	0	4
0	0	0	$-\frac{1}{3}$	$\frac{5}{6}$	1	$\frac{4}{3}$	0	$\frac{25}{6}$
0	1	0	0	0	1	0	0	12
0	0	1	0	0	0	1	0	6
0	0	0	72	0	−16	12	1	4920

$\xrightarrow[R_5+16R_2]{R_1+\frac{6}{5}R_2 \\ R_3-R_2}$

x	y	z	s	t	u	v	P	Const.
1	0	0	0	1	0	0	0	9
0	0	0	$-\frac{1}{3}$	$\frac{5}{6}$	1	$\frac{4}{3}$	0	$\frac{25}{6}$
0	1	0	$\frac{1}{3}$	$-\frac{5}{6}$	0	$-\frac{4}{3}$	0	$\frac{47}{6}$
0	0	1	0	0	0	1	0	6
0	0	0	$\frac{200}{3}$	$\frac{40}{3}$	0	$\frac{100}{3}$	1	$4986\frac{2}{3}$

Therefore, $x = 9$, $y = 47/6$, $z = 6$, and $P \approx 4986.67$; that is, the company should manufacture 9000 bottles of formula *I*, 7833 bottles of formula *II*, and 6000 bottles of formula *III* for a maximum profit of \$4986.67.

USING TECHNOLOGY EXERCISES 4.2, page 233

1. $x = 1.2, y = 0, z = 1.6, w = 0$, and $P = 8.8$

3. $x = 1.6, y = 0, z = 0, w = 3.6$, and $P = 12.4$

EXERCISES 4.2, page 246

1. We solve the associated regular problem:

   Maximize $P = -C = 2x - y$ subject to

$$x + 2y \le 6$$
$$3x + 2y \le 12$$
$$x \ge 0,\ y \ge 0$$

Using the simplex method where u and v are slack variables, we have

	x	y	u	v	P	Const.	Ratio	
	1	2	1	0	0	6	6	$\xrightarrow{\frac{1}{3}R_2}$
p.r.→	3	2	0	1	0	12	4	
	−2	1	0	0	1	0		
	↑ p.c.							

x	y	u	v	P	Const.	
1	2	1	0	0	6	$\xrightarrow[R_3+2R_2]{R_1-R_2}$
1	$\frac{2}{3}$	0	$\frac{1}{3}$	0	4	
−2	1	0	0	1	0	

x	y	u	v	P	Const.
0	$\frac{4}{3}$	1	$-\frac{1}{3}$	0	2
1	$\frac{2}{3}$	0	$\frac{1}{3}$	0	4
0	$\frac{7}{3}$	0	$\frac{2}{3}$	1	8

Therefore, $x = 4$, $y = 0$, and $C = -P = -8$.

3. We maximize $P = -C = 3x + 2y$. Using the simplex method, we obtain

	x	y	u	v	P	Const.	Ratio	
	3	4	1	0	0	24	8	$\xrightarrow{\frac{1}{7}R_2}$
p.r.→	7	−4	0	1	0	16	$\frac{16}{7}$	
	−3	−2	0	0	1	0		
	↑ p.c.							

x	y	u	v	P	Const.	
3	4	1	0	0	24	$\xrightarrow[R_3+3R_2]{R_1-3R_2}$
1	$-\frac{4}{7}$	0	$\frac{1}{7}$	0	$\frac{16}{7}$	
−3	−2	0	0	1	0	

	x	y	u	v	P	Const.	Ratio
$p.r. \to$	0	$\frac{40}{7}$	1	$-\frac{3}{7}$	0	$\frac{120}{7}$	3
	1	$-\frac{4}{7}$	0	$\frac{1}{7}$	0	$\frac{16}{7}$	--
	0	$-\frac{26}{7}$	0	$\frac{3}{7}$	1	$\frac{48}{7}$	
		↑ $p.c.$					

$\xrightarrow{\frac{7}{40}R_1}$

x	y	u	v	P	Const.
0	1	$\frac{7}{40}$	$-\frac{3}{40}$	0	3
1	$-\frac{4}{7}$	0	$\frac{1}{7}$	0	$\frac{16}{7}$
0	$-\frac{26}{7}$	0	$\frac{3}{7}$	1	$\frac{48}{7}$

$\xrightarrow[R_3+\frac{26}{7}R_1]{R_2+\frac{4}{7}R_1}$

x	y	u	v	P	Const.
0	1	$\frac{7}{40}$	$-\frac{3}{40}$	0	3
1	0	$\frac{1}{10}$	$\frac{1}{10}$	0	4
0	0	$\frac{13}{20}$	$\frac{3}{20}$	1	18

The last tableau is in final form. We find $x = 4$, $y = 3$, and $C = -P = -18$.

5. We maximize $P = -C = -2x + 3y + 4z$ subject to the given constraints. Using the simplex method we obtain

	x	y	z	u	v	v	P	Const.	Ratio
$p.r. \to$	–1	2	–1	1	0	0	0	8	--
	1	–2	2	0	1	0	0	10	5
	2	4	–3	0	0	1	0	12	--
	2	–3	–4	0	0	0	1	0	
			↑ $p.c.$						

$\xrightarrow{\frac{1}{2}R_2}$

x	y	z	u	v	v	P	Const.
–1	2	–1	1	0	0	0	8
$\frac{1}{2}$	–1	1	0	$\frac{1}{2}$	0	0	5
2	4	–3	0	0	1	0	12
2	–3	–4	0	0	0	1	0

R_1+R_2
R_3+3R_2
$\xrightarrow{R_4+4R_2}$

	x	y	z	u	v	v	P	Const.	Ratio	
p.r.→	$-\frac{1}{2}$	1	0	1	$\frac{1}{2}$	0	0	13	13	R_2+R_1
	$\frac{1}{2}$	−1	1	0	$\frac{1}{2}$	0	0	5	--	R_3-R_1
	$\frac{7}{2}$	1	0	0	$\frac{3}{2}$	1	0	27	27	R_4+7R_1 →
	4	−7	0	0	2	0	1	20		
		↑ p.c.								

x	y	z	u	v	v	P	Const.
$-\frac{1}{2}$	1	0	1	$\frac{1}{2}$	0	0	13
0	0	1	1	1	0	0	18
4	0	0	−1	1	1	0	14
$\frac{1}{2}$	0	0	7	$\frac{11}{2}$	0	1	111

The last tableau is in final form. We see that $x = 0$, $y = 13$, $z = 18$, $w = 14$, and $C = -P = -111$.

7. $x = 5/4$, $y = 1/4$, $u = 2$, $v = 3$, and $C = P = 13$.

9. $x = 5$, $y = 10$, $z = 0$, $u = 1$, $v = 2$, and $C = P = 80$.

11. We first write the tableau

x	y	Const.
1	2	4
3	2	6
2	5	

Then obtain the following by interchanging rows and columns:

u	v	Const.
1	3	2
2	2	5
4	6	

From this table we construct the dual problem:

Maximize the objective function

$$P = 4u + 6v \text{ subject to}$$
$$u + 3v \le 2$$
$$2u + 2v \le 5$$
$$u \ge 0,\ v \ge 0$$

Solving the dual problem using the simplex method with x and y as the slack variables, we obtain

	u	v	x	y	P	Const.	Ratio
p.r. →	1	3	1	0	0	2	$\frac{2}{3}$
	2	2	0	1	0	5	$\frac{5}{2}$
	−4	−6	0	0	1	0	
		↑ p.c.					

$\xrightarrow{\frac{1}{3}R_1}$

u	v	x	y	P	Const.
$\frac{1}{3}$	1	$\frac{1}{3}$	0	0	$\frac{2}{3}$
2	2	0	1	0	5
−4	−6	0	0	1	0

$\xrightarrow[R_3+6R_1]{R_2-2R_1}$

	u	v	x	y	P	Const.	Ratio
p.r. →	$\frac{1}{3}$	1	$\frac{1}{3}$	0	0	$\frac{2}{3}$	2
	$\frac{4}{3}$	0	$-\frac{2}{3}$	1	0	$\frac{11}{3}$	$2\frac{3}{4}$
	−2	0	2	0	1	4	
	↑ p.c.						

$\xrightarrow{3R_1}$

u	v	x	y	P	Const.
1	3	1	0	0	2
$\frac{4}{3}$	0	$-\frac{2}{3}$	1	0	$\frac{11}{3}$
−2	0	2	0	1	4

$\xrightarrow[R_3+2R_1]{R_2-\frac{4}{3}R_1}$

u	v	x	y	P	*Const.*
1	3	1	0	0	2
0	–4	–2	1	0	1
0	6	4	0	1	8

Interpreting the final tableau, we see that $x = 4$, $y = 0$, and $P = C = 8$.

13. We first write the tableau

x	y	*Const.*
6	1	60
2	1	40
1	1	30
6	4	

Then obtain the following by interchanging rows and columns:

u	v	w	*Const.*
6	2	1	6
1	1	1	4
60	40	30	

From this table we construct the dual problem:

Maximize $P = 60u + 40v + 30w$ subject to

$$6u + 2v + w \le 6$$
$$u + v + w \le 4$$
$$u \ge 0,\ v \ge 0,\ w \ge 0$$

We solve the problem as follows.

	u	v	w	x	y	P	*Const.*	*Ratio*
$p.r. \rightarrow$	6	2	1	1	0	0	6	1
	1	1	1	0	1	0	4	4
	–60	–40	–30	0	0	1	0	--
	$\uparrow$ $p.c.$							

$\xrightarrow{\frac{1}{6}R_1}$

u	v	w	x	y	P	Const.
1	$\frac{1}{3}$	$\frac{1}{6}$	$\frac{1}{6}$	0	0	1
1	1	1	0	1	0	4
−60	−40	−30	0	0	1	0

$\xrightarrow[R_3+60R_1]{R_2-R_1}$

	u	v	w	x	y	P	Const.	Ratio
	1	$\frac{1}{3}$	$\frac{1}{6}$	$\frac{1}{6}$	0	0	1	6
p.r. →	0	$\frac{2}{3}$	$\frac{5}{6}$	$-\frac{1}{6}$	1	0	3	18 / 5
	0	−20	−20	10	0	1	60	--
			↑ p.c.					

$\xrightarrow{\frac{6}{5}R_2}$

u	v	w	x	y	P	Const.
1	$\frac{1}{3}$	$\frac{1}{6}$	$\frac{1}{6}$	0	0	1
0	$\frac{4}{5}$	1	$-\frac{1}{5}$	$\frac{6}{5}$	0	$\frac{18}{5}$
0	−20	−20	10	0	1	60

$\xrightarrow[R_3+20R_2]{R_1-\frac{1}{6}R_2}$

	u	v	w	x	y	P	Const.	Ratio
p.r. →	1	$\frac{1}{5}$	0	$\frac{1}{5}$	$-\frac{1}{5}$	0	$\frac{2}{5}$	2
	0	$\frac{4}{5}$	1	$-\frac{1}{5}$	$\frac{6}{5}$	0	$\frac{18}{5}$	9 / 2
	0	−4	0	6	24	1	132	--
		↑ p.c.						

$\xrightarrow{5R_1}$

u	v	w	x	y	P	Const.
5	1	0	1	−1	0	2
1	$\frac{4}{5}$	1	$-\frac{1}{5}$	$\frac{6}{5}$	0	$\frac{18}{5}$
0	−4	0	6	24	1	132

$\xrightarrow[R_3+4R_1]{R_2-\frac{4}{5}R_1}$

u	v	w	x	y	P	Const.
5	1	0	1	−1	0	2
−4	0	1	−1	2	0	2
20	0	0	10	20	1	140

The last tableau is in final form. We find that $x = 10$, $y = 20$, and $C = 140$.

15. We first write the tableau

x	y	z	*Const.*
20	10	1	10
1	1	2	20
200	150	120	

Then obtain the following by interchanging rows and columns:

u	v	*Const.*
20	1	200
10	1	150
1	2	120
10	20	

From this table we construct the dual problem:

Maximize $P = 10u + 20v$ subject to

$$20u + v \le 200$$

$$10u + v \le 150$$

$$u + 2v \le 120$$

$$u \le 0,\ v \le 0$$

Solving this problem, we obtain the following tableaus:

	u	v	x	y	z	P	*Const.*	*Ratio*
	20	1	1	0	0	0	200	200
	10	1	0	1	0	0	150	150
$p.r. \rightarrow$	1	2	0	0	1	0	120	60
	−10	−20	0	0	0	1	0	
		$\uparrow$ $p.c$						

$\xrightarrow{\frac{1}{2}R_3}$

u	v	x	y	z	P	Const.
20	1	1	0	0	0	200
10	1	0	1	0	0	150
$\frac{1}{2}$	1	0	0	$\frac{1}{2}$	0	60
−10	−20	0	0	0	1	0

$\xrightarrow{\substack{R_1-R_3 \\ R_2-R_3 \\ R_4+20R_3}}$

u	v	x	y	z	P	Const.
$\frac{39}{2}$	0	1	0	$-\frac{1}{2}$	0	140
$\frac{19}{2}$	0	0	1	$-\frac{1}{2}$	0	90
$\frac{1}{2}$	1	0	0	$\frac{1}{2}$	0	60
0	0	0	0	10	1	1200

This last tableau is in final form. We find that $x = 0$, $y = 0$, $z = 10$, and $C = 1200$.

17. We first write the tableau

x	y	z	Const.
1	2	2	10
2	1	1	24
1	1	1	16
6	8	4	

Then obtain the following by interchanging rows and columns:

u	v	w	Const.
1	2	1	6
2	1	1	8
2	1	1	4
10	24	16	

From this table we construct the dual problem:

Maximize the objective function
$P = 10u + 24v + 16w$ subject to

$$u + 2v + w \le 6$$
$$2u + v + w \le 8$$
$$2u + v + w \le 4$$
$$u \ge 0, v \ge 0, w \ge 0$$

Solving the dual problem using the simplex method with x, y, and z as slack variables, we obtain

	u	v	w	x	y	z	P	Const.	Ratio
p.r.→	1	2	1	1	0	0	0	6	3
	2	1	1	0	1	0	0	8	8
	2	1	1	0	0	1	0	4	4
	−10	−24	−16	0	0	0	1	0	
		↑ p.c.							

$\xrightarrow{\frac{1}{2}R_1}$

u	v	w	x	y	z	P	Const.
$\frac{1}{2}$	1	$\frac{1}{2}$	$\frac{1}{2}$	0	0	0	3
2	1	1	0	1	0	0	8
2	1	1	0	0	1	0	4
−10	−24	−16	0	0	0	1	0

$\xrightarrow[R_4+24R_1]{R_2-R_1,\ R_3-R_1}$

	u	v	w	x	y	z	P	Const.	Ratio
	$\frac{1}{2}$	1	$\frac{1}{2}$	$\frac{1}{2}$	0	0	0	3	6
	$\frac{3}{2}$	0	$\frac{1}{2}$	$-\frac{1}{2}$	1	0	0	5	10
p.r.→	$\frac{3}{2}$	0	$\frac{1}{2}$	$-\frac{1}{2}$	0	1	0	1	2
	2	0	−4	12	0	0	1	72	
			↑ p.c.						

$\xrightarrow{2R_3}$

u	v	w	x	y	z	P	Const.
$\frac{1}{2}$	1	$\frac{1}{2}$	$\frac{1}{2}$	0	0	0	3
$\frac{3}{2}$	0	$\frac{1}{2}$	$-\frac{1}{2}$	1	0	0	5
3	0	1	−1	0	2	0	2
2	0	−4	12	0	0	1	72

$\xrightarrow[R_4+4R_3]{R_1-\frac{1}{2}R_3,\ R_2-\frac{1}{2}R_3}$

u	v	w	x	y	z	P	Const.
−1	1	0	1	0	−1	0	2
0	0	0	0	1	−1	0	4
3	0	1	−1	0	2	0	2
14	0	0	8	0	8	1	80

The solution to the primal problem is $x = 8$, $y = 0$, $z = 8$, and $C = 80$.

19. We first write

Maximize $P = 6u + 2v + 4w$ subject to the constraints

x	y	z	*Const.*
2	4	3	6
5	0	1	2
0	6	2	4
30	12	20	

Then obtain the following by interchanging rows and columns:

u	v	w	*Const.*
2	6	0	30
4	0	6	12
3	1	2	20
6	2	4	

From this table we construct the dual problem:

Maximize $P = 6u + 2v + 4w$ subject to

$$\begin{aligned} 2u + 6v \quad\quad &\leq 30 \\ 4u + \quad\quad 6w &\leq 12 \\ 3u + v + 2w &\leq 20 \\ u \geq 0,\ v \geq 0,\ w &\geq 0 \end{aligned}$$

Using the simplex method, we obtain

	u	v	w	x	y	z	P	*Const.*	*Ratio*
	2	6	0	1	0	0	0	30	15
$p.r. \rightarrow$	4	0	6	0	1	0	0	12	3
	3	1	2	0	0	1	0	20	20 / 8
	−6	−2	−4	0	0	0	1	0	
	$\uparrow$ p.c.								

$\xrightarrow{\frac{1}{4}R_2}$

u	v	w	x	y	z	P	Const.
2	6	0	1	0	0	0	30
1	0	$\frac{3}{2}$	0	$\frac{1}{4}$	0	0	3
3	1	2	0	0	1	0	20
−6	−2	−4	0	0	0	1	0

$\xrightarrow{\substack{R_1-2R_2 \\ R_3-3R_2 \\ R_4+6R_2}}$

	u	v	w	x	y	z	P	Const.	Ratio
p.r.→	0	6	−3	1	$-\frac{1}{2}$	0	0	24	4
	1	0	$\frac{3}{2}$	0	$\frac{1}{4}$	0	0	3	---
	0	1	$-\frac{5}{2}$	0	$-\frac{3}{4}$	1	0	11	11
	0	−2	5	0	$\frac{3}{2}$	0	1	18	
		↑ p.c.							

$\xrightarrow{\frac{1}{6}R_1}$

u	v	w	x	y	z	P	Const.
0	1	$-\frac{1}{2}$	$\frac{1}{6}$	$-\frac{1}{12}$	0	0	4
1	0	$\frac{3}{2}$	0	$\frac{1}{4}$	0	0	3
0	1	$-\frac{5}{2}$	0	$-\frac{3}{4}$	1	0	11
0	−2	5	0	$\frac{3}{2}$	0	1	18

$\xrightarrow{\substack{R_3-R_1 \\ R_4+2R_1}}$

u	v	w	x	y	z	P	Const.
0	1	$-\frac{1}{2}$	$\frac{1}{6}$	$-\frac{1}{12}$	0	0	4
1	0	$\frac{3}{2}$	0	$\frac{1}{4}$	0	0	3
0	0	−2	$-\frac{1}{6}$	$-\frac{2}{3}$	1	0	7
0	0	4	$\frac{1}{3}$	$\frac{4}{3}$	0	1	26

The last tableau is in final form. We find $x = 1/3$, $y = 4/3$, $z = 0$, and $C = 26$.

21. This problem was formulated in Exercise 12, Section 3.2, page 218. We rewrite the constraints in the form

$$
\begin{aligned}
-x_1 - x_2 - x_3 \qquad\qquad\qquad\qquad & \geq -800 \\
-x_4 - x_5 - x_6 & \geq -600 \\
x_1 \qquad\qquad + x_4 \qquad\qquad & \geq \ 500 \\
x_2 \qquad\qquad + x_5 \qquad & \geq \ 400 \\
x_3 \qquad\qquad + x_6 & \geq \ 400
\end{aligned}
$$

We solve this problem using duality. We first write

x_1	x_2	x_3	x_4	x_5	x_6	*Const.*
−1	−1	−1	0	0	0	−800
0	0	0	−1	−1	−1	−600
1	0	0	1	0	0	500
0	1	0	0	1	0	400
0	0	1	0	0	1	400
16	20	22	18	16	14	

Interchanging the rows with the columns, we obtain

u_1	u_2	u_3	u_4	u_5	*Const.*
−1	0	1	0	0	16
−1	0	0	1	0	20
−1	0	0	0	1	22
0	−1	1	0	0	18
0	−1	0	1	0	16
0	−1	0	0	1	14
−800	−600	500	400	400	

from which we obtain the dual problem:

Maximize $P = -800u_1 - 600u_2 + 500u_3 + 400u_4 + 400u_5$ subject to

$$
\begin{aligned}
-u_1 \quad + u_3 \quad\quad\quad &\le 16 \\
-u_1 \quad\quad + u_4 \quad &\le 20 \\
-u_1 \quad\quad\quad + u_5 &\le 22 \\
-u_2 + u_3 \quad\quad\quad &\le 18 \\
-u_2 \quad + u_4 \quad &\le 16 \\
-u_2 + u_5 \quad\quad\quad &\le 14 \\
u_1 \ge 0,\ u_2 \ge 0, \ldots, u_5 \ge 0 &
\end{aligned}
$$

The initial simplex tableau is

u_1	u_2	u_3	u_4	u_5	x_1	x_2	x_3	x_4	x_5	x_6	P	*Const.*
–1	0	1	0	0	1	0	0	0	0	0	0	16
–1	0	0	1	0	0	1	0	0	0	0	0	20
–1	0	0	0	1	0	0	1	0	0	0	0	22
0	–1	1	0	0	0	0	0	1	0	0	0	18
0	–1	0	1	0	0	0	0	0	1	0	0	16
0	–1	0	0	1	0	0	0	0	0	1	0	14
800	600	–500	–400	–400	0	0	0	0	0	0	1	0

Using the sequence of row operations

1. $R_4 - R_1$, $R_7 + 500R_1$ 2. $R_2 - R_5$, $R_7 + 400R_5$ 3. $R_3 - R_6$, $R_7 + 400R_6$

4. $R_3 - R_2$, $R_4 + R_2$, $R_5 + R_2$, $R_6 + R_2$, $R_7 + 200R_2$

we obtain the final tableau

u_1	u_2	u_3	u_4	u_5	x_1	x_2	x_3	x_4	x_5	x_6	P	*Const.*
–1	0	1	0	0	1	0	0	0	0	0	0	16
–1	1	0	0	0	0	1	0	0	–1	0	0	4
0	0	0	0	0	0	–1	1	0	1	–1	0	4
0	0	0	0	0	–1	1	0	1	–1	0	0	6
–1	0	0	1	0	0	1	0	0	0	0	0	20
–1	0	0	0	1	0	1	0	0	–1	1	0	18
100	0	0	0	0	500	200	0	0	200	400	1	20,800

We find $x_1 = 500$, $x_2 = 200$, $x_3 = 0$, $x_4 = 0$, $x_5 = 200$, $x_6 = 400$, and $C = 20{,}800$. So the schedule is

Location *I*: 500 to warehouse *A*, 200 to warehouse *B*
Location *II*: 200 to warehouse *B*, 400 to warehouse *C*
Shipping costs: $20,800.

23. The given data may be summarized as follows:

	Orange Juice	**Grapefruit Juice**
Vitamin A	60 I.U.	120 I.U.
Vitamin C	16 I.U.	12 I.U.
Calories	14	11

Suppose x ounces of orange juice and y ounces of pink-grapefruit juice are required for each glass of the blend. Then the problem is

$$\begin{aligned} \text{Minimize } C = 14x + 11y \text{ subject to} \\ 60x + 120y \geq 1200 \\ 16x + 12y \geq 200 \\ x \geq 0, y \geq 0 \end{aligned}$$

To construct the dual problem, we first write down the tableau

x	y	*Const.*
60	120	1200
16	12	200
14	11	

Then obtain the following by interchanging rows and columns:

u	v	*Const.*
60	16	14
120	12	11
1200	200	

From this table we construct the dual problem:

$$\begin{aligned} \text{Maximize } P = 1200u + 200v \text{ subject to} \\ 60u + 16v \leq 14 \\ 120u + 12v \leq 11 \\ u \geq 0, v \geq 0 \end{aligned}$$

The initial tableau is

u	v	x	y	P	*Const.*
60	16	1	0	0	14
120	12	0	1	0	11
−1200	−200	0	0	1	0

Using the following sequence of row operations,

1. $\frac{1}{120}R_2$ 2. $R_1 - 60R_2$, $R_3 + 1200R_2$ 3. $\frac{1}{10}R_1$ 4. $R_2 - \frac{1}{10}R_1$, $R_3 + 80R_1$

we obtain the final tableau

u	v	x	y	P	Const.
0	1	$\frac{1}{10}$	$-\frac{1}{20}$	0	$\frac{17}{20}$
0	0	$-\frac{1}{100}$	$\frac{1}{75}$	0	$\frac{1}{150}$
0	0	8	6	1	178

We conclude that the owner should use 8 ounces of orange juice and 6 ounces of grapefruit juice per glass of the blend for a minimal calorie count of 178.

USING TECHNOLOGY EXERCISES 4.2, page 252

1. $x = \frac{4}{3}$, $y = \frac{10}{3}$, $z = 0$, and $C = \frac{14}{3}$

3. $x = 0.9524$, $y = 4.2857$, $z = 0$, and $C = 6.0952$.

EXERCISES 4.3, page 265

1. Maximize $C = -P = -2x + 3y$ subject to

$$\begin{aligned} -3x - 5y &\le -20 \\ 3x + y &\le 16 \\ -2x + y &\le 1 \\ x \ge 0, y &\ge 0 \end{aligned}$$

3. Maximize $P = -C = -5x - 10y - z$ subject to

$$\begin{aligned} -2x - y - z &\le -4 \\ -x - 2y - 2z &\le -2 \\ 2x + 4y + 3z &\le 12 \\ x \ge 0,\ y \ge 0, \text{ and } z &\ge 0 \end{aligned}$$

5. We set up the tableau and solve the problem using the simplex method:

	x	y	u	v	P	Const	Ratio
	2	5	1	0	0	20	4
p.r.→	1	−5	0	1	0	−5	1
	−1	−2	0	0	1	0	
		↑ p.c.					

$\xrightarrow{-\frac{1}{5}R_2}$

x	y	u	v	P	Const
2	5	1	0	0	20
$-\frac{1}{5}$	1	0	$-\frac{1}{5}$	0	1
−1	−2	0	0	1	0

$\xrightarrow[R_3+2R_2]{R_1-5R_2}$

x	y	u	v	P	Const	Ratio
3	0	1	1	0	15	5
$-\frac{1}{5}$	1	0	$-\frac{1}{5}$	0	1	--
$-\frac{7}{5}$	0	0	$-\frac{2}{5}$	1	2	

$\xrightarrow{\frac{1}{3}R_1}$

x	y	u	v	P	Const
1	0	$\frac{1}{3}$	$\frac{1}{3}$	0	5
$-\frac{1}{5}$	1	0	$-\frac{1}{5}$	0	1
$-\frac{7}{5}$	0	0	$-\frac{2}{5}$	1	2

$\xrightarrow[R_3+\frac{7}{5}R_1]{R_2+\frac{1}{5}R_1}$

x	y	u	v	P	Const
1	0	$\frac{1}{3}$	$\frac{1}{3}$	0	5
0	1	$\frac{1}{15}$	$-\frac{2}{15}$	0	2
0	0	$\frac{7}{15}$	$\frac{1}{15}$	1	9

The maximum value of P is 9 when $x = 5$ and $y = 2$.

7. We first rewrite the problem as a maximization problem with inequality constraints using $\leq$, obtaining the following equivalent problem:

Maximize $P = -C = 2x - y$ subject to

$$x + 2y \leq 6$$

$$3x + 2y \leq 12$$

$$x \geq 0, y \geq 0$$

Following the procedure outlined for nonstandard problems, we have

	x	y	u	v	P	Const	Ratio
	1	2	1	0	0	6	6
$p.r. \rightarrow$	3	2	0	1	0	12	4
	−2	1	0	0	1	0	
	$\uparrow$ $p.c.$						

$\xrightarrow{\frac{1}{3}R_2}$

x	y	u	v	P	Const
1	2	1	0	0	6
1	$\frac{2}{3}$	0	$\frac{1}{3}$	0	4
−2	1	0	0	1	0

$\xrightarrow[R_3+2R_2]{R_1-R_2}$

x	y	u	v	P	Const
0	$\frac{4}{3}$	1	$-\frac{1}{3}$	0	6
1	$\frac{2}{3}$	0	$\frac{1}{3}$	0	4
0	$\frac{7}{3}$	0	$\frac{2}{3}$	1	8

We conclude that C attains a minimum value of -8 when $x = 4$ and $y = 0$.

9. Using the simplex method we have

	x	y	u	v	P	Const	Ratio
	1	3	1	0	0	6	6
p.r. →	−2	3	0	1	0	−6	3
	−1	−4	0	0	1	0	
	↑ p.c.						

$\xrightarrow{-\frac{1}{2}R_2}$

x	y	u	v	P	Const
1	3	1	0	0	6
1	$-\frac{3}{2}$	0	$-\frac{1}{2}$	0	3
−1	−4	0	0	1	0

$\xrightarrow{\substack{R_1 - R_2 \\ R_3 + R_2}}$

x	y	u	v	P	Const
0	$\frac{9}{2}$	1	$\frac{1}{2}$	0	3
1	$-\frac{3}{2}$	0	$-\frac{1}{2}$	0	3
0	$-\frac{11}{2}$	0	$-\frac{1}{2}$	1	3

$\xrightarrow{\frac{2}{9}R_1}$

x	y	u	v	P	Const
0	1	$\frac{2}{9}$	$\frac{1}{9}$	0	$\frac{2}{3}$
1	$-\frac{3}{2}$	0	$-\frac{1}{2}$	0	3
0	$-\frac{11}{2}$	0	$-\frac{1}{2}$	1	3

$\xrightarrow{\substack{R_2 + \frac{3}{2}R_1 \\ R_3 + \frac{11}{2}R_1}}$

x	y	u	v	P	Const
0	1	$\frac{2}{9}$	$\frac{1}{9}$	0	$\frac{2}{3}$
1	0	$\frac{1}{3}$	$-\frac{1}{3}$	0	4
0	0	$\frac{11}{9}$	$\frac{1}{9}$	1	$\frac{20}{3}$

We conclude that P attains a maximum value of 20/3, when $x = 4$ and $y = 2/3$.

11. We rewrite the problem as

Maximize $P = x + 2y$ subject to

$$2x + 3y \le 12$$
$$-x + 3y \le 3$$
$$-x + 3y \ge 3$$
$$x \ge 0, y \ge 0$$

The initial tableau is

x	y	u	v	w	P	Const.
2	3	1	0	0	0	12
–1	3	0	1	0	0	3
1	–3	0	0	1	0	–3
–1	–2	0	0	0	1	0

Using the following sequence of row operations

1. $-\frac{1}{3}R_3$ 2. $R_1 - 3R_3$, $R_2 - 3R_3$, $R_4 + 2R_3$ 3. $\frac{1}{3}R_1$

4. $R_3 + \frac{1}{3}R_1$, $R_4 + \frac{5}{3}R_1$ 5. $R_1 - \frac{1}{3}R_2$, $R_3 + \frac{2}{9}R_2$, $R_4 + \frac{1}{9}R_2$

we obtain the final tableau

x	y	u	v	w	P	Const.
1	0	$\frac{1}{3}$	$-\frac{1}{3}$	0	0	3
0	0	0	1	1	0	0
0	1	$\frac{1}{9}$	$\frac{2}{9}$	0	0	2
0	0	$\frac{5}{9}$	$\frac{1}{9}$	0	1	7

We conclude that P attains a maximum value of 7 when $x = 3$ and $y = 2$.

13. We rewrite the problem as

Maximize $P = 5x + 4y + 2z$ subject to

$$x + 2y + 3z \le 24$$
$$-x + y - z \le -6$$
$$x \ge 0,\ y \ge 0,\ z \ge 0$$

The initial tableau is

x	y	z	u	v	P	Const.
1	2	3	1	0	0	24
–1	1	–1	0	1	0	–6
–5	–4	–2	0	0	1	0

Using the following sequence of row operations

1. $-R_2$ 2. $R_1 - 2R_2$, $R_3 + 5R_2$ 3. $\frac{1}{3}R_1$ 4. $R_2 + R_1$, $R_3 + 9R_1$

5. $3R_1$ 6, $R_2 + \frac{2}{3}R_1$; $R_3 + 2R_1$

we obtain the final tableau

x	y	z	u	v	P	Const.
0	3	2	1	1	0	18
1	2	3	1	0	0	24
0	6	13	5	0	1	120

from which we deduce that P attains a maximum value of 120 when $x = 24$, $y = 0$, and $z = 0$.

15. The problem is to maximize $P = -C = -x + 2y - z$ subject to the given constraints. The initial tableau is

x	y	z	u	v	w	P	Const.
1	−2	3	1	0	0	0	10
2	1	−2	0	1	0	0	15
2	1	3	0	0	1	0	20
1	−2	1	0	0	0	1	0

Using the following sequence of row operations,
1. $R_1 + 2R_2$, $R_3 - R_2$, $R_4 + 2R_2$ 2. $\frac{1}{5}R_3$ 3. $R_1 + R_3$, $R_2 + 2R_3$, $R_4 + 3R_3$
we obtain the final tableau

x	y	z	u	v	w	P	Const.
5	0	0	1	$\frac{9}{5}$	$\frac{1}{5}$	0	41
2	1	0	0	$\frac{3}{5}$	$\frac{2}{5}$	0	17
0	0	1	0	$-\frac{1}{5}$	$\frac{1}{5}$	0	1
5	0	0	0	$\frac{7}{5}$	$\frac{3}{5}$	1	33

We conclude that C attains a minimum value of -33 when $x = 0$, $y = 17$, $z = 1$, and $C = -P = -33$.

17. Rewriting the third constraint as $-x + 2y - z \leq -4$, we obtain the following initial tableau

x	y	z	u	v	w	P	Const.
1	2	3	1	0	0	0	28
2	3	−1	0	1	0	0	6
−1	2	−1	0	0	1	0	−4
−2	−1	−1	0	0	0	1	0

Using the following sequence of row operations,
1. $\frac{1}{2}R_2$ 2. $R_1 - R_2$, $R_3 + R_2$, $R_4 + 2R_2$ 3. $-\frac{2}{3}R_3$
4. $R_1 - \frac{7}{2}R_3$, $R_2 + \frac{1}{2}R_3$, $R_4 + 2R_3$ 5. $\frac{3}{26}R_1$ 6. $R_2 - \frac{1}{3}R_1$; $R_3 + \frac{7}{3}R_1$, $R_4 + \frac{8}{3}R_1$

7. $\frac{26}{7}R_1$ 8. $R_2 + \frac{11}{26}R_1$, $R_3 + \frac{1}{26}R_1$, $R_4 + \frac{8}{13}R_1$

we obtain the final tableau

x	y	z	u	v	w	P	$Const.$
0	$\frac{26}{7}$	0	$\frac{3}{7}$	$\frac{2}{7}$	1	0	$\frac{68}{7}$
1	$\frac{11}{7}$	0	$\frac{1}{7}$	$\frac{3}{7}$	1	0	$\frac{46}{7}$
0	$\frac{1}{7}$	1	$\frac{2}{7}$	$-\frac{1}{7}$	0	0	$\frac{50}{7}$
0	$\frac{16}{7}$	0	$\frac{4}{7}$	$\frac{5}{7}$	0	1	$\frac{142}{7}$

from which we deduce that P attains a maximum value of 142/7 when $x = 46/7$, $y = 0$, and $z = 50/7$.

19. Rewriting the third constraint ($2x + y + z = 10$) in the form

$$2x + y + z \geq 10 \quad \text{and} \quad -2x - y - z \leq -10,$$

we obtain the following initial tableau.

x	y	z	t	u	v	w	P	$Const.$
1	2	1	1	0	0	0	0	20
3	1	0	0	1	0	0	0	30
2	1	1	0	0	1	0	0	10
−2	−1	−1	0	0	0	1	0	−10
−1	−2	−3	0	0	0	0	1	0

Using the following sequence of row operations
1. $-R_4$ 2. $R_1 - R_4$, $R_3 - R_4$, $R_5 + 3R_4$ 3. $R_1 - R_3$, $R_4 + R_3$, $R_5 + 3R_3$
we obtain the final tableau

x	y	z	t	u	v	w	P	$Const.$
−1	1	0	1	0	−1	0	0	10
3	1	0	0	1	0	0	0	30
0	0	0	0	0	1	1	0	0
2	1	1	0	0	1	0	0	10
5	1	0	0	0	3	0	1	30

We conclude that P attains a maximum value of 30 when $x = 0$, $y = 0$, and $z = 10$.

21. Let x and y denote the number of acres of crops A and B, respectively to be planted. Then the problem is

Maximize $P = 150x + 200y$ subject to the constraints

$$\begin{aligned} x + \quad y &\le 150 \\ 40x + 60y &\le 7400 \\ 20x + 25y &\le 3300\,. \\ x &\ge 80 \\ x \ge 0, y &\ge 0 \end{aligned}$$

Using the simplex method, we obtain

x	y	u	v	w	z	P	*Const.*	*Ratio*
1	1	1	0	0	0	0	150	150
40	60	0	1	0	0	0	7400	185
20	25	0	0	1	0	0	3300	165
–1	0	0	0	0	1	0	–80	–80
–150	–200	0	0	0	0	1	0	

$\xrightarrow{-R_4}$

x	y	u	v	w	z	P	*Const.*
1	1	1	0	0	0	0	150
40	60	0	1	0	0	0	7400
20	25	0	0	1	0	0	3300
1	0	0	0	0	–1	0	80
–150	–200	0	0	0	0	1	0

$\xrightarrow[R_5+150R_4]{R_1-R_4,\ R_2-40R_4,\ R_3-20R_4}$

x	y	u	v	w	z	P	*Const.*
0	1	1	0	0	1	0	70
0	60	0	1	0	40	0	4200
0	25	0	0	1	20	0	1700
1	0	0	0	0	–1	0	80
0	–200	0	0	0	–150	1	12,000

$\xrightarrow{\frac{1}{25}R_3}$

x	y	u	v	w	z	P	Const.
0	1	1	0	0	1	0	70
0	60	0	1	0	40	0	4200
0	1	0	0	$\frac{1}{25}$	$\frac{4}{5}$	0	68
1	0	0	0	0	-1	0	80
0	-200	0	0	0	-150	1	12,000

$\xrightarrow{\substack{R_1 - R_3 \\ R_2 - 60R_3 \\ R_5 + 200R_3}}$

x	y	u	v	w	z	P	Const.
0	0	1	0	$-\frac{1}{25}$	$\frac{1}{5}$	0	2
0	0	0	1	$-\frac{12}{25}$	-8	0	120
0	1	0	0	$\frac{1}{25}$	$\frac{4}{5}$	0	68
1	0	0	0	0	-1	0	80
0	0	0	0	8	10	1	25,600

We conclude that the farmer should plant 80 acres of crop A and 68 acres of crop B to realize a maximum profit of $25,600.

23. Let x and y denote the amount (in dollars) invested in company A and company B, respectively. Then the problem is

$$\text{Maximize } P = 0.08x + 0.06y \text{ subject to}$$
$$-x + 3y \le 0$$
$$y \ge 10{,}000{,}000$$
$$x + y = 60{,}000{,}000$$
$$x \ge 0,\ y \ge 0$$

Substituting $x = 60{,}000{,}000 - y$ into the first equation and the first and second inequalities, we have

$$\text{Maximize } P = 0.08(60{,}000{,}000 - y) + 0.06y$$
$$= 4{,}800{,}000 - 0.02y \text{ subject to}$$
$$y \le 15{,}000{,}000$$
$$y \ge 10{,}000{,}000$$
$$x \ge 0, y \ge 0.$$

Using the simplex method, we have

	y	u	v	P	Const.
	1	1	0	0	15,000,000
p.r.→	−1	0	1	0	−10,000,000
	0.02	0	0	1	4,800,000
	↑ p.c.				

$\xrightarrow{-R_2}$

y	u	v	P	Const.
1	1	0	0	15,000,000
1	0	−1	0	10,000,000
0.02	0	0	1	4,800,000

$\xrightarrow[R_5-0.02R_2]{R_1-R_2}$

y	u	v	P	Const.
0	1	1	0	5,000,000
1	0	−1	0	10,000,000
0	0	0.02	1	4,600,000

We conclude that the bank should extend \$50 million in home loans, \$10 million of commercial-development loans to attain a maximum return of \$4.6 million.

25. Let x, y, and z denote the number of units of products A, B, and C manufactured by the company. Then the linear programming problem is

Maximize $P = 18z + 12y + 15z$ subject to

$$
\begin{aligned}
2x + y + 2z &\le 900 \\
3x + y + 2z &\le 1080 \\
2x + 2y + z &\le 840 \\
x - y + z &\le 0 \\
x \ge 0, \; y \ge 0, \; z &\ge 0
\end{aligned}
$$

The initial tableau is

x	y	z	t	u	v	w	P	Const.
2	1	2	1	0	0	0	0	900
3	1	2	0	1	0	0	0	1080
2	2	1	0	0	1	0	0	840
1	−1	1	0	0	0	1	0	0
−18	−12	−15	0	0	0	0	1	0

Using the following sequence of row operations,

1. $R_1 - 2R_4,\ R_2 - 3R_4,\ R_3 - 2R_4,\ R_5 + 18R_4$ 2. $\frac{1}{4}R_3$

3. $R_1 - 3R_3,\ R_2 - 4R_3,\ R_4 + R_3,\ R_5 + 30R_3$ 4. $\frac{4}{3}R_4$

5. $R_1 - \frac{3}{4}R_4,\ R_3 + \frac{1}{4}R_4,\ R_5 + \frac{9}{2}R_4$

we obtain the final tableau

x	y	z	t	u	v	w	P	Const.
-1	0	0	1	0	-1	-1	0	60
0	0	0	0	1	-1	-1	0	240
$\frac{1}{3}$	1	0	0	0	$\frac{1}{3}$	$-\frac{1}{3}$	0	280
$\frac{4}{3}$	0	1	0	0	$\frac{1}{3}$	$\frac{2}{3}$	0	280
-6	0	0	0	0	9	6	1	7560

and conclude that the company should produce 0 units of product A, 280 units of product B, and 280 units of product C to realize a maximum profit of \$7,560.

27. Let x denote the number of ounces of food A and y denote the number of ounces of food B used in the meal. Then the problem is to minimize the amount of cholesterol in the meal. Thus, the linear programming problem is

$$\text{Maximize } P = -C = -2x - 5y \text{ subject to}$$
$$30x + 25y \geq 400$$
$$x + \tfrac{1}{2}y \geq 10$$
$$2x + 5y \geq 40$$
$$x \geq 0, y \geq 0$$

The initial tableau is

x	y	u	v	w	C	Const.
-30	-25	1	0	0	0	-400
-1	$-\frac{1}{2}$	0	1	0	0	-10
-2	-5	0	0	1	0	-40
2	5	0	0	0	1	0

Using the following sequence of row operations

1. $-R_2$ 2. $R_1 + 30R_2;\ R_3 + 2R_2;\ R_4 - 2R_2$ 3. $-\frac{1}{4}R_3$

4. $R_1 + 10R_2;\ R_2 - \frac{1}{2}R_3;\ R_4 - 4R_3$ 5. $-\frac{1}{25}R_1$ 6. $R_2 + \frac{5}{4}R_1;\ R_3 - \frac{1}{2}R_1$

we obtain the final tableau

x	y	u	v	w	C	Const.
0	0	$-\frac{1}{25}$	1	$\frac{1}{10}$	0	2
1	0	$-\frac{1}{20}$	1	$\frac{1}{4}$	0	10
0	1	$\frac{1}{50}$	0	$-\frac{3}{10}$	0	4
0	0	0	0	1	1	–40

Thus, the minimum content of cholesterol is 40mg when 10 ounces of food A and 4 ounces of food B are used.(Since the u-column is not in unit form, we see that the problem has multiple solutions.)

CHAPTER 4 REVIEW EXERCISES, page 270

1. This is a regular linear programming problem. Using the simplex method with u and v as slack variables, we obtain the following sequence of tableaus:

	x	y	u	v	P	Const	Ratio
p.r.→	1	3	1	0	0	15	5
	4	1	0	1	0	16	16
	–3	–4	0	0	1	0	
		↑ p.c.					

$\xrightarrow{\frac{1}{3}R_1}$

x	y	u	v	P	Const.
$\frac{1}{3}$	1	$\frac{1}{3}$	0	0	5
4	1	0	1	0	16
–3	–4	0	0	1	0

$\xrightarrow[R_3+4R_1]{R_2-R_1}$

	x	y	u	v	P	Const.
	$\frac{1}{3}$	1	$\frac{1}{3}$	0	0	5
p.c.→	$\frac{11}{3}$	0	$-\frac{1}{3}$	1	0	11
	$-\frac{5}{3}$	0	$\frac{4}{3}$	0	1	20
	↑ p.c.					

$\xrightarrow{\frac{3}{11}R_2}$

x	y	u	v	P	Const.
$\frac{1}{3}$	1	$\frac{1}{3}$	0	0	5
1	0	$-\frac{1}{11}$	$\frac{3}{11}$	0	3
$-\frac{5}{3}$	0	$\frac{4}{3}$	0	1	20

$\xrightarrow{\substack{R_1-\frac{1}{3}R_2 \\ R_3+\frac{5}{3}R_2}}$

x	y	u	v	P	Const.
0	1	$\frac{4}{11}$	$-\frac{1}{11}$	0	4
1	0	$-\frac{1}{11}$	$\frac{3}{11}$	0	3
0	0	$\frac{13}{11}$	$\frac{5}{11}$	1	25

and conclude that $x = 3$, $y = 4$, and $P = 25$.

3. Using the simplex method to solve this regular linear programming problem we have

	x	y	z	u	v	P	Const.	Ratio
p.r. →	1	2	3	1	0	0	12	4
	1	–3	2	0	1	0	10	5
	–2	–3	–5	0	0	1	0	
			↑ p.c.					

$\xrightarrow{\frac{1}{3}R_1}$

x	y	z	u	v	P	Const.
$\frac{1}{3}$	$\frac{2}{3}$	1	$\frac{1}{3}$	0	0	4
1	–3	2	0	1	0	10
–2	–3	–5	0	0	1	0

$\xrightarrow{\substack{R_2-2R_1 \\ R_3+5R_1}}$

	x	y	z	u	v	P	Const.	Ratio
	$\frac{1}{3}$	$\frac{2}{3}$	1	$\frac{1}{3}$	0	0	4	12
p.r. →	$\frac{1}{3}$	$-\frac{13}{3}$	0	$-\frac{2}{3}$	1	0	2	6
	$-\frac{1}{3}$	$\frac{1}{3}$	0	$\frac{5}{3}$	0	1	20	
	↑ p.c.							

$\xrightarrow{3R_2}$

x	y	z	u	v	P	Const.
$\frac{1}{3}$	$\frac{2}{3}$	1	$\frac{1}{3}$	0	0	4
0	–13	0	–2	3	0	6
$-\frac{1}{3}$	$\frac{1}{3}$	0	$\frac{5}{3}$	0	1	20

$\xrightarrow{\substack{R_1-\frac{1}{3}R_2 \\ R_3+\frac{1}{3}R_2}}$

	x	y	z	u	v	P	Const.	Ratio
p.r.→	0	5	1	1	−1	0	2	2/5
	1	−13	0	−2	3	0	6	--
	0	−4	0	1	1	1	22	
		↑ p.c.						

$\xrightarrow{\frac{1}{5}R_1}$

x	y	z	u	v	P	Const.
0	1	$\frac{1}{5}$	$\frac{1}{5}$	$-\frac{1}{5}$	0	$\frac{2}{5}$
0	−13	0	−2	3	0	6
0	−4	0	1	1	1	22

$\xrightarrow[R_3+4R_1]{R_2+13R_1}$

x	y	z	u	v	P	Const.
0	1	$\frac{1}{5}$	$\frac{1}{5}$	$-\frac{1}{5}$	0	$\frac{2}{5}$
1	0	$\frac{13}{5}$	$\frac{3}{5}$	$\frac{2}{5}$	0	$\frac{56}{5}$
0	0	$\frac{4}{5}$	$\frac{9}{5}$	$\frac{1}{5}$	1	$\frac{118}{5}$

We conclude that the P attains a maximum value of 23.6 when $x = 11.2$, $y = 0.4$ and $z = 0$.

5. We first write the tableau

x	y	Const.
2	3	6
2	1	4
3	2	

Then obtain the following by interchanging rows and columns:

u	v	Const.
2	2	3
3	1	2
6	4	

From this table we construct the dual problem:

Maximize the objective function $P = 6u + 4v$ subject to the constraints

$$2u + 2v \le 3$$

$$3u + v \le 2$$

$$u \ge 0,\ v \ge 0$$

Using the simplex method, we have

	u	v	x	y	P	Const	Ratio
	2	2	1	0	0	3	3/2
p.r.→	3	1	0	1	0	2	2/3
	−6	−4	0	0	1	0	
	↑ p.c.						

$\xrightarrow{\frac{1}{3}R_2}$

u	v	x	y	P	Const
2	2	1	0	0	3
1	$\frac{1}{3}$	0	$\frac{1}{3}$	0	$\frac{2}{3}$
−6	−4	0	0	1	0

$\xrightarrow{R_1-2R_2,\ R_3+6R_2}$

u	v	x	y	P	Const	Ratio
0	$\frac{4}{3}$	1	$-\frac{2}{3}$	0	$\frac{5}{3}$	3/4
1	$\frac{1}{3}$	0	$\frac{1}{3}$	0	$\frac{2}{3}$	2
0	−2	0	2	1	4	

$\xrightarrow{\frac{3}{4}R_1}$

u	v	x	y	P	Const
0	1	$\frac{3}{4}$	$-\frac{1}{2}$	0	$\frac{5}{4}$
1	$\frac{1}{3}$	0	$\frac{1}{3}$	0	$\frac{2}{3}$
0	−2	0	2	1	4

$\xrightarrow{R_2-\frac{1}{3}R_1,\ R_3+2R_1}$

u	v	x	y	P	Const
0	1	$\frac{3}{4}$	$-\frac{1}{2}$	0	$\frac{5}{4}$
1	0	$-\frac{1}{4}$	$\frac{1}{2}$	0	$\frac{1}{4}$
0	0	$\frac{3}{2}$	1	1	$\frac{13}{2}$

Therefore, C attains a minimum value of 13/2 when $x = 3/2$ and $y = 1$.

7. We first write the tableau

x	y	z	Const.
3	2	1	4
1	1	3	6
24	18	24	

Then obtain the following by interchanging rows and columns:

u	v	Const.
3	2	24
2	1	18
1	3	24
4	6	

From this table we construct the dual problem:

Maximize the objective function $P = 4u + 6v$ subject to

$$3u + v \le 24$$
$$2u + v \le 18$$
$$u + 3v \le 24$$
$$u \ge 0,\ v \ge 0$$

The initial tableau is

u	v	x	y	z	P	Const.
3	1	1	0	0	0	24
2	1	0	1	0	0	18
1	3	0	0	1	0	24
–4	–6	0	0	0	1	0

Using the following sequence of row operations

1. $\frac{1}{3}R_3$ 2. $R_1 - R_3$, $R_2 - R_3$, $R_4 + 6R_3$ 3. $\frac{3}{8}R_1$ 4. $R_2 - \frac{5}{3}R_1$, $R_3 - \frac{1}{3}R_1$, $R_4 + 2R_1$

we obtain the final tableau

u	v	x	y	z	P	Const.
1	0	$\frac{3}{8}$	0	$-\frac{1}{8}$	0	6
0	0	$-\frac{5}{8}$	1	$-\frac{1}{8}$	0	0
0	1	$-\frac{1}{8}$	0	$\frac{3}{8}$	0	6
0	0	$\frac{3}{4}$	0	$\frac{7}{4}$	0	60

We conclude that C attains a minimum value of 60 when $x = 3/4$, $y = 0$, and $z = 7/4$.

9. Rewriting the problem, we have

Maximize $P = 3x - 4y$ subject to

$$x + y \le 45$$
$$-x + 2y \le -10$$
$$x \ge 0,\ y \ge 0$$

Using the simplex method, we have

x	y	u	v	P	Const	Ratio
1	1	1	0	0	45	45
–1	2	0	1	0	–10	10
–3	4	0	0	1	0	

$\xrightarrow{-R_2}$

x	y	u	v	P	Const
1	1	1	0	0	45
1	–2	0	–1	0	10
–3	4	0	0	1	0

$\xrightarrow{R_1 - R_2,\ R_3 + 3R_2}$

x	y	u	v	P	Const
0	3	1	1	0	35
1	–2	0	–1	0	10
0	–2	0	–3	1	30

$\xrightarrow{R_2 + R_1,\ R_3 + 3R_1}$

x	y	u	v	P	Const
0	3	1	1	0	35
1	1	1	–0	0	45
0	7	3	0	1	135

We conclude that P attains a maximum value of 135 when $x = 45$ and $y = 0$.

11. We first write the problem in the form

Maximize $P = 2x + 3y$ subject to

$$2x + 5y \le 20$$
$$x - 5y < -5$$
$$x \ge 0,\ y \ge 0$$

The initial tableau is

x	y	u	v	P	Const
2	5	1	0	0	20
1	–5	0	1	0	–5
–2	–3	0	0	1	0

Using the sequence of row operations

1. $\frac{1}{5}R_1$ 2. $R_2 + 5R_1$, $R_3 + 3R_1$ 3. $\frac{1}{3}R_2$ 4. $R_1 - \frac{2}{5}R_2$, $R_3 + \frac{4}{5}R_2$

we obtain the final tableau

x	y	u	v	P	Const
0	1	$\frac{1}{15}$	$-\frac{2}{15}$	0	2
1	0	$\frac{1}{3}$	$\frac{1}{3}$	0	5
0	0	$\frac{13}{15}$	$\frac{4}{15}$	1	16

We conclude that P attains a maximum value of 16 when $x = 5$ and $y = 2$.

13. Let x, y, and z denote the number of units of products A, B, and C made, respectively. Then the problem is to maximize the profit

$$P = 4x + 6y + 8z \text{ subject to}$$
$$9x + 12y + 18z \le 360$$
$$6x + 6y + 10z \le 240$$
$$x \ge 0,\ y \ge 0,\ z \ge 0$$

The initial tableau is

x	y	z	u	v	P	Const.
9	12	18	1	0	0	360
6	6	10	0	1	0	240
–4	–6	–8	0	0	1	0

Using the sequence of row operations

1. $\frac{1}{18}R_1$ 2. $R_2 - 10R_1$ 3. $R_3 + 8R_1$ 4. $\frac{3}{2}R_1$ 5. $R_2 + \frac{2}{3}R_1$, $R_3 + \frac{2}{3}R_1$

we obtain the final tableau

x	y	z	u	v	P	Const.
$\frac{3}{4}$	1	$\frac{3}{2}$	$\frac{1}{12}$	0	0	30
$\frac{3}{2}$	0	1	$-\frac{1}{2}$	1	0	60
$\frac{1}{2}$	0	1	$\frac{1}{2}$	0	1	180

and conclude that the company should produce 0 units of product A, 30 units of product B, and 0 units of product C to realize a maximum profit of $180.

15. Let x, y, and z, denote the amount invested in stocks, bonds, and money-market funds, respectively. Then the problem is

Maximize $P = 0.15x + 0.10y + 0.08z$ subject to

$$x + y + z \le 200{,}000$$
$$-z \le -50{,}000$$
$$y - x + z \le 0$$
$$x \ge 0,\ y \ge 0,\ z \ge 0$$

The initial tableau is

x	y	z	u	v	w	P	Const.
1	1	1	1	0	0	0	200,000
0	0	−1	0	1	0	0	−50,000
1	−1	1	0	0	1	0	0
−0.15	−0.10	−0.08	0	0	0	1	0

Using the following sequence of row operations,

1. $-R_2$ 2. $R_1 - R_2$, $R_3 - R_2$, $R_4 + 0.08R_2$ 3. $-R_3$ 4. $R_1 - R_3$, $R_4 + \frac{1}{10}R_3$ 5. $\frac{1}{2}R_1$ 6. $R_3 + R_1$, $R_4 + \frac{1}{4}R_1$

we obtain the final tableau

x	y	z	u	v	w	P	Const.
1	0	0	$\frac{1}{2}$	1	$\frac{1}{2}$	0	50,000
0	0	1	0	−1	0	0	50,000
0	1	0	$\frac{1}{2}$	0	$-\frac{1}{2}$	0	100,000
0	0	0	$\frac{1}{8}$	$\frac{7}{100}$	$\frac{1}{40}$	1	21,500

and we conclude that Sandra should invest $50,000 in stocks, $100,000 in bonds, and $50,000 in money-market funds to realize a maximum return of $21,500 per year on her investments.

CHAPTER 5

EXERCISES 5.1, page 283

1. The interest is given by $I = (500)(2)(0.08) = 80$, or \$80.
The accumulated amount is 500 + 80, or \$580.

3. The interest is given by $I = (800)(0.06)(0.75) = 36$, or \$36.
The accumulated amount is 800 + 36, or \$836.

5. We are given that $A = 1160$, $t = 2$, and $r = 0.08$, and we are asked to find P. Since
$$A = P(1 + rt)$$
we see that
$$P = \frac{A}{1+rt} = \frac{1160}{1+(0.08)(2)} = 1000, \text{ or } \$1000.$$

7. We use the formula $I = Prt$ and solve for t when $I = 20$, $P = 1000$, and $r = 0.05$.
Thus,
$$20 = 1000(0.05)(\frac{t}{365})$$
and
$$t = \frac{365(20)}{50} = 146, \quad \text{or 146 days.}$$

9. We use the formula $A = P(1 + rt)$ with $A = 1075$, $P = 1000$, $t = 0.75$, and solve for r.
Thus,
$$1075 = 1000(1 + 0.75r)$$
$$75 = 750r$$
or
$$r = 0.10.$$
Therefore, the interest rate is 10 percent per year.

11. $A = 1000(1 + 0.07)^8 = 1718.19$, or \$1718.19.

13. $A = 2500\left(1+\frac{0.07}{2}\right)^{20} = 4974.47$, or \$4974.47.

15. $A = 12000\left(1+\frac{0.08}{4}\right)^{42} = 27{,}566.93$, or \$27,566.93.

17. $A = 150{,}000\left(1+\dfrac{0.14}{12}\right)^{48} = 261{,}751.04$, or \$261,751.04.

19. $A = 150{,}000\left(1+\dfrac{0.12}{365}\right)^{1095} = 214{,}986.69$, or \$214,986.69.

21. Using the formula

$$r_{eff} = \left(1+\frac{r}{m}\right)^{m} - 1$$

with $r = 0.10$ and $m = 2$, we have

$$r_{eff} = \left(1+\frac{0.10}{2}\right)^{2} - 1 = 0.1025, \quad \text{or 10.25 percent..}$$

23. Using the formula

$$r_{eff} = \left(1+\frac{r}{m}\right)^{m} - 1$$

with $r = 0.08$ and $m = 12$, we have

$$r_{eff} = \left(1+\frac{0.08}{4}\right)^{4} - 1 = 0.08300, \quad \text{or 8.3 percent per year.}$$

25. The present value is given by

$$P = 40{,}000\left(1+\frac{0.08}{2}\right)^{-8} = 29{,}227.61, \quad \text{or \$29,227.61.}$$

27. The present value is given by

$$P = 40{,}000\left(1+\frac{0.07}{12}\right)^{-48} = 30{,}255.95, \quad \text{or \$30,255.95.}$$

29. Think of \$4300 as the principal and \$306 as the accumulated amount at the end of 30 days. If r denotes the simple interest rate per annum, then we have $P = 300$, $A = 306$, $n = 1/12$, and we are required to find r. Using (8b) we have

$$306 = 300\left(1+\frac{r}{12}\right) = 300 + r\left(\frac{300}{12}\right)$$

and $r = \left(\dfrac{12}{300}\right)6 = 0.24.$

31. The rate that you would expect to pay is

$$A = 380(1 + 0.08)^5 = 558.34, \text{ or } \$558.34 \text{ per day.}$$

33. The amount that they can expect to pay is given by

$$A = 150{,}000(1 + 0.05)^4 = 182{,}325.94, \quad \text{or approximately } \$182{,}326.$$

35. The investment will be worth

$$A = 1.5\left(1 + \frac{0.095}{2}\right)^{20} = 3.794651$$

or approximately $3.8 million dollars.

37. Using the formula

$$P = A\left(1 + \frac{r}{m}\right)^{-mt}$$

we have

$$P = 40{,}000\left(1 + \frac{0.085}{4}\right)^{-20} = 26{,}267.49, \text{ or } \$26{,}267.49.$$

39. a. They should set aside

$$P = 100{,}000(1 + 0.085)^{-13} = 34{,}626.88, \text{ or } \$34{,}626.88.$$

b. They should set aside

$$P = 100{,}000\left(1 + \frac{0.085}{2}\right)^{-26} = 33{,}886.16, \quad \text{or } \$33{,}886.16.$$

c. They should set aside

$$P = 100{,}000\left(1 + \frac{0.085}{4}\right)^{-52} = 33{,}506.76, \text{ or } \$33{,}506.76.$$

41. The present value of the $8000 loan due in 3 years is given by

$$P = 8000\left(1 + \frac{0.10}{2}\right)^{-6} = 5969.72, \text{ or } \$5969.72.$$

The present value of the $15,000 loan due in 6 years is given by

$$P = 15{,}000\left(1 + \frac{0.10}{2}\right)^{-12} = 8352.56, \text{ or } \$8352.56.$$

Therefore, the amount the proprietors of the inn will be required to pay at the end of 5 years is given by

$$A = 14{,}322.28\left(1+\frac{0.10}{2}\right) = 23{,}329.48, \quad \text{or } \$23{,}329.48.$$

43. Using the compound interest formula with A = 128,000, P = 100,000 and t = 6, we have

$$128{,}000 = 100{,}000(1+R)^6$$
$$(1+R)^{1/6} = (1.28)^{1/6}$$
$$1+R = 1.042,$$
$$R = 0.042, \quad \text{or 4.2 percent.}$$

45. Let the effective rate of interest be R. Then R satisfies

$$A = P(1+R)^t$$

or

$$10{,}000 = 6595.37\left(1+\frac{R}{2}\right)^{10}$$

$$1+\frac{R}{2} = (1.51621516)^{1/10}$$
$$= 1.0425$$

and

$$R = 0.085, \quad \text{or 8.5 percent.}$$

47. a. We obtain a family of straight lines with varying slope and P-intercept as P increases. For a fixed rate of interest, the accumulated amount A grows at the rate of Pr units per year starting initially with an amount of $\$P$.
b. We obtain a family of straight lines emanating from the point (0, P) and with varying slope as r increases. For a fixed principal, the accumulated amount A grows at the rate Pr units per year starting initially with an amount of $\$P$.

USING TECHNOLOGY EXERCISES, page 287

1. \$5872.78 3. \$475.49 5. 8.95%/yr 7. 10.20%/yr

9. :PROGRAM: PREVAL

```
:Disp "A"
:Input A
:Disp "r"
:Input r
:Disp "t"
:Input t
:Disp "m"
```

```
:Input m
:A(1 + r/m)^(−m*t) → P
:Disp "PRESENT VALUE IS"
:Disp P
```

11. \$94,038.74

13. \$62,244.96

EXERCISES 5.2, page 295

1. $S = 1000\left[\dfrac{(1+0.1)^{10}-1}{0.1}\right] = 15{,}937.42$, or \$15,937.42.

3. $S = 1500\left[\dfrac{\left(1+\dfrac{0.08}{4}\right)^{24}-1}{\dfrac{0.08}{4}}\right] = 54{,}759.35$, or \$54,759.35.

5. $S = 600\left[\dfrac{\left(1+\dfrac{0.12}{4}\right)^{36}-1}{\dfrac{0.12}{4}}\right] = 37{,}965.57$, or \$37,965.57.

7. $P = 5000\left[\dfrac{1-(1+0.08)^{-8}}{0.08}\right] = 28{,}733.19$, or \$28,733.19.

9. $P = 4000\left[\dfrac{1-(1+0.09)^{-5}}{0.09}\right] = 15{,}558.61$, or \$15,558.61.

11. $P = 800\left[\dfrac{1-\left(1+\dfrac{0.12}{4}\right)^{-28}}{\dfrac{0.12}{4}}\right] = 15{,}011.29$, or \$15,011.29

13. She will have

$$S = 1500\left[\frac{(1+0.08)^{25}-1}{0.08}\right] = 109{,}658.91, \text{ or } \$109{,}658.91.$$

15. On October 31, Mrs Lynde's account will be worth

$$S = 40\left[\frac{\left(1+\frac{0.07}{12}\right)^{11}-1}{\frac{0.07}{12}}\right] = 453.06, \text{ or } \$453.06.$$

One month later, this account will be worth $A = (453.06)\left(1+\frac{0.07}{12}\right) = 455.70$, or

$455.70.

17. The amount in Mr. Collin's employee retirement account is given by

$$S = 100\left[\frac{\left(1+\frac{0.07}{12}\right)^{144}-1}{\frac{0.07}{12}}\right] = 22{,}469.60, \text{ or } \$22{,}469.60.$$

The amount in Mr. Collin's IRA is given by

$$S = 2000\left[\frac{(1+0.09)^{8}-1}{0.09}\right] = 22{,}056.85, \text{ or } \$22{,}056.85.$$

Therefore, the total amount in his retirement fund is given by

$$22{,}469.60 + 22{,}056.85 = 44{,}526.45, \text{ or } \$44{,}526.45.$$

19. The equivalent cash payment is given by

$$P = 450\left[\frac{1-\left(1+\frac{0.09}{12}\right)^{-24}}{\frac{0.09}{12}}\right] = 9850.12, \text{ or } \$9850.12.$$

21. We use the formula for the present value of an annuity obtaining

$$P = 22\left[\frac{1-\left(1+\frac{0.18}{12}\right)^{-36}}{\frac{0.18}{12}}\right] = 608.54, \text{ or } \$608.54.$$

23. With an \$800 monthly payment, the present value of their loan would be

$$P = 800\left[\frac{1-\left(1+\frac{0.095}{12}\right)^{-360}}{\frac{0.095}{12}}\right] = 95{,}141.34, \text{ or } \$95{,}141.34.$$

With a \$1000 monthly payment, the present value of their loan would be

$$P = 1000\left[\frac{1-\left(1+\frac{0.095}{12}\right)^{-360}}{\frac{0.095}{12}}\right] = 118{,}926.68, \text{ or } \$118{,}926.68$$

Since they intend to make a \$25,000 down payment, the range of homes they should consider is \$120,141 to \$143,927.

25. The lower limit of their investment is

$$A = 800\left[\frac{1-\left(1+\frac{0.09}{12}\right)^{-180}}{\frac{0.09}{12}}\right] + 25{,}000 = 103{,}874.73$$

or approximately \$103,875. The upper limit of their investment is

$$A = 1000\left[\frac{1-\left(1+\frac{0.09}{12}\right)^{-180}}{\frac{0.09}{12}}\right] + 25{,}000 = 123{,}593.41$$

or approximately \$123,593. Therefore, the price range of houses they should consider is \$103,875 to \$123,593.

USING TECHNOLOGY EXERCISES 5.2, page 297

1. \$59,622.15

3. \$8453.59

5.
```
:PROGRAM: PVAN
:Disp "R"
:Input R
:Disp "i"
:Input i
:Disp "N"
:Input N
:(R/i)(1-(1+i)^(-N))→ P
:Disp "AMOUNT IS"
:Disp P
```

7. \$45,983.53

9. \$18,344.08

EXERCISES 5.3, page 305

1. The size of each installment is given by

$$R = \frac{100{,}000(0.08)}{1-(1+0.08)^{-10}} = 14{,}902.95, \text{ or } \$14{,}902.95.$$

3. The size of each installment is given by

$$R = \frac{50{,}000(0.01)}{1-(1+0.01)^{-12}} - = 444.24, \text{ or } \$444.24.$$

5. The size of each installment is given by

$$R = \frac{25{,}000(0.0075)}{1-(1+0.0075)^{-48}} = 622.13, \text{ or } \$622.13.$$

7. The size of each installment is

$$R = \frac{80{,}000(0.00875)}{1-(1+0.00875)^{-360}} = 731.79, \text{ or } \$731.79.$$

9. The periodic payment that is required is

$$R = \frac{20{,}000(0.02)}{(1+0.02)^{12}-1} = 1491.19, \text{ or } \$1491.19.$$

11. The periodic payment that is required is

$$R = \frac{100{,}000(0.0075)}{(1+0.0075)^{120}-1} = 516.76, \text{ or } \$516.76$$

13. The periodic payment that is required is

$$R = \frac{250{,}000(0.00875)}{(1+0.00875)^{300}-1} = 172.95, \text{ or } \$172.95.$$

15. The size of each installment is given by

$$R = \frac{100{,}000(0.10)}{1-(1+0.10)^{-10}} = 16{,}274.54, \text{ or } \$16{,}274.54.$$

17. The monthly payment in each case is given by

$$R = \frac{100{,}000\left(\frac{r}{12}\right)}{1-\left(1+\frac{r}{12}\right)^{-360}}$$

Thus, if $r = 0.08$, then $R = \dfrac{100{,}000\left(\frac{0.08}{12}\right)}{1-\left(1+\frac{0.08}{12}\right)^{-360}} = 733.76$, or \$733.76

If $r = 0.09$, then $R = \dfrac{100{,}000\left(\frac{0.09}{12}\right)}{1-\left(1+\frac{0.09}{12}\right)^{-360}} = 804.62$, or \$804.62

If $r = 0.10$, then $R = \dfrac{100{,}000\left(\dfrac{0.10}{12}\right)}{1-\left(1+\dfrac{0.10}{12}\right)^{-360}} = 877.57$, or \$877.57.

If $r = 0.11$, then $R = \dfrac{100{,}000\left(\dfrac{0.11}{12}\right)}{1-\left(1+\dfrac{0.11}{12}\right)^{-360}} = \$952.32$.

a. The difference in monthly payments in the two loans is
$877.57 - \$665.30 = \$212.27.

b. The monthly mortgage payment on a \$150,000 mortgage would be
1.5(\$877.57) = \$1316.36.

The monthly mortgage payment on a \$50,000 mortgage would be
0.5(\$877.57) = \$438.79.

19. a. The amount of the loan required is 16000 - (0.25)(12000) or 13,000 dollars. If the car is financed over 36 months, the payment will be

$$R = \frac{12{,}000\left(\frac{0.10}{12}\right)}{1-\left(1+\frac{0.10}{12}\right)^{-36}} \approx 387.21, \text{ or } \$387.21 \text{ per month.}$$

If the car is financed over 48 months, the payment will be

$$R = \frac{12{,}000\left(\frac{0.10}{12}\right)}{1-\left(1+\frac{0.10}{12}\right)^{-48}} \approx 304.35, \text{ or } \$304.35 \text{ per month.}$$

b. The interest charges for the 36-month plan are
36(387.21) - 12000 = 1939.56,
or \$1939.56. The interest charges for the 48-month plan are
48(304.35) - 12000 = 2608.80, or \$2608.80.

21. The amount borrowed is 180,000 - 20,000 = 160,000 dollars. The size of the monthly installment is

$$R = \frac{160{,}000\left(\frac{0.08}{12}\right)}{1-\left(1+\frac{0.08}{12}\right)^{-360}} \approx 1174.0233, \text{ or } \$1174.02.$$

To find their equity after five years, we compute

$$P = 1174.0233\left[\frac{1-\left(1+\frac{0.08}{12}\right)^{-300}}{\frac{0.08}{12}}\right] \approx 152{,}112$$

or $152,112, and so their equity is

$$180{,}000 - 152{,}112 = 27{,}888, \text{ or } \$27{,}888.$$

To find their equity after ten years, we compute

$$P = 1174.0233\left[\frac{1-\left(1+\frac{0.08}{12}\right)^{-240}}{\frac{0.08}{12}}\right] \approx 140{,}360, \text{ or } \$140{,}360.$$

and their equity is 180,000 - 140,360 = 39,640, or $39,640.
To find their equity after twenty years, we compute

$$P = 1174.0233\left[\frac{1-\left(1+\frac{0.08}{12}\right)^{-120}}{\frac{0.08}{12}}\right] \approx 96{,}765, \text{ or } \$96{,}765,$$

and their equity is 180,000 - 96,765, or $83,235.

23. The amount that must be deposited quarterly into this fund is

$$R = \frac{\left(\frac{0.09}{4}\right)200{,}000}{\left(1+\frac{0.09}{4}\right)^{40} - 1} = 3{,}135.48, \text{ or } \$3{,}135.48.$$

25. The size of each quarterly installment is given by

$$R = \frac{\left(\frac{0.10}{4}\right)20{,}000}{\left(1+\frac{0.10}{4}\right)^{12} - 1} = 1449.74, \text{ or } \$1449.74.$$

27. The value of the IRA account after 20 years is

$$S = 375\left[\frac{\left(1+\frac{0.08}{4}\right)^{80} - 1}{\frac{0.08}{4}}\right] = 72{,}664.48, \text{ or } \$72{,}664.48.$$

The payment he would receive at the end of each quarter for the next 15 years is given by

$$R = \frac{\left(\frac{0.08}{4}\right)72{,}664.48}{1-\left(1+\frac{0.08}{4}\right)^{-60}} = 2090.41, \text{ or } \$2090.41.$$

If he continues working and makes quarterly payments until age 65, the value of the IRA account would be

$$S = 375\left[\frac{\left(1+\frac{0.08}{4}\right)^{100} - 1}{\frac{0.08}{4}}\right] = 117{,}087.11, \text{ or } \$117{,}087.11.$$

The payment he would receive at the end of each quarter for the next 10 years is given by

$$R = \frac{\left(\frac{0.08}{4}\right)117{,}087.11}{1-\left(1+\frac{0.08}{4}\right)^{-40}} = 4280.21, \text{ or } \$4280.21.$$

29. The monthly payment the Sandersons are required to make under the terms of their original loan is given by

$$R = \frac{100{,}000\left(\frac{0.10}{12}\right)}{1-\left(1+\frac{0..10}{12}\right)^{-240}} \approx 965.02, \text{ or } \$965.02.$$

The monthly payment the Sandersons are required to make under the terms of their new loan is given by

$$R = \frac{100{,}000\left(\frac{0.078}{12}\right)}{1-\left(1+\frac{0.078}{12}\right)^{-240}} \approx 824.04, \text{ or } \$824.04.$$

The amount of money that the Sandersons can expect to save over the life of the loan by refinancing is given by

$$240(965.02 - 824.04) = 33{,}835.20, \text{ or } \$33{,}835.20.$$

USING TECHNOLOGY EXERCISES 5.3, page 309

1. $3645.40

3. $18,443.75

5.
```
:PROGRAM: SINKFD
:Disp "S"
:Input S
:Disp "i"
:Input i
:Disp "N"
:Input N
:S*i/((1+i)^N-1)→R
:Disp "R is"
:Disp R
```

7. $916.26

9. $809.31

11. \$45,069.31. The amortization schedule follows.

End of Period	Interest charged	Repayment made	Payment toward Principal	Outstanding Principal
0				265,000.00
1	19,610.00	45,069.31	25,459.31	239,540.69
2	17,726.01	45,069.31	27,343.30	212,197.39
3	15,702.61	45,069.31	29,366.70	182,830.69
4	13,529.47	45,069.31	31,539.84	151,290.85
5	11,195.52	45,069.31	33,873.79	117,417.06
6	8,688.86	45,069.31	36,380.45	81,036.61
7	5,996.71	45,069.31	39,072.60	41,964.01
8	3,105.34	45,069.35	41,964.01	.00

EXERCISES 5.4, page 316

1. $a_9 = 6 + (9 - 1)3 = 30$

3. $a_8 = -15 + (8 - 1)\left(\dfrac{3}{2}\right) = -\dfrac{9}{2} = -4.5.$

5. $a_{11} - a_4 = (a_1 + 10d) - (a_1 + 3d) = 7\text{d}$. Also, $a_{11} - a_4 = 107 - 30 = 77$.

Therefore, $7d = 77$, and $d = 11$. Next,

$$a_4 = a + 3\text{d} = a + 3(11) = a + 33 = 30.$$

and $a = -3$. Therefore, the first five terms are -3, 8, 19, 30, 41.

7. Here $a = x$, $n = 7$, and $d = y$. Therefore, the required term is

$a_7 = x + (7 - 1)y = x + 6y.$

9. Using the formula for the sum of the terms of an arithmetic progression with $a = 4$, $d = 7$ and $n = 15$, we have

$$S_n = \frac{n}{2}[2a + (n-1)d]]$$

$$S_{15} = \frac{15}{2}[2(4) + (15-1)7] = \frac{15}{2}(106) = 795.$$

11. The common difference is $d = 2$ and the first term is $a = 15$. Using the formula for the nth term

$$a_n = a + (n - 1)d,$$

we have $57 = 15 + (n - 1)(2) = 13 + 2n$

$2n = 44,$ and $n = 22.$

Using the formula for the sum of the terms of an arithmetic progression with $a = 15$, $d = 2$ and $n = 22$, we have

$$S_n = \frac{n}{2}[2a + (n-1)d]]$$

$$S_{22} = \frac{22}{2}[2(15) + (22-1)2] = 11(72) = 792.$$

13.

$$\begin{aligned} &f(1) + f(2) + f(3) + \cdots + f(20) \\ &= [3(1) - 4] + [3(2) - 4] + [3(3) - 4] + \cdots + [3(20) - 4] \\ &= 3(1 + 2 + 3 + \cdots + 20) + 20(-4) \\ &= 3\left(\frac{20}{2}\right)[2(1) + (20-1)1] - 80 \\ &= 550. \end{aligned}$$

15.

$$S_n = \frac{n}{2}[2a_1 + (n-1)d]] = \frac{n}{2}(a_1 + a_1 + (n-1)d]$$

$$= \frac{n}{2}(a_1 + a_n)$$

a. $S_{11} = \frac{11}{2}(3 + 47) = 275.$

b. $S_{20} = \frac{20}{2}[5 + (-33)] = -280.$

17. Let n be the number of weeks till she reaches 10 miles. Then

$$a_n = 1+(n-1)\frac{1}{4} = 1+\frac{1}{4}n-\frac{1}{4} = \frac{1}{4}n+\frac{3}{4} = 10$$

Therefore, $n + 3 = 40$, and $n = 37$; that is, it will take Karen 37 weeks to meet her goal.

19. To compute the tourist's fare by taxi, take $a = 1$, $d = 0.60$, and $n = 25$. Then the required fare is given by

$$a_{25} = 1 + (25 - 1)0.60 = 15.4,$$

or \$15.40. Therefore, by taking the airport limousine, the tourist will save

$$15.40 - 7.50 = 7.90, \text{ or } \$7.90.$$

21. a. Using the formula for the sum of an arithmetic progression, we have

$$\begin{aligned} S_n &= \frac{n}{2}[2a+(n-1)d]] \\ &= \frac{N}{2}[2(1)+(N-1)(1)] \\ &= \frac{N}{2}(N+1). \end{aligned}$$

b. $S_{10} = \frac{10}{2}(10+1) = 5(11) = 55$

$$D_3 = (C-S)\frac{N-(n-1)}{55} = (6000-500)\frac{10-(3-1)}{55} = 5500\left(\frac{8}{55}\right)$$

$= 800$, or \$800.

23. This is a geometric progression with $a = 4$ and $r = 2$. Next, $a_7 = 4(2)^6 = 256$,

and $\quad S_7 = \dfrac{4(1-2^7)}{1-2} = 508.$

25. If we compute the ratios

$$\frac{a_2}{a_1} = \frac{-\frac{3}{8}}{\frac{1}{2}} = -\frac{3}{4} \quad \text{and} \quad \frac{a_3}{a_2} = \frac{\frac{1}{4}}{-\frac{3}{8}} = -\frac{2}{3},$$

we see that the given sequence is not geometric since the ratios are not equal.

27. This is a geometric progression with $a = 243$, and $r = 1/3$.

$$a_7 = 243(\tfrac{1}{3})^6 = \tfrac{1}{3}.$$

$$S_7 = \frac{243(1-(\frac{1}{3})^7)}{1-\frac{1}{3}} = 364\tfrac{1}{3}.$$

29. First, we compute

$$r = \frac{a_2}{a_1} = \frac{-3}{3} = -1.$$

Next, $a_{20} = -3(-1)^{19} = 3$

and so $S_{20} = \dfrac{-3[1-(-1)^{20}]}{[1-(-1)]} = 0$.

31. The population in five years is expected to be

$$200{,}000(1.08)^{6-1} = 200{,}000\,(1.08)^5 = 293{,}866.$$

33. The salary of a union member whose salary was $22,000 six years ago is given by the 7th term of a geometric progression whose first term is 22,000 and whose common ratio is 1.11. Thus

$$a_7 = (22{,}000)(1.11)^6 = 41{,}149.12, \text{ or } \$41{,}149.12.$$

35. With 8 percent raises per year, the employee would make

$$S_4 = 28{,}000\left[\frac{1-(1.08)^4}{1-1.08}\right] = 126{,}171.14,$$

or $126,171.14 over the next four years.
With $1500 raises per year, the employee would make

$$S_4 = \frac{4}{2}[2(28{,}000)+(4-1)1500] = 121.000$$

or $121,000 over the next four years. We conclude that the employee should choose the 8 percent per year raises.

37. *a*. During the sixth year, she will receive

$$a_6 = 10{,}000(1.15)^5 = 20{,}113.57, \text{ or } \$20{,}113.57.$$

b. The total amount of the six payments will be given by

$$S_6 = \frac{10{,}000[1-(1.15)^6]}{1-1.15} = 87{,}537.38, \text{ or } \$87{,}537.38.$$

39. The book value of the office equipment at the end of the eighth year is given by

$$V(8) = 150{,}000\left(1 - \frac{2}{10}\right)^8 = 25{,}165.82, \text{ or } \$25{,}165.82.$$

41. The book value of the restaurant equipment at the end of six years is given by

$$V(6) = 150{,}000(0.8)^6 = 39{,}321.60,$$

or $39,321.60. By the end of the sixth year, the equipment will have depreciated by

$$D(n) = 150{,}000 - 39{,}321.60 = 110{,}678.40, \text{ or } \$110{,}678.40.$$

CHAPTER 5, REVIEW EXERCISES, page 320

1. a. Here $P = 5000$, $r = 0.1$, and $m = 1$. Thus, $i = r = 0.1$ and $n = 4$. So

$$A = 5000(1.1)^4 = 7320.5, \text{ or } \$7320.50.$$

b. Here $m = 2$ so that $i = 0.1/2 = 0.05$ and $n = (4)(2) = 8$. So

$$A = 5000(1.05)^8 = 7387.28 \text{ or } \$7387.28.$$

c. Here $m = 4$, so that $i = 0.1/4 = 0.025$ and $n = (4)(4) = 16$. So

$$A = 5000(1.025)^{16} = 7{,}422.53, \text{ or } \$7422.53.$$

d. Here $m = 12$, so that $i = 0.1/12$ and $n = (4)(12) = 48$. So

$$A = 5000\left(1 + \frac{0.10}{12}\right)^{48} = 7446.77, \text{ or } \$7446.77.$$

3. a. The effective rate of interest is given by

$$r_{eff} = \left(1 + \frac{r}{m}\right)^m - 1 = (1 + 0.12) - 1 = 0.12, \text{ or 12 percent.}$$

b. The effective rate of interest is given by

$$r_{eff} = \left(1 + \frac{r}{m}\right)^m - 1 = \left(1 + \frac{0.12}{2}\right)^2 - 1 = 0.1236, \text{ or 12.36 percent.}$$

c. The effective rate of interest is given by

$$r_{eff} = \left(1 + \frac{r}{m}\right)^m - 1 = \left(1 + \frac{0.12}{4}\right)^2 - 1 = 0.125509, \text{ or 12.5509 percent.}$$

d. The effective rate of interest is given by

$$r_{eff} = \left(1+\frac{r}{m}\right)^m - 1 = \left(1+\frac{0.12}{12}\right)^2 - 1 \quad 0.126825, \text{ or } 12.6825 \text{ percent.}$$

5. The present value is given by

$$P = 41{,}413\left(1+\frac{0.065}{4}\right)^{-20} = 30{,}000.29, \quad \text{or approximately } \$30{,}000.$$

7.

$$S = 150\left[\frac{\left(1+\frac{0.08}{4}\right)^{28} - 1}{\frac{0.08}{4}}\right] = 5557.68, \text{ or } \$5557.68.$$

9. Using the formula for the present value of an annuity with $R = 250$, $n = 36$, $i = 0.09/12 = 0.0075$, we have

$$P = 250\left[\frac{1-(1.0075)^{-36}}{0.0075}\right] = 7861.70, \text{ or } \$7861.70.$$

11. Using the amortization formula with $P = 22{,}000$, $n = 36$, and $i = 0.085/12$, we find

$$R = \frac{22{,}000\left(\frac{0.085}{12}\right)}{1-\left(1+\frac{0.085}{12}\right)^{-36}} = 694.49, \text{ or } \$694.49.$$

13. Using the sinking fund formula with $S = 18{,}000$, $n = 48$, and $i = 0.06/12$, we have

$$R = \frac{\left(\frac{0.06}{12}\right)18{,}000}{\left(1+\frac{0.06}{12}\right)^{48} - 1} = 332.73, \text{ or } \$332.73.$$

15. We are asked to find r such that

$$\left(1+\frac{r}{365}\right)^{365}=\left(1+\frac{0.072}{12}\right)^{12}=1.074424168.$$

Then

$$1+\frac{r}{365}=(1.074424168)^{1/365}$$

$$\frac{r}{365}=(1.074424168)^{1/365}-1$$

and

$$r=365[1.074424168)^{1/365}-1]$$
$$=0.071791886, \text{ or approximately } 7.179 \text{ percent.}$$

17. Let a_n denote the sales during the nth year of operation. Then

$$\frac{a_{n+1}}{a_n}=1.14=r.$$

Therefore, the sales during the fourth year of operation were

$$a_4=a_1r^{n-1}=1{,}750{,}000(1.14)^{4-1}=2{,}592{,}702, \text{ or } \$2{,}592{,}702.$$

The total sales over the first four years of operation are given by

$$S_4=\frac{a(1-r^4)}{1-r}=\frac{1{,}750{,}000[1-(1.14)^4]}{1-(1.14)}=8{,}612{,}002.$$

or \$8,612,002.

19. Using the present value formula for compound interest, we have

$$P=A\left(1+\frac{r}{m}\right)^{-mt}=19{,}440.31\left(1+\frac{0.065}{12}\right)^{-12(4)}=15{,}000.00, \text{ or } \$15{,}000.$$

21. Using the sinking fund formula with $S=40{,}000$, $n=120$, and $i=0.08/12$, we find

$$R=\frac{\left(\frac{0.08}{12}\right)40{,}000}{\left(1+\frac{0.08}{12}\right)^{120}-1}=218.64, \text{ or } \$218.64.$$

23. Using the formula for the present value of an annuity, we see that the equivalent cash payment of Ms. Lemsky's auto lease is

$$P = 300\frac{1-\left(1+\frac{0.05}{12}\right)^{-48}}{\frac{0.05}{12}} = 13{,}026.89, \text{ or } \$13{,}026.89.$$

25. a. The monthly payment is given by

$$P = \frac{(120{,}000)(0.0075)}{1-(1+0.0075)^{-360}} = 965.55, \text{ or } \$965.55.$$

b. We can find the total interest payment by computing

$$360(965.55) - 120{,}000 = 227{,}598, \text{ or } \$227{,}598.$$

c. We first compute the present value of their remaining payments. Thus,

$$P = 965.55\left[\frac{1-(1+0.0075)^{-240}}{0.0075}\right] = 107{,}316.01.$$

or \$107,316.01. Then their equity is

$$150{,}000 - 107{,}316.01, \text{ or approximately } \$42{,}684.$$

27. Using the sinking fund formula with $S = 500{,}000$, $n = 20$, and $i = 0.10/4$, we find that the amount of each installment should be

$$R = \frac{\left(\frac{0.10}{4}\right)500{,}000}{\left(1+\frac{0.10}{4}\right)^{20} - 1} = 19{,}573.56, \text{ or } \$19{,}573.5629.$$

29. Using the amortization formula, we find that Mr. Baker's monthly payment will be

$$R = \frac{32{,}000\left(\frac{0.186}{12}\right)}{1-\left(1+\frac{0.186}{12}\right)^{-18}} \approx 205.09, \text{ or } \$205.09.$$

CHAPTER 6

EXERCISES 6.1, page 332

1. $\{x \mid x$ is gold medalist in the 1996 Summer Olympic Games$\}$

3. $\{x \mid x$ is an integer greater than 2 and less than 8$\}$

5. $\{2,3,4,5,6\}$

7. $\{-2\}$

9. a. True--the order in which the elements are listed is not important.
   b. False-- A is a set, not an element.

11. a. False. The empty set has no elements.
    b. False. 0 is an element and $\varnothing$ is a set.

13. True.

15. a. True. 2 belongs to A.
    b. False. For example, 5 belongs to A but $5 \notin \{2,4,6\}$.

17. a. and b.

19. a. $\varnothing$, $\{1\}$, $\{2\}$, $\{1,2\}$

    b. $\varnothing$, $\{1\}$, $\{2\}$, $\{3\}$, $\{1,2\}$, $\{1,3\}$, $\{2,3\}$, $\{1,2,3\}$

    c. $\varnothing$, $\{1\}$, $\{2\}$, $\{3\}$, $\{4\}$, $\{1,2\}$, $\{1,3\}$, $\{1,4\}$, $\{2,3\}$, $\{2,4\}$, $\{3,4\}$, $\{1,2,3\}$, $\{1,2,4\}$, $\{2,3,4\}$, $\{1,3,4\}$, $\{1,2,3,4\}$

21. $\{1, 2, 3, 4, 6, 8, 10\}$

23. {Jill, John, Jack, Susan, Sharon}

25. a.

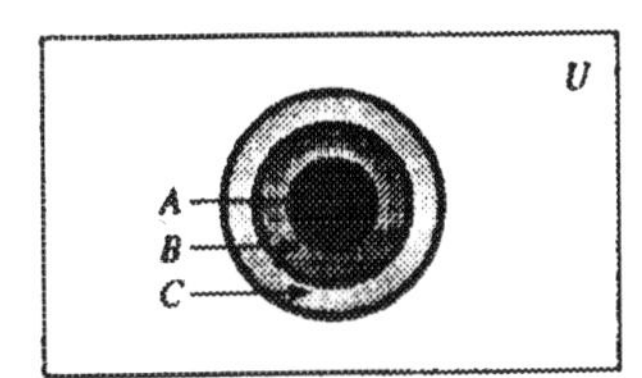

b.

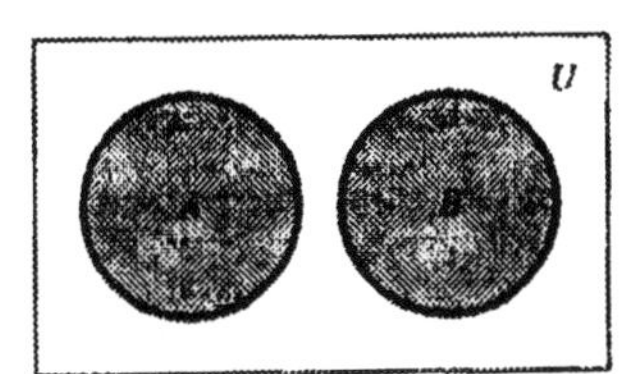

c.

27. a.

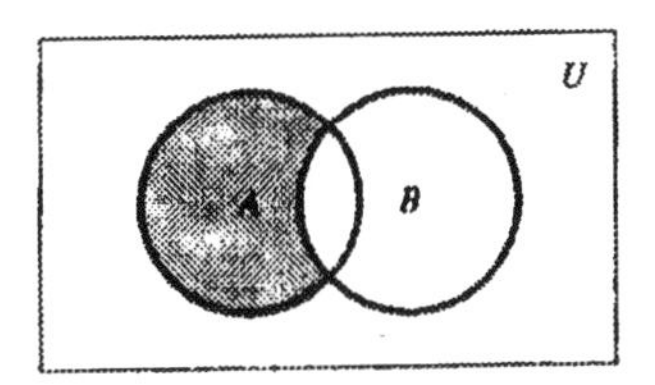

b.

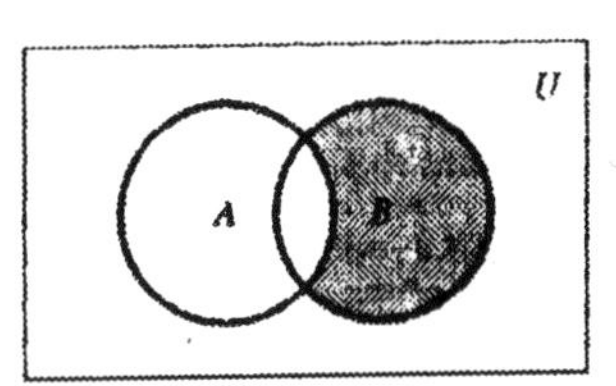

29. a.

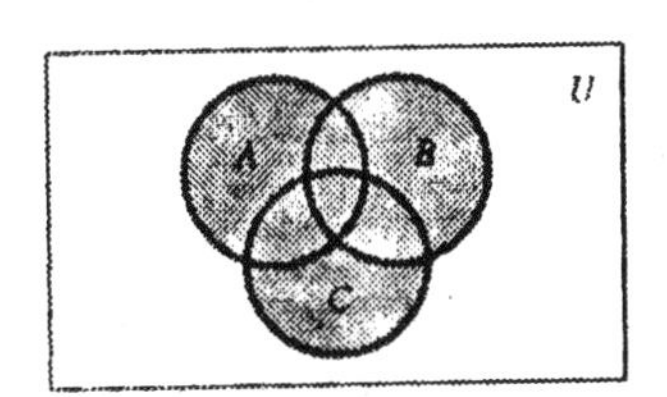

b.

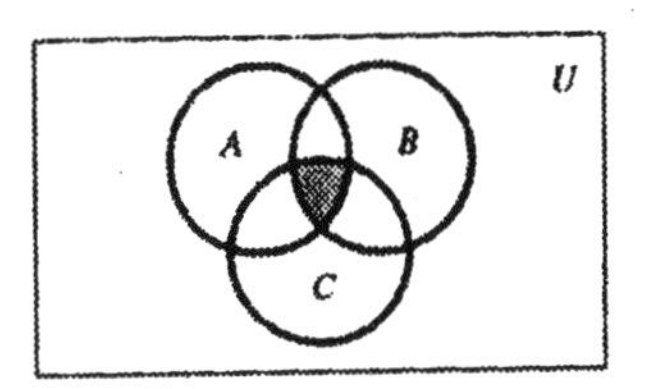

31. a. b.

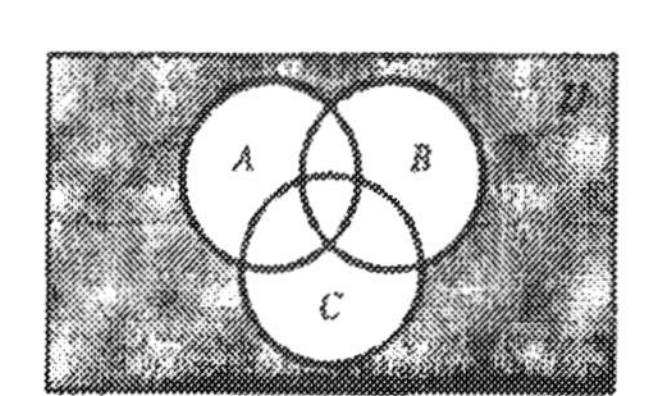

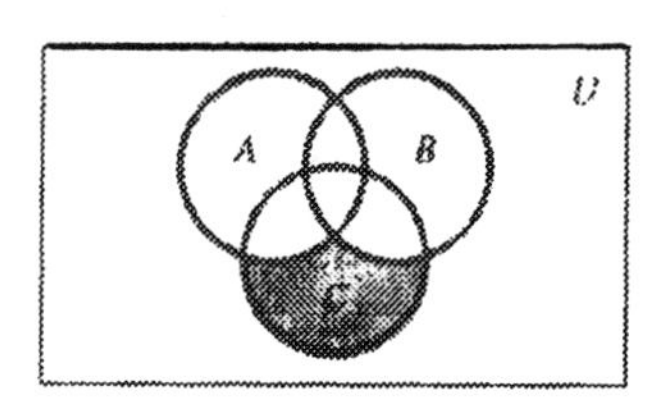

33. a. $A^C = \{2,4,6,8,10\}$
 b. $B \cup C = \{2,4,6,8,10\} \cup \{1,2,4,5,8,9\} = \{1,2,4,5,6,8,9,10\}$
 c. $C \cup C' = U = \{1,2,3,4,5,6,7,8,9,10\}$

35. a. $(A \cap B) \cup C = C = \{1,2,4,5,8,9\}$
 b. $(A \cup B \cup C)^C = \varnothing$
 c. $(A \cap B \cap C)^C = = U = \{1,2,3,4,5,6,7,8,9,10\}$

37. a. The sets are not disjoint. 4 is an element of both sets.
 b. The sets are disjoint as they have no common elements.

39. a. The set of all employees at the Universal Life Insurance Company who do not drink tea.
 b. The set of all employees at the Universal Life Insurance Company who do not drink coffee.

41. a. The set of all employees at the Universal Life Insurance Company who drink tea but not coffee.
 b. The set of all employees at the Universal Life Insurance Company who drink coffee but not tea.

43. a. The set of all employees at the hospital who are not doctors
 b. The set of all employees at the hospital who are not nurses

45. a. The set of all employees at the hospital who are female doctors.
 b. The set of all employees at the hospital who are both doctors and administrators.

47. a. $D \cap F$ b. $R \cap F^C \cap L^C$

49. a. B^C b. $A \cap B$ c. $A \cap B \cap C^C$

51. a. Region 1: $A \cap B \cap C$ is the set of tourists who used all three modes of transportation over a 1-week period in London.
b. Regions 1 and 4: $A \cap C$ is the set of tourists who have taken the underground and a bus over a 1-week period in London.
c. Regions 4, 5, 7, and 8: B^c is the set of tourists who have not taken a cab over a 1-week period in London.

53. $A \subset A \cup B$ $B \subset A \cup B$

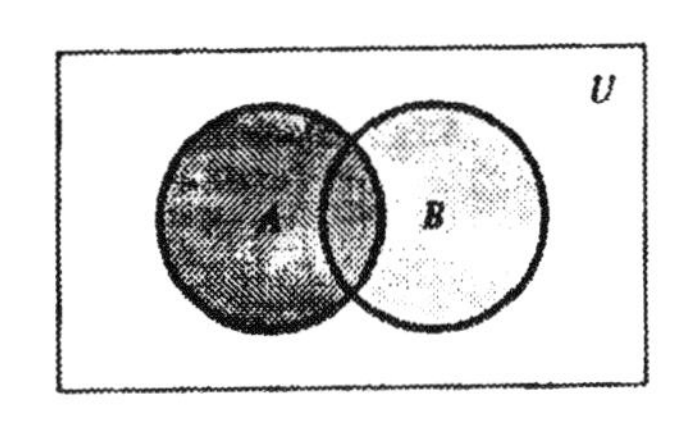

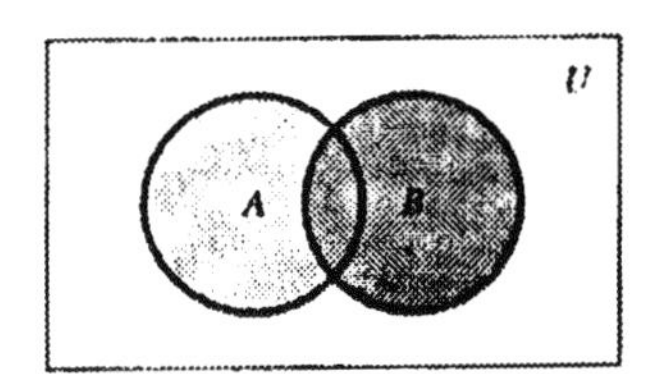

55. $A \cup (B \cup C) = (A \cup B) \cup C$

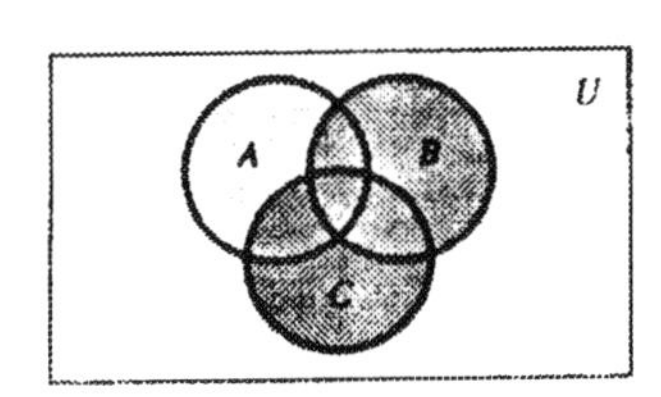

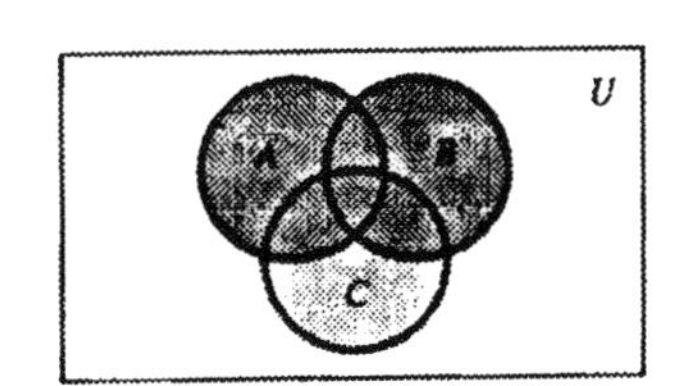

57. $A \cap (B \cup C) = (A \cap B) \cup (A \cap C)$

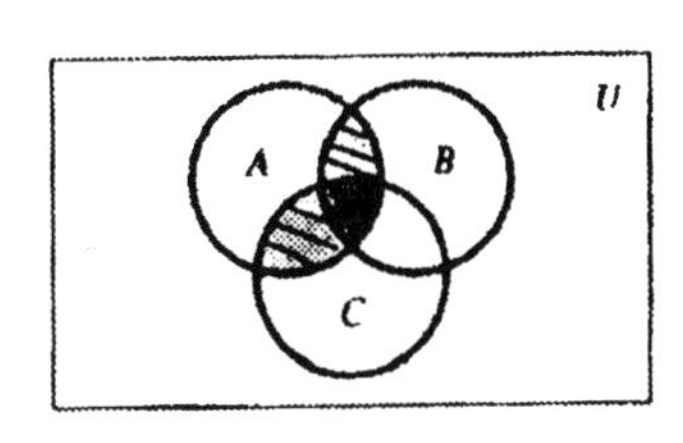

=

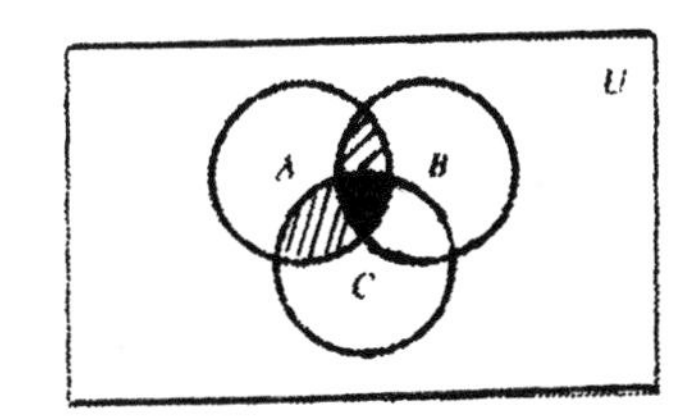

59. a. $A \cup (B \cup C) = \{1,3,5,7,9\} \cup (\{1,2,4,7,8\} \cup \{2,4,6,8\})$
$= \{1\ 3,5,7,9\} \cup \{1,2,4,6,7,8\}$
$= \{1,2,3,4,5,6,7,8,9\}$

$$A \cup (B \cup C) = (\{1,3,5,7,9\} \cup (\{1,2,4,7,8\}) \cup \{2,4,6,8\})$$
$$= \{1,2,3,4,5,7,8,9\} \cup \{2,4,6,8\}$$
$$= \{1,2,3,4,5,6,7,8,9\}$$

b. $A \cap (B \cap C) = \{1,3,5,7,9\} \cap (\{1,2,4,7,8\} \cap \{2,4,6,8\})$
$$= \{1,3,5,7,9\} \cap (\{2,4,8\}$$
$$= \varnothing$$

$$(A \cap B) \cap C = (\{1,3,5,7,9\} \cap \{1,2,4,7,8\}) \cap \{2,4,6,8\}$$
$$= \{1,7\} \cap \{2,4,6,8\}$$
$$= \varnothing.$$

61. a. r, u, v, w, x, y b. v, r

63. a. t, y, s b. t, s, w, x, z

65. $A \subset C$

EXERCISES 6.2, page 339

1. $A \cup B = \{a,e,g,h,i,k,l,m,o,u\}$, and so $n(A \cup B) = 10$. Next, $n(A) + n(B) = 5 + 5 = 10$.

3. a. $A = \{2,4,6,8\}$ and $n(A) = 4$.
   b. $B = \{6,7,8,9,10\}$ and $n(B) = 5$
   c. $A \cup B = \{2,4,6,7,8,9,10\}$ and $n(A \cup B) = 7$.
   d. $A \cap B = \{6,8\}$ and $n(A \cap B) = 2$.

5. $A \cup B = \{a,e,i,o,u\} \cup \{b,d,e,o,u\} = \{a,b,d,e,i,o,u\}$ and $n(A \cup B) = 7$
   $A = \{a,e,i,o,u\}$ so $n(A) = 5$, $B = \{b,d,e,o,u\}$ and $n(B) = 5$, and
   $A \cap B = \{a,e,i,o,u\} \cap \{b,d,e,o,u\} = \{e,o,u\}$ so that $n(A \cap B) = 3$.
   Therefore, $n(A \cup B) = n(A) + n(B) - n(A \cap B) = 5 + 5 - 3 = 7$.

7. $n(A \cap B) = n(A) + n(B) - n(A \cup B) = 10 + 8 - 15 = 3$.

9. Refer to the Venn diagram at the right.

   a. $n(A^C \cap B) = 40$.

   b. $n(B^C) = 60 + 60 = 120$.

   c. $n(A^C \cap B^C) = n(A \cup B)^C = 60$.

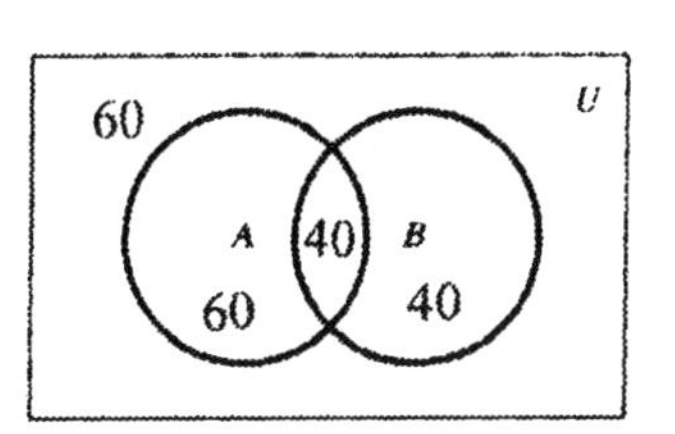

11. $n(A \cup B) = n(A) + n(B) - n(A \cap B)$ so

   $n(A) = n(A \cup B) + n(A \cap B) - n(B)$

   $= 14 + 3 - 6 = 11$.

13. $n(A \cap B \cap C)$

   $= n(A \cup B \cup C) - n(A) - n(B) - n(C) + n(A \cap B) + n(A \cap C) + n(B \cap C)$

   $= 31 - 16 - 16 - 14 + 6 + 5 + 6 = 2$.

15. Let $A = \{x \mid x$ is a subscriber to the Sunday *L.A. Times*$\}$. Then, we are given that $n(A) = 900$, $n(A \cap B) = 500$, and $n(A \cup B) = 1000$. Refer to the Venn diagram at the right. Since

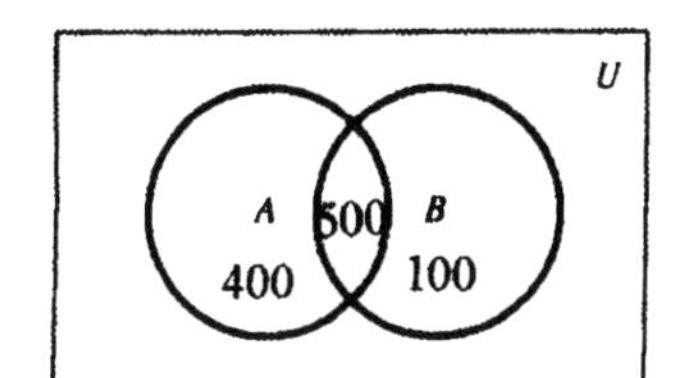

   $n(A \cup B) = n(A) + n(B) - n(A \cap B)$,

   we see that

   $n(B) = n(A \cup B) + n(A \cap B) - n(A)$

   $= 1000 + 500 - 900 = 600$

   Next, $(B \cap A^C) = n(B) - n(A \cap B) = 600 - 500 = 100$.

17. Let A denote the set of prisoners in the Wilton County Jail who were accused of a felony and B the set of prisoners in that jail who were accused of a misdemeanor. Then we are given that

   $n(A \cup B) = 190$

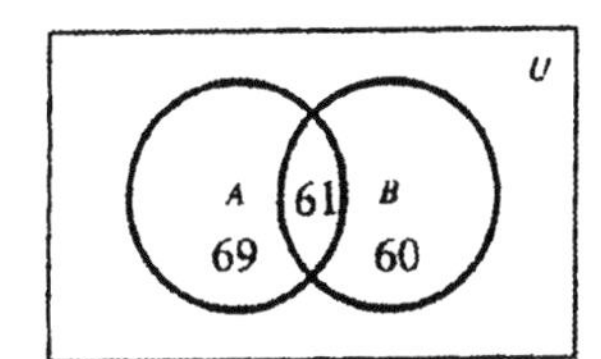

   Refer to the diagram at the right.

   Then the number of prisoners who were accused of both a felony and a misdemeanor is given by $(A \cap B) = n(A) + n(B) - n(A \cup B) = 130 + 121 - 190 = 61$.

19. Let U denote the set of all customers surveyed, and let

   $A = \{x \in U \mid x$ buys brand $A\}$

   $B = \{x \in U \mid x$ buys brand $B\}$.

Then $n(U) = 120$, $n(A) = 80$,
$n(B) = 68$, and $n(A \cap B) = 42$.

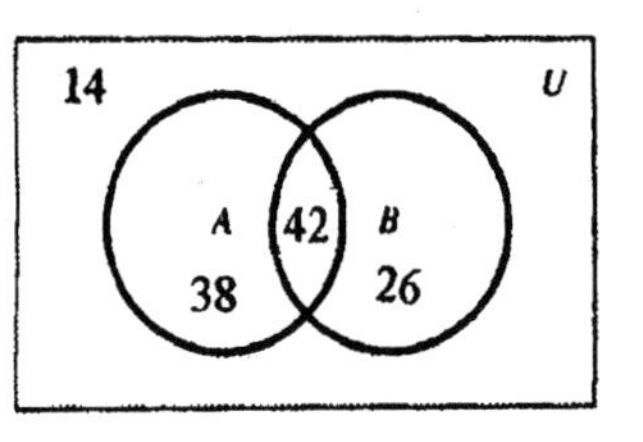

Refer to the diagram at the right.

a. The number of customers who buy at least one of these brands is

$$n(A \cup B) = 80 + 68 - 42 = 106.$$

b. The number who buy exactly one of these brands is

$$n(A \cap B^C) + n(A^C \cap B) = 38 + 26 = 64$$

c. The number who buy only brand A is $n(A \cap B^C) = 38$.

d. The number who buy none of these brands is

$$n[(A \cup B)^C] = 120 - 106 = 14.$$

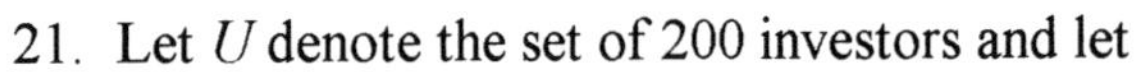

21. Let U denote the set of 200 investors and let
$A = \{x \in U \mid x \text{ uses a discount broker}\}$
$B = \{x \in U \mid x \text{ uses a full-service broker}\}$.

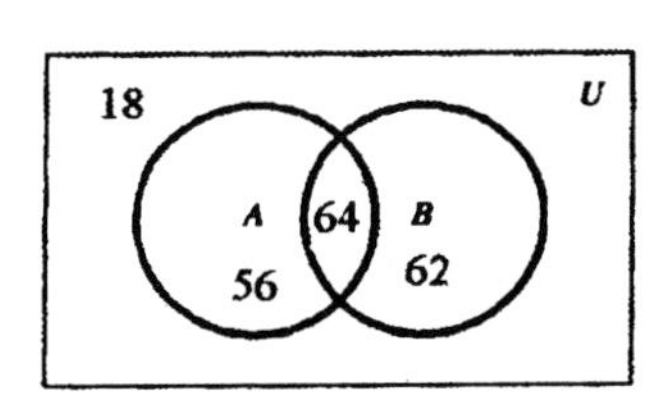

Refer to the diagram at the right.

a. The number of investors who use at least one kind of broker is

$$n(A \cup B) = n(A) + n(B) - n(A \cap B)$$
$$= 120 + 126 - 64 = 182.$$

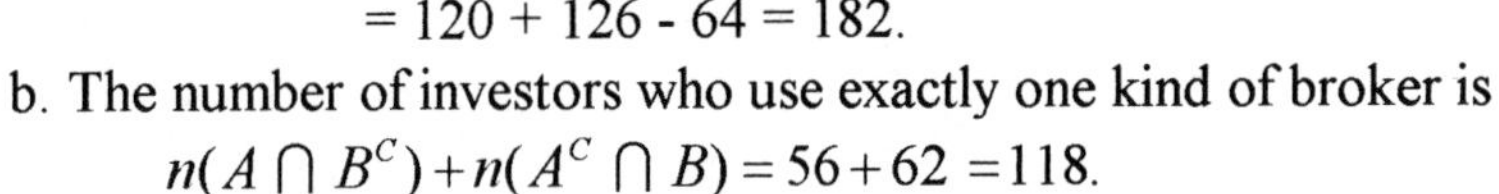

b. The number of investors who use exactly one kind of broker is

$$n(A \cap B^C) + n(A^C \cap B) = 56 + 62 = 118.$$

c. The number of investors who use only discount brokers is

$$n(A \cap B^C) = 56.$$

d. The number of investors who don't use a broker is

$$n(A \cup B)^C = n(U) - n(A \cup B) = 200 - 182 = 18.$$

In Exercises 23 - 25, refer to the figure that follows.

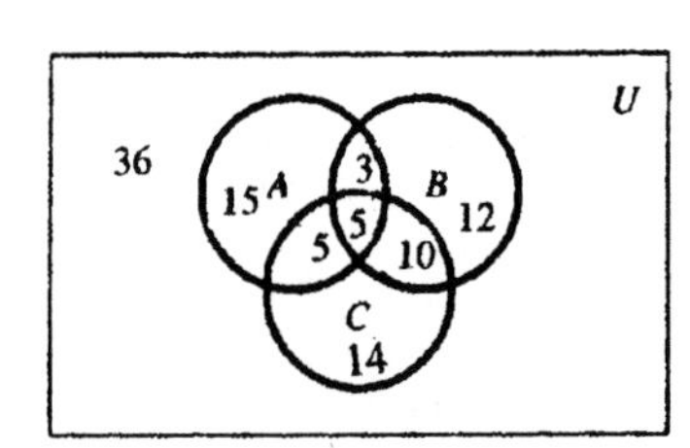

23. a. $n[A \cap (B \cup C)] = 13$ b. $n[A \cap (B \cup C)^C] = 15$

25. a. $n[A \cup (B \cap C)] = 38$
b. $n(A^C \cap B^C \cap C^C)^C = n[(A \cup B \cup C)] = 64.$

27. Let U denote the set of all economists surveyed, and let
$$A = \{ x \in U \mid x \text{ had lowered his estimate of the consumer inflation rate} \}$$
$$B = \{ x \in U \mid x \text{ had raised his estimate of the } GDP \text{ growth rate} \}.$$
Refer to the diagram at the right.
Then $n(U) = 10$, $n(A) = 7$, $n(B) = 8$, and $n(A \cap B^C) = 2$. Then the number of economists who had both lowered their estimate of the consumer inflation rate and raised their estimate of the *GDP* rate is given by
$$n(A \cap B) = 5.$$

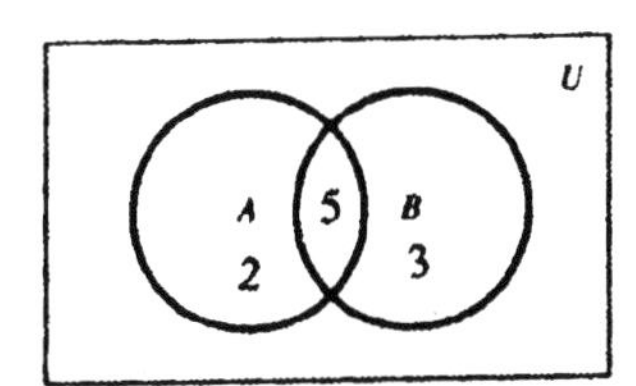

29. Let U denote the set of 100 college students who were surveyed and let
$$A = \{ x \in U \mid x \text{ is a student who reads } Time \text{ magazine} \}$$
$$B = \{ x \in U \mid x \text{ is a student who reads } Newsweek \text{ magazine} \}$$
and
$$C = \{ x \in U \mid x \text{ is a student who reads } U.S. \text{ } News \text{ and } World \text{ } Report \text{ magazine} \}$$
Then $n(A) = 40$, $n(B) = 30$, $n(C) = 25$, $n(A \cap B) = 15$,
$n(A \cap C) = 12$, $n(B \cap C) = 10$, and $n(A \cap B \cap C) = 4$.

Refer to the diagram on the right.

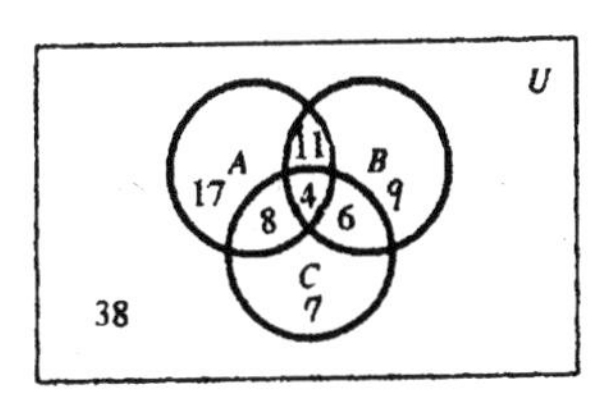

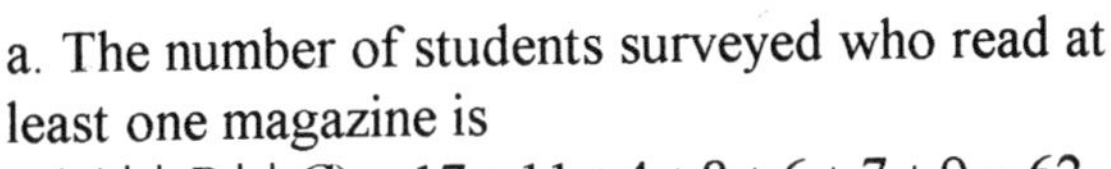

a. The number of students surveyed who read at least one magazine is
$n(A \cup B \cup C) = 17 + 11 + 4 + 8 + 6 + 7 + 9 = 62$

b. The number of students surveyed who read exactly one magazine is
$$n(A \cap B^C \cap C^C) + n(A^C \cap B \cap C^C) + n(A^C \cap B^C \cap C)$$
$$= 17 + 9 + 7 = 33.$$

c. The number of students surveyed who read exactly two magazines is

$$n(A \cap B \cap C^C) + n(A^C \cap B \cap C) + n(A \cap B^C \cap C)$$
$$= 11 + 6 + 8 = 25.$$

d. The number of students surveyed who did not read any of these magazines is

$$n(A \cup B \cup C)^C = 100 - 62 = 38.$$

31. Let U denote the set of all customers surveyed, and let

$$A = \{ x \in U \mid x \text{ buys brand } A \}$$
$$B = \{ x \in U \mid x \text{ buys brand } B \}.$$
$$C = \{ x \in U \mid x \text{ buys brand } C \}.$$

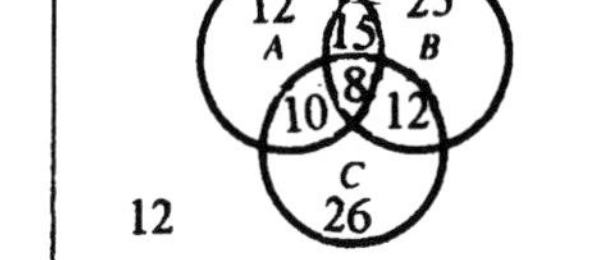

Refer to the figure at the right.
Then

$n(U) = 120$, $n(A \cap B \cap C^C) = 15$,
$n(A^C \cap B \cap C^C) = 25$, $n(A^C \cap B^C \cap C) = 26$,
$n(A \cap B \cap C^C) = 15$, $n(A \cap B^C \cap C) = 10$,
$n(A^C \cap B \cap C) = 12$, and $n(A \cap B \cap C) = 8$.

a. The number of customers who buy at least one of these brands is
$n(A \cup B \cup C) = 12 + 15 + 25 + 12 + 8 + 10 + 26 = 108.$

b. The number who buy labels A and B but not C is
$n(A \cap B \cap C^C) = 15$

c. The number who buy brand A is
$n(A) = 12 + 10 + 15 + 8 = 45.$

d. The number who buy none of these brands is
$n[(A \cup B \cup C)^C] = 120 - 108 = 12.$

EXERCISES 6.3, page 347

1. By the multiplication principle, the number of rates is given by (4)(3) = 12.

3. By the multiplication principle, the number of ways that a blackjack hand can be dealt is (4)(16) = 64.

5. By the multiplication principle, she can create (2)(4)(3) = 24 different ensembles.

7. The number of paths is $2 \times 4 \times 3$, or 24.

9. By the multiplication principle, we see that the number of ways a health-care plan can be selected is $(10)(3)(2) = 60$.

11. The number of different responses is

$$\underbrace{(5)(5)\ldots(5)}_{50 \text{ terms}} = 5^{50}.$$

13. The number of selections is given by $(2)(5)(3)$, or 30 selections.

15. The number of different selections is $(10)(10)(10)(10) - 10 = 10000 - 10 = 9990$.

17. a. The number of possible classifications is $(2)(3)(6) = 36$.

b.

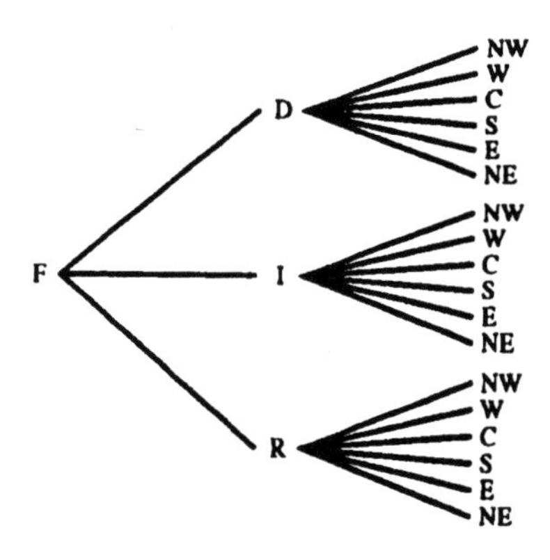

19. If every question is answered, there are 2^{10}, or 1024, ways.In the second case, there are 3 ways to answer each question, and so we have 3^{10}, or 59,049, ways.

21. The number of ways the first, second and third prizes can be awarded is

$$(15)(14)(13) = 2730.$$

23. The number of ways in which the nine symbols on the wheels can appear in the window slot is $(9)(9)(9)$, or 729. The number of ways in which the eight symbols other than the "lucky dollar" can appear in the window slot is $(8)(8)(8)$ or 512. Therefore, the number of ways in which the "lucky dollars" can appear in the window slot is $729 - 512$, or 217.

EXERCISES 6.4, page 360

1. $3(5!) = 3(5)(4)(3)(2)(1) = 360.$

3. $\frac{5!}{2!3!} = 5(2) = 10.$

5. $P(5,5) = \frac{5!}{(5-2)!} = \frac{5!}{0!} = 120$

7. $P(5,2) = \frac{5!}{(5-2)!} = \frac{5!}{3!} = (5)(4) = 20$

9. $P(n,1) = \frac{n!}{(n-1)!} = n$

11. $C(6,6) = \frac{6!}{6!0!} = 1$

13. $C(7,4) = \frac{7!}{4!3!} = \frac{7 \cdot 6 \cdot 5}{3 \cdot 2} = 35$

15. $C(5,0) = \frac{5!}{5!0!} = 1$

17. $C(9,6) = \frac{9!}{3!6!} = \frac{9 \cdot 8 \cdot 7}{3 \cdot 2} = 84$

19. $C(n,2) = \frac{n!}{(n-2)!2!} = \frac{n(n-1)}{2}$

21. $P(n,n-2) = \frac{n!}{(n-(n-2))!} = \frac{n!}{(n-n+2)!} = \frac{n!}{2}$

23. Order is important here since the word "*glacier*" is different from "*reicalg*", so this is a permutation.

25. Order is not important here. Therefore, we are dealing with a combination. If we consider a sample of three record-o-phones of which one is defective, it does not matter whether the defective record-o-phone is the first member of our sample, the second member of our sample, or the third member of our sample. The net result is a sample of three record-o-phones of which one is defective.

27. The order is important here. Therefore, we are dealing with a permutation. Consider, for example, 9 books on a library shelf. Each of the 9 books would have a call number, and the books would be placed in order of their call numbers; that is, a call number of 902 would come before a call number of 910.

29. The order is not important here, and consequently we are dealing with a combination. It would not matter if the hand *Q Q Q 5 5* were dealt or the hand *5 5 Q Q Q*. In each case the hand would consist of three queens and a pair.

31. The number of 4-letter permutations is $P(4,4)=\dfrac{4!}{0!}=4\cdot3\cdot2\cdot1=24$.

33. The number of seating arrangements is $P(4,4)=\dfrac{4!}{0!}=24$.

35. The number of different batting orders is $P(9,9)=\dfrac{9!}{0!}=362{,}880$.

37. The number of different ways the 3 candidates can be selected is

$$C(12,3)=\frac{12!}{9!3!}=\frac{12\cdot11\cdot10}{3\cdot2\cdot1}=220.$$

39. There are 10 letters in the word *ANTARCTICA*, 3*A*s, 1*N*, 2*T*s, 1*R*, 2*C*s, and 1*I*. Therefore, we use the formula for the permutation of n objects, not all distinct:

$$\frac{n!}{n_1!n_2!\cdots n_r!}=\frac{10!}{3!2!2!}=151{,}200$$

41. The number of ways the 3 sites can be selected is

$$C(12,3)=\frac{12!}{9!3!}=\frac{12\cdot11\cdot10}{3\cdot2\cdot1}=220$$

43. The number of ways in which the sample of 3 transistors can be selected is

$$C(100,3)=\frac{100!}{97!3!}=\frac{100\cdot99\cdot98}{3\cdot2\cdot1}=161{,}700.$$

45. In this case order is important, as it makes a difference whether a commercial is shown first, last, or in between. The number of ways that the director can schedule the commercials is given by

$$P(6,6)=6!=720.$$

47. The inquiries can be directed in

$$P(12,6)=\frac{12!}{6!}=12\cdot11\cdot10\cdot9\cdot8\cdot7=665{,}280$$

or 665,280 ways.

49. a. The ten books can be arranged in

$$P(10,10)=10!=3{,}628{,}800 \text{ ways.}$$

b. If books on the same subject are placed together, then they can be arranged on the shelf

$$P(3,3) \times P(4,4) \times P(3,3) \times P(3,3) = 5184 \text{ ways.}$$

Here we have computed the number of ways the mathematics books can be arranged times the number of ways the social science books can be arranged times the number of ways the biology books can be arranged times the number of ways the 3 sets of books can be arranged.

51. Notice that order is certainly important here.

a. The number of ways that the 20 featured items can be arranged is given by

$$P(20,20) = 20! = 2.43 \times 10^{18}.$$

b. If items from the same department must appear in the same row, then the number of ways they can be arranged on the page is

Number of ways of arranging the rows x Number of ways of arranging the items in each of the 5 rows

$$P(5,5) \bullet P(4,4) \times P(4,4) \times P(4,4) \times P(4,4) \times P(4,4)$$
$$= 5! \times (4!)^5 = 955{,}514{,}880.$$

53. The number of ways is given by

$$2\{C(2,2) + [C(3,2) - C(2,2)]\} = 2[1 + (3 - 1)] = 2 \times 3 = 6$$

(number of players)[number of ways to win in exactly 2 sets - number of ways to win in exactly 3 sets]

55. The number of ways the measure can be passed is

$$C(3,3) \times [C(8,6) + C(8,7) + C(8,8)] = 37.$$

Here three of the three permanent members must vote for passage of the bill and this can be done in $C(3,3) = 1$ way. Of the 8 nonpermanent members who are voting 6 can vote for passage of the bill, or 7 can vote for passage, or 8 can vote for passage. Therefore, there are

$$C(8,6) + C(8,7) + C(8,8) = 37 \text{ ways}$$

that the nonpermanent members can vote to ensure passage of the measure. This gives $1 \times 37 = 37$ ways that the members can vote so that the bill is passed.

57. a. If no preference is given to any student, then the number of ways of awarding the 3 teaching assistantships is

$$C(12,3) = \frac{12!}{3!9!} = 220.$$

b. If it is stipulated that one particular student receive one of the assistantships, then the remaining two assistantships must be awarded to two of the remaining 11 students. Thus, the number of ways is

$$C(11,2) = \frac{11!}{2!9!} = 55.$$

c. If at least one woman is to be awarded one of the assistantships, and the group of students consists of seven men and five women, then the number of ways the assistantships can be awarded is given by

$$C(5,1) \times C(7,2) \; + \; C(5,2) \times C(7,1) \; + \; C(5,3)$$
$$= \frac{5!}{4!1!} \cdot \frac{7!}{5!2!} + \frac{5!}{3!2!} \cdot \frac{7!}{6!1!} + \frac{5!}{3!2!} = 105 + 70 + 10 = 185.$$

59. The number of ways of awarding the 7 contracts to 3 different firms is given by

$$P(7,3) = \frac{7!}{4!} = 210.$$

The number of ways of awarding the 3 contracts to 2 different firms is

$$C(7,2) \times P(3,2) = 126.$$ (First pick the two firms, and then award the 3 contracts.)

Therefore, the number of ways the contracts can be awarded if no firm is to receive more than 2 contracts is given by

$$210 + 126 = 336.$$

61. The number of different curricula that are available for the student's consideration is given by

$$C(5,1) \times C(3,1) \times C(6,2) \times C(4,1) \; + \; C(5,1) \times C(3,1) \times C(6,2) \times C(3,1)$$
$$= \frac{5!}{4!1!} \cdot \frac{3!}{2!1!} \cdot \frac{6!}{4!2!} \cdot \frac{4!}{3!1!} + \frac{5!}{4!1!} \cdot \frac{3!}{2!1!} \cdot \frac{6!}{4!2!} \cdot \frac{3!}{2!1!}$$
$$= (5)(3)(15)(4) + (5)(3)(15)(3) = 900 + 675 = 1575.$$

63. The number of ways is given by

$$P(10,2) + P(10,1) + P(10,0) = 45 + 10 + 1 = 56$$

or

$$C(10,8) + C(10,9) + C(10,10) = 56.$$

65. The number of ways of dealing a straight flush (5 cards in sequence in the same suit is given by

the number of ways of selecting 5 cards in sequence in the same suit × the number of ways of selecting a suit

$$10 \quad \bullet \quad C(4,1) = \quad 40.$$

67. The number of ways of dealing a flush (5 cards in one suit that are not all in sequence) is given by

the number of ways of selecting 5 cards in one suit - the number of ways of selecting 5 cards in one suit in sequence

$$4C(13,5) \quad - \quad 4(10)$$
$$= 5148 - 40 = 5108.$$

69. The number of ways of dealing a full house (3 of a kind and a pair) is given by

the number of ways of picking 3 of a kind from a given rank × the number of ways of picking a pair from the 12 remaining ranks

$$13C(4,3) \bullet 12C(4,2)$$
$$= 13(4) \bullet (12)(6) = 3744.$$

71. The bus will travel a total of 6 blocks. Each route must include 2 blocks running north and south and 4 blocks running east and west. To compute the total number of possible routes, it suffices to compute the number of ways the 2 blocks running north and south can be selected from the six blocks. Thus,

$$C(6,2) = \frac{6!}{2!4!} = 15.$$

73. The number of ways that the quorum can be formed is given by

$$C(12,6) + C(12,7) + C(12,8) + C(12,9) + C(12,10) + C(12,11) + C(12,12)$$
$$= \frac{12!}{6!6!} + \frac{12!}{7!5!} + \frac{12!}{8!4!} + \frac{12!}{9!3!} + \frac{12!}{10!2!} + \frac{12!}{11!1!} + \frac{12!}{12!0!}$$
$$= 924 + 792 + 495 + 220 + 66 + 12 + 1 = 2510.$$

75. Using the formula given in Exercise 74, we see that the number of ways of seating the 5 commentators at a round table is $(5 - 1)! = 4! = 24$.

77. The number of possible corner points is $C(8,3) = \frac{8!}{5!3!} = 56$.

USING TECHNOLOGY EXERCISES, page 365

1. $1.307674368 \times 10^{12}$

3. $2.56094948229 \times 10^{16}$

5. 674,274,182,400

7. 133,784,560

9. 4,656,960

11. Using the multiplication principle, the number of 10-question exams she can set is given by

$$C(25,3) \times C(40,5) \times C(30,2) = 658,337,004,000.$$

CHAPTER 6, REVIEW EXERCISES, page 367

1. $\{3\}$. The set consists of all solutions to the equation $3x - 2 = 7$.

3. $\{4,6,8,10\}$

5. Yes.

7. Yes.

9.

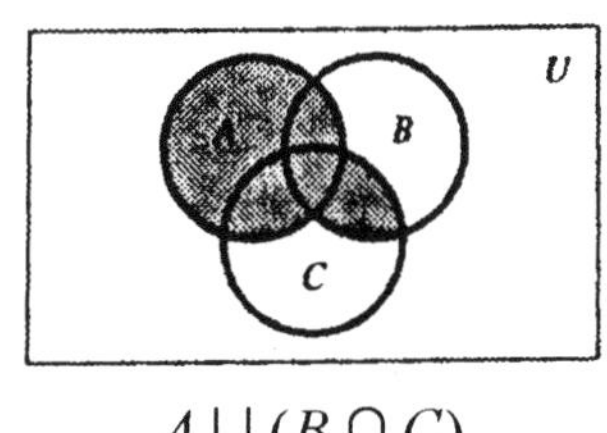

$A \cup (B \cap C)$

11.

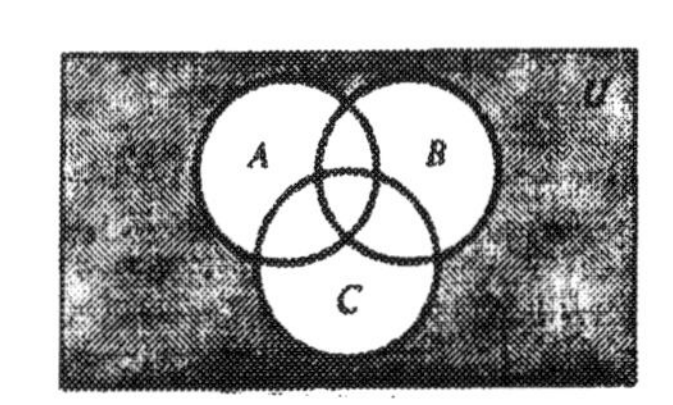

$A^c \cap B^c \cap C^c$

13. $A \cup (B \cup C) = \{a,b\} \cup [\{b,c,d\} \cup \{a,d,e\}]$
$= \{a,b\} \cup \{a,b,c,d,e\} = \{a,b,c,d,e\}.$
$(A \cup B) \cup C = [\{a,b\} \cup \{b,c,d\}] \cup \{a,d,e\}$
$= \{a,b,c,d\} \cup \{a,d,e\} = \{a,b,c,d,e\}.$

15. $A \cap (B \cup C) = \{a,b\} \cap [\{b,c,d\} \cup \{a,d,e\}] = \{a,b\} \cap \{a,b,c,d,e\} = \{a,b\}.$
$(A \cap B) \cup (A \cap C) = [\{a,b\} \cap \{b,c,d\}] \cup [\{a,b\} \cap \{a,d,e\} = \{b\} \cup \{a\} = \{a,b\}.$

17. The set of all participants in a consumer behavior survey who both avoided buying a product because it is not recyclable and boycotted a company's products because of its record on the environment.

19. The set of all participants in a consumer behavior survey who both did not use cloth diapers rather than disposable diapers and voluntarily recycled their garbage.

In Exercises 21-25, refer to the following Venn diagram.

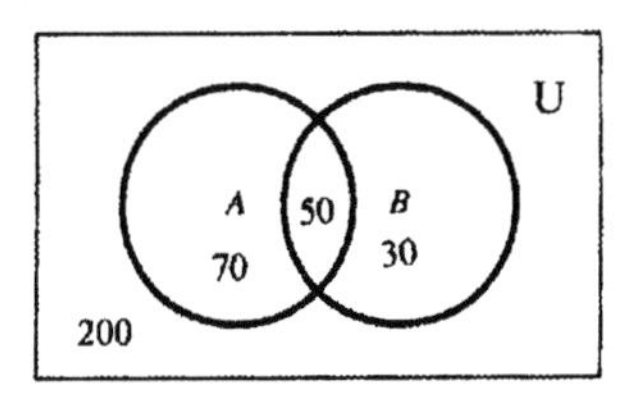

21. $n(A \cup B) = n(A) + n(B) - n(A \cap B) = 120 + 80 - 50 = 150.$

23. $n(B^c) = n(U) - n(B) = 350 - 80 = 270.$

25. $n(A \cap B^c) = n(A) - n(A \cap B) = 120 - 50 = 70.$

27. $C(20,18) = \dfrac{20!}{18!2!} = 190.$

29. $C(5,3) \cdot P(4,2) = \dfrac{5!}{3!2!} \cdot \dfrac{4!}{2!} = 10 \cdot 12 = 120.$

31. Let U denote the set of 5 major cards, and let

$A = \{x \in U \mid x \text{ offered cash advances}\}$

$B = \{x \in U \mid x \text{ offered extended payments for all goods and services purchased}\}$

$C = \{x \in U \mid x \text{ required an annual fee that was less than \$35}\}$

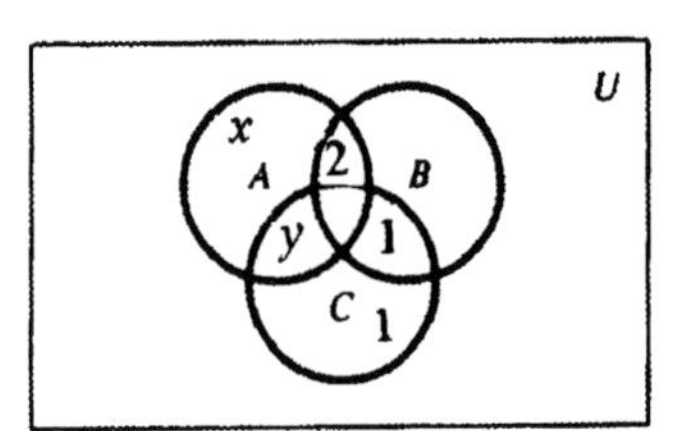

Thus, $n(A) = 3, n(B) = 3, n(C) = 2$

$n(A \cap B) = 2, n(B \cap C) = 1$

and $n(A \cap B \cap C) = 0$

Using the Venn diagram on the right, we have

$$x + y + 2 = 3$$
$$y + 2 = 2$$

Solving the system, we find $x = 1$, and $y = 0$. Therefore, the number of cards that offer cash advances and have an annual fee that is less than \$35 is given by

$$n(A \cap C) = y = 0.$$

33. The number of ways the compact discs can be arranged on a shelf is

$$P(6,6) = 6! = 720 \text{ ways.}$$

35. The number of possible outcomes is $(6)(4)(5)(6) = 720$.

37. a. Since there is repetition of the letters C, I, and N, we use the formula for the permutation of n objects, not all distinct, with $n = 10$, $n_1 = 2$, $n_2 = 3$, and $n_3 = 3$. Then the number of permutations that can be formed is given by

$$\frac{10!}{2!3!3!} = 50{,}400.$$

b. Here, again, we use the formula for the permutation of n objects, not all distinct, this time with $n = 8$, $n_1 = 2$, $n_2 = 2$, and $n_3 = 2$. Then the number of permutations is given by

$$\frac{8!}{2!2!2!} = 5040.$$

39. a. The number of ways the 7 students can be assigned to seats is

$$P(7,7) = 7! = 5040.$$

b. The number of ways 2 specified students can be seated next to each other is

$$2(6) = 12.$$

(Think of seven numbered seats. Then the students can be seated in seats 1-2, or 2-3, or 3-4, or 4-5, or 5-6, or 6-7. Since there are 6 such possibilities and the pair of students can be seated in 2 ways, we conclude that there are 2(6) possible arrangements.)Then the remaining 5 students can be seated in $P(5,5) = 5!$ ways.Therefore, the number of ways the 7 students can be seated if two specified students sit next to each other is

$$2(6)5! = 1440.$$

Finally, the number of ways the students can be seated if the two students do not sit next to each other is

$$P(7,7) - 2(6)5! = 5040 - 1440 = 3600.$$

41. a. The number of samples that can be selected is

$$C(15,4) = \frac{15!}{4!11!} = 1365.$$

b. There are

$$C(10,4) = \frac{10!}{4!6!} = 210$$

ways of selecting 4 balls none of which are white. Therefore, there are

$$1365 - 210,$$

or 1155 ways of selecting 4 balls of which at least one is white.

CHAPTER 7

EXERCISES 7.1, page 379

1. $E \cup F = \{a,b,d,f\}$; $E \cap F = \{a\}$.

3. $F^C = \{b,c,e\}$; $E \cap G^C = \{a,b\} \cap \{a,d,f\} = \{a\}$.

5. Since $E \cap F = \{a\}$ is not a null set, we conclude that E and F are not mutually exclusive.

7. $E \cup F \cup G = \{2,4,6\} \cup \{1,3,5\} \cup \{5,6\} = \{1,2,3,4,5,6\}$.

9. $(E \cup F \cup G)^C = \{1,2,3,4,5,6\}^C = \varnothing$.

11. Yes, $E \cap F = \varnothing$; that is, E and F do not contain any common elements.

13. $E \cup F$ 15. G^C 17. $(E \cup F \cup G)^C$

19. $\varnothing$, $\{a\}$, $\{b\}$, $\{c\}$, $\{a,b\}$, $\{a,c\}$, $\{b,c\}$, $\{a,b,c\}$.

21. a. $S = \{R,B\}$ b. $\varnothing$, $\{B\}$, $\{R\}$, $\{B,R\}$

23. a. $S = \{(H,1), (H,2), (H,3), (H,4), (H,5), (H,6), (T,1), (T,2), (T,3), (T,4), (T,5), (T,6)\}$
 b. $E = \{(H,2), (H,4), (H,6)\}$

25. $S = \{(d,d,d), (d,d,n), (d,n,d), (n,d,d), (d,n,n), (n,d,n), (n,n,d), (n,n,n)\}$

27. a. $\{ABC, ABD, ABE, ACD, ACE, ADE, BCD, BCE, BDE, CDE\}$;
 b. 6 c. 3 d. 6

29. a. E^C b. $E^C \cap F^C$ c. $E \cup F$ d. $(E \cap F^C) \cup (E^C \cap F)$

31. a. $S = \{x \mid x > 0\}$ b. $E = \{x \mid 0 < x \leq 2\}$ c. $F = \{x \mid x > 2\}$

33. a. $S = \{0,1,2,3,...,10\}$ b. $E = \{0,1,2,3\}$ c. $F = \{5,6,7,8,9,10\}$

35. a. $S = \{0,1,2,...,20\}$ b. $E = \{0,1,2,...,9\}$ c. $F = \{20\}$

37. If E is an event of an experiment then E^c is the event containing the elements in S that are not in E. Therefore $E \cap E^C = \varnothing$ and the two sets are mutually exclusive.

39. The number of events of this experiment is 2^n.

EXERCISES 7.2, page 388

1. $\{(H,H)\}$, $\{(H,T)\}$, $\{(T,H)\}$, $\{(T,T)\}$.

3. $\{(D,m)\}$, $\{(D,f)\}$, $\{(R,m)\}$, $\{(R,f)\}$, $\{(I,m)\}$, $\{(I,f)\}$

5. $\{(1,i)\}$, $\{(1,d)\}$, $\{(1,s)\}$, $\{(2,i)\}$, $\{(2,d)\}$, $\{(2,s)\}$, ..., $\{(5,i)\}$, $\{(5,d)\}$, $\{(5,s)\}$

7. $\{(A,Rh^+)\}$, $\{(A,Rh^-)\}$, $\{(B,Rh^+)\}$, $\{B, Rh^-)\}$, $\{(AB, Rh^+)\}$, $\{(AB,Rh^-)\}$, $\{(O,Rh^+)\}$, $\{(O,Rh^-)\}$

9. The probability distribution associated with this data is

Grade	A	B	C	D	F
Probability	0.10	0.25	0.45	0.15	0.05

11. a. $S = \{(0 < x \leq 200), (200 < x \leq 400), (400 < x \leq 600), (600 < x \leq 800), (800 < x \leq 1000), (x > 1000)\}$

b.

Number of cars (x)	Probability
$0 < x \leq 200$	0.075
$200 < x \leq 400$	0.1
$400 < x \leq 600$	0.175
$600 < x \leq 800$	0.35
$800 < x \leq 1000$	0.225
$x > 1000$	0.075

13. The probability distribution associated with this data is

Rating	*A*	*B*	*C*	*D*	*F*
Probability	0.026	0.199	0.570	0.193	0.012

15. The probability distribution is

Number of figures produced (in dozens)	30	31	32	33	34	35	36
Probability	0125	0	0.1875	0.25	0.1875	0.125	0.125

17. The probability is

$$\frac{84,000,000}{179,000,000} = 0.469.$$

19. a. The probability that a person killed by lightning is a male is

$$\frac{376}{439} \approx 0.856.$$

b. The probability that a person killed by lightning is a female is

$$\frac{439 - 376}{439} = \frac{63}{439} \approx 0.144.$$

21. The probability that the retailer uses electronic tags as antitheft devices is

$$\frac{81}{179} \approx 0.46.$$

23. a. $P(D) = \frac{13}{52} = \frac{1}{4}$ b. $P(B) = \frac{26}{52} = \frac{1}{2}$ c. $P(A) = \frac{4}{52} = \frac{1}{13}$

25. The probability of arriving at the traffic light when it is red is

$$\frac{30}{80} = 0.375.$$

27. The probability is

$$P(D) + P(C) + P(B) + P(A) = 0.15 + 0.45 + 0.25 + 0.10 = 0.95.$$

29. a. $P(E) = \frac{62}{9+62+27} = \frac{62}{98} \approx 0.633$ b. $P(E) = \frac{27}{98} \approx 0.276$

31. There are two ways of getting a 7, one die showing a 3 and the other die showing a 4, and vice versa.

33. No, the outcomes are not equally likely.

35. No. Since the coin is weighted, the outcomes are not equally likely.

37. a. $P(A) = P(s_1) + P(s_2) = \frac{1}{12} + \frac{1}{12} = \frac{1}{6}$

b. $P(B) = P(s_2) + P(s_4) + P(s_5) + P(s_6) = \frac{1}{4} + \frac{1}{6} + \frac{1}{12} = \frac{5}{6}$

c. $P(C) = 1.$

39. Let G denote a female birth and let B denote a male birth. Then the eight equally likely outcomes of this experiment are

$$GGG \quad GGB \quad GBG \quad BGG \quad BGB \quad BBG \quad GBB \quad BBB.$$

a. The event that there are two girls and a boy in the family is

$$E = \{GGB, GBG, BGG\}.$$

Since there are three favorable outcomes, $P(E) = 3/8$.

b. The event that the oldest child is a girl is

$$F = \{GGG, GGB, GBG, GBB\}.$$

Since there are 4 favorable outcomes, $P(F) = 1/2$.

c. The event that the oldest child is a girl and the youngest child is a boy is
$G = \{GGB, GBB\}$.
Since there are two favorable outcomes, $P(F) = 1/4$.

EXERCISES 7.3, page 398

1. Refer to Example 3, page 386. Let E denote the event of interest. Then
$$P(E) = \frac{18}{36} = \frac{1}{2}$$

3. Refer to Example 3, page 386. The event of interest is $E = \{1,1\}$, and $P(E) = 1/36$.

5. Let E denote the event of interest. Then $E = \{(6,2),(6,1),(1,6),(2,6)\}$
and $$P(E) = \frac{4}{36} = \frac{1}{9}.$$

7. Let E denote the event that the card drawn is a king, and let F denote the event that the card drawn is a diamond. Then the required probability is
$$P(E \cap F) = \frac{1}{52}.$$

9. Let E denote the event that a face card is drawn. Then
$$P(E) = \frac{12}{52} = \frac{3}{13}.$$

11. Let E denote the event that an ace is drawn. Then $P(E) = 1/13$. Then E^c is the event that an ace is not drawn and
$$P(E^C) = 1 - P(E) = \frac{12}{13}.$$

13. Let E denote the event that a ticket holder will win first prize, then
$$P(E) = \frac{1}{500} = 0.002,$$
and the probability of the event that a ticket holder will not win first prize is
$P(E^c) = 1 - 0.002 = 0.998$.

15. Property 2 of the laws of probability is violated. The sum of the probabilities must add up to 1. In this case $P(S) = 1.1$, which is not possible.

17. The five events are not mutually exclusive; the probability of winning at least one purse is

$$1 \text{ - probability of losing all 5 times } = 1 - \frac{9^5}{10^5} = 1 - 0.5905 = 0.4095.$$

19. The two events are not mutually exclusive; hence, the probability of the given event is

$$\frac{1}{6} + \frac{1}{6} - \frac{1}{36} = \frac{11}{36}.$$

21. $E^C \cap F^C = \{c, d, e\} \cap \{a, b, e\} = \{e\} \neq \varnothing$.

23. Let G denote the event that a customer purchases a pair of glasses and let C denote the event that the customer purchases a pair of contact lenses. Then

$$P(G \cup C)^C \neq 1 - P(G) - P(C).$$

Mr. Owens has not considered the case in which the customer buy both glasses and contact lenses.

25. a. $P(E \cap F) = 0$ since E and F are mutually exclusive.

b. $P(E \cup F) = P(E) + P(F) - P(E \cap F) = 0.2 + 0.5 = 0.7$.

c. $P(E^c) = 1 - P(E) = 1 - 0.2 = 0.8$.

d. $P(E^C \cap F^C) = P[(E \cup F)^C] = 1 - P(E \cup F) = 1 - 0.7 = 0.3$.

27. a. $P(A) = P(s_1) + P(s_2) = \frac{1}{8} + \frac{3}{8} = \frac{1}{2}$

$P(B) = P(s_1) + P(s_3) = \frac{1}{8} + \frac{1}{4} = \frac{3}{8}$

b. $P(A^C) = 1 - P(A) = 1 - \frac{1}{2} = \frac{1}{2}$

$P(B^C) = 1 - P(B) = 1 - \frac{3}{8} = \frac{5}{8}$

c. $P(A \cap B) = P(s_1) = \frac{1}{8}$

d. $P(A \cup B) = P(A) + P(B) - P(A \cap B) = \frac{1}{2} + \frac{3}{8} - \frac{1}{8} = \frac{3}{4}$

29. Referring to the following diagram we see that

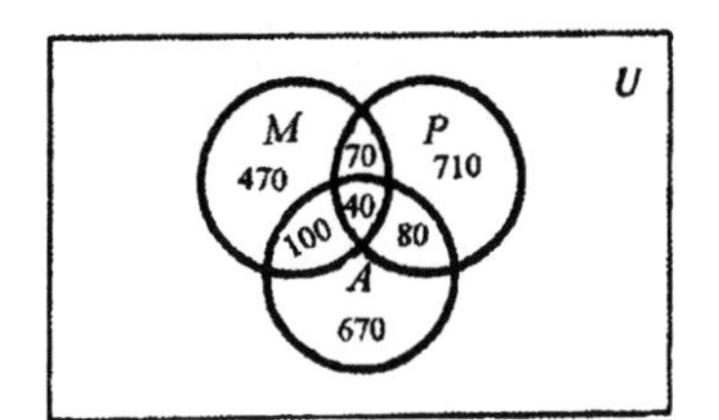

P: lack of parental support
M: malnutrition
A: abused or neglected

the probability that a teacher selected at random from this group said that lack of parental support is the only problem hampering a student's schooling is

$$\frac{710}{2140} = 0.33.$$

31. Let E and F denote the events that the person surveyed learned of the products from *Good Housekeeping* and *The Ladies Home Journal*, respectively. Then

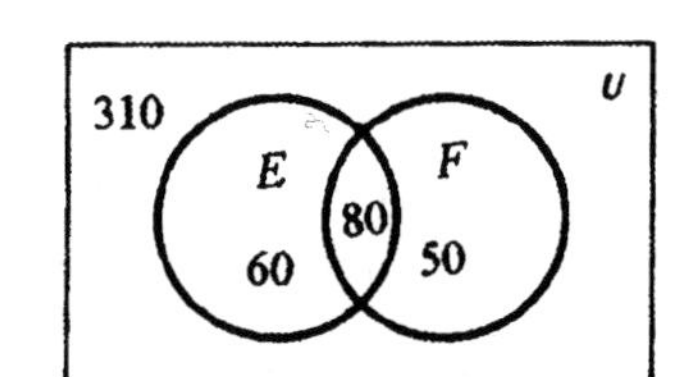

$$P(E) = \frac{140}{500} = \frac{7}{25}, \quad P(F) = \frac{130}{500} = \frac{13}{50}$$

and $P(E \cap F) = \frac{80}{500} = 0.16$

a. $P(E \cap F) = \frac{80}{500} = 0.16$

b. $P(E \cup F) = \frac{14}{50} + \frac{13}{50} - \frac{8}{50} = \frac{19}{50} = 0.38$

c. $P(E \cap F^C) + P(E^C \cap F) = \frac{60}{500} + \frac{50}{500} = \frac{110}{500} = 0.22$.

33. Let $A = \{t \mid t < 3\}$, $B = \{t \mid t \leq 4\}$, $C = \{t \mid t > 5\}$

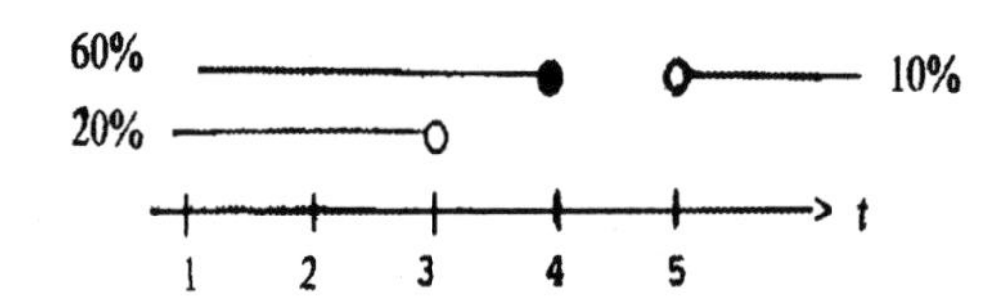

a. $D = \{t \mid t \leq 5\}$ and $P(D) = 1 - P(C) = 1 - 0.1 = 0.9$.

b. $E = \{t \mid t > 4\}$ and $P(E) = 1 - P(B) = 1 - 0.6 = 0.4$.

c. $F = \{t \mid 3 \le t \le 4\}$ and $P(F) = P(A^C \cap B) = 0.4$.

35. a. The probability that the participant favors tougher gun-control laws is

$$\frac{150}{250} = 0.6.$$

b. The probability that the participant owns a handgun is

$$\frac{58+25}{250} = 0.332.$$

c. The probability that the participant owns a handgun but not a rifle is

$$\frac{58}{250} = 0.232.$$

d. The probability that the participant favors tougher gun-control laws and does not own a handgun is

$$\frac{12+138}{250} = 0.6.$$

37. The event E that a multiple of 5 is drawn and the event F that a multiple of 7 is drawn are not mutually exclusive. In fact, the correct answer is

$$\frac{16}{50} = \frac{8}{25}.$$

39. True. As in Exercise 38, we have $P(A) = P(B) - P(B - A)$. Since $P(B - A) \ge 0$, we see that $P(A) \le P(B)$.

EXERCISES 7.4, page 407

1. Let E denote the event that the coin lands heads all five times. Then

$$P(E) = \frac{1}{2^5} = \frac{1}{32}.$$

3. Let E denote the event that the coin lands tails all 5 times, then

$$P(E^c) = 1 - P(E) = 1 - \frac{1}{32} = \frac{31}{32},$$

where E^c is the event that the coin lands heads at least once.

5. $P(E)=\dfrac{13\cdot C(4,2)}{C(52,2)}=\dfrac{78}{1326}\approx 0.0588$.

7. $P(E)=\dfrac{C(26,2)}{C(52,2)}=\dfrac{325}{1326}\approx 0.2451$.

9. The probability of the event that two of the balls will be white and two will be blue is

$$P(E)=\frac{n(E)}{n(S)}=\frac{C(3,2)\cdot C(5,2)}{C(8,4)}=\frac{(3)(10)}{70}=\frac{3}{7}.$$

11. The probability of the event that exactly three of the balls are blue is

$$P(E)=\frac{n(E)}{n(S)}=\frac{C(5,3)C(3,1)}{C(8,4)}=\frac{30}{70}=\frac{3}{7}.$$

13. $$P(E)=\frac{C(3,2)\cdot C(1,1)}{8}=\frac{3}{8}.$$

15. $$P(E)=1-\frac{C(3,3)}{8}=\frac{1}{8}.$$

17. The number of elements in the sample space is 2^{10}. There are

$$C(10,6)=\frac{10!}{6!4!},$$

or 210 ways of answering exactly six questions correctly. Therefore, the required probability is

$$\frac{210}{2^{10}}=\frac{210}{1024}\approx 0.205.$$

19. a. Let E denote the event that both of the bulbs are defective. Then

$$P(E)=\frac{C(4,2)}{C(24,2)}=\frac{\frac{4!}{2!2!}}{\frac{24!}{22!2!}}=\frac{4\cdot 3}{24\cdot 23}=\frac{1}{46}\approx 0.022.$$

b. Let F denote the event that none of the bulbs are defective. Then

$$P(F)=\frac{C(20,2)}{C(24,2)}=\frac{20!}{18!2!}\cdot\frac{22!2!}{24!}=\frac{20}{24}\cdot\frac{19}{23}=0.6884.$$

Therefore, the probability that at least one of the light bulbs is defective is given by

$$1-P(F)=1-0.6884=0.3116.$$

21. a. The probability that both of the cartridges are defective is

$$P(E)=\frac{C(6,2)}{C(80,2)}=\frac{30}{6320}=0.0048.$$

b. Let F denote the event that none of the cartridges are defective. Then

$$P(F)=\frac{C(74,2)}{C(80,2)}=\frac{5402}{6320}=0.855,$$

and $P(F^c)=1-P(F)=1-0.855=0.145$ is the probability that at least 1 of the cartridges is defective.

23. a. The probability that Mary's name will be selected is

$$P(E)=\frac{12}{100}=0.12;$$

The probability that both Mary's and John's names will be selected is

$$P(F)=\frac{C(98,10)}{C(100,12)}=\frac{\frac{98!}{88!10!}}{\frac{100!}{88!12!}}=\frac{12\cdot 11}{100\cdot 99}\approx 0.013.$$

b. The probability that Mary's name will be selected is

$$P(M)=\frac{6}{40}=0.15.$$

The probability that both Mary's and John's names will be selected is

$$P(M)\cdot P(J)=\frac{6}{60}\cdot\frac{6}{40}=\frac{36}{2400}=0.015.$$

25. The probability is given by

$$\frac{C(12,8)\cdot C(8,2)}{C(20,10)}+\frac{C(12,9)C(8,1)}{C(20,10)}+\frac{C(12,10)}{C(20,10)}$$
$$=\frac{(28)(495)+(220)(8)+66}{184,756}\approx 0.085$$

27. a. The probability that he will select brand B is

$$\frac{C(4,2)}{C(5,3)}=\frac{6}{15}=\frac{3}{5}. \qquad \left(\frac{\text{the number of selections that include brand } B}{\text{the number of possible selections}}\right)$$

b. The probability that he will select brands B and C is

$$\frac{C(3,1)}{C(5,3)}=0.3$$

c. The probability that he will select at least one of the two brands, B and C is

$$1-\frac{C(3,3)}{C(5,3)}=0.9.$$ (1 - probability that he does not select brands *B* and *C*.)

29. The probability that the three "Lucky Dollar" symbols will appear in the window of the slot machine is

$$P(E)=\frac{n(E)}{n(S)}=\frac{(1)(1)(1)}{C(1,9)C(1,9)C(1,9)}=\frac{1}{729}.$$

31. The probability of a ticket holder having all four digits in exact order is

$$\frac{1}{C(10,1)\cdot C(10,1)\cdot C(10,1)\cdot C(10,1)}=\frac{1}{10,000}=0.0001.$$

33. The probability of a ticket holder having a specified digit in exact order is

$$\frac{C(1,1)C(10,1)C(10,1)C(10,1)}{10^4}=0.10.$$

35. The number of ways of selecting a 5-card hand from 52 cards is given by

$$C(52,5)=2,598,960.$$

The number of straight flushes that can be dealt in each suit is 10, so there are 4(10) possible straight flushes. Therefore, the probability of being dealt a straight flush is

$$\frac{4(10)}{C(52,5)}=\frac{40}{2,598,960}=0.0000154.$$

37. The number of ways of being dealt a flush in one suit is $C(13,5)$, and, since there are four suits, the number of ways of being dealt a flush is $4\cdot C(13,5)$. Since we wish to exclude the hands that are straight flushes we subtract the number of possible straight flushes from $4\cdot C(13,5)$. Therefore, the probability of being drawn a flush, but not a straight flush, is

$$\frac{4\cdot C(13,5)-40}{C(52,5)}=\frac{5108}{2,598,960}=0.0019654.$$

39. The total number of ways to select three cards of one rank is

$$13\cdot C(4,3).$$

The remaining two cards must form a pair of another rank and there are

$$12\cdot C(4,2)$$

ways of selecting these pairs. Next, the total number of ways to be dealt a full house is

$$13 \cdot C(4,3) \cdot 12 \cdot C(4,2) = 3744.$$

Hence, the probability of being dealt a full house is

$$\frac{3{,}744}{2{,}598{,}960} \approx 0.0014406.$$

41. Let E denote the event that in a group of 5, no two will have the same sign. Then

$$P(E) = \frac{12 \cdot 11 \cdot 10 \cdot 9 \cdot 8}{12^5} \approx 0.381944.$$

Therefore, the probability that at least two will have the same sign is given by

$$1 - P(E) = 1 - 0.381944 \approx 0.618.$$

b. $P(\text{no Aries}) = \dfrac{11 \cdot 11 \cdot 11 \cdot 11 \cdot 11}{12^5} \approx 0.647228.$

$$P(1 \text{ Aries}) = \frac{C(5,1) \cdot (1)(11)(11)(11)(11)}{12^5} \approx 0.2941945.$$

Therefore, the probability that at least two will have the sign Aries is given by

$$1 - [P(\text{no Aries}) + P(1 \text{ Aries})] = 1 - 0.9414225 \approx 0.059.$$

43. Referring to the table on page 406, we see that in a group of 50 people, the probability that none of the people will have the same birthday is $1 - 0.970 = 0.03$.

EXERCISES 7.5, page 422

1. a. $P(A|B) = \dfrac{P(A \cap B)}{P(B)} = \dfrac{0.2}{0.5} = \dfrac{2}{5}.$

   b. $P(B|A) = \dfrac{P(A \cap B)}{P(A)} = \dfrac{0.2}{0.6} = \dfrac{1}{3}.$

3. $P(A \cap B) = P(A)P(B|A) = (0.6)(0.5) = 0.3.$

5. $P(A) \cdot P(B) = (0.3)(0.6) = 0.18 = P(A \cap B)$. Therefore the events are independent.

7. $P(A \cap B) = P(A) + P(B) - P(A \cup B) = 0.5 + 0.7 - 0.85 = 0.35$

$$= P(A) \cdot P(B)$$

so they are independent events.

9. a. $P(A \cap B) = P(A)P(B) = (0.4)(0.6) = 0.24$.

   b. $P(A \cup B) = P(A) + P(B) - P(A \cap B) = 0.4 + 0.6 - 0.24 = 0.76$.

11. a. $P(A) = 0.5$ b. $P(E|A) = 0.4$
    c. $P(A \cap E) = P(A)P(E|A) = (0.5)(0.4) = 0.2$
    d. $P(E) = (0.5)(0.4) + (0.5)(0.3) = 0.35$.
    e. No. $P(A \cap E) \neq P(A) \cdot P(E) = (0.5)(0.35)$
    f. A and E are not independent events.

13.

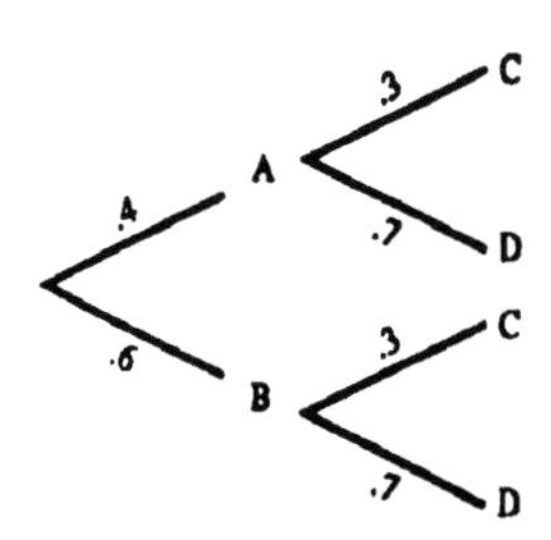

a. $P(A) = 0.4$ b. $P(C|A) = 0.3$
c. $P(A \cap C) = P(A)P(C|A) = (0.4)(0.3) = 0.12$
d. $P(C) = (0.4)(0.3) + (0.6)(0.3) = 0.3$
e. Yes. $P(A \cap C) = 0.12 = P(A)P(C)$
f. Yes.

15. Let A denote the event that the sum of the numbers is less than 9 and let B denote the event that one of the numbers is a 6.

$$\text{Then, } P(A|B) = \frac{P(A \cap B)}{P(B)} = \frac{\frac{4}{36}}{\frac{11}{36}} = \frac{4}{11}.$$

17. a. The probability that the first card drawn is a heart is

$$P(H) = \frac{13}{52} = \frac{1}{4}.$$

b. The probability that the second card drawn is a heart given that the first card drawn was not a heart is

$$P(H|H^C) = \frac{13}{51}.$$

c. The probability that the second card drawn is a heart given that the first card drawn was a heart is

$$P(H|H) = \frac{12}{51}.$$

19. Referring to the following tree diagram, we see that

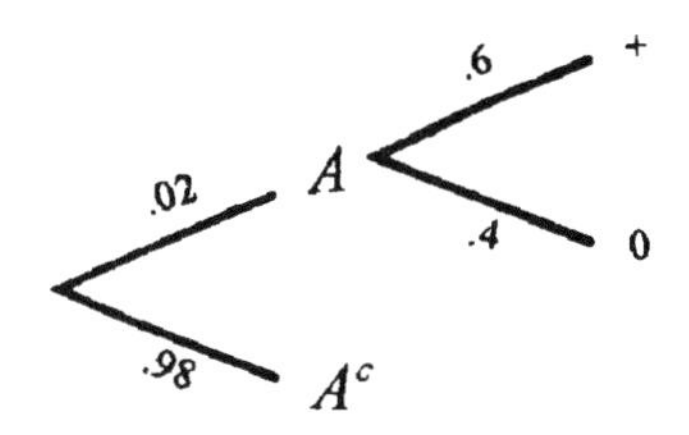

the probability that a tax return selected at random will result in additional assessments being levied on the taxpayer is

$$(0.02)(0.6) = 0.012.$$

21. Let $C = \{x|x$ is an eligible voter with a college degree$\}$ and
$V = \{x|x$ is an eligible voter who voted$\}$.
Referring to the following tree diagram, we see that

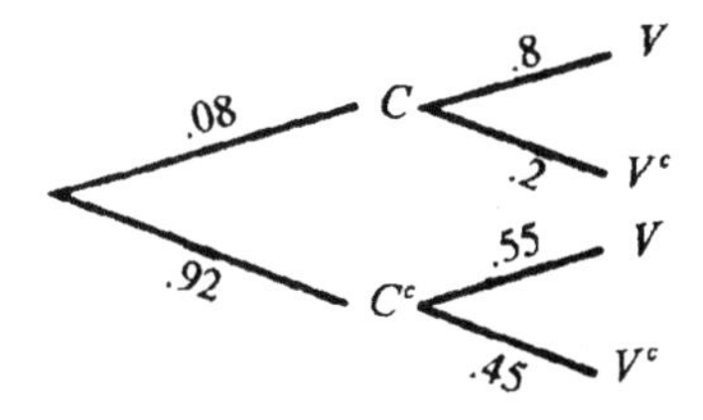

a. the probability that an eligible voter selected at random has a college degree and voted in the last presidential election is

$$P(C \cap V) = (0.08)(0.8) = 0.064$$

b. the probability that an eligible voter selected at random does not have a college degree and did not vote in the last presidential election is

$$P(C^C \cap V^C) = (0.92)(0.45) = 0.414.$$

c. the probability that an eligible voter selected at random voted in the last presidential election is

$$P(C \cap V) + P(C^C \cap V) = (0.08)(0.8) + (0.92)(0.55) = 0.57.$$

d. the probability that an eligible voter selected at random did not vote in the last presidential election is

$$P(V^c) = 1 - P(V) = 1 - 0.57 = 0.43.$$

23. Refer to the tree diagram that follows.

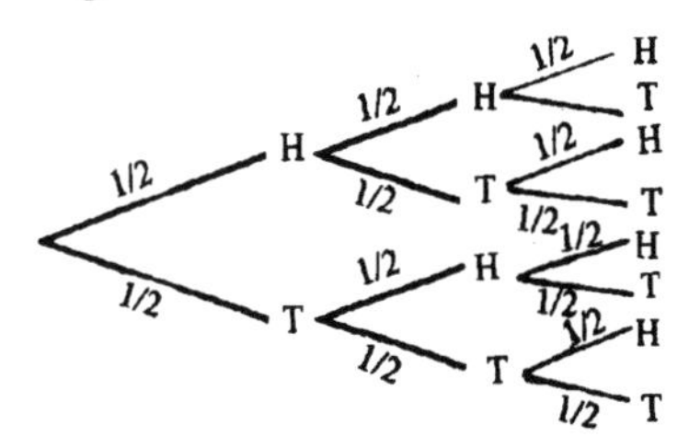

a. The probability that the coin will land heads at least twice is

$$\frac{4}{8} = \frac{1}{2}.$$

b. The probability that the coin will land heads on the second toss given that heads were thrown on the first toss is

$$\frac{2}{4} = \frac{1}{2}.$$

c. The probability that the coin will land heads on the third toss given that tails were thrown on the first toss is

$$\frac{2}{4} = \frac{1}{2}.$$

25. $P(D) = P(A \cap D) + P(B \cap D) + P(C \cap D)$

$= (0.45)(0.01) + (0.25)(0.02) + (0.30)(0.015) = 0.014.$

27. a.

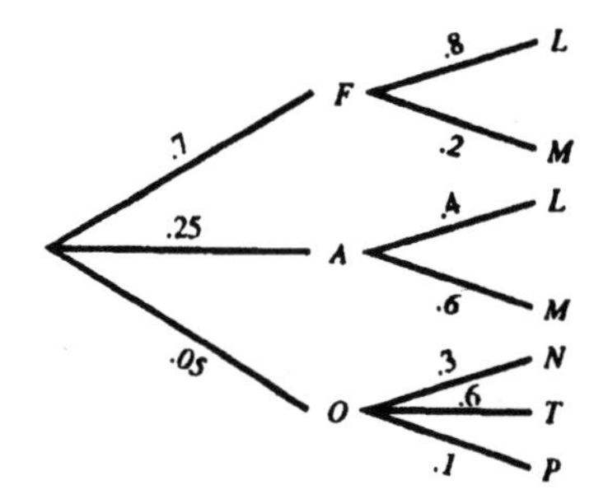

L = 30-year loan

M = 15-year loan

N = 20-year loan

T = 10-year loan

P = "5-year or less" loan

b. The required probability is $P(A \cap M) = (0.25)(0.6) = 0.15$.

c. The required probability is

$$P(F \cap M) + P(A \cap M) = (0.7)(0.2) + (0.25)(0.6) = 0.29.$$

29. Let D denote the event that a light is defective and D^c the event that a light is nondefective. Referring to the following tree diagram

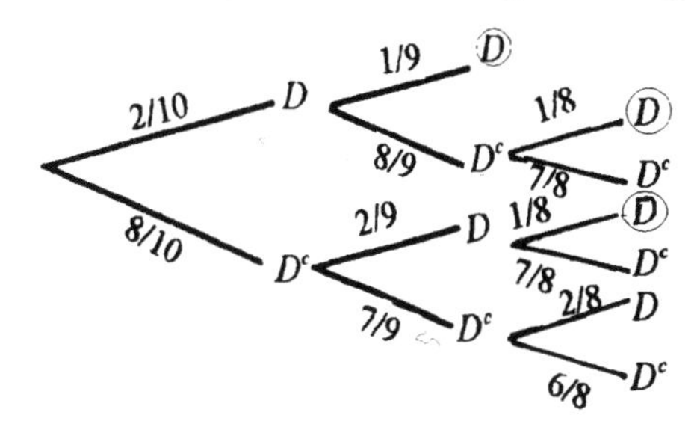

we see that the probability that both defective lights will be found after three trials is

$$P(D \cap D) + P(D \cap D^C \cap D) + P(D^C \cap D \cap D)$$

$$= \frac{2}{10} \cdot \frac{1}{9} + \frac{2}{10} \cdot \frac{8}{9} \cdot \frac{1}{8} + \frac{8}{10} \cdot \frac{2}{9} \cdot \frac{1}{8} = \frac{16+16+16}{720} = \frac{1}{15}.$$

31. a. $P(A) = \dfrac{8120}{10730} = 0.757$; $P(B) = \dfrac{6101}{10730} = 0.569$;

$$P(A \cap B) = \frac{4222}{10730} = 0.393$$

$$P(B|A) = \frac{P(A \cap B)}{P(A)} = \frac{n(A \cap B)}{n(A)} = \frac{4222}{8120} = 0.520$$

$$P(B|A^C) = \frac{P(A^C \cap B)}{P(A^C)} = \frac{n(A^C \cap B)}{n(A^C)} = \frac{1879}{2610} = 0.720$$

b. $P(B|A) \neq P(B)$, so A and B are not independent events.

33. $P(A \cap B) = \dfrac{1}{4} \cdot \dfrac{25}{51} = \dfrac{25}{204}$; $P(A) \cdot P(B) = \dfrac{1}{4} \cdot \dfrac{1}{2} = \dfrac{1}{8}$

Since $P(A \cap B) \neq P(A) \cdot P(B)$, A and B are dependent events.

35. a. These are independent events. Therefore, the probability is

$(0.9)(0.9) = 0.81$.

b. The probability that the mail will be delivered before 2 P.M. is

$(0.9)(0.9)(0.9) = 0.729$.

37. The probability that on a particular day

a. all four machines will break down is

$$P(A)P(B)P(C)P(D) = \frac{1}{50} \cdot \frac{1}{60} \cdot \frac{1}{75} \cdot \frac{1}{40} = 0.0000001.$$

b. none of the machines will break down is

$$P(A^C)P(B^C)P(C^C)P(D^C) = \frac{49}{50} \cdot \frac{59}{60} \cdot \frac{74}{75} \cdot \frac{39}{40} = 0.9270473 \approx 0.927.$$

c. exactly one machine will break down is

$$P(A)P(B^C)P(C^C)P(D^C) + P(A^C)P(B)P(C^C)P(D^C)$$
$$+ P(A^C)P(B^C)P(C)P(D^C) + P(A^C)P(B^C)P(C^C)P(D)$$
$$= \frac{1}{50} \cdot \frac{59}{60} \cdot \frac{74}{75} \cdot \frac{39}{40} + \frac{49}{50} \cdot \frac{1}{60} \cdot \frac{74}{75} \cdot \frac{39}{40} + \frac{49}{50} \cdot \frac{59}{60} \cdot \frac{1}{75} \cdot \frac{39}{40} + \frac{49}{50} \cdot \frac{59}{60} \cdot \frac{74}{75} \cdot \frac{1}{40}$$
$$= 0.189193 + 0.0157126 + 0.0125276 + 0.2737704$$
$$= 0.0709299 \approx 0.071.$$

39. a. No, because $E \cap S = E \neq \varnothing$ unless $E = \varnothing$.

b. Yes, because $E \cap \varnothing = \varnothing$.

41. Using Formula (2), we find

$$P(A|A \cup B) = \frac{P(A \cap (A \cup B))}{P(A \cup B)}.$$

Since A and B are mutually exclusive, $A \cap (A \cup B) = A$

$$P(A \cup B) = P(A) + P(B).$$

Therefore, $P(A|A \cup B) = \dfrac{P(A)}{P(A) + P(B)}$.

EXERCISES 7.6, page 431

1.

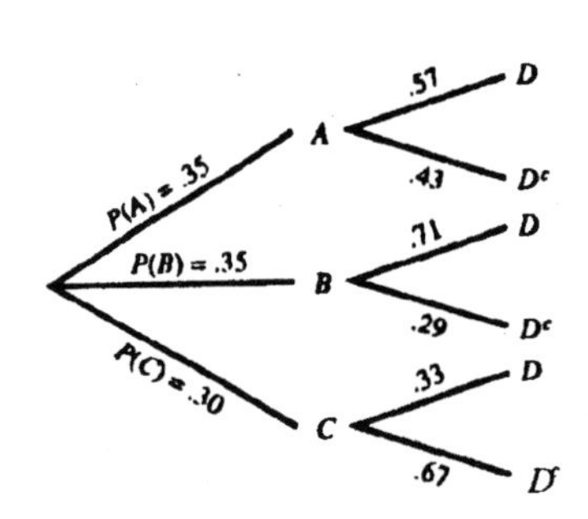

3. a. $P(D^C) = \dfrac{15+10+20}{35+35+30} = 0.45$

b. $P(B|D^C) = \dfrac{10}{15+10+20} = 0.22$.

5. a. $P(D) = \dfrac{25+20+15}{50+40+35} = 0.48$ b. $P(B|D) = \dfrac{20}{25+20+15} = 0.33$

7. a. $P(A)\cdot P(D|A) = (0.4)(0.2) = 0.08$

b. $P(B)\cdot P(D|B) = (0.6)(0.25) = 0.15$

c. $P(A|D) = \dfrac{P(A)\cdot P(D|A)}{P(A)\cdot P(D|A) + P(B)\cdot P(D|B)} = \dfrac{(0.4)(0.2)}{0.08+0.15} \approx 0.35$

9. a. $P(A)\cdot P(D|A) = \dfrac{1}{3}\cdot\dfrac{1}{4} = \dfrac{1}{12}$

b. $P(B)\cdot P(D|B) = \dfrac{1}{2}\cdot\dfrac{1}{2} = \dfrac{1}{4}$

c. $P(C)\cdot P(D|C) = \dfrac{1}{6}\cdot\dfrac{1}{3} = \dfrac{1}{18}$

d. $P(A|D) = \dfrac{P(A)\cdot P(D|A)}{P(A)\cdot P(D|A) + P(B)\cdot P(D|B) + P(C)\cdot P(C|B)}$

$$= \frac{\frac{1}{12}}{\frac{1}{12}+\frac{1}{4}+\frac{1}{18}} \approx \frac{1}{12}\cdot\frac{36}{14} = \frac{3}{14}$$

11. Let A denote the event that the first card drawn is a heart and B the event that the second card drawn is a heart. Then

$$P(A|B) = \frac{P(A)\cdot P(B|A)}{P(A)\cdot P(B|A) + P(A^C)\cdot P(B|A^C)}$$

$$= \frac{\frac{1}{4}\cdot\frac{12}{51}}{\frac{1}{4}\cdot\frac{12}{51}+\frac{3}{4}\cdot\frac{13}{51}} = \frac{4}{17}.$$

13. Using the following tree diagram, we see that

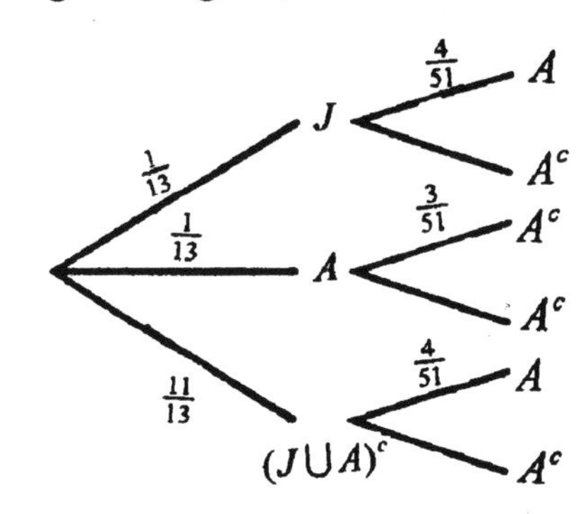

$$P(J|A) = \frac{\frac{1}{13} \cdot \frac{4}{51}}{\frac{1}{13} \cdot \frac{4}{51} + \frac{1}{13} \cdot \frac{3}{51} + \frac{11}{13} \cdot \frac{4}{51}} = \frac{\frac{4}{13 \cdot 51}}{\frac{51}{13 \cdot 51}} = 0.0784 .$$

15. The probabilities associated with this experiment are represented in the following tree diagram.

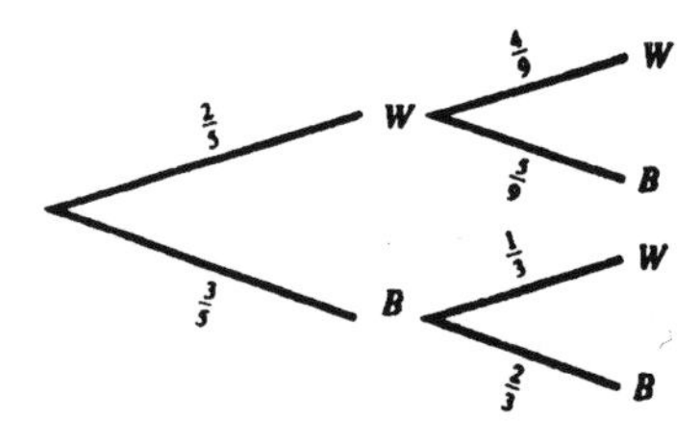

17. Referring to the tree diagram in Exercise 15, we see that the probability that the transferred ball was black given that the second ball was white is

$$P(B|W) = \frac{\frac{3}{5} \cdot \frac{1}{3}}{\frac{2}{5} \cdot \frac{4}{9} + \frac{3}{5} \cdot \frac{1}{3}} = \frac{9}{17}.$$

19. Let D denote the event that a senator selected at random is a Democrat, R denote the event that a senator selected at random is a Republican, and M the event that a senator has served in the military. From the following tree diagram

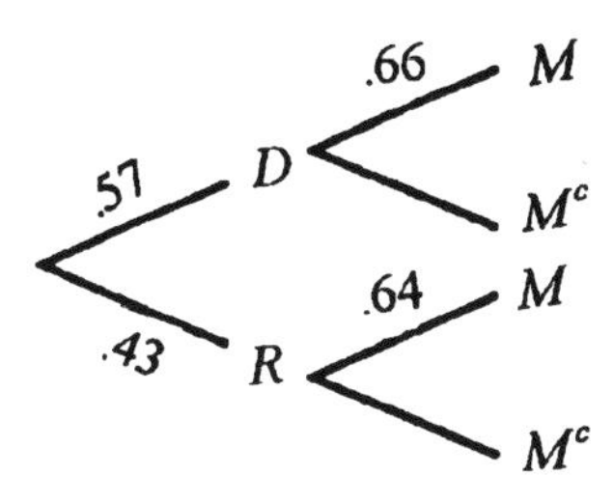

we see that the probability that a senator selected at random who has served in the military is a Republican is

$$P(R|M) = \frac{P(R)P(M|R)}{P(M)} = \frac{(0.64)(0.43)}{(0.66)(0.57)+(0.64)(0.43)}$$
$$= 0.4225.$$

21. Let H_2 denote the event that the coin tossed is the two-headed coin, H_B denote the event that the coin tossed is the biased coin, and H_F denote the event that the coin tossed is the fair coin. Referring to the following tree diagram, we see that

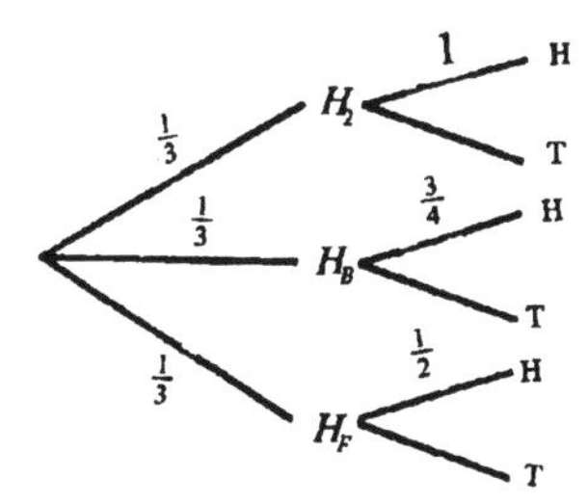

a. $P(H) = \frac{1}{3}\cdot 1 + \frac{1}{3}\cdot\frac{3}{4} + \frac{1}{3}\cdot\frac{1}{2} = \frac{1}{3}\cdot\frac{1}{4}\cdot\frac{1}{6} = \frac{9}{12} = \frac{3}{4}$

b. $P(H_F|H) = \frac{\frac{1}{3}\cdot\frac{1}{2}}{\frac{3}{4}} = \frac{2}{9}$

23. Let D denote the event that the person has the disease, and let Y denote the event that the test is positive. Referring to the following tree diagram, we see that the required probability is

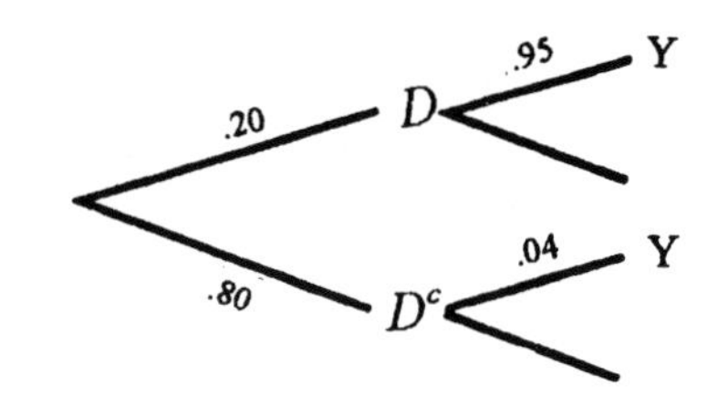

$$P(D|Y) = \frac{P(D) \cdot P(Y|D)}{P(D) \cdot P(Y|D) + P(D^C) \cdot P(Y|D^C)}$$
$$= \frac{(0.2)(0.95)}{(0.2)(0.95) + (0.8)(0.04)} \approx 0.856.$$

25. Let x denote the age of an insured driver, and let A denote the event that an insured driver is in an accident. Using the tree diagram we find,

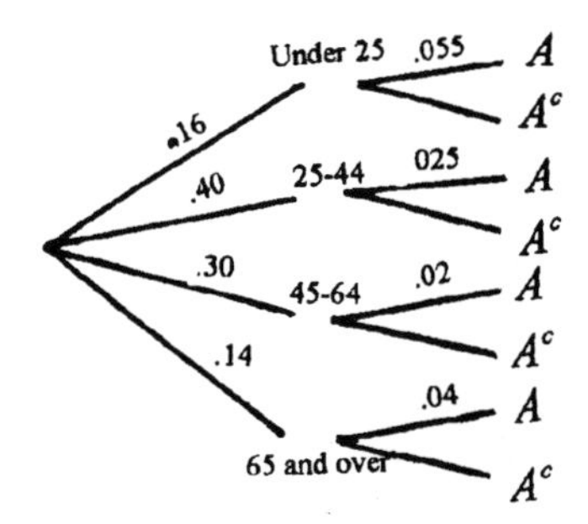

a. $P(A) = (0.16)(0.055) + (0.4)(0.025) + (0.3)(0.02) + (0.14)(0.04)$
$\approx 0.03.$

b. $P(x < 25|A) = \dfrac{(0.16)(0.055)}{(0.03)} \approx 0.29$

27. Let E, F, and G denote the events that the child selected at random is 12 years old, 13 years old, or 14 years old, respectively; and let C^c denote the even that the child does not have a cavity. Using the tree diagram at the right, we see that

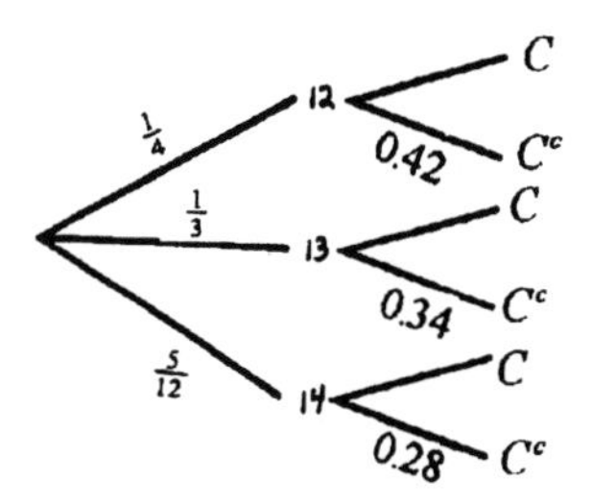

$$P(G|C^C) = \frac{\frac{5}{12}(0.28)}{\frac{1}{4}(0.42) + \frac{1}{3}(0.34) + \frac{5}{12}(0.28)} = 0.348$$

29. Let M and F denote the events that a person arrested for crime in 1988 was male or female, respectively; and let U denote the event that the person was under the age of 18. Using the following tree diagram, we have

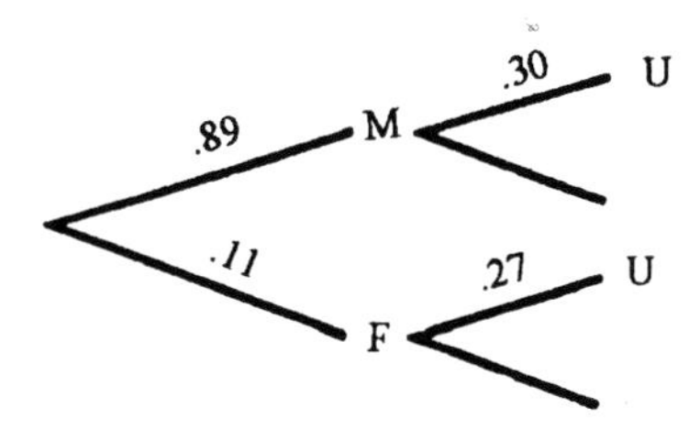

a. $P(U) = (0.89)(0.30) + (0.11)(0.27) = 0.2967.$

b. $P(F \mid U) = \dfrac{(0.11)(0.27)}{(0.89)(0.30)+(0.11)(0.27)} = 0.1001.$

31. Let I and II denote a customer who purchased the drug in capsule or tablet form, respectively; and let E denote a customer in Group I and II who purchased the extra-strength dosage of the drug. Then using the tree diagram that follows, we see that

$$P(I \mid E) = \frac{(0.57)(0.38)}{(0.57)(0.38)+(0.43)(0.31)} = 0.619$$

.57 I .38 E
E^c
.43 II .31 E
E^c

33. Let D and N denote the events that a employee was placed by Ms.Dwyer or Ms. Newberg, respectively; and let S denote the event that the employee placed by one of these women was satisfactory. Using the tree diagram shown at the right, we see that

$$P(N \mid S^C) = \frac{(0.55)(0.3)}{(0.45)(0.2)+(0.55)(0.3)}$$
$$= 0.647$$

.45 D .80 S
.20 S^c
.55 N .70 S
.30 S^c

35. Using the tree diagram shown at the right, we see that

$$P(S_2 \mid S) = \frac{(0.95)(0.8)}{(0.95)(0.8)+(0.2)(0.3)} \approx 0.93.$$

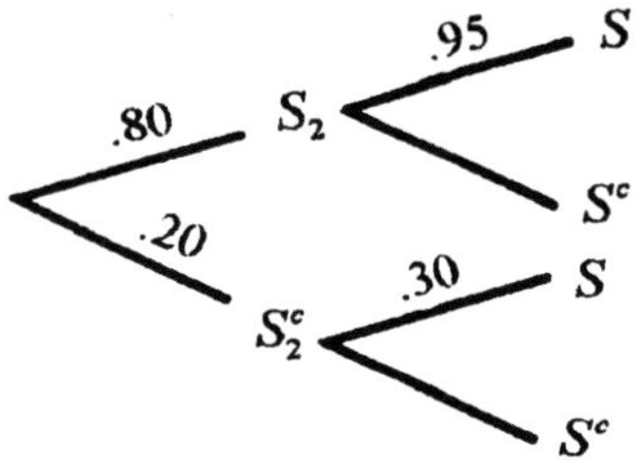

EXERCISES 7.7, page 449

1. Yes. All entries are nonnegative and the sum of the entries in each row is equal to 1.

3. Yes.

5. No. The sum of the entries of the third row is not 1.

7. Yes.

9. No. It is not a square ($n \times n$) matrix.

11. a. The conditional probability that the outcome state 1 will occur given that the outcome state 1 has occurred is 0.3.
    b. 0.7
    c. We compute $X_1 = X_0T = \begin{bmatrix} .4 & .6 \end{bmatrix}\begin{bmatrix} .3 & .7 \\ .6 & .4 \end{bmatrix} = \begin{bmatrix} .48 & .52 \end{bmatrix}.$

13. We compute $X_0T = \begin{bmatrix} .5 & .5 \end{bmatrix}\begin{bmatrix} .6 & .4 \\ .2 & .8 \end{bmatrix} = \begin{bmatrix} .4 & .6 \end{bmatrix}$

    Thus, after 1 stage of the experiment, the probability of state 1 occurring is 0.4 and the probability of state 2 occurring is 0.6. The tree diagram describing this process follows.

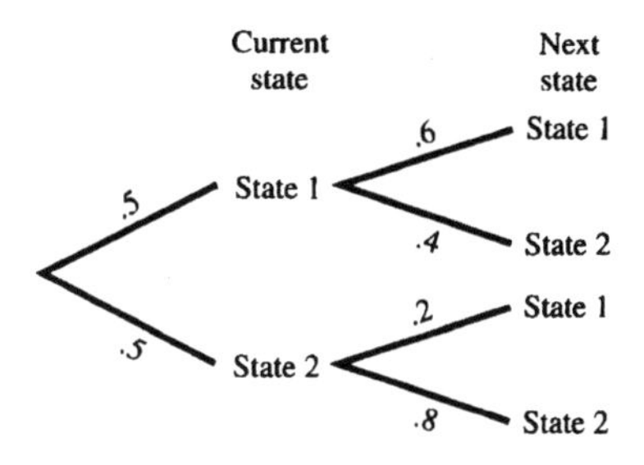

Using this diagram, we see that the probabilities of state 1 and state 2 occurring in the next stage of the experiment are given by

$$P(S_1) = (0.5)(0.6) + (0.5)(0.2) = 0.4$$
$$P(S_2) = (0.5)(0.4) + (0.5)(0.8) = 0.6$$

Observe that these probabilities are precisely those represented in the probability distribution vector X_0T.

15. $X_1 = X_0T = \begin{bmatrix} .6 & .4 \end{bmatrix}\begin{bmatrix} .4 & .6 \\ .8 & .2 \end{bmatrix} = \begin{bmatrix} .56 & .44 \end{bmatrix}$.

$X_2 = X_1T = \begin{bmatrix} .56 & .44 \end{bmatrix}\begin{bmatrix} .4 & .6 \\ .8 & .2 \end{bmatrix} = \begin{bmatrix} .576 & .424 \end{bmatrix}$.

17. $X_1 = \begin{bmatrix} \frac{1}{4} & \frac{1}{2} & \frac{1}{4} \end{bmatrix}\begin{bmatrix} \frac{1}{4} & \frac{1}{4} & \frac{1}{2} \\ \frac{1}{4} & \frac{1}{2} & \frac{1}{4} \\ \frac{1}{2} & \frac{1}{2} & 0 \end{bmatrix} = \begin{bmatrix} \frac{5}{16} & \frac{7}{16} & \frac{1}{4} \end{bmatrix}$

$X_2 = \begin{bmatrix} \frac{5}{16} & \frac{7}{16} & \frac{1}{4} \end{bmatrix}\begin{bmatrix} \frac{1}{4} & \frac{1}{4} & \frac{1}{2} \\ \frac{1}{4} & \frac{1}{2} & \frac{1}{4} \\ \frac{1}{2} & \frac{1}{2} & 0 \end{bmatrix} = \begin{bmatrix} \frac{5}{16} & \frac{27}{64} & \frac{17}{64} \end{bmatrix}$.

19. Since all entries in the matrix are positive, it is regular.

21. $T^2 = \begin{bmatrix} 1 & 0 \\ .8 & .2 \end{bmatrix}\begin{bmatrix} 1 & 0 \\ .8 & .2 \end{bmatrix} = \begin{bmatrix} 1 & 0 \\ .96 & .04 \end{bmatrix}$

$T^3 = \begin{bmatrix} 1 & 0 \\ .96 & .04 \end{bmatrix}\begin{bmatrix} 1 & 0 \\ .8 & .2 \end{bmatrix} = \begin{bmatrix} 1 & 0 \\ .992 & .008 \end{bmatrix}$

and we see that the a_{12} entry will always be zero, so T is not regular.

23. $$T^2 = \begin{bmatrix} \frac{1}{2} & \frac{1}{2} & 0 \\ \frac{3}{4} & 0 & \frac{1}{4} \\ 0 & \frac{1}{2} & \frac{1}{2} \end{bmatrix} \begin{bmatrix} \frac{1}{2} & \frac{1}{2} & 0 \\ \frac{3}{4} & 0 & \frac{1}{4} \\ 0 & \frac{1}{2} & \frac{1}{2} \end{bmatrix} = \begin{bmatrix} \frac{5}{8} & \frac{1}{4} & \frac{1}{8} \\ \frac{3}{8} & \frac{1}{2} & \frac{1}{8} \\ \frac{3}{8} & \frac{1}{4} & \frac{3}{8} \end{bmatrix}$$
and so the matrix is regular.

25. $$T^2 = \begin{bmatrix} .7 & .3 & 0 \\ .2 & .8 & 0 \\ .3 & .3 & .4 \end{bmatrix} \begin{bmatrix} .7 & .3 & 0 \\ .2 & .8 & 0 \\ .3 & .3 & .4 \end{bmatrix} = \begin{bmatrix} .55 & .45 & 0 \\ .3 & .7 & 0 \\ .39 & .45 & .16 \end{bmatrix}$$
and so forth. Continuing, we see that $T^3, T^4, \ldots,$ will have the a_{13} and a_{23} entries equal to zero and T is not regular.

27. We solve the matrix equation
$$\begin{bmatrix} x & y \end{bmatrix} = \begin{bmatrix} \frac{1}{3} & \frac{2}{3} \\ \frac{1}{4} & \frac{3}{4} \end{bmatrix} = \begin{bmatrix} x & y \end{bmatrix}$$
or equivalently, the system of equations
$$\begin{aligned} \tfrac{1}{3}x + \tfrac{1}{4}y &= x \\ \tfrac{2}{3}x + \tfrac{3}{4}y &= y \\ x + \ \ y &= 1. \end{aligned}$$
Solving this system of equations, we find the required vector to be
$$\begin{bmatrix} \frac{3}{11} & \frac{8}{11} \end{bmatrix}.$$

29. We have $XT = X$, that is,
$$\begin{bmatrix} x & y \end{bmatrix} = \begin{bmatrix} .5 & .5 \\ .2 & .8 \end{bmatrix} = \begin{bmatrix} x & y \end{bmatrix}$$
or equivalently, the sytem of equations
$$\begin{aligned} .5x + .2y &= x \\ .5x + .8y &= y. \end{aligned}$$
These two equations are equivalent to the single equation $0.5x - 0.2y = 0$. We must also have $x + y = 1$. So we have the system
$$\begin{aligned} .5x - .2y &= x \\ x + \ \ y &= 1 \end{aligned}$$

The second equation gives $y = 1 - x$, which when substituted into the first equation yields,

$$0.5x - 0.2(1 - x) = 0, \;\; 0.7x - 0.2 = 0, \;\; \text{or} \;\; x = 2/7.$$

Therefore, $y = 5/7$ and the steady-state distribution vector is

$$\begin{bmatrix} \frac{2}{7} & \frac{5}{7} \end{bmatrix}.$$

31. We solve the system

$$\begin{bmatrix} x & y & x \end{bmatrix} = \begin{bmatrix} 0 & 1 & 0 \\ \frac{1}{8} & \frac{5}{8} & \frac{1}{4} \\ 1 & 0 & 0 \end{bmatrix} = \begin{bmatrix} x & y & z \end{bmatrix}$$

together with the equation $x + y + z = 1$; that is, the system

$$\begin{aligned} -x + \tfrac{1}{8}y + z &= 0 \\ x - \tfrac{3}{8}y &= 0 \\ \tfrac{1}{4}y - z &= 0 \\ x + y + z &= 1. \end{aligned}$$

Using the Gauss-Jordan method, we find that the required steady-state vector is $\begin{bmatrix} \frac{3}{13} & \frac{8}{13} & \frac{2}{13} \end{bmatrix}$.

33. We solve the system

$$\begin{bmatrix} x & y & z \end{bmatrix} = \begin{bmatrix} .2 & 0 & .8 \\ 0 & .6 & .4 \\ .3 & .4 & .3 \end{bmatrix} = \begin{bmatrix} x & y & z \end{bmatrix}$$

together with the equation $x + y + z = 1$, or equivalently, the system

$$\begin{aligned} -0.8x + 0.3z &= 0 \\ -0.4y + 0.4z &= 0 \\ 0.8x + 0.4y - 0.7z &= 0 \\ x + y + z &= 1. \end{aligned}$$

Using the Gauss-Jordan method, we find that the required steady-state vector is

$$\begin{bmatrix} \frac{3}{19} & \frac{8}{19} & \frac{8}{19} \end{bmatrix}.$$

35. a.

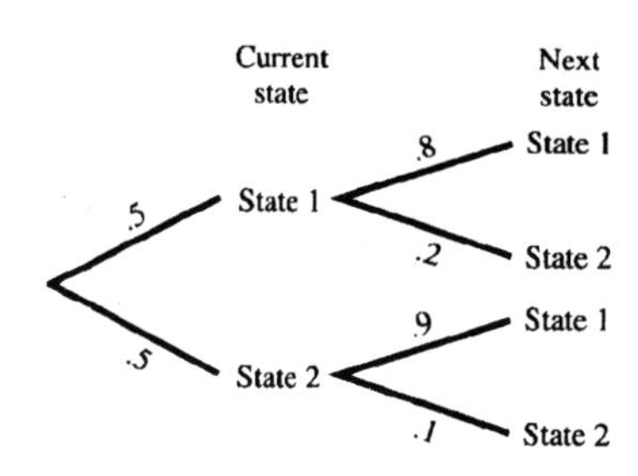

b. $T = \begin{array}{c} \\ L \\ R \end{array}\!\!\begin{array}{c} \begin{array}{cc} L & R \end{array} \\ \begin{bmatrix} .8 & .2 \\ .9 & .1 \end{bmatrix} \end{array}$

c. $X_0 = \begin{array}{c} \begin{array}{cc} L & R \end{array} \\ \begin{bmatrix} .5 & .5 \end{bmatrix} \end{array}$

d. $X_1 = \begin{bmatrix} .5 & .5 \end{bmatrix} \begin{array}{c} \\ L \\ R \end{array}\!\!\begin{array}{c} \begin{array}{cc} L & R \end{array} \\ \begin{bmatrix} .8 & .2 \\ .9 & .1 \end{bmatrix} \end{array} = \begin{array}{c} \begin{array}{cc} L & R \end{array} \\ \begin{bmatrix} .85 & .15 \end{bmatrix} \end{array}$

so the required probability is 0.85.

37. a. The transition matrix for the Markov chain is given by

$$T = \begin{array}{c} \\ A \\ U \\ N \end{array}\!\!\begin{array}{c} \begin{array}{ccc} A & U & N \end{array} \\ \begin{bmatrix} .85 & .10 & .05 \\ 0 & .95 & .05 \\ .10 & .05 & .85 \end{bmatrix} \end{array}.$$

b. The probability vector describing the distribution of land 10 years ago is given by

$$\begin{array}{c} \begin{array}{ccc} A & U & N \end{array} \\ \begin{bmatrix} .50 & .15 & .35 \end{bmatrix} \end{array}.$$

c. To find the required probability vector, we compute

$$X_0 T = \begin{bmatrix} .50 & .15 & .35 \end{bmatrix} \begin{bmatrix} .85 & .10 & .05 \\ 0 & .95 & .05 \\ .10 & .05 & .85 \end{bmatrix} = \begin{bmatrix} .46 & .21 & .33 \end{bmatrix}$$

$$X_1 T = \begin{bmatrix} .46 & .21 & .33 \end{bmatrix} \begin{bmatrix} .85 & .10 & .05 \\ 0 & .95 & .05 \\ .10 & .05 & .85 \end{bmatrix} = \begin{bmatrix} .424 & .262 & .314 \end{bmatrix}.$$

Thus, the probability vector describing the distribution of land 10 years from now is

$$\begin{bmatrix} .424 & .262 & .314 \end{bmatrix}.$$

39. $X_1 = X_0T = [.6 \quad .3 \quad .1]\begin{bmatrix} .75 & .15 & .10 \\ .05 & .90 & .05 \\ .05 & .10 & .85 \end{bmatrix} = [.47 \quad .37 \quad .16].$

$$X_2 = X_1T = [.47 \quad .37 \quad .16]\begin{bmatrix} .75 & .15 & .10 \\ .05 & .90 & .05 \\ .05 & .10 & .85 \end{bmatrix} = [.379 \quad .420 \quad .202].$$

After 1 year A has 47%, B has 37%, and C has 16%; after 2 years A has 38%, B has 42%, and C has 20%.

41. $X_1 = X_0T = [.3 \quad .3 \quad .2 \quad .2]\begin{bmatrix} .80 & .10 & .05 & .05 \\ .10 & .70 & .10 & .10 \\ .20 & .10 & .60 & .10 \\ .10 & .05 & .05 & .80 \end{bmatrix} = [.33 \quad .27 \quad .175 \quad .225.]$

Similarly, $X_2 = X_1T = [.3485 \quad .25075 \quad .15975 \quad .241]$

and $X_3 = X_2T = [.3599 \quad .2384 \quad .1504 \quad .2513]$.

Assuming that the present trend continues, 36% of the students in their senior year will major in business, 23.8% will major in the humanities, 15% will major in education, and 25.1% will major in the natural sciences.

43. a. We compute

$$X_1 = [.48 \quad .52]\begin{bmatrix} .72 & .28 \\ .12 & .88 \end{bmatrix} = [.408 \quad .592],$$

and conclude that, ten years from now, there will be 40.8 percent 1-wage-earners and 59.2% 2-wage earners.

b. We solve the system

$$[x \quad y]\begin{bmatrix} .72 & .28 \\ .12 & .88 \end{bmatrix} = [x \quad y].$$

together with the equation $x + y = 1$.

$$\begin{aligned} -0.28x + 0.12y &= 0 \\ 0.28x - 0.12y &= 0 \\ x + \quad y &= 1 \end{aligned}$$

Solving, we find $x = 0.3$ and $y = 0.7$, and conclude that in the long run, there will be 30% 1-wage earners and 70% 2-wage earners.

45. a. If this trend continues, the percentage of homeowners in this city who will own single-family homes or condominiums two years from now will be given by $X_2 = X_1T$. Thus,

$$X_1 = X_0T = \begin{bmatrix} .8 & .2 \end{bmatrix} \begin{bmatrix} .85 & .15 \\ .35 & .65 \end{bmatrix} = \begin{bmatrix} .75 & .25 \end{bmatrix}$$

$$X_2 = X_1T = \begin{bmatrix} .75 & .25 \end{bmatrix} \begin{bmatrix} .85 & .15 \\ .35 & .65 \end{bmatrix} = \begin{bmatrix} .725 & .275 \end{bmatrix}$$

and we conclude that 72.5% will own single-family home and 27.5% will own condominiums at that time.

b. We solve the system

$$\begin{bmatrix} x & y \end{bmatrix} \begin{bmatrix} .85 & .15 \\ .35 & .65 \end{bmatrix} = \begin{bmatrix} x & y \end{bmatrix}$$

together with the equation $x + y = 1$. Thus,

$$\begin{aligned} -0.15x + 0.35y &= 0 \\ 0.15x - 0.35y &= 0 \\ x + \quad y &= 1 \end{aligned}$$

Solving, we find $x = 0.7$ and $y = 0.3$, and conclude that in the long run 70% will own single family homes and 30% will own condominiums.

47. In the long-run each network will command 33.333% of the audience. Observe that the same steady state is reached regardless of the initial state of the system.

49. We wish to solve the system

$$\begin{bmatrix} x & y & z \end{bmatrix} = \begin{bmatrix} .75 & .15 & .10 \\ .05 & .90 & .05 \\ .05 & .10 & .85 \end{bmatrix} = \begin{bmatrix} x & y & z \end{bmatrix},$$

or, equivalently,

$$\begin{aligned} .75x + .05y + .05z &= x \\ .15x + .90y + .10z &= y \\ .10x + .05y + .85z &= z \\ x + \quad y + \quad z &= 1 \end{aligned}$$

Using the Gauss-Jordan method to solve this system of equations, we find

$$\begin{bmatrix} .167 & .542 & .292 \end{bmatrix}$$

and conclude that in the long run manufacturer A will have 16.7% of the market, manufacturer B will have 54.2% of the market, and manufacturer C will have 29.2% of the market.

USING TECHNOLOGY EXERCISES 7.7, page 456

1. $X_5 = [.204489 \quad .131869 \quad .261028 \quad .186814 \quad .2158]$

3. Manufacturer A will have 23.95% of the market, manufacturer B willl have 49.71% of the market share, and manufacturer C will have 26.34% of the market share.

CHAPTER 7, REVIEW EXERCISES, page 458

1. a. $P(E \cap F) = 0$ since E and F are mutually exclusive.

   b. $P(E \cup F) = P(E) + P(F) - P(E \cap F)$
   $= 0.4 + 0.2 = 0.6$

   c. $P(E^c) = 1 - P(E) = 1 - 0.4 = 0.6.$

   d. $P(E^C \cap F^C) = P(E \cup F)^C = 1 - P(E \cup F) = 1 - 0.6 = 0.4.$

   e. $P(E^C \cup F^C) = P(E \cap F)^C = 1 - P(E \cap F) = 1 - 0 = 1.$

3. a. The probability of the number being even is
   $P(2) + P(4) + P(6) = 0.12 + 0.18 + 0.19 = 0.49.$
   b. The probability that the number is either a 1 or a 6 is
   $P(1) + P(6) = 0.20 + 0.19 = 0.39.$
   c. The probability that the number is less than 4 is
   $P(1) + P(2) + P(3) = 0.20 + 0.12 + 0.16 = 0.48.$

5. Let A denote the event that a video-game cartridge has an audio defect and let V denote the event that a video-game cartridge has a video defect. Then using the following Venn diagram, we have

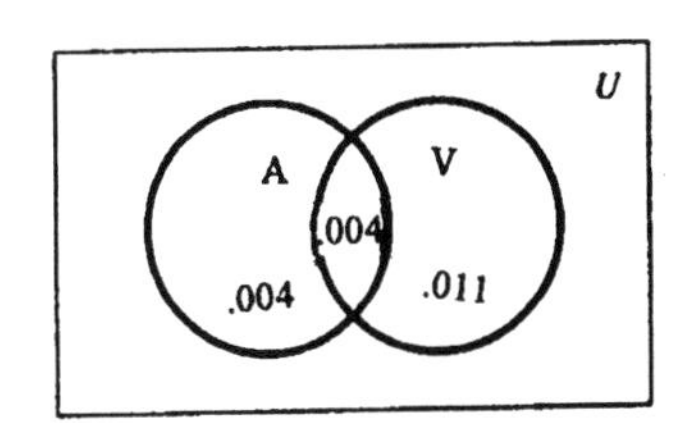

a. the probability that a cartridge purchased by a customer will have a video or audio defect is

$$P(A \cup V) = P(A \cap V^C) + P(V \cap A^C) + P(A \cap V)$$
$$= 0.004 + 0.004 + 0.011 = 0.019$$

b. the probability that a cartridge purchased by a customer will have not have a video or audio defect is

$$P(A \cup V)^C = 1 - P(A \cup V) = 1 - 0.019 = 0.981.$$

7. $P(A \cap E) = (0.3)(0.6) = 0.18.$

9. $P(C \cap E) = (0.2)(0.3) = 0.06.$

11. $P(E) = 0.18 + 0.25 + 0.06 = 0.49$

13. a. The probability that none of the pens in the sample are defective is

$$\frac{C(18,3)}{C(20,3)} = \frac{\frac{18!}{15!3!}}{\frac{20!}{17!3!}} = \frac{18!}{15!} \cdot \frac{17!}{20!} = \frac{68}{95} \approx 0.7159.$$

Therefore, the probability that at least one is defective is given by

$$1 - 0.71579 \approx 0.284.$$

b. The probability that two are defective is given by

$$\frac{C(2,2) \cdot C(18,1)}{C(20,3)} = \frac{18}{\frac{20!}{17!3!}} = \frac{6}{380} \approx 0.0158 .$$

Therefore, the probability that no more than 1 is defective is given by

$$1 - \frac{6}{380} = \frac{374}{380} \approx 0.984 .$$

15. Let E denote the event that the sum of the numbers is 8 and let D denote the event that the numbers appearing on the face of the two dice are different. Then

$$P(E|D) = \frac{P(E \cap D)}{P(D)} = \frac{4}{30} = \frac{2}{15}.$$

17. The probability that all three cards are face cards is

$$\frac{C(12,3)}{C(52,3)} = 0.00995.$$

19. The probability that the second card is black, given that the first card was red is

$$\frac{26}{51} = 0.510.$$

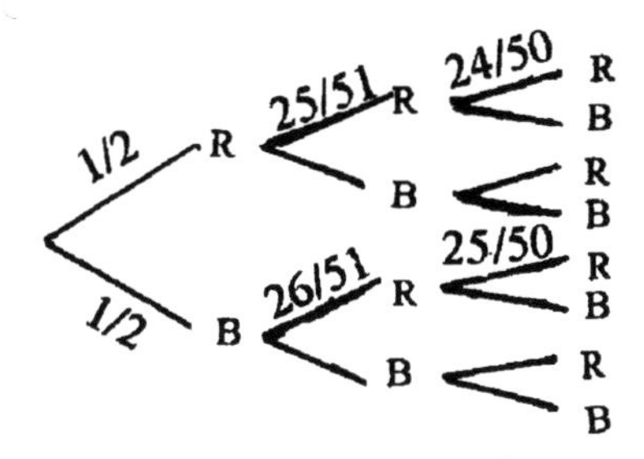

21. Since the entries $a_{21} = -2$ and $a_{22} == -8$ are negative, the given matrix is not stochastic and is hence not a regular stochastic matrix.

23.
$$T^2 = \begin{bmatrix} \frac{1}{2} & 0 & \frac{1}{2} \\ 0 & 0 & 1 \\ \frac{1}{3} & \frac{1}{3} & \frac{1}{3} \end{bmatrix} \begin{bmatrix} \frac{1}{2} & 0 & \frac{1}{2} \\ 0 & 0 & 1 \\ \frac{1}{3} & \frac{1}{3} & \frac{1}{3} \end{bmatrix} = \begin{bmatrix} \frac{5}{12} & \frac{1}{6} & \frac{5}{12} \\ \frac{1}{3} & \frac{1}{3} & \frac{1}{3} \\ \frac{5}{18} & \frac{1}{9} & \frac{11}{18} \end{bmatrix}$$

and so the matrix is regular.

25.
$$X_1 = \begin{bmatrix} \frac{1}{2} & \frac{1}{2} & 0 \end{bmatrix} \begin{bmatrix} 0 & \frac{2}{5} & \frac{3}{5} \\ \frac{1}{4} & \frac{1}{2} & \frac{1}{4} \\ \frac{3}{5} & \frac{1}{5} & \frac{1}{5} \end{bmatrix} = \begin{bmatrix} \frac{1}{8} & \frac{9}{20} & \frac{17}{40} \end{bmatrix}$$

$$X_2 = \begin{bmatrix} \frac{1}{8} & \frac{9}{20} & \frac{17}{40} \end{bmatrix} \begin{bmatrix} 0 & \frac{2}{5} & \frac{3}{5} \\ \frac{1}{4} & \frac{1}{2} & \frac{1}{4} \\ \frac{3}{5} & \frac{1}{5} & \frac{1}{5} \end{bmatrix} = \begin{bmatrix} \frac{147}{400} & \frac{9}{25} & \frac{109}{400} \end{bmatrix}$$
$$= \begin{bmatrix} 0.3675 & 0.36 & 0.2725 \end{bmatrix}.$$

27. We solve the matrix equation

$$\begin{bmatrix} x & y \end{bmatrix} \begin{bmatrix} .6 & .4 \\ .3 & .7 \end{bmatrix} = \begin{bmatrix} x & y \end{bmatrix}$$

or equivalently, the system of equations

$$-.4x + .3y = 0$$
$$.4x - .3y = 0$$
$$x + y = 1.$$

Solving this system of equations, we find the steady-state distribution vector to be

$$\begin{bmatrix} \frac{3}{7} & \frac{4}{7} \end{bmatrix},$$

and the steady-state matrix to be

$$\begin{bmatrix} \frac{3}{7} & \frac{4}{7} \\ \frac{3}{7} & \frac{4}{7} \end{bmatrix}.$$

29. We solve the system

$$\begin{bmatrix} x & y & z \end{bmatrix} \begin{bmatrix} .6 & .2 & .2 \\ .4 & .2 & .4 \\ .3 & .2 & .5 \end{bmatrix} = \begin{bmatrix} x & y & z \end{bmatrix}$$

together with the equation $x + y + z = 1$, or equivalently, the system

$$\begin{aligned} .6x+.4y+.3z &= x \\ .2x+.2y+.2z &= y \\ .2x+.4y+.5z &= z \\ x+\ \ y+\ \ z &= 1 \end{aligned}$$

Upon solving the system, we find that

$x = .457$, $y = .20$ and $z = .343$,

and the steady-state distribution vector is given by

$$\begin{bmatrix} .457 & .20 & .343 \end{bmatrix}.$$

Therefore, the steady-state matrix is

$$\begin{bmatrix} .457 & .20 & .343 \\ .457 & .20 & .343 \\ .457 & .20 & .343 \end{bmatrix}.$$

31. Let M denote the event that an employee at the insurance company is a male and let F denote the event that an employee at the insurance company is on flex time. Then

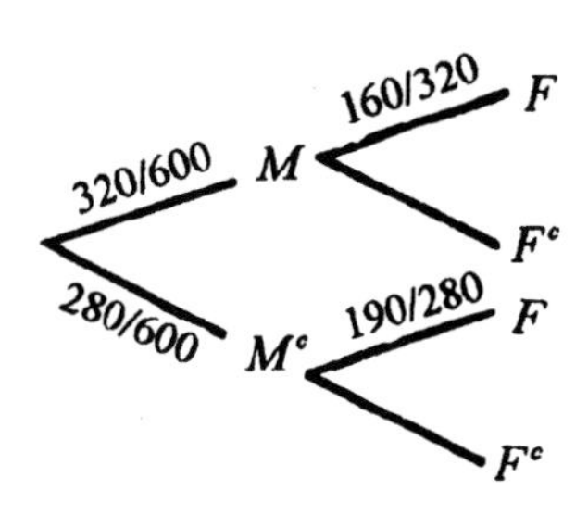

$$P(M|F) = \frac{\frac{320}{600} \cdot \frac{160}{320}}{\frac{320}{600} \cdot \frac{160}{320} + \frac{280}{600} \cdot \frac{190}{280}} = 0.4571.$$

33. Let E denote the event that an applicant selected at random is eligible for admission, and let Pa denote the event that an applicant selected at random passes the admission exam. Using the tree diagram that follows, we see that

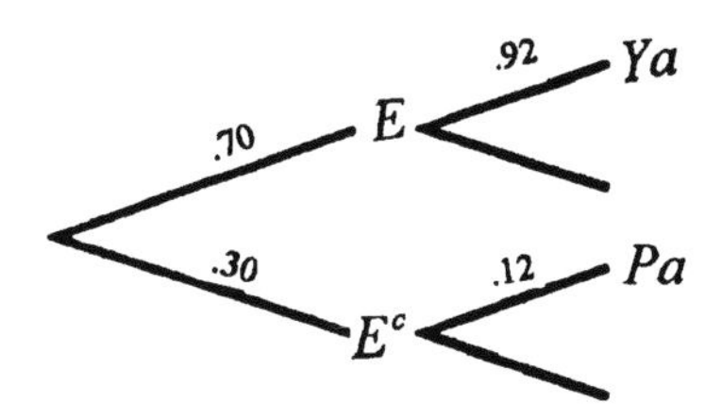

a. $P(Pa) = (0.70)(0.92) + (0.30)(0.12) = 0.68$.

b. $P(E^C \mid Pa) = \dfrac{(0.30)(0.12)}{(0.70)(0.92) + (0.30)(0.12)} = 0.053$.

CHAPTER 8

EXERCISES 8.1, page 468

1. a. See part (b).
   b.
   c. $\{GGG\}$

Outcome	GGG	GGR	GRG	RGG
Value	3	2	2	2

Outcome	GRR	RGR	RRG	RRR
Value	1	1	1	0

3. X may assume the values in the set $S = \{1,2,3,...\}$.

5. The event that the sum of the dice is 7 is

$$E = \{(1,6),(2,5),(3,4),(4,3),(5,2),(6,1)\}$$

and $P(E) = \frac{6}{36} = \frac{1}{6}$.

7. X may assume the value of any positive integer. The random variable is infinite discrete.

9. $\{d \mid d \geq 0\}$. The random variable is continuous.

11. X may assume the value of any positive integer. The random variable is infinite discrete.

13. a. $P(X = -10) = 0.20$
    b. $P(X \geq 5) = 0.1 + 0.25 + 0.1 + 0.15 = 0.60$
    c. $P(-5 \leq X \leq 5) = 0.15 + 0.05 + 0.1 = 0.30$
    d. $P(X \leq 20) = 0.20 + 0.15 + 0.05 + 0.1 + 0.25 + 0.1 + 0.15 = 1$

15.

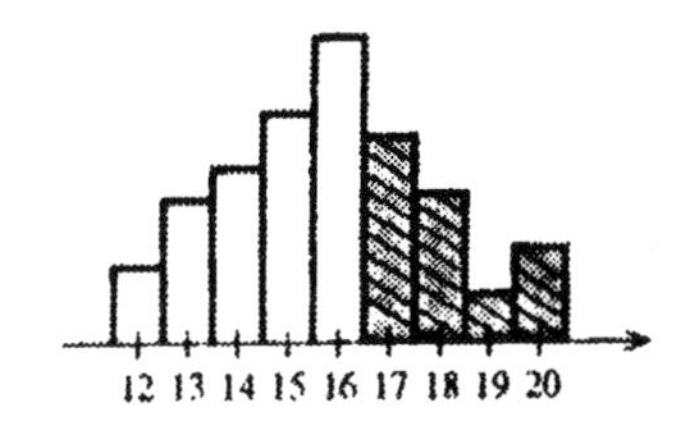

17. a.

x	1	2	3	4	5	6
$P(X=x)$	$\frac{1}{6}$	$\frac{1}{6}$	$\frac{1}{6}$	$\frac{1}{6}$	$\frac{1}{6}$	$\frac{1}{6}$

y	1	2	3	4	5	6
$P(Y=y)$	$\frac{1}{6}$	$\frac{1}{6}$	$\frac{1}{6}$	$\frac{1}{6}$	$\frac{1}{6}$	$\frac{1}{6}$

b.

$x+y$	2	3	4	5	6	7	8	9	10	11	12
$P(X+Y)=x+y)$	$\frac{1}{36}$	$\frac{2}{36}$	$\frac{3}{36}$	$\frac{4}{36}$	$\frac{5}{36}$	$\frac{6}{36}$	$\frac{5}{36}$	$\frac{4}{36}$	$\frac{3}{36}$	$\frac{2}{36}$	$\frac{1}{36}$

19. a.

x	0	1	2	3	4	5
$P(X=x)$	0.017	0.067	0.033	0.117	0.233	0.133

x	6	7	8	9	10
$P(X=x)$	0.167	0.1	0.05	0.067	0.017

b.

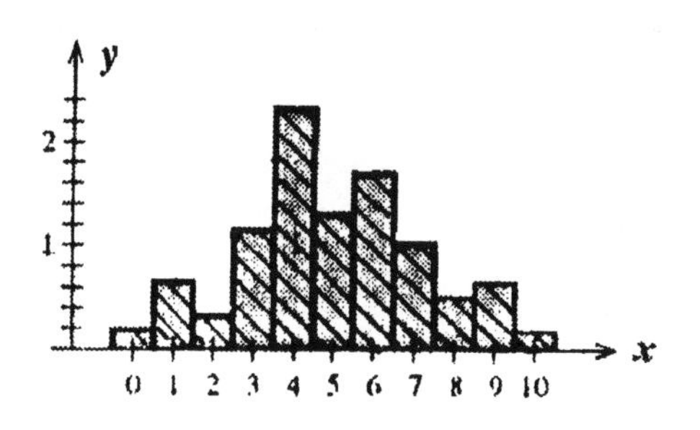

21.

x	1	2	3	4	5
$P(X=x)$	0.007	0.029	0.021	0.079	0.164

x	6	7	8	9	10
$P(X=x)$	0.15	0.20	0.207	0.114	0.029

USING TECHNOLOGY EXERCISES, page 451

1.

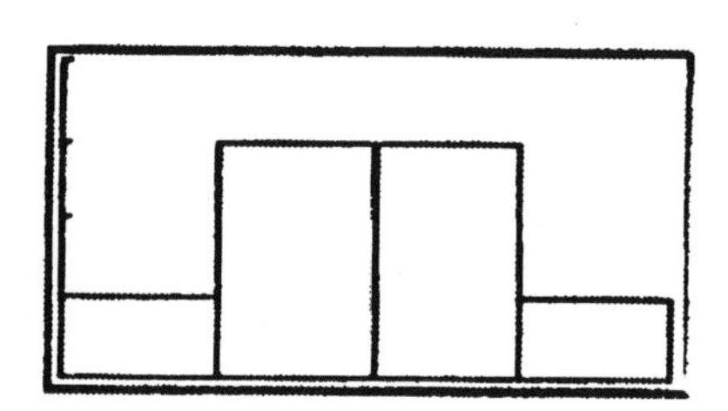

3.

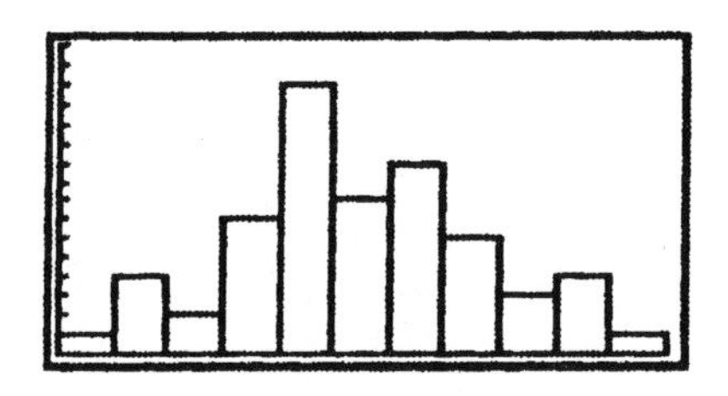

EXERCISES 8.2, page 483

1. a. The student's grade-point average is given by

$$\frac{(2)(4)(3)+(3)(3)(3)+(4)(2)(3)+(1)(1)(3)}{(10)(3)}$$

or 2.6.

b.

x	0	1	2	3	4
$P(X=x)$	0	0.1	0.4	0.3	0.2

$E(X) = 1(0.1) + 2(0.4) + 3(0.3) + 4(0.2) = 2.6.$

3. $E(X) = -5(0.12) + -1(0.16) + 0(0.28) + 1(0.22) + 5(0.12) + 8(0.1)$
$= 0.86.$

5. $E(X) = 0(0.07) + 25(0.12) + 50(0.17) + 75(0.14) + 100(0.28)$
$+ 125(0.18) + 150(0.04)$
$= 78.5,$ or \$78.50.

7. A customer entering the store is expected to buy

$E(X) = (0)(0.42) + (1)(0.36) + (2)(0.14) + (3)(0.05) + (4)(0.03)$

$= 0.91$, or 0.91 videocassettes.

9. The expected number of accidents is given by
$E(X) = (0)(0.935) + (1)(0.03) + (2)(0.02) + (3)(0.01) + (4)(0.005)$
$= 0.12.$

11. The expected number of machines that will break down on a given day is given by
$E(X) = (0)(0.43) + (1)(0.19) + (2)(0.12) + (3)(0.09) + (4)(0.04)$
$+ (5)(0.03) + (6)(0.03) + (7)(0.02) + (8)(0.05)$
$= 1.73.$

13. The expected gain of the insurance company is given by
$E(X) = 0.96P - (20{,}000 - P)(0.04),$
where P is the amount the man can expect to pay. We want $E(X) = 0$, that is,
$P - (0.04)(20{,}000) = 0.$
Solving for P, we find
$P = 800$
so the minimum he can expect to pay is \$800.

15. The expected gain of the insurance company is given by
$E(X) = 0.9935P - (25{,}000 - P)0.0065 \geq 0,$
where P is the amount that Mr. Taylor can expect to pay. Since
$E(X) \geq 0,$

$P - 0.0065(25{,}000) \geq 0.$

Solving for P, we find $P \geq 162.50$, so the minimum Mr. Taylor can expect to pay is $162.50.

17. The expected value of the first project is

$(0.7)(180{,}000) + (0.3)(150{,}000) = 171{,}000$, or \$171,000

and the expected value of the second project is

$(0.6)(220{,}000) + (0.4)(80{,}000) = 164{,}000$, or \$164,000.

He should choose the first project if he wants to maximize his expected profit.

19. a. DAHL MOTORS

$$E(X) = 5(0.05) + 6(0.09) + 7(0.14) + 8(0.24) + 9(0.18) + 10(0.14) + 11(0.11) + 12(0.05) = 8.52.$$

FARTHINGTON AUTO SALES

$$E(X) = 5(0.08) + 6(0.21) + 7(0.31) + 8(0.24) + 9(0.10) + 10(0.06) = 7.25.$$

b. The expected weekly profit from Dahl Motors is

$8.52(362) = 3084.24$,

or \$3084.24. The expected weekly profit from Farthington Auto Sales is

$7.25(436)$, or \$3161.

We conclude that Mr. Hunt should purchase Farthington Auto Sales.

21. a. The odds that it will rain tomorrow are 3 to 7.

b. The odds that it will not rain tomorrow are 7 to 3.

23. The expected value of the player's winnings are given by

$$E(X) = (1)(\frac{18}{38}) + (-1)(\frac{18}{38}) + (-2)(\frac{2}{38}) = -\frac{4}{38} \approx -0.105,$$

or a loss of $10\frac{1}{2}$ cents.

25. The odds in favor of E occurring are

$$\frac{P(E)}{P(E^C)} = \frac{0.8}{0.2},$$

or 4 to 1. The odds against E occurring are 1 to 4.

27. The probability of E occurring is given by

$$P(E)=\frac{9}{9+7}=\frac{9}{16}\approx 0.5625.$$

29. The probability that Joan will make the sale is given by

$$P(E)=\frac{8}{8+5}=\frac{8}{13}\approx 0.6154.$$

31. The probability that the boxer will win the match is given by

$$P(E)=\frac{4}{3+4}=\frac{4}{7}\approx 0.5714.$$

33. a. $E(cx)=(cx_1)p_1+(cx_2)p_2+\cdots+(cx_n)p_n$

$$=c(x_1p_1+x_2p_2+\cdots+x_np_n)$$
$$=cE(X).$$

b. The expected loss is $(300)(-\frac{2}{38})=-15.79$, or a loss of \$15.79.

35. The mean-wage rate is given by

$$(\tfrac{60}{450})(10.70)+(\tfrac{90}{450})(10.80)+(\tfrac{75}{450})(10.90)+(\tfrac{120}{450})(11.00)+(\tfrac{60}{450})(11.10)+(\tfrac{45}{450})(11.20)$$
$$=10.94$$

or \$10.94. The mode is \$11.00, and the median is

$$\frac{10.90+11.00}{2}=10.95, \text{ or } \$10.95.$$

EXERCISES 8.3, page 494

1. $\mu=(1)(0.4)+(2)(0.3)+3(0.2)+(4)(0.1)=2.$

$\text{Var}(X)=(0.4)(1-2)^2+(0.3)(2-2)^2+(0.2)(3-2)^2+(0.1)(4-2)^2$
$=0.4+0+0.2+0.4=1$

$\sigma=\sqrt{1}=1.$

3. $\mu=-2(\frac{1}{16})+-1(\frac{4}{16})+0(\frac{6}{16})+1(\frac{4}{16})+2(\frac{1}{16})=\frac{0}{16}=0.$

$\text{Var}(X)=\frac{1}{16}(-2-0)^2+\frac{4}{16}(-1-0)^2+\frac{6}{16}(0-0)^2+\frac{4}{16}(1-0)^2+\frac{1}{16}(2-0)^2$

$$= 1$$

$$\sigma = \sqrt{1} = 1.$$

5. $\mu = 0.1(430) + (0.2)(480) + (0.4)(520) + (0.2)(565) + (0.1)(580)$
$= 518.$

$$\text{Var}(X) = 0.1(430 - 518)^2 + (0.2)(480 - 518)^2 + (0.4)(520 - 518)^2 + (0.2)(565 - 518)^2 + (0.1)(580 - 518)^2 = 1891.$$

$$\sigma = \sqrt{1891} = 43.49.$$

7. The mean of the histogram in Figure (b) is more concentrated about its mean than the histogram in Figure (a). Therefore, the histogram in Figure (a) has the larger variance.

9. $E(X) = 1(0.1) + 2(0.2) + 3(0.3) + 4(0.2) + 5(0.2) = 3.2.$

$$\text{Var}(X) = (0.1)(1 - 3.2)^2 + (0.2)(2 - 3.2)^2 + (0.3)(3 - 3.2)^2 + (0.2)(4 - 3.2)^2 + (0.2)(5 - 3.2)^2 = 1.56$$

11. $\mu = \dfrac{1+2+3+\cdots+8}{8} = 4.5$

$$V(X) = \tfrac{1}{8}(1-4.5)^2 + \tfrac{1}{8}(2-4.5)^2 + \cdots + \tfrac{1}{8}(8-4.5)^2 = 5.25$$

13. a. Let X be the annual birth rate during the years 1981 - 1990.

b.

x	15.5	15.6	15.7	15.9	16.2	16.7
$P(X=x)$	0.2	0.1	0.3	0.2	0.1	0.1

c. $E(X) = (0.2)(15.5) + (0.1)(15.6) + (0.3)(15.7) + (0.2)(15.9) + (0.1)(16.2) + (0.1)(16.7)$
$= 15.84.$

$$V(X) = (0.2)(15.5 - 15.84)^2 + (0.1)(15.6 - 15.84)^2 + (0.3)(15.7 - 15.84)^2 + (0.2)(15.9 - 15.84)^2 + (0.1)(16.2 - 15.84)^2 + (0.1)(16.7 - 15.84)^2 = 0.1224$$

$\sigma = \sqrt{0.1224} \approx 0.350$.

15. a. Mutual Fund A

$\mu = (0.2)(-4) + (0.5)(8) + (0.3)(10) = 6.2$, or \$620.

$V(X) = (0.2)(-4 - 6.2)^2 + (0.5)(8 - 6.2)^2 + (0.3)(10 - 6.2)^2$

$= 26.76$, or \$267,600.

Mutual Fund B

$\mu = (0.2)(-2) + (0.4)(6) + (0.4)(8)$

$= 5.2$, or \$520.

$V(X) = (0.2)(-2 - 5.2)^2 + (0.4)(6 - 5.2)^2 + (0.4)(8 - 5.2)^2$

$= 13.76$, or \$137,600.

b. Mutual Fund A

c. Mutual Fund B

17. $\text{Var}(X) = (0.4)(1)^2 + (0.3)(2)^2 + (0.2)(3)^2 + (0.1)(4)^2 - (2)^2 = 1$.

19. $\mu = [\frac{10}{500}(80) + \frac{20}{500}(90) + \cdots + \frac{5}{500}(200)] = 139.6$, or \$139,600.

$V(X) = [\frac{10}{500}(80 - 13.96)^2 + \frac{20}{500}(90 - 13.96)^2 + \cdots + \frac{5}{500}(200 - 13.96)^2][(1000)^2]$

$= 1443.84 \times 10^6$ dollars.

$\sigma = \sqrt{1443.84 \times 10^6} = 37.998 \times 10^3$, or \$37,998.

21. a. Using Chebychev's inequality we have

$P(\mu - k\sigma \leq X \leq \mu + k\sigma) \geq 1 - 1/k^2$.

$\mu - k\sigma = 42 - k(2) = 38$, and $k = 2$,

and $P(\mu - k\sigma \leq X \leq \mu + k\sigma) \geq 1 - 1/(2)^2$

$\geq 1 - 1/4$

$\geq 3/4$, or at least 0.75.

b. Using Chebychev's inequality we have

$P(\mu - k\sigma \leq X \leq \mu + k\sigma) \geq 1 - 1/k^2$.

$\mu - k\sigma = 42 - k(2) = 32$, and $k = 5$,

and $P(\mu - k\sigma \leq X \leq \mu + k\sigma) \geq 1 - 1/(5)^2$

$\geq 1 - 1/25$

$\geq 24/25$, or at least 0.96.

23. Using Chebychev's inequality we have

$P(\mu - k\sigma \leq X \leq \mu + k\sigma) \geq 1 - 1/k^2$.

Here $k = 2$, so

$P(\mu - k\sigma \leq X \leq \mu + k\sigma) \geq 1 - 1/2^2 = 3/4$.

This means that at least 75 percent of the values are expected to lie between $\mu - 2\sigma$ and $\mu + 2\sigma$.

25. Using Chebychev's inequality we have

$$P(\mu - k\sigma \leq X \leq \mu + k\sigma) \geq 1 - 1/k^2.$$

Here, $\mu - k\sigma = 24 - k(3) = 20$, and $k = 4/3$.

So $\quad P(\mu - k\sigma \leq X \leq \mu + k\sigma) \geq 1 - 1/(4/3)^2 \geq 1 - 9/16 = 7/16.$

or at least 0.4375.

27. Here $\mu = 5$ and $\sigma = 0.02$. Next, we require that $c = k\sigma$, or $k = \dfrac{c}{0.02}$. Next, solve

$$0.96 = 1 - \left(\frac{0.02}{c}\right)^2$$

$$0.04 = \frac{0.0004}{c^2} = 0.01$$

$$c^2 = \frac{0.0004}{0.04} = 0.01 \quad \text{and} \quad c = 0.1.$$

We conclude that for $P(5 - c \leq X \leq 5 + c) \geq 0.96$. We require that $c \geq 0.1$.

USING TECHNOLOGY EXERCISES, page 479

1. a.

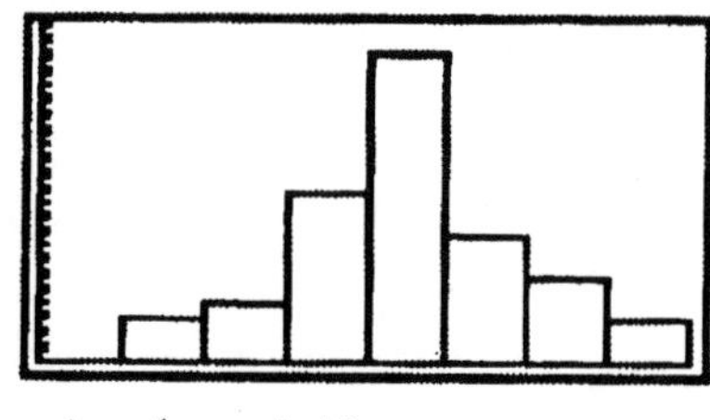

b. $\mu = 4$ and $\sigma = 1.40$

3. a.

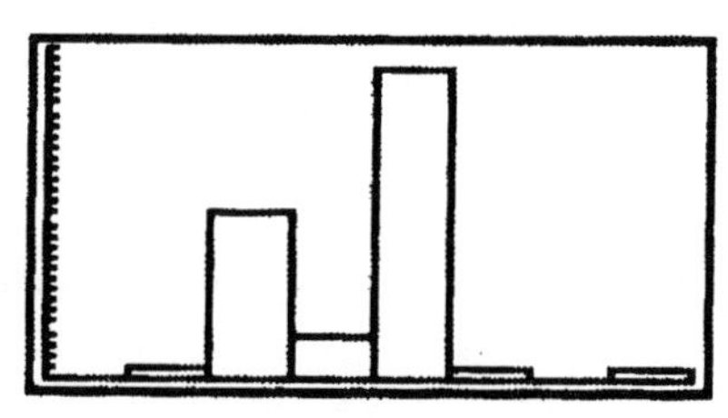

b. $\mu = 17.34$ and $\sigma = 1.11$

5. a. Let X denote the random variable that gives the weight of a carton of sugar.
b. The probability distribution for the random variable X is

x	4.96	4.97	4.98	4.99	5.00	5.01	5.02	5.03
$P(X = x)$	$\frac{3}{30}$	$\frac{4}{30}$	$\frac{4}{30}$	$\frac{1}{30}$	$\frac{1}{30}$	$\frac{5}{30}$	$\frac{3}{30}$	$\frac{3}{30}$

x	5.04	5.05	5.06
$P(X=x)$	$\frac{4}{30}$	$\frac{1}{30}$	$\frac{1}{30}$

$\mu = 5.00467 \approx 5.00.$ $V(X) = 0.0009.$ $\sigma = \sqrt{0.0009} = 0.03.$

EXERCISES 8.4, page 509

1. Yes. The number of trials is fixed, there are two outcomes of the experiment, the probability in each trial is fixed $(p = \frac{1}{6})$, and the trials are independent of each other.

3. No. There are more than 2 outcomes in each trial.

5. No. There are more than 2 outcomes in each trial and the probability of success (an accident) in each trial is not the same.

7. $C(4,2)(\frac{1}{2})^2(\frac{2}{3})^2 = \frac{4!}{2!2!}(\frac{4}{81}) \approx 0.296.$

9. $C(5,3)(0.2)^3(0.8)^2 = (\frac{5!}{2!3!})(0.2)^3(0.8)^2 \approx 0.0512.$

11. The required probability is given by
$$P(X=0) = C(5,0)(\tfrac{1}{3})^0(\tfrac{2}{3})^5 \approx 0.132.$$

13. The required probability is given by
$$\begin{aligned}P(X \geq 3) &= C(6,3)(\tfrac{1}{2})^3(\tfrac{1}{2})^{6-3} + C(6,4)(\tfrac{1}{2})^4(\tfrac{1}{2})^{6-4} + C(6,5)(\tfrac{1}{2})^5(\tfrac{1}{2})^{6-5}\\ &\quad + C(6,6)(\tfrac{1}{2})^6(\tfrac{1}{2})^{6-6}\\ &= \tfrac{6!}{3!3!}(\tfrac{1}{2})^6 + \tfrac{6!}{4!2!}(\tfrac{1}{2})^6 + \tfrac{6!}{5!1!}(\tfrac{1}{2})^6 + \tfrac{6!}{6!0!}(\tfrac{1}{2})^6\\ &= \tfrac{1}{64}(20+15+6+1) = \tfrac{21}{32}.\end{aligned}$$

15. The probability of no failures, or, equivalently, the probability of five successes is
$$P(X=5) = C(5,5)(\tfrac{1}{3})^5(\tfrac{2}{3})^{5-5} = \tfrac{1}{243} \approx 0.00412.$$

17. Here $n = 4$, and $p = 1/6$. Then

$$P(X=2) = C(4,2)(\tfrac{1}{6})^2(\tfrac{5}{6})^2 = (\tfrac{25}{216}) \approx 0.116.$$

19. Here $n = 5$, $p = 0.4$, and therefore, $q = 1 - 0.4 = 0.6$.

a. Using Table 2 in the Appendix, we see that

$P(X=0) = 0.078$, $P(X=1) = 0.259$, $P(X=2) = 0.346$,
$P(X=3) = 0.230$, $P(X=4) = 0.077$, $P(X=5) = 0.010$.

b.

x	0	1	2	3	4	5
$P(X=x)$	0.078	0.259	0.346	0.230	0.077	0.010

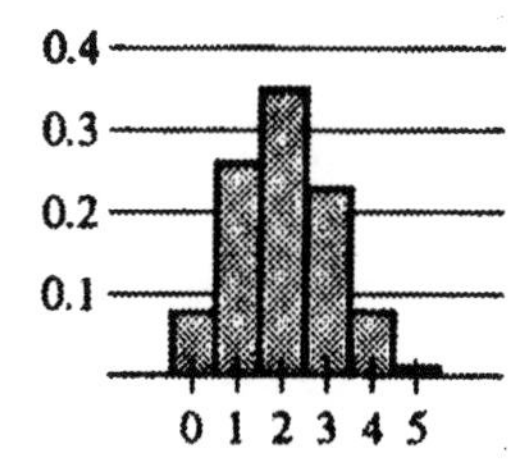

c. $\mu = np = 5(0.4) = 2$

$\sigma = \sqrt{npq} = \sqrt{5(0.4)(0.6)} = \sqrt{1.2} = 1.095.$

21. Here $1 - p = 1/50$ or $p = 49/50$. So the probability of obtaining 49 or 50 nondefective fuses is

$$P(X=49) + P(X=50) = C(50,49)(\tfrac{49}{50})^{49}(\tfrac{1}{50}) + C(50,50)(\tfrac{49}{50})^{50}(\tfrac{1}{50})^0 \approx 0.74.$$

This is also the probability of at most one defective fuse. So the inference is incorrect.

23. The required probability is given by

$$P(X=6) = C(6,6)(\tfrac{1}{4})^6(\tfrac{3}{4})^0 = 0.0002.$$

25. a. The probability that six or more people stated a preference for brand A is

$$\begin{aligned} P(X \geq 6) &= C(10,6)(.6)^6(.4)^4 + C(10,7)(.6)^7(.4)^3 \\ &\quad + C(10,8)(.6)^8(.4)^2 + C(10,9)(.6)^9(.4)^1 \\ &\quad + C(10,10)(.6)^{10}(.4)^0 \\ &\approx 0.251 + 0.215 + 0.121 + 0.040 + 0.006 = 0.633. \end{aligned}$$

(Use Table 2 in Appendix C.)

b. The required probability is 1 - 0.633 = 0.367.

27. This is a binomial experiment with $n = 9$, $p = 1/3$, and $q = 2/3$.
a. The probability is given by
$$P(X = 3) = C(9,3)(\tfrac{1}{3})^3(\tfrac{2}{3})^6 = (\tfrac{9!}{6!3!})(\tfrac{1}{3})^3(\tfrac{2}{3})^6 \approx 0.273.$$
b. The probability is given by
$$\begin{aligned} &P(X=0) + P(X=1) + P(X=2) + P(X=3) \\ &= C(9,0)(\tfrac{1}{3})^0(\tfrac{2}{3})^9 + C(9,1)(\tfrac{1}{3})(\tfrac{2}{3})^8 + C(9,2)(\tfrac{1}{3})^2(\tfrac{2}{3})^7 + C(9,3)(\tfrac{1}{3})^3(\tfrac{2}{3})^6 \\ &= 0.026 + 0.117 + 0.234 + 0.273 \approx 0.650. \end{aligned}$$

29. In order to obtain a score of at least 90 percent the student needs to answer 3 or 4 of the remaining questions correctly. The probability of doing this is
$$P(X \geq 3) = C(4,3)(0.5)^3(0.5) + C(4,4)(0.5)^4(0.5)^0 = 0.3125.$$

31. Here $1 - p = 0.015$ and $p = 0.985$. The probability that the sample will be accepted, or, equivalently, that the sample contains all six nondefective cartridges is given by
$$P(X = 6) = C(6,6)(0.985)^6(0.015)^0 \approx 0.913.$$

33. This is a binomial experiment with $n = 4$, $p = 0.001$, and $q = 0.999$.
a. The probability that exactly one engine will fail is given by
$$P(X = 1) = C(4,1)(0.001)(0.999)^3 = 4(0.001)(0.999)^3 = 0.003988.$$
b. The probability that exactly two engines will fail is given by
$$\begin{aligned} P(X = 2) &= C(4,2)(0.001)^2(0.999)^2 = 6(0.001)^2(0.999)^2 \\ &= 0.000006. \end{aligned}$$
c. The probability that more than two engines will fail is given by
$$\begin{aligned} P(X > 2) &= P(X = 3) + P(X = 4) \\ &= C(4,3)(0.001)^3(0.999) + C(4,4)(0.001)^4(0.999)^0 \\ &= 3.996 \times 10^{-9} + 1 \times 10^{-12} \approx 3.997 \times 10^{-9}. \end{aligned}$$

35. The required probability is
$$\begin{aligned} P(X \leq 1) &= P(X = 0) + P(X = 1) \\ &= C(20,0)(0.1)^0(0.9)^{20} + C(20,1)(0.1)^1(0.9)^{19} \approx 0.3917. \end{aligned}$$

37. Take $p = 1/2$. The probability of obtaining no heads in n tosses is
$$P(X = n) = C(n,n)(\tfrac{1}{2})^n(\tfrac{1}{2})^0 = (\tfrac{1}{2})^n.$$
The probability of obtaining at least one head is $1 - (\frac{1}{2})^n$. We want this to exceed 0.99. Thus,

$$1-\left(\frac{1}{2}\right)^n \geq 0.99$$

$$\frac{1}{2^n} \geq 0.01$$

$$2^n \geq 100$$

or $$n \geq \frac{\ln 100}{\ln 2} \approx 6.64$$

So one must toss the coin at least 7 times.

39. a. The expected number of students who will graduate within four years is

$$\mu = np = (0.6)(2000) = 1200.$$

b. The standard deviation of the number of students who will graduate within four years is

$$\sigma = \sqrt{npq} = \sqrt{(2000)(0.6)(0.4)} \approx 21.91.$$

EXERCISES 8.5, page 520

1. $P(Z < 1.45) = 0.9265.$

3. $P(Z < -1.75) = 0.0401.$

5. $P(-1.32 < Z < 1.74) = P(Z < 1.74) - P(Z < -1.32)$
$= 0.9591 - 0.0934 = 0.8657.$

7. $P(Z < 1.37) = 0.9147.$

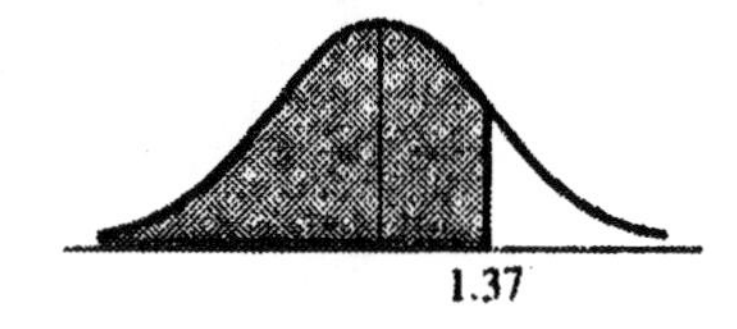

9. $P(Z < -0.65) = 0.2578.$

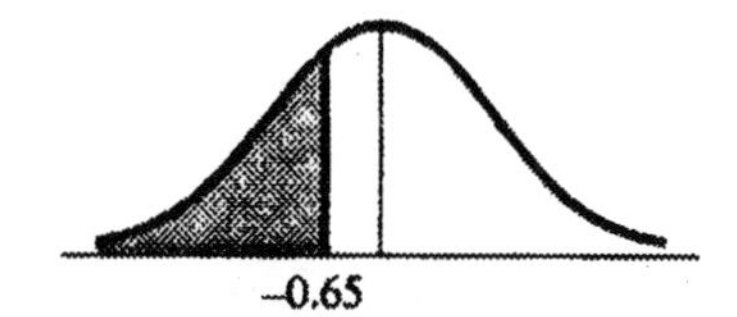

11. $P(Z > -1.25) = 1 - P(Z < -1.25) = 1 - 0.1056 = 0.8944$

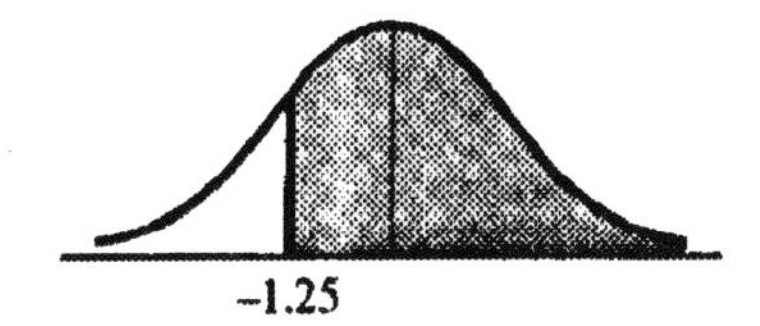

13. $P(0.68 < Z < 2.02) = P(Z < 2.02) - P(Z < 0.68)$
$= 0.9783 - 0.7517 = 0.2266.$

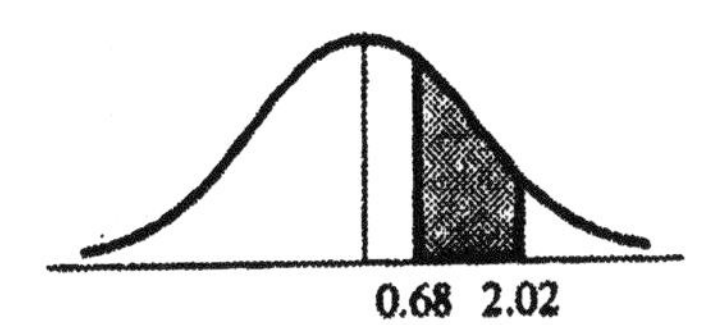

15. a. Referring to Table 3, we see that $P(Z < z) = 0.8907$ implies that $z = 1.23$.

b. Referring to Table 3, we see that $P(Z < z) = 0.2090$ implies that $z = -0.81$.

17. a. $P(Z > -z) = 1 - P(Z < -z) = 1 - 0.9713 = 0.0287$ implies $z = 1.9$.
b. $P(Z < -z) = 0.9713$ implies that $z = -1.9$.

19. a. $P(X < 60) = P(Z < \frac{60-50}{5}) = P(Z < 2) = 0.9772.$

b. $P(X > 43) = P(Z > \frac{43-50}{5}) = P(Z > -1.4) = P(Z < 1.4) = 0.9192.$

c. $P(46 < X < 58) = P(\frac{46-50}{5} < Z < \frac{58-50}{5}) = P(-0.8 < Z < 1.6)$
$= P(Z < 1.6) - P(Z < -0.8)$
$= 0.9452 - 0.2119 = 0.7333.$

EXERCISES 8.6, page 530

1. $\mu = 20$ and $\sigma = 2.6$.

a. $P(X > 22) = P(Z > \frac{22-20}{2.6}) = P(Z > 0.77) = P(Z < -0.77) = 0.2206.$

b. $P(X < 18) = P(Z < \frac{18-20}{2.6}) = P(Z < -0.77) = 0.2206.$

c. $P(19 < X < 21) = P(\frac{19-20}{2.6} < Z < \frac{21-20}{2.6}) = P(-0.39 < Z < 0.39)$

$= P(Z < 0.39) - P(Z < -0.39)$

$= 0.6517 - 0.3483 = 0.3034.$

3. $\mu = 750$ and $\sigma = 75$.

a. $P(X > 900) = P(Z > \frac{900-750}{75}) = P(Z > 2) = P(Z < -2) = 0.0228.$

b. $P(X < 600) = P(Z < \frac{600-750}{75}) = P(Z < -2) = 0.0228.$

c. $P(750 < X < 900) = P(Z < \frac{750-750}{75} < Z < \frac{900-750}{75})$

$= P(0 < Z < 2) = P(Z < 2) - P(Z < 0)$

$= 0.9772 - 0.5000 = 0.4772.$

d. $P(600 < X < 800) = P(\frac{600-750}{75} < Z < \frac{800-750}{75})$

$= P(-2 < Z < .667) = P(Z < .667) - P(Z < -2)$

$= 0.7486 - 0.0228 = 0.7258.$

5. $\mu = 100$ and $\sigma = 15$.

a. $P(X > 140) = P(Z > \frac{140-100}{15}) = P(Z > 2.667) = P(Z < -2.667)$

$= 0.0038.$

b. $P(X > 120) = P(Z > \frac{120-100}{15}) = P(Z > 1.33) = P(Z < -1.33)$

$= 0.0918.$

c. $P(100 < X < 120) = P(\frac{100-100}{15} < Z < \frac{120-100}{15}) = P(0 < Z < 1.333)$

$= P(Z < 0) - P(Z < 1.333)$

$= 0.9082 - 0.5000 = 0.4082.$

d. $P(X < 90) = P(Z < \frac{90-100}{15}) = P(Z < -0.667) = 0.2514.$

7. Here $\mu = 375$ and $\sigma = 50$.

$$P(350 < X < 450) = P(\frac{350-375}{50} < Z < \frac{450-375}{50}) = P(-0.5 < Z < 1.5)$$
$$= P(Z < 1.5) - P(Z < -0.5)$$
$$= 0.9332 - 0.3085 = 0.6247.$$

9. Here $\mu = 22$ and $\sigma = 4$.

$$P(X < 12) = P(Z < \frac{12-22}{4}) = P(Z < -2.5) = 0.0062, \text{ or } 0.62 \text{ percent.}$$

11. $\mu = 70$ and $\sigma = 10$.

To find the cut-off point for an A, we solve $P(Y < y) = 0.85$ for y. Now

$$P(Y < y) = P\left(Z < \frac{y-70}{10}\right) = 0.85 \text{ implies } \frac{y-72}{10} = 1.04$$

or $y = 80.4 \approx 80$.

For a B: $P(Y < y) = P\left(Z < \frac{y-70}{10}\right) = 0.60$ implies $\frac{y-70}{10} = 0.25$, or $y \approx 73$.

For a C: $P\left(Z \le \frac{y-70}{10}\right) = 0.2$ implies $\frac{y-70}{10} = -0.84$ or $y \approx 62$.

For a D: $P\left(Z < \frac{y-70}{10}\right) = 0.05$ implies $\frac{y-70}{10} = -1.65$, or $y \approx 54$.

13. Let X denote the number of heads in 25 tosses of the coin. Then X is a binomial random variable. Also, $n = 25$, $p = 0.4$, and $q = 0.6$. So

$$\mu = (25)(0.4) = 10$$
$$\sigma = \sqrt{(25)(0.4)(0.6)} \approx 2.45.$$

Approximating the binomial distribution by a normal distribution with a mean of 10 and a standard deviation of 2.45, we find upon letting Y denote the associated normal random variable,

a. $P(X < 10) \approx P(Y < 9.5)$

$$= P\left(Z < \frac{9.5-10}{2.45}\right) = P(Z < -0.20) = 0.4207.$$

b. $P(10 \le X \le 12) \approx P(9.5 < Y < 12.5)$

$$= P\left(\frac{9.5-10}{2.45} < Z < \frac{12.5-10}{2.45}\right) = P(Z < 1.02) - P(Z < -0.20)$$
$$= P(Z < 1.02) - P(Z < -0.20) = 0.8461 - 0.4207 = 0.4254.$$

c. $P(X > 15) \approx P(Y \geq 15)$

$$= P(Z > \frac{15.5-10}{2.45}) = P(Z > 2.25) = P(Z < -2.25) = 0.0122.$$

15. Let X denote the number of times the marksman hits his target. Then X has a binomial distribution with $n = 30$, $p = 0.6$ and $q = 0.4$. Therefore,

$$\mu = (30)(0.6) = 18, \quad \sigma = \sqrt{(30)(0.6)(0.4)} = 2.68.$$

a. $P(X \geq 20) \approx P(Y \geq 19.5)$

$$= P\left(Z > \frac{19.5-18}{2.68}\right) = P(Z > 0.56) = P(Z < -0.56) = 0.2877.$$

b. $P(X < 10) \approx P(Y < 9.5) = P\left(Z < \frac{9.5-18}{2.68}\right) = P(Z < -3.17) = 0.0008.$

c. $P(15 \leq X \leq 20) \approx P(14.5 < Y < 20.5) = P\left(\frac{14.5-18}{2.68} < Z < \frac{20.5-18}{2.68}\right)$

$$= P(Z < 0.93) - P(Z < -1.31)$$

$$= 0.8238 - 0.0951 = 0.7287.$$

17. Let X denote the number of "seconds." Then X has a binomial distribution with $n = 200$, $p = 0.03$, and $q = 0.97$. Then

$$\mu = (200)(0.03) = 6$$

$$\sigma = \sqrt{(200)(0.03)(0.97)} \approx 2.41,$$

and $P(X < 10) \approx P(Y < 9.5) = P(Z < \frac{9.5-6}{2.41}) = \text{P}(Z < 1.45) = 0.9265.$

19. a. Let X denote the number of mice that recovered from the disease. Then X has a binomial distribution with $n = 50$, $p = 0.5$, and $q = 0.5$, so

$$\mu = (50)(0.5) = 25$$

$$\sigma = \sqrt{(50)(0.5)(0.5)} \approx 3.54,$$

Approximating the binomial distribution by a normal distribution with a mean of 25 and a standard deviation of 3.54, we find that the probability that 35 or more of the mice would recover from the disease without benefit of the drug is

$$P(X \geq 35) \approx P(Y > 34.5)$$

$$= P(Z > \frac{34.5-25}{3.54})$$

$= P(Z > 2.68) = P(Z < -2.68) = 0.0037.$

b. The drug is very effective.

21. Let n denote the number of reservations the company should accept. Then we need to find

$$P(X \geq 2000) \approx P(Y > 1999.5) = 0.01$$

or equivalently,

$$P(Z \geq \frac{1999.5 - np}{\sqrt{npq}}) = 0.01 \qquad \text{[Here p = 0.92 and q = 0.08.]}$$

Using Table 3 in Appendix C, we find

$$\frac{1999.5 - 0.92n}{\sqrt{0.0736n}} = -2.33$$

$$(1999.5 - 0.92n)^2 = (-2.33)^2(0.0736n)$$

$$3{,}998{,}000.25 - 3679.08n + 0.8464n^2 = 0.39956704n,$$

or $$0.8464n^2 - 3679.479567n + 3{,}998{,}000.25 = 0.$$

Using the quadratic formula, we obtain

$$n = \frac{3679.479567 \pm \sqrt{2940.2375}}{1.6928} \approx 2142,$$

or 2142. [You can verify that 2206 is not a root of the original equation (before squaring).] Therefore, the company should accept no more than 2142 reservations.

CHAPTER 8, REVIEW EXERCISES, page 534

1. a. $S = \{WWW, WWB, WBW, WBB, BWW, BWB, BBW, BBB\}$

b.

Outcome	*WWW*	*WWB*	*WBW*	*WBB*
Value	0	1	1	2

Outcome	*BWW*	*BWB*	*BBW*	*BBB*
Value	1	2	2	3

c.

x	0	1	2	3
$P(X = x)$	$\frac{1}{35}$	$\frac{12}{35}$	$\frac{18}{35}$	$\frac{4}{35}$

d.

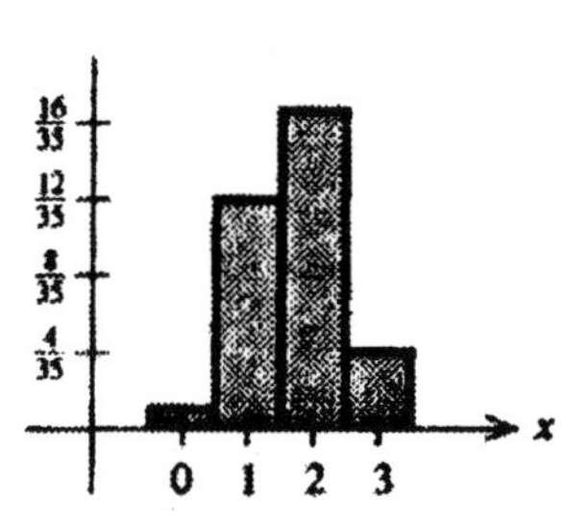

3. a. $P(1 \le X \le 4) = 0.1 + 0.2 + 0.3 + 0.2 = 0.8.$

b. $\mu = 0(0.1) + 1(0.1) + 2(0.2) + 3(0.3) + 4(0.2) + 5(0.1) = 2.7.$

$$V(X) = 0.1(0 - 2.7)^2 + 0.1(1 - 2.7)^2 + 0.2(2 - 2.7)^2 + 0.3(3 - 2.7)^2 + 0.2(4 - 2.7)^2 + .1(5 - 2.7)^2$$
$$= 2.01$$

$$\sigma = \sqrt{2.01} \approx 1.418.$$

5. $P(Z < 0.5) = 0.6915.$

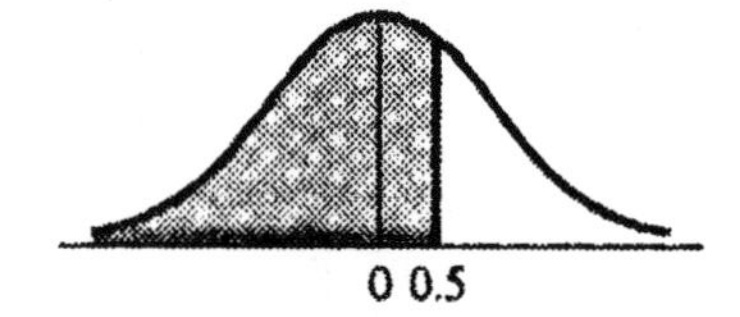

7. $P(-0.75 < Z < 0.5) = P(Z < 0.5) - P(Z < -0.75)$
$= 0.6915 - 0.2266 = 0.4649.$

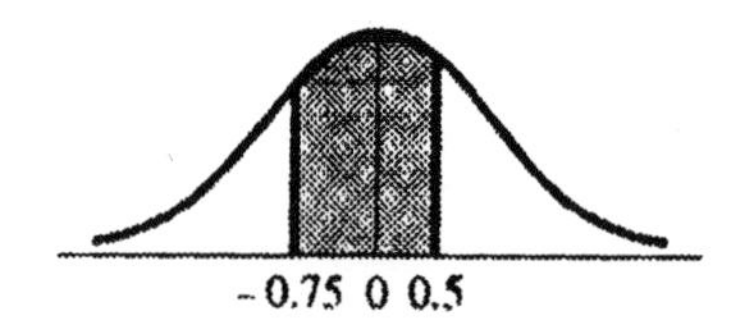

9. If $P(Z < z) = 0.9922$, then $z = 2.42$.

11. If $P(Z > z) = 0.9788$, then $P(Z < -z) = 0.9788$, and $-z = 2.03$, or $z = -2.03$.

13. $P(X < 11) = P\left(Z < \frac{11-10}{2}\right) = P(Z < 0.5) = 0.6915.$

15. $P(7 < X < 9) = P\left(\frac{7-10}{2} < Z < \frac{9-10}{2}\right) = P(-1.5 < Z < -0.5)$

$= P(Z < -0.5) - P(Z < -1.5)$

$= 0.3085 - 0.0668 = 0.2417.$

17. This is a binomial experiment with $p = 0.7$, and so $q = 0.3$. The probability that he will get exactly two strikes in four attempts is given by

$$P(X = 2) = C(4,2)(0.7)^2(0.3)^2 \approx 0.2646.$$

The probability that he will get at least two strikes in four attempts is given by

$P(X = 2) + P(X = 3) + P(X = 4)$

$= C(4,2)(0.7)^2(0.3)^2 + C(4,3)(0.7)^3(0.3) + C(4,4)(0.7)^4(0.3)^0$

$= 0.2646 + 0.4116 + 0.2401 \approx 0.9163.$

19. Here $\mu = 64.5$ and $\sigma = 2.5$. Next,

$$64.5 - 2.5k = 59.5 \text{ and } 64.5 + 2.5k = 69.5$$

and $k = 2$. Therefore, the required probability is given by

$$P(59.5 \leq X \leq 69.5) \geq 1 - \frac{1}{2^2} = 0.75.$$

21. Let the random variable X be the number of people for whom the drug is effective. Then

$$\mu = (0.15)(800) = 120 \text{ and } \sigma = \sqrt{(800)(0.15)(0.85)} = \sqrt{102} \approx 10.1.$$

23. Here $\mu = (0.6)(100) = 60$ and $\sigma = \sqrt{100(0.6)(0.4)} = 4.899$.

Then $P(X > 50) \approx P(Y > 50.5) = P\left(Z > \frac{50.5-60}{4.899}\right) = P(Z > -1.94)$

$= P(Z < 1.94) = 0.9738.$

CHAPTER 9

EXERCISES 9.1, page 541

1. $27^{2/3} = (3^3)^{2/3} = 3^2 = 9.$

3. $\left(\dfrac{1}{\sqrt{3}}\right)^0 = 1.$ Recall that any number raised to the zero power is 1.

5. $\left[\left(\dfrac{1}{8}\right)^{1/3}\right]^{-2} = \left(\dfrac{1}{2}\right)^{-2} = (2^2) = 4.$

7. $\left(\dfrac{7^{-5}\cdot 7^2}{7^{-2}}\right)^{-1} = (7^{-5+2+2})^{-1} = (7^{-1})^{-1} = 7^1 = 7.$

9. $(125^{2/3})^{-1/2} = 125^{(2/3)(-1/2)} = 125^{-1/3} = \dfrac{1}{125^{1/3}} = \dfrac{1}{5}.$

11. $\dfrac{\sqrt{32}}{\sqrt{8}} = \sqrt{\dfrac{32}{8}} = \sqrt{4} = 2.$

13. $\dfrac{16^{5/8}16^{1/2}}{16^{7/8}} = 16^{(5/8+1/2-7/8)} = 16^{1/4} = 2.$

15. $16^{1/4}\cdot 8^{-1/3} = 2\cdot\left(\dfrac{1}{8}\right)^{1/3} = 2\cdot\dfrac{1}{2} = 1.$

17. True.

19. False. $x^3 \times 2x^2 = 2x^{3+2} = 2x^5 \neq 2x^6.$

21. False. $\dfrac{2^{4x}}{1^{3x}} = \dfrac{2^{4x}}{1} = 2^{4x}.$

23. False. $\dfrac{1}{4^{-3}} = 4^3 = 64.$

25. False. $(1.2^{1/2})^{-1/2} = (1.2)^{-1/4} \neq 1.$

27. $(xy)^{-2} = \dfrac{1}{(xy)^2}.$

29. $\dfrac{x^{-1/3}}{x^{1/2}} = x^{(-1/3)-(1/2)} = x^{-5/6} = \dfrac{1}{x^{5/6}}.$

31. $12^0(s+t)^{-3} = 1\cdot\frac{1}{(s+t)^3} = \frac{1}{(s+t)^3}.$

33. $\frac{x^{7/3}}{x^{-2}} = x^{(7/3)+2} = x^{(7/3)+(6/3)} = x^{13/3}.$

35. $(x^2y^{-3})(x^{-5}y^3) = (x^{2-5}y^{-3+3}) = x^{-3}y^0 = x^{-3} = \frac{1}{x^3}.$

37. $\frac{x^{3/4}}{x^{-1/4}} = x^{(3/4)-(-1/4)} = x^{4/4} = x.$

39. $\left(\frac{x^3}{-27y^6}\right)^{-2/3} = x^{3(-2/3)}\left(-\frac{1}{27}\right)^{-2/3} y^{6(-2/3)} = x^{-2}\left(-\frac{1}{3}\right)^{-2} y^{-4} = \frac{9}{x^2y^4}.$

41. $\left(\frac{x^{-3}}{y^{-2}}\right)\left(\frac{y}{x}\right)^4 = \frac{x^{-3(2)}y^4}{y^{-2(2)}x^4} = \left(\frac{y^{4+4}}{x^{4+6}}\right) = \frac{y^8}{x^{10}}.$

43. $\sqrt[3]{x^{-2}}\cdot\sqrt{4x^5} = x^{-2/3}\cdot 4^{1/2}\cdot x^{5/2} = x^{-(2/3)+(5/2)}\cdot 2 = 2x^{11/6}.$

45. $-\sqrt[4]{16x^4y^8} = -(16^{1/4}\cdot x^{4/4}\cdot y^{8/4}) = -2xy^2.$

47. $\sqrt[6]{64x^8y^3} = (64)^{1/6}\cdot x^{8/6}y^{3/6} = 2x^{4/3}y^{1/2}.$

49. $2^{3/2} = (2)(2^{1/2}) = 2(1.414) = 2.828.$

51. $9^{3/4} = (3^2)^{3/4} = 3^{6/4} = 3^{3/2} = 3\cdot 3^{1/2} = 3(1.732) = 5.196.$

53. $10^{3/2} = 10^{1/2}\cdot 10 = (3.162)(10) = 31.62.$

55. $10^{2.5} = 10^2\cdot 10^{1/2} = 100(3.162) = 316.2.$

57. $\frac{3}{2\sqrt{x}}\cdot\frac{\sqrt{x}}{\sqrt{x}} = \frac{3\sqrt{x}}{2x}.$

59. $\frac{2y}{\sqrt{3y}} \cdot \frac{\sqrt{3y}}{\sqrt{3y}} = \frac{2y\sqrt{3y}}{3y} = \frac{2}{3}\sqrt{3y}.$

61. $\frac{1}{\sqrt[3]{x}} \cdot \frac{\sqrt[3]{x^2}}{\sqrt[3]{x^2}} = \frac{\sqrt[3]{x^2}}{\sqrt[3]{x^3}} = \frac{\sqrt[3]{x^2}}{x}.$

63. $\frac{2\sqrt{x}}{3} \cdot \frac{\sqrt{x}}{\sqrt{x}} = \frac{2x}{3\sqrt{x}}.$

65. $\sqrt{\frac{2y}{x}} = \frac{\sqrt{2y}}{\sqrt{x}} \cdot \frac{\sqrt{2y}}{\sqrt{2y}} = \frac{2y}{\sqrt{2xy}}.$

67. $\frac{\sqrt[3]{x^2 z}}{y} \cdot \frac{\sqrt[3]{xz^2}}{\sqrt[3]{xz^2}} = \frac{\sqrt[3]{x^3 z^3}}{y\sqrt[3]{xz^2}} = \frac{xz}{y\sqrt[3]{xz^2}}.$

EXERCISES 9.2, page 550

1. $(7x^2 - 2x + 5) + (2x^2 + 5x - 4) = 7x^2 - 2x + 5 + 2x^2 + 5x - 4$
$= 9x^2 + 3x + 1.$

3. $(5y^2 - 2y + 1) - (y^2 - 3y - 7) = 5y^2 - 2y + 1 - y^2 + 3y + 7$
$= 4y^2 + y + 8.$

5. $x - \{2x - [-x - (1 - x)]\} = x - \{2x - [-x - 1 + x]\}$
$= x - \{2x + 1\}$
$= x - 2x - 1$
$= -x - 1.$

7. $(\frac{1}{3} - 1 + e) - (-\frac{1}{3} - 1 + e^{-1}) = \frac{1}{3} - 1 + e + \frac{1}{3} + 1 - \frac{1}{e}$
$= \frac{2}{3} + e - \frac{1}{e}$
$= \frac{3e^2 + 2e - 3}{3e}.$

9. $3\sqrt{8}+8-2\sqrt{y}+\frac{1}{2}\sqrt{x}-\frac{3}{4}\sqrt{y}=3\sqrt{4\cdot 2}+8+\frac{1}{2}\sqrt{x}-\frac{11}{4}\sqrt{y}$

$=6\sqrt{2}+8+\frac{1}{2}\sqrt{x}-\frac{11}{4}\sqrt{y}.$

11. $(x + 8)(x - 2) = x(x - 2) + 8(x - 2) = x^2 - 2x + 8x - 16 = x^2 + 6x - 16.$

13. $(a + 5)^2 = (a + 5)(a + 5) = a(a + 5) + 5(a + 5) = a^2 + 5a + 5a + 25$
$= a^2 + 10a + 25.$

15. $(x + 2y)^2 = (x + 2y)(x + 2y) = x(x + 2y) + 2y(x + 2y)$
$= x^2 + 2xy + 2yx + 4y^2 = x^2 + 4xy + 4y^2.$

17. $(2x + y)(2x - y) = 2x(2x - y) + y(2x - y) = 4x^2 - 2xy + 2xy - y^2$
$= 4x^2 - y^2.$

19. $(x^2 - 1)(2x) - x^2(2x) = 2x^3 - 2x - 2x^3 = -2x.$

21. $2\left(t+\sqrt{t}\right)^2-2t^2=2(t+\sqrt{t})(t+\sqrt{t})-2t^2$

$=2(t^2+2t\sqrt{t}+t)-2t^2$

$=2t^2+4t\sqrt{t}+2t-2t^2$

$=4t\sqrt{t}+2t=2t(2\sqrt{t}+1).$

23. $4x^5 - 12x^4 - 6x^3 = 2x^3(2x^2 - 6x - 3).$

25. $7a^4 - 42a^2b^2 + 49a^3b = 7a^2(a^2 - 6b^2 + 7ab).$

27. $e^{-x} - xe^{-x} = e^{-x}(1 - x).$

29. $2x^{-5/2}-\frac{3}{2}x^{-3/2}=\frac{1}{2}x^{-5/2}(4-3x).$

31. $6ac + 3bc - 4ad - 2bd = 3c(2a + b) - 2d(2a + b) = (2a + b)(3c - 2d).$

33. $4a^2 - b^2 = (2a + b)(2a - b).$ (Difference of two squares)

35. $10 - 14x - 12x^2 = -2(6x^2 + 7x - 5) = -2(3x + 5)(2x - 1).$

37. $3x^2 - 6x - 24 = 3(x^2 - 2x - 8) = 3(x - 4)(x + 2).$

39. $12x^2 - 2x - 30 = 2(6x^2 - x - 15) = 2(3x - 5)(2x + 3)$.

41. $9x^2 - 16y^2 = (3x)^2 - (4y)^2 = (3x - 4y)(3x + 4y)$.

43. $x^6 + 125 = (x^2)^3 + (5)^3 = (x^2 + 5)(x^4 - 5x^2 + 25)$.

45. $(x^2 + y^2)x - xy(2y) = x^3 + xy^2 - 2xy^2 = x^3 - xy^2$.

47. $$\begin{aligned} &2(x - 1)(2x + 2)^3[4(x - 1) + (2x + 2)] \\ &\quad = 2(x - 1)(2x + 2)^3[4x - 4 + 2x + 2] \\ &\quad = 2(x - 1)(2x + 2)^3[6x - 2] \\ &\quad = 4(x - 1)(3x - 1)(2x + 2)^3. \end{aligned}$$

49. $$\begin{aligned} &4(x - 1)^2(2x + 2)^3(2) + (2x + 2)^4(2)(x - 1) \\ &\quad = 2(x - 1)(2x + 2)^3[4(x - 1) + (2x + 2)] \\ &\quad = 2(x - 1)(2x + 2)^3(6x - 2) \\ &\quad = 4(x - 1)(3x - 1)(2x + 2)^3. \end{aligned}$$

51. $$\begin{aligned} (x^2 + 2)^2[5(x^2 + 2)^2 - 3](2x) &= (x^2 + 2)^2[5(x^4 + 4x^2 + 4) - 3](2x) \\ &= (2x)(x^2 + 2)^2(5x^4 + 20x^2 + 17). \end{aligned}$$

53. $x^2 + x - 12 = 0$, or $(x + 4)(x - 3) = 0$, so that $x = -4$ or $x = 3$. We conclude that the roots are $x = -4$ and $x = 3$.

55. $4t^2 + 2t - 2 = (2t - 1)(2t + 2) = 0$. Thus, $t = 1/2$ and $t = -1$ are the roots.

57. $\frac{1}{4}x^2 - x + 1 = (\frac{1}{2}x - 1)(\frac{1}{2}x - 1) = 0$. Thus $\frac{1}{2}x = 1$, and $x = 2$ is a double root of the equation.

59. Here we use the quadratic formula to solve the equation $4x^2 + 5x - 6 = 0$. Then, $a = 4$, $b = 5$, and $c = -6$. Therefore,

$$\begin{aligned} x &= \frac{-b \pm \sqrt{b^2 - 4ac}}{2a} = \frac{-(5) \pm \sqrt{(5)^2 - 4(4)(-6)}}{2(4)} = \frac{-5 \pm \sqrt{121}}{8} \\ &= \frac{-5 \pm 11}{8}. \end{aligned}$$

Thus, $x = -\frac{16}{8} = -2$ and $x = \frac{6}{8} = \frac{3}{4}$ are the roots of the equation.

61. We use the quadratic formula to solve the equation $8x^2 - 8x - 3 = 0$. Here $a = 8$, $b = -8$, and $c = -3$. Therefore,

$$x = \frac{-b \pm \sqrt{b^2 - 4ac}}{2a} = \frac{-(-8) \pm \sqrt{(-8)^2 - 4(8)(-3)}}{2(8)} = \frac{8 \pm \sqrt{160}}{16}$$

$$= \frac{8 \pm 4\sqrt{10}}{16} = \frac{-2 \pm \sqrt{10}}{4}.$$

Thus, $x = \frac{1}{2} + \frac{1}{4}\sqrt{10}$ and $x = \frac{1}{2} - \frac{1}{4}\sqrt{10}$ are the roots of the equation.

63. We use the quadratic formula to solve $2x^2 + 4x - 3 = 0$. Here, $a = 2$, $b = 4$, and $c = -3$. Therefore

$$x = \frac{-b \pm \sqrt{b^2 - 4ac}}{2a} = \frac{-(4) \pm \sqrt{(4)^2 - 4(2)(-3)}}{2(2)} = \frac{-4 \pm \sqrt{40}}{4}$$

$$= \frac{-4 \pm 2\sqrt{10}}{4} = \frac{-2 \pm \sqrt{10}}{2}.$$

Thus, $x = -1 + \frac{1}{2}\sqrt{10}$ and $x = -1 - \frac{1}{2}\sqrt{10}$ are the roots of the equation.

EXERCISES 9.3, page 558

1. $$\frac{x^2 + x - 2}{x^2 - 4} = \frac{(x+2)(x-1)}{(x+2)(x-2)} = \frac{x-1}{x-2}.$$

3. $$\frac{12t^2 + 12t + 3}{4t^2 - 1} = \frac{3(4t^2 + 4t + 1)}{4t^2 - 1} = \frac{3(2t+1)(2t+1)}{(2t+1)(2t-1)} = \frac{3(2t+1)}{2t-1}.$$

5. $$\frac{(4x-1)(3) - (3x+1)(4)}{(4x-1)^2} = \frac{12x - 3 - 12x - 4}{(4x-1)^2} = -\frac{7}{(4x-1)^2}.$$

7. $$\frac{(2x+3)(1) - (x+1)(2)}{(2x+3)^2} = \frac{2x + 3 - 2x - 2}{(2x+3)^2} = \frac{1}{(2x+3)^2}.$$

9. $$\frac{(e^x + 1)e^x - e^x(2)(e^x + 1)(e^x)}{(e^x + 1)^4} = \frac{e^x(e^x + 1)(1 - 2e^x)}{(e^x + 1)^4} = \frac{e^x(1 - 2e^x)}{(e^x + 1)^3}.$$

11. $\dfrac{2a^2-2b^2}{b-a}\cdot\dfrac{4a+4b}{a^2+2ab+b^2}=\dfrac{2(a+b)(a-b)4(a+b)}{-(a-b)(a+b)(a+b)}=-8.$

13. $\dfrac{3x^2+2x-1}{2x+6}\div\dfrac{x^2-1}{x^2+2x-3}=\dfrac{(3x-1)(x+1)}{2(x+3)}\cdot\dfrac{(x+3)(x-1)}{(x+1)(x-1)}=\dfrac{3x-1}{2}.$

15. $\dfrac{58}{3(3t+2)}+\dfrac{1}{3}=\dfrac{58+3t+2}{3(3t+2)}=\dfrac{3t+60}{3(3t+2)}=\dfrac{t+20}{3t+2}.$

17. $\dfrac{2x}{2x-1}-\dfrac{3x}{2x+5}=\dfrac{2x(2x+5)-3x(2x-1)}{(2x-1)(2x+5)}=\dfrac{4x^2+10x-6x^2+3x}{(2x-1)(2x+5)}$

$$=\frac{-2x^2+13x}{(2x-1)(2x+5)}=-\frac{x(2x-13)}{(2x-1)(2x+5)}.$$

19. $(x+\dfrac{1}{x})(x^2-1)=x^3+x-x-\dfrac{1}{x}=\dfrac{x^4-1}{x}.$

21. $\dfrac{4}{x^2-9}-\dfrac{5}{x^2-6x+9}=\dfrac{4}{(x+3)(x-3)}-\dfrac{5}{(x-3)^2}$

$$=\frac{4(x-3)-5(x+3)}{(x-3)^2(x+3)}=-\frac{x+27}{(x-3)^2(x+3)}.$$

23. $2+\dfrac{1}{a+2}-\dfrac{2a}{a-2}=\dfrac{2(a+2)(a-2)+a-2-2a(a+2)}{(a+2)(a-2)}$

$$=\frac{2a^2-8+a-2-2a^2-4a}{(a+2)(a-2)}=-\frac{3a+10}{(a+2)(a-2)}.$$

25. $\dfrac{1+\dfrac{1}{x}}{1-\dfrac{1}{x}}=\dfrac{\dfrac{x+1}{x}}{\dfrac{x-1}{x}}=\dfrac{x+1}{x}\cdot\dfrac{x}{x-1}=\dfrac{x+1}{x-1}.$

27. $$\frac{x^{-2}-y^{-2}}{x+y}=\frac{\frac{1}{x^2}-\frac{1}{y^2}}{x+y}=\frac{\frac{y^2-x^2}{x^2y^2}}{x+y}=\frac{(y+x)(y-x)}{x^2y^2}\cdot\frac{1}{x+y}=\frac{y-x}{x^2y^2}.$$

29. $$\frac{4x^2}{2\sqrt{2x^2+7}}+\sqrt{2x^2+7}=\frac{4x^2+2\sqrt{2x^2+7}\sqrt{2x^2+7}}{2\sqrt{2x^2+7}}=\frac{4x^2+4x^2+14}{2\sqrt{2x^2+7}}$$
$$=\frac{4x^2+7}{\sqrt{2x^2+7}}.$$

31. $$\frac{2x(x+1)^{-1/2}-(x+1)^{1/2}}{x^2}=\frac{(x+1)^{-1/2}(2x-x-1)}{x^2}=\frac{(x+1)^{-1/2}(x-1)}{x^2}$$
$$=\frac{x-1}{x^2\sqrt{x+1}}.$$

33. $$\frac{(2x+1)^{1/2}-(x+2)(2x+1)^{-1/2}}{2x+1}=\frac{(2x+1)^{-1/2}(2x+1-x-2)}{2x+1}$$
$$=\frac{(2x+1)^{-1/2}(x-1)}{2x+1}=\frac{x-1}{(2x+1)^{3/2}}.$$

35. (b) is equivalent to $\dfrac{a}{1+\frac{b}{x}}=\dfrac{a}{\frac{x+b}{x}}=\dfrac{ax}{x+b}$ which is (a).

37. (b) is equivalent to $\dfrac{x+\frac{1}{x^3}}{1-\frac{1}{x^3}}=\dfrac{x^4+1}{x^3}\cdot\dfrac{x^3}{x^3-1}=\dfrac{x^4+1}{x^3-1}$ which is (a).

39. $$\frac{1}{\sqrt{3}-1}\cdot\frac{\sqrt{3}+1}{\sqrt{3}+1}=\frac{\sqrt{3}+1}{3-1}=\frac{\sqrt{3}+1}{2}.$$

41. $$\frac{1}{\sqrt{x}-\sqrt{y}}\cdot\frac{\sqrt{x}+\sqrt{y}}{\sqrt{x}+\sqrt{y}}=\frac{\sqrt{x}+\sqrt{y}}{x-y}.$$

43. $$\frac{\sqrt{a}+\sqrt{b}}{\sqrt{a}-\sqrt{b}}\cdot\frac{\sqrt{a}+\sqrt{b}}{\sqrt{a}+\sqrt{b}}=\frac{(\sqrt{a}+\sqrt{b})^2}{a-b}.$$

45. $\frac{\sqrt{x}}{3} \cdot \frac{\sqrt{x}}{\sqrt{x}} = \frac{x}{3\sqrt{x}}$.

47 $\frac{1-\sqrt{3}}{3} \cdot \frac{1+\sqrt{3}}{1+\sqrt{3}} = \frac{1^2-(\sqrt{3})^2}{3(1+\sqrt{3})} = -\frac{2}{3(1+\sqrt{3})}$.

49. $\frac{1+\sqrt{x+2}}{\sqrt{x+2}} \cdot \frac{1-\sqrt{x+2}}{1-\sqrt{x+2}} = \frac{1-(x+2)}{\sqrt{x+2}(1-\sqrt{x+2})} = -\frac{x+1}{\sqrt{x+2}(1-\sqrt{x+2})}$.

EXERCISES 9.4, page 564

1. The statement is false because -3 is greater than -20. (See the number line that follows).

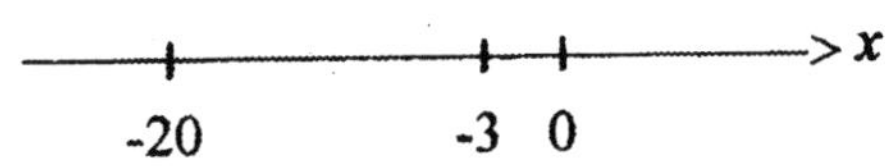

3. The statement is false because 2/3 [which is equal to (4/6)] is less than 5/6.

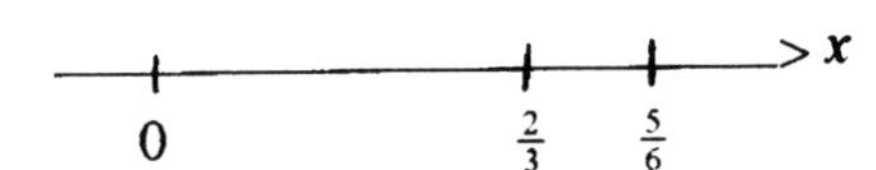

5. The interval (3,6) is shown on the number line that follows. Note that this is an open interval indicated by **(** and **)**

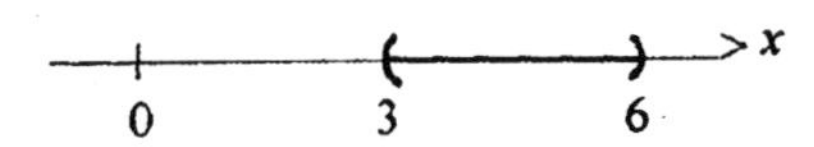

7. The interval [-1,4) is shown on the number line that follows. Note that this is a half-open interval indicated by **[** (closed) and **)** (open).

9. The infinite interval $(0,\infty)$ is shown on the number line that follows.

11. First, $2x + 4 < 8$ (Add -4 to each side of the inequality.)
Next, $2x < 4$ (Multiply each side of the inequality by 1/2)
and $x < 2$.

13. We are given the inequality $-4x \geq 20$.
Then $x \leq -5$. (Multiply both sides of the inequality by -1/4 and reverse the sign of the inequality.)

We write this in interval notation as $(-\infty,-5]$.

15. We are given the inequality $-6 < x - 2 < 4$.
First $-6 + 2 < x < 4 + 2$ (Add +2 to each member of the inequality.)
and $-4 < x < 6$,
so the solution set is the open interval $(-4,6)$.

17. We want to find the values of x that satisfy the inequalities
$$x + 1 > 4 \text{ or } x + 2 < -1.$$
Adding -1 to both sides of the first inequality, we obtain
$$x + 1 - 1 > 4 - 1,$$
or $x > 3$.
Similarly, adding -2 to both sides of the second inequality, we obtain
$$x + 2 - 2 < -1 - 2,$$
or $x < -3$.
Therefore, the solution set is $(-\infty,-3) \cup (3,\infty)$.

19. We want to find the values of x that satisfy the inequalities
$$x + 3 > 1 \text{ and } x - 2 < 1.$$
Adding -3 to both sides of the first inequality, we obtain
$$x + 3 - 3 > 1 - 3,$$
or $x > -2$.
Similarly, adding 2 to each side of the second inequality, we obtain
$$x < 3,$$
and the solution set is $(-2,3)$.

21. We want to find the values of x that satisfy the inequalities $(x + 3)(x - 5) \le 0$. From the sign diagram

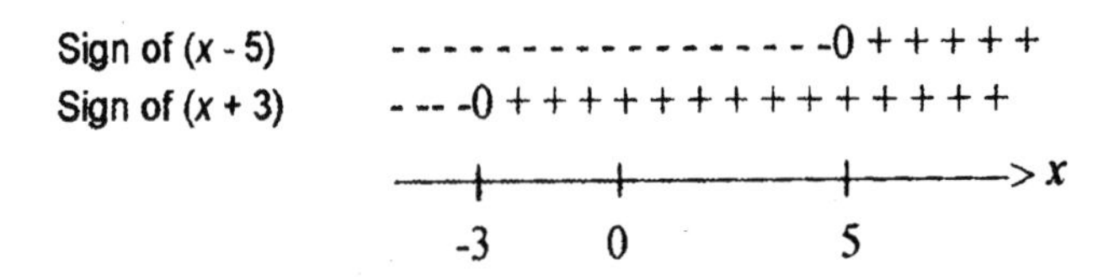

we see that the given inequality is satisfied when $-3 \le x \le 5$, that is, when the signs of the two factors are different or when one of the factors is equal to zero.

23. We want to find the values of x that satisfy the inequalities $(2x - 3)(x - 1) \ge 0$. From the sign diagram

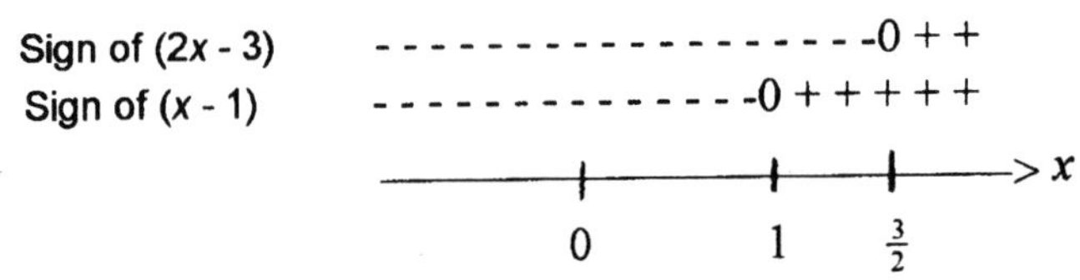

we see that the given inequality is satisfied when $x \le 1$ or $x \ge \frac{3}{2}$; that is,when the signs of both factors are the same, or one of the factors is equal to zero.

25. We want to find the values of x that satisfy the inequalities $\frac{x+3}{x-2} \ge 0$. From the sign diagram

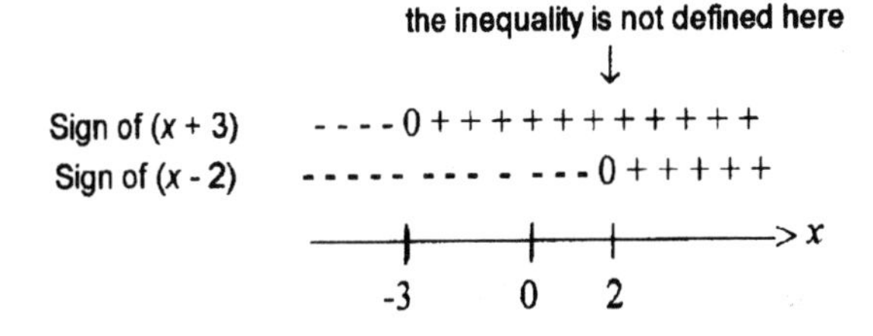

we see that the given inequality is satisfied when $x \le -3$ or $x > 2$, that is, when the signs of the two factors are the same. Notice that $x = 2$ is not included because the inequality is not defined at that value of x.

27. We want to find the values of x that satisfy the inequalities $\frac{x-2}{x-1} \le 2$. Subtracting from each side of the given inequality gives

$$\frac{x-2}{x-1} - 2 \le 0, \quad \frac{x-2-2(x-1)}{x-1} \le 0, \text{ or } \quad -\frac{x}{x-1} \le 0.$$

From the sign diagram

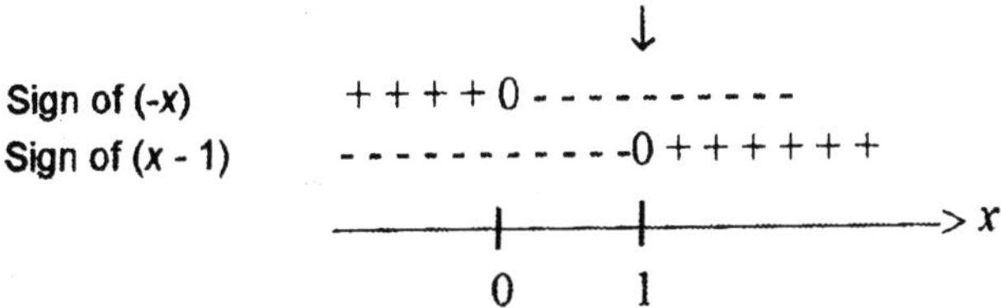

we see that the given inequality is satisfied when $x \leq 0$ or $x > 1$; that is, when the signs of the two factors differ. Notice that $x = 1$ is not included because the inequality is undefined at that value of x.

29. $|-6+2| = 4.$

31. $\dfrac{|-12+4|}{|16-12|} = \dfrac{|-8|}{4} = 2.$

33. $\sqrt{3}|-2| + 3|-\sqrt{3}| = \sqrt{3}(2) + 3\sqrt{3} = 5\sqrt{3}.$

35. $|\pi - 1| + 2 = \pi - 1 + 2 = \pi + 1.$

37. $|\sqrt{2} - 1| + |3 - \sqrt{2}| = \sqrt{2} - 1 + 3 - \sqrt{2} = 2.$

39. False. If $a > b$, then $-a < -b$, $-a + b < -b + b$, and $b - a < 0$.

41. False. Let $a = -2$ and $b = -3$. Then $a^2 = 4$ and $b^2 = 9$, and $4 < 9$. Note that we only need to provide a counterexample to show that the statement is not always true.

43. True. There are three possible cases.

Case 1 If $a > 0$, $b > 0$, then $a^3 > b^3$, since $a^3 - b^3 = (a - b)(a^2 + ab + b^2) > 0$.

Case 2 If $a > 0$, $b < 0$, then $a^3 > 0$ and $b^3 < 0$ and it follows that $a^3 > b^3$.

Case 3 If $a < 0$ and $b < 0$, then $a^3 - b^3 = (a - b)(a^2 + ab + b^2) > 0$, and we see that $a^3 > b^3$. (Note that $(a - b) > 0$ and $ab > 0$.)

45. False. Take $a = -2$, then $|-a| = |-(-2)| = |2| = 2 \neq a$.

47. True. If $a - 4 < 0$, then $|a - 4| = 4 - a = |4 - a|$. If $a - 4 > 0$, then

$$|4 - a| = a - 4 = |a - 4|.$$

49. False. Take $a = 3$, $b = -1$. Then $|a+b| = |3-1| = 2 \neq |a|+|b| = 3+1 = 4$.

51. If the car is driven in the city, then it can be expected to cover

$$(18.1)(20) = 362 \qquad \text{(miles/gal} \cdot \text{gal)}$$

or 362 miles on a full tank. If the car is driven on the highway, then it can be expected to cover

$$(18.1)(27) = 488.7 \qquad \text{(miles/gal} \cdot \text{gal)}$$

or 488.7 miles on a full tank. Thus, the driving range of the car may be described by the interval [362, 488.7].

53.

$$6(P - 2500) \leq 4(P + 2400)$$

$$6P - 15000 \leq 4P + 9600$$

$$2P \leq 24600, \text{ or } P \leq 12300.$$

Therefore, the maximum profit is \$12,300.

55. Let x represent the salesman's monthly sales in dollars. Then

$$0.15(x - 12000) \geq 1000$$
$$15(x - 12000) \geq 100000$$
$$15x - 180000 \geq 100000$$
$$15x \geq 280000$$
$$x \geq 18666.67.$$

We conclude that the salesman must earn at least \$18,666.67 to reach his goal.

57. The rod is acceptable if $0.49 < 0.51$ or $-0.01 < x - 0.5 < 0.01$. This gives the required inequality

$$|x - 0.5| \leq 0.01.$$

59. We want to solve the inequality

$$-6x^2 + 30x - 10 \geq 14. \qquad \text{(Remember } x \text{ is expressed in thousands.)}$$

Adding -14 to both sides of this inequality, we have

$$-6x^2 + 30x - 10 - 14 \geq 14 - 14,$$

or $\quad -6x^2 + 30x - 24 \geq 0.$

Dividing both sides of the inequality by -6 (which reverses the sign of the inequality), we have

$$x^2 - 5x + 4 \leq 0.$$

Factoring this last expression, we have

$$(x - 4)(x - 1) \leq 0.$$

From the following sign diagram,

Sign of (x - 4)	- - - - - - - - - - - - - - - - -0+ + + + + + + + +
Sign of (x - 1)	- - - - - -0 + + + + + + + + + + + + + + + + +

(number line with x-axis marked at 0, 1, 4)

we see that x must lie between 1 and 4. (The inequality is only satisfied when the two factors have opposite signs.) Since x is expressed in thousands of units, we see that the manufacturer must produce between 1000 and 4000 units of the commodity.

CHAPTER 9 REVIEW EXERCISES, page 566

1. $\left(\frac{9}{4}\right)^{3/2} = \frac{9^{3/2}}{4^{3/2}} = \frac{27}{8}$.

3. $(3 \cdot 4)^{-2} = 12^{-2} = \frac{1}{12^2} = \frac{1}{144}$.

5. $\frac{(3 \cdot 2^{-3})(4 \cdot 3^5)}{2 \cdot 9^3} = \frac{3 \cdot 2^{-3} \cdot 2^2 \cdot 3^5}{2 \cdot (3^2)^3} = \frac{2^{-1} \cdot 3^6}{2 \cdot 3^6} = \frac{1}{4}$.

7. $\frac{4(x^2+y)^3}{x^2+y} = 4(x^2+y)^2$.

9. $\frac{\sqrt[4]{16x^5yz}}{\sqrt[4]{81xyz^5}} = \frac{(2^4x^5yz)^{1/4}}{(3^4xyz^5)^{1/4}} = \frac{2x^{5/4}y^{1/4}z^{1/4}}{3x^{1/4}y^{1/4}z^{5/4}} = \frac{2x}{3z}$.

11. $\left(\frac{3xy^2}{4x^3y}\right)^{-2}\left(\frac{3xy^3}{2x^2}\right)^3 = \left(\frac{3y}{4x^2}\right)^{-2}\left(\frac{3y^3}{2x}\right)^3 = \left(\frac{4x^2}{3y}\right)^2\left(\frac{3y^3}{2x}\right)^3 = \frac{(16x^4)(27y^9)}{(9y^2)(8x^3)} = 6xy^7$.

13. $\sqrt[3]{81x^5y^{10}} \cdot \sqrt[3]{9xy^2} = 3^{4/3}x^{5/3}y^{10/3} \cdot 3^{2/3}x^{1/3}y^{2/3} = 3^2x^2y^4 = 9x^2y^4$.

15. $-2\pi^2r^3 + 100\pi r^2 = -2\pi r^2(\pi r - 50)$.

17. $16 - x^2 = 4^2 - x^2 = (4 - x)(4 + x)$.

19. $-2x^2 - 4x + 6 = (-2x - 6)(x - 1) = -2(x + 3)(x - 1)$.

21. $\dfrac{(t+6)(60)-(60t+180)}{(t+6)^2}=\dfrac{60t+360-60t-180}{(t+6)^2}=\dfrac{180}{(t+6)^2}.$

23. $\dfrac{2}{3}\left(\dfrac{4x}{2x^2-1}\right)+3\left(\dfrac{3}{3x-1}\right)=\dfrac{8x}{3(2x^2-1)}+\dfrac{9}{3x-1}=\dfrac{8x(3x-1)+27(2x^2-1)}{3(2x^2-1)(3x-1)}$

$$=\frac{78x^2-8x-27}{3(2x^2-1)(3x-1)}.$$

25. $8x^2 + 2x - 3 = (4x + 3)(2x - 1) = 0$ and $x = -3/4$ and $x = 1/2$ are the roots of the equation.

27. $-x^3 - 2x^2 + 3x = -x(x^2 + 2x - 3) = -x(x + 3)(x - 1) = 0$ and the roots of the equation are $x = 0$, $x = -3$, and $x = 1$.

29. Adding x to both sides yields $3 \leq 3x + 9$ or $3x \geq -6$, $\Rightarrow x \geq -2$. We conclude that the solution set is $[-2,\infty)$.

31. The inequalities imply $x > 5$ or $x < -4$. So the solution is $(-\infty,-4) \cup (5,\infty)$.

33. $|-5+7|+|-2|=|2|+|-2|=2+2=4.$

35. $|2\pi-6|-\pi=2\pi-6-\pi=\pi-6.$

37. Factoring the given expression, we have $(2x - 1)(x + 2) \leq 0$. From the sign diagram we conclude that the given inequality is satisfied when $-2 \leq x \leq \frac{1}{2}$.

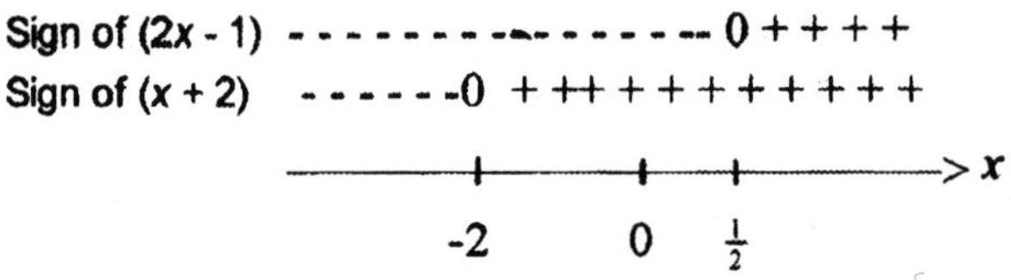

Sign of (2x - 1) - - - - - - - - - - - - - - - - - 0 + + + + +
Sign of (x + 2) - - - - - - 0 + + + + + + + + + + + + +

-2 0 $\frac{1}{2}$ → x

39. The given inequality is equivalent to $|2x-3|<5$ or $-5 < 2x - 3 < 5$. Thus, $-2 < 2x < 8$, or $-1 < x < 4$.

41. $\dfrac{\sqrt{x}-1}{x-1}=\dfrac{\sqrt{x}-1}{x-1}\cdot\dfrac{\sqrt{x}+1}{\sqrt{x}+1}=\dfrac{(\sqrt{x})^2-1}{(x-1)(\sqrt{x}+1)}=\dfrac{x-1}{(x-1)(\sqrt{x}+1)}=\dfrac{1}{\sqrt{x}+1}.$

43. Here we use the quadratic formula to solve the equation $x^2 - 2x - 5$. Then $a = 1$, $b = -2$, and $c = -5$. Thus,

$$x = \frac{-b \pm \sqrt{b^2 - 4ac}}{2a} = \frac{-(-2) \pm \sqrt{(-2)^2 - 4(1)(-5)}}{2(1)} = \frac{2 \pm \sqrt{24}}{2} = 1 \pm \sqrt{6}.$$

45. $2(1.5C + 80) \leq 2(2.5C - 20) \Rightarrow 1.5C + 80 \leq 2.5C - 20$, so $C \geq 100$ and the minimum cost is \$100.

CHAPTER 10

EXERCISES 10.1, page 578

1. $f(x) = 5x + 6$. Therefore

$f(3) = 5(3) + 6 = 21$

$f(-3) = 5(-3) + 6 = -9$

$f(a) = 5(a) + 6 = 5a + 6$

$f(-a) = 5(-a) + 6 = -5a + 6$

$f(a + 3) = 5(a + 3) + 6 = 5a + 15 + 6 = 5a + 21.$

3. $g(x) = 3x^2 - 6x - 3$

$g(0) = 3(0) - 6(0) - 3 = -3$

$g(-1) = 3(-1)^2 - 6(-1) - 3 = 3 + 6 - 3 = 6$

$g(a) = 3(a)^2 - 6(a) - 3 = 3a^2 - 6a - 3$

$g(-a) = 3(-a)^2 - 6(-a) - 3 = 3a^2 + 6a - 3$

$g(x + 1) = 3(x + 1)^2 - 6(x + 1) - 3 = 3(x^2 + 2x + 1) - 6x - 6 - 3$

$= 3x^2 + 6x + 3 - 6x - 9 = 3x^2 - 6.$

5. $s(t) = \dfrac{2t}{t^2 - 1}$. Therefore,

$$s(4) = \frac{2(4)}{(4)^2 - 1} = \frac{8}{15}.$$

$$s(0) = \frac{2(0)}{0^2 - 1} = 0$$

$$s(a) = \frac{2(a)}{a^2 - 1} = \frac{2a}{a^2 - 1}$$

$$s(2 + a) = \frac{2(2 + a)}{(2 + a)^2 - 1} = \frac{2(2 + a)}{a^2 + 4a + 4 - 1} = \frac{2(2 + a)}{a^2 + 4a + 3}$$

$$s(t+1)=\frac{2(t+1)}{(t+1)^2-1}=\frac{2(t+1)}{t^2+2t+1-1}=\frac{2(t+1)}{t(t+2)}.$$

7. $f(t)=\dfrac{2t^2}{\sqrt{t-1}}$. Therefore, $f(2)=\dfrac{2(2^2)}{\sqrt{2-1}}=8$

$$f(a)=\frac{2a^2}{\sqrt{a-1}}$$

$$f(x+1)=\frac{2(x+1)^2}{\sqrt{(x+1)-1}}=\frac{2(x+1)^2}{\sqrt{x}}$$

$$f(x-1)=\frac{2(x-1)^2}{\sqrt{(x-1)-1}}=\frac{2(x-1)^2}{\sqrt{x-2}}.$$

9. Since $x = -2 \le 0$, we see that $f(-2) = (-2)^2 + 1 = 4 + 1 = 5$

   Since $x = 0 \le 0$, we see that $f(0) = (0)^2 + 1 = 1$

   Since $x = 1 > 0$, we see that $f(1)=\sqrt{1}=1$.

11. Since $x = -1 < 1$, $f(-1)=-\frac{1}{2}(-1)^2+3=\frac{5}{2}$.

    Since $x = 0 < 1$, $f(0)=-\frac{1}{2}(0)^2+3=3$.

    Since $x = 1 \ge 1$, $f(1) = 2(1^2) + 1 = 3$.

    Since $x = 2 \ge 1$, $f(2) = 2(2^2) + 1 = 9$.

13. $g(2)=\sqrt{2^2-1}=\sqrt{3}$ and the point $(2, \sqrt{3})$ lies on the graph of g.

15. $f(-2)=\dfrac{|-2-1|}{-2+1}=\dfrac{|-3|}{-1}=-3$ and the point (-2,-3) does lie on the graph of f.

17. Since $f(x)$ is a real number for any value of x, the domain of f is $(-\infty, \infty)$.

19. $f(x)$ is not defined at $x = 0$ and so the domain of f is $(-\infty,0)\cup(0,\infty)$.

21. $f(x)$ is a real number for all values of x. Note that $x^2 + 1 \geq 1$ for all x. Therefore, the domain of f is $(-\infty, \infty)$.

23. Since the square root of a number is defined for all real numbers greater than or equal to zero, we have

$$5 - x \geq 0, \text{ or } -x \geq -5$$

and $\quad x \leq 5$. (Recall that multiplying by -1 reverses the sign of an inequality.) Therefore, the domain of g is $(-\infty,5]$.

25. The denominator of f is zero when

$$x^2 - 1 = 0 \text{ or } x = \pm 1.$$

Therefore, the domain of f is $(-\infty,-1) \cup (-1,1) \cup (1,\infty)$.

27. f is defined when $x + 3 \geq 0$, that is, when $x \geq -3$. Therefore, the domain of f is $[-3,\infty)$.

29. The numerator is defined when

$$1 - x \geq 0, \quad -x \geq -1 \quad \text{or} \quad x \leq 1.$$

Furthermore, the denominator is zero when $x = \pm 2$. Therefore, the domain is the set of all real numbers in $(-\infty,-2) \cup (-2,1]$.

31. a. The domain of f is the set of all real numbers.

b. $f(x) = x^2 - x - 6$. Therefore,

$$f(-3) = (-3)^2 - (-3) - 6 = 9 + 3 - 6 = 6.$$

$$f(-2) = (-2)^2 - (-2) - 6 = 4 + 2 - 6 = 0.$$

$$f(-1) = (-1)^2 - (-1) - 6 = 1 + 1 - 6 = -4.$$

$$f(0) = (0)^2 - (0) - 6 = -6.$$

$$f\left(\tfrac{1}{2}\right) = \left(\tfrac{1}{2}\right)^2 - \left(\tfrac{1}{2}\right) - 6 = \tfrac{1}{4} - \tfrac{2}{4} - \tfrac{24}{4} = -\tfrac{25}{4}.$$

$$f(1) = (1)^2 - 1 - 6 = -6.$$

$$f(2) = (2)^2 - 2 - 6 = 4 - 2 - 6 = -4.$$

$$f(3) = (3)^2 - 3 - 6 = 9 - 3 - 6 = 0.$$

c.

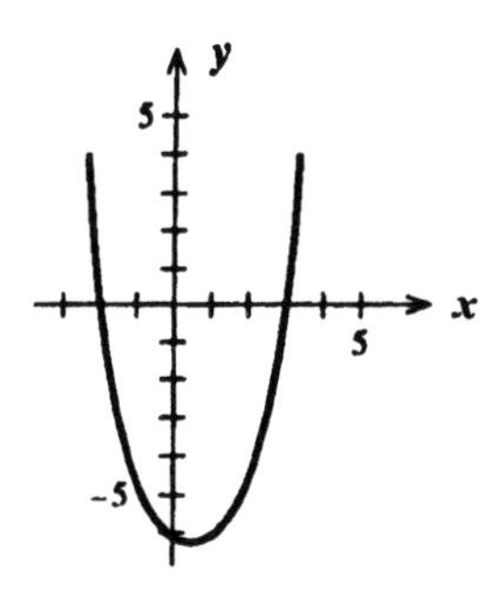

33.

x	-3	-2	-1	0	1	2	3
$f(x)$	19	9	3	1	3	9	19

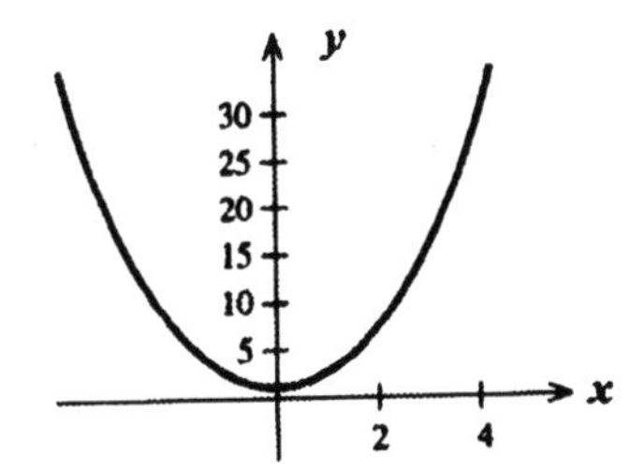

35.

x	0	1	2	4	9	16
$f(x)$	2	3	3.41	4	5	6

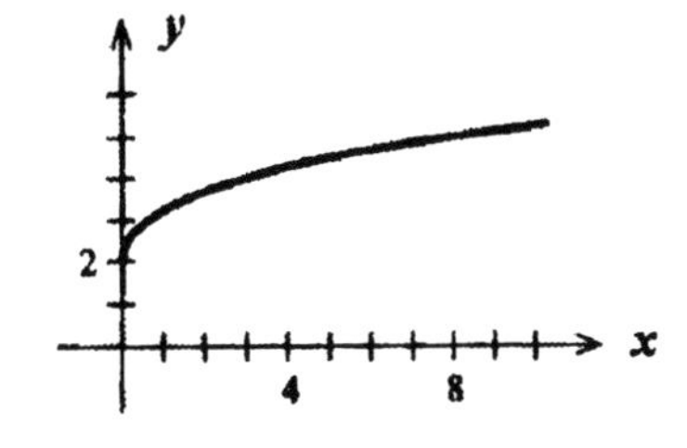

37.

x	0	-1	-3	-8	-15
$f(x)$	1	1.4	2	3	4

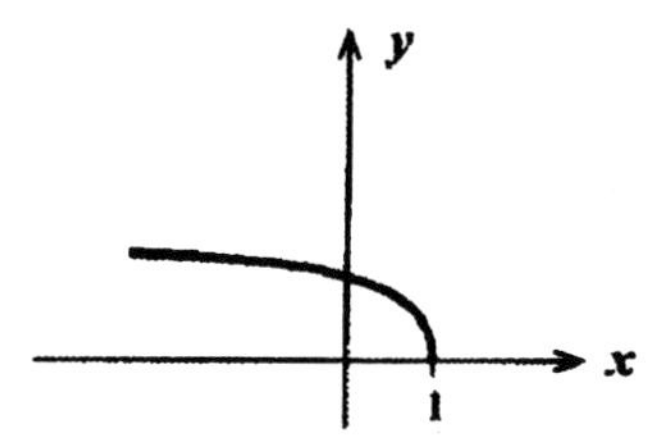

39.

x	-3	-2	-1	0	1	2	3
$f(x)$	2	1	0	-1	0	1	2

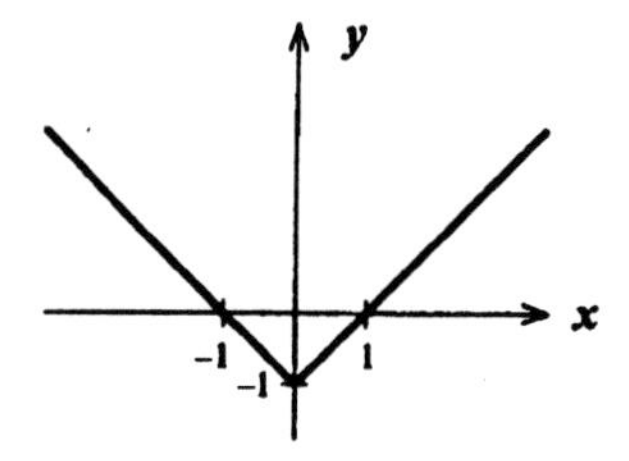

41.

x	-3	-2	-1	0	1	2	3
$f(x)$	-3	-2	-1	1	3	5	7

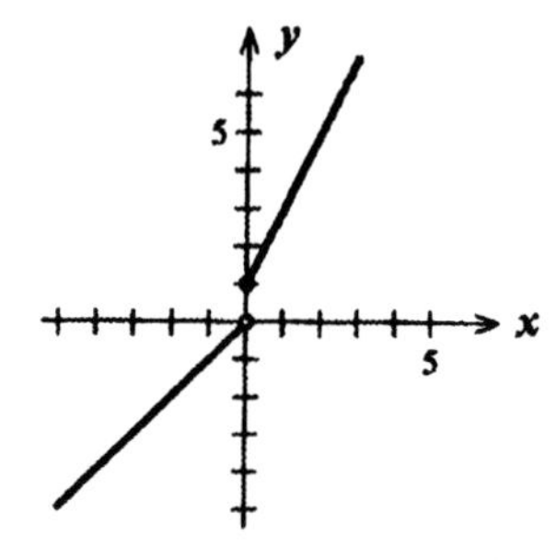

43. If $x \leq 1$, the graph of f is the half-line $y = -x + 1$. For $x > 1$, use the table

x	2 3 4
$f(x)$	3 8 15

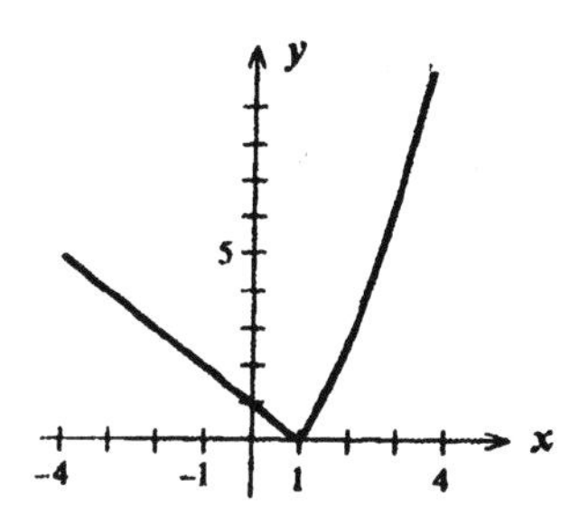

45. Each vertical line cuts the given graph at exactly one point, and so the graph represents y as a function of x.

47. Since there is a vertical line that intersects the graph at three points, the graph does not represent y as a function of x.

49. Each vertical line intersects the graph of f at exactly one point, and so the graph represents y as a function of x.

51. Each vertical line intersects the graph of f at exactly one point, and so the graph represents y as a function of x.

53. The circumference of a circle with a 5-inch radius is given by

$$C(5) = 2\pi(5) = 10\pi, \text{ or } 10\pi \text{ inches.}$$

55. a. From $t = 0$ to $t = 5$, the graph for cassettes lies above that for CDs so from 1985 to 1990, sales of prerecorded cassettes were greater than that of CDs.
b. Sales of prerecorded CDs were greater than that of prerecorded cassettes from 1990 on.
c. The graphs intersect at the point with coordinates $x = 5$ and $y \approx 3.5$, and this tells us that the sales of the two formats were the same in 1990 with the level of sales at approximately \$3.5 billion.

57. a. The slope of the straight line passing through the points (0, 0.58) and (20, 0.95) is

$$m = \frac{0.95 - 0.58}{20 - 0} = 0.0185,$$

and so an equation of the straight line passing through these two points is

$$y - 0.58 = 0.0185(t - 0) \text{ or } y = 0.0185t + 0.58$$

Next, the slope of the straight line passing through the points (20, 0.95) and (30, 1.1) is

$$m = \frac{1.1 - 0.95}{30 - 20} = 0.015$$

and so an equation of the straight line passing through the two points is

$$y - 0.95 = 0.015(x - 20) \text{ or } y = 0.015t + 0.65.$$

Therefore, the rule for f is

$$f(t) = \begin{cases} 0.0185t + 0.58 & 0 \le t \le 20 \\ 0.015t + 0.65 & 20 < t \le 30 \end{cases}$$

b. The ratios were changing at the rates of 0.0185/yr and 0.015/yr from 1960 through 1980, and from 1980 through 1990, respectively.

c. The ratio was 1 when $t \approx 20$. This shows that the number of bachelor's degrees earned by women equaled the number earned bymen for the first time around 1980.

59. a. $T(x) = 0.06x$

b. $T(200) = 0.06(200) = 12$, or \$12.00.

$T(5.65) = 0.06(5.65) = 0.34$, or \$0.34.

61. The child should receive

$$D(4) = \tfrac{2}{25}(500)(4) = 160, \text{ or } 160 \text{ mg.}$$

63. a. The daily cost of leasing from Ace is

$$C_1(x) = 30 + 0.15x,$$

while the daily cost of leasing from Acme is

$$C_2(x) = 25 + 0.20x,$$

where x is the number of miles driven.

b.

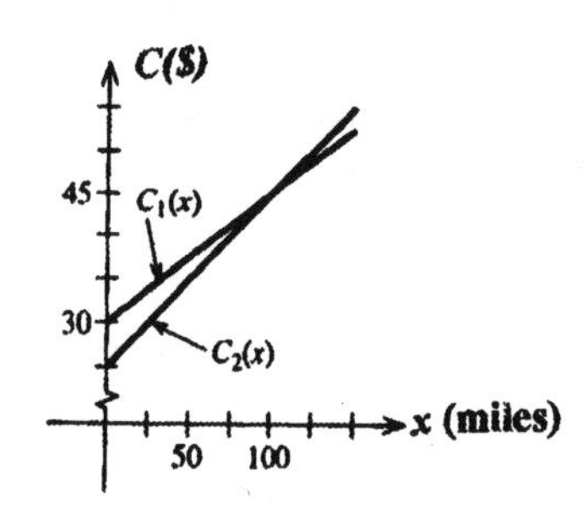

c. The costs will be the same when $C_1(x) = C_2(x)$, that is, when

$$30 + 0.15x = 25 + 0.20x$$
$$0.05x = 5, \text{ or } x = 100.$$

Since $\quad C_1(70) = 30 + 0.15(70) = 40.5$

and $\quad C_2(70) = 25 + 0.20(70) = 39,$

and the customer plans to drive less than 100 miles, she should rent from Acme.

65. Here $V = -20{,}000n + 1{,}000{,}000$.
The book value in 2004 will be

$$V = -20{,}000(19) + 1{,}000{,}000, \text{ or } \$620{,}000.$$

The book value in 2006 will be

$$V = -20{,}000(23) + 1{,}000{,}000, \text{ or } \$540{,}000.$$

67. a. We require that $0.04 - r^2 \geq 0$ and $r \geq 0$. This is true if $0 \leq r \leq 0.2$. Therefore, the domain of v is $[0,0.2]$.
b. Compute

$$v(0) = 1000[0.04 - (0)^2] = 1000(0.04) = 40.$$
$$v(0.1) = 1000[0.04 - (0.1)^2] = 1000(0.04 - .01)$$
$$= 1000(0.03) = 30.$$
$$v(0.2) = 1000[0.04 - (0.2)^2] = 1000(0.04 - 0.04) = 0.$$

As the distance r increases, the velocity of the blood decreases.

69. Between 8 A.M. and 9 A.M., the average worker can be expected to assemble

$$N(1) - N(0) = (-1 + 6 + 15) - 0 = 20,$$

or 20 walkie-talkies. Between 9 A.M. and 10 A.M., we can expect

$$N(2) - N(1) = [-2^3 + 6(2^2) + 15(2)] - (-1 + 6 + 15)$$
$$= 46 - 20,$$

or 26 walkie-talkies can be assembled by the average worker.

71. When the proportion of popular votes won by the Democratic presidential candidate is 0.60, the proportion of seats in the House of Representatives won by Democratic candidates is given by

$$s(0.6) = \frac{(0.6)^3}{(0.6)^3 + (1-0.6)^3} = \frac{0.216}{0.216+0.064} = \frac{0.216}{0.280} \approx 0.77.$$

73.

$$f(x) = \begin{cases} 78 & \text{if } 2 < x \le 3 \\ 101 & \text{if } 3 < x \le 4 \\ 124 & \text{if } 4 < x \le 5 \\ 147 & \text{if } 5 < x \le 6 \\ 170 & \text{if } 6 < x \le 7 \\ 193 & \text{if } 7 < x \le 8 \\ 216 & \text{if } 8 < x \le 9 \\ 239 & \text{if } 9 < x \le 10 \\ 262 & \text{if } 10 < x \le 11 \\ 285 & \text{if } 11 < x \le 12. \end{cases}$$

a. The domain of f is $(0,12]$.

b.

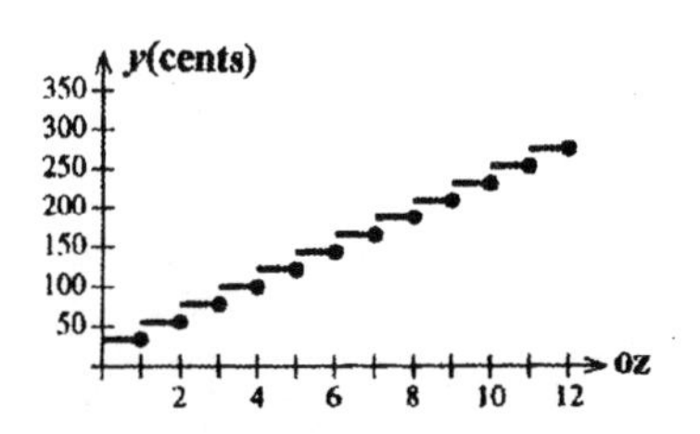

USING TECHNOLOGY EXERCISES 10.1, page 587

1.

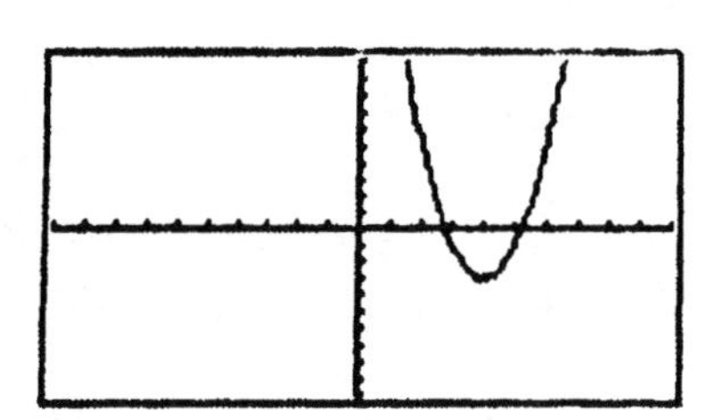

3.

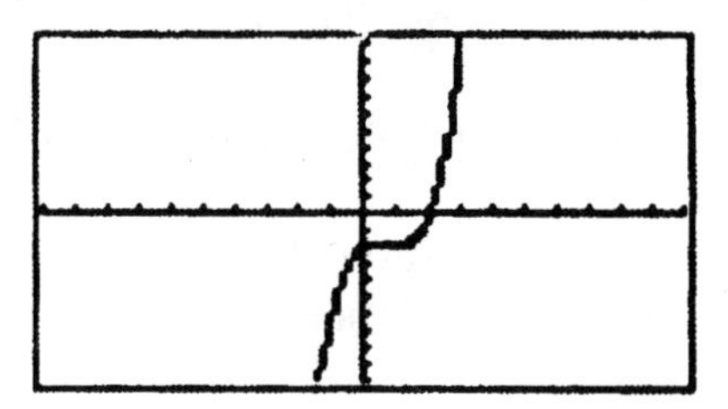

5.

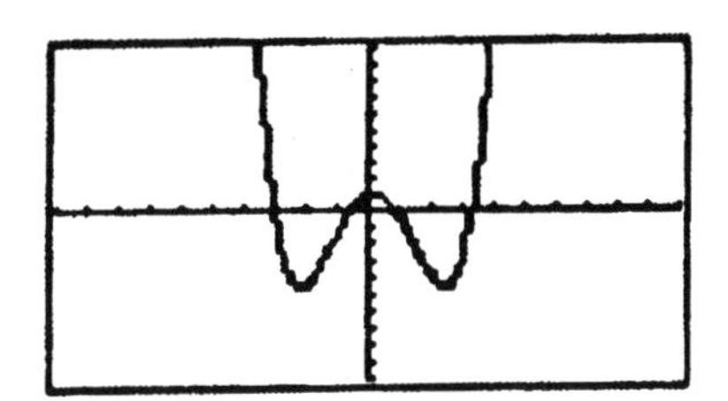

7.

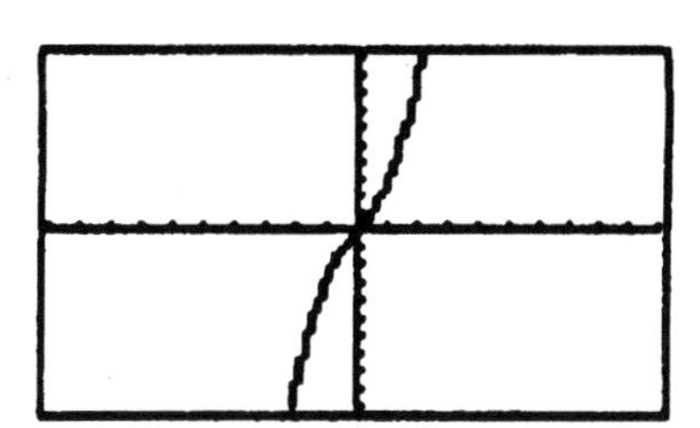

9. a.

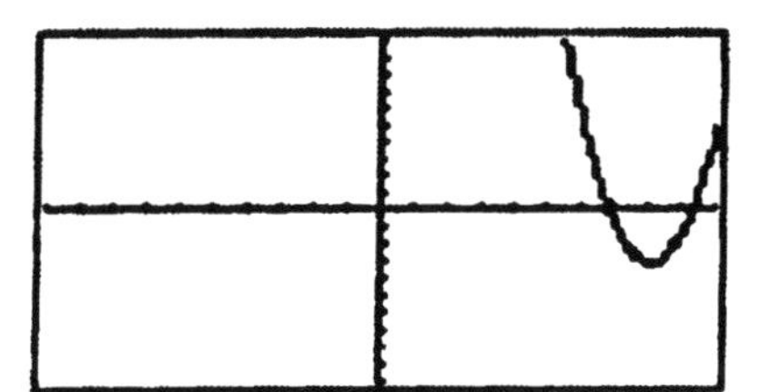

b.

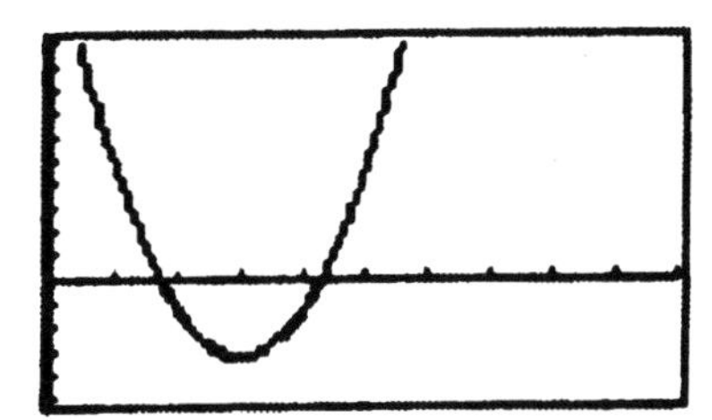

11. a.

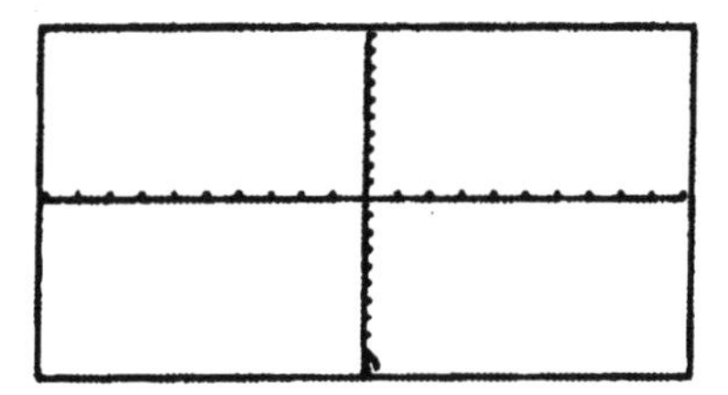

b.

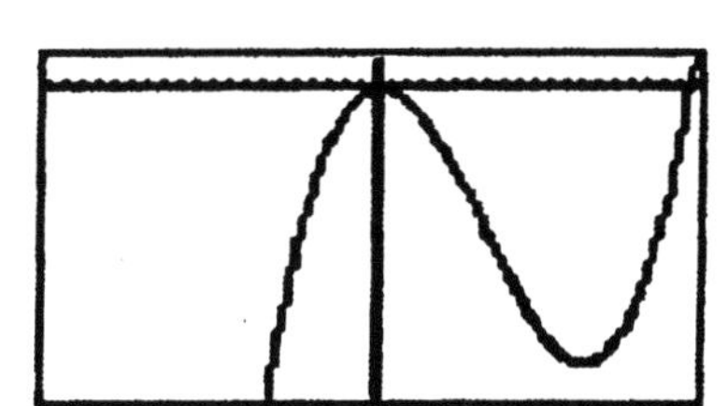

13. a.

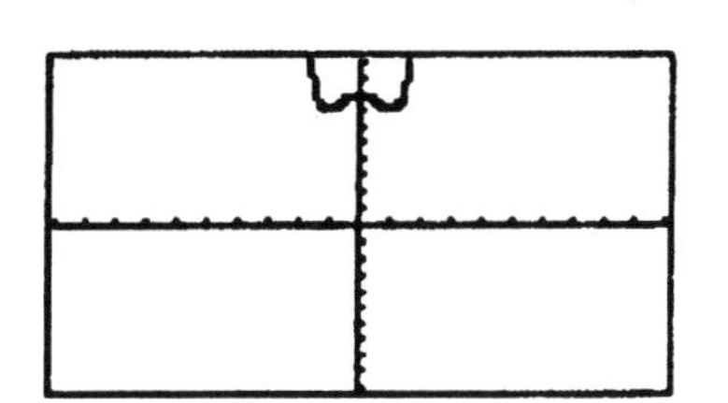

b.

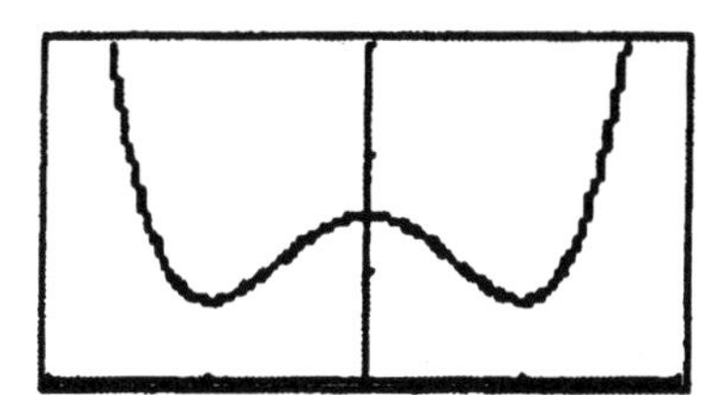

15. a.

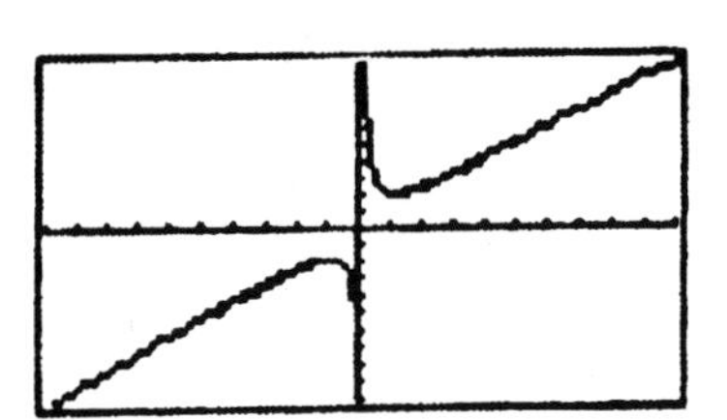

b.

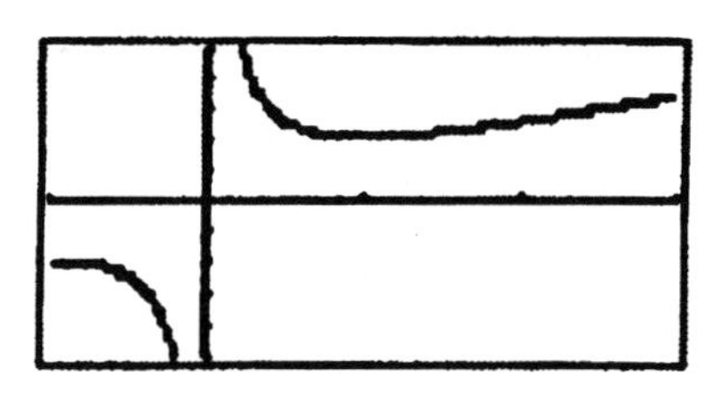

17. a

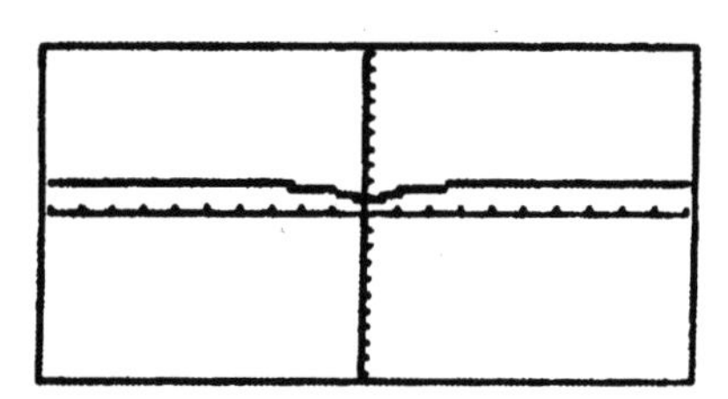

b.

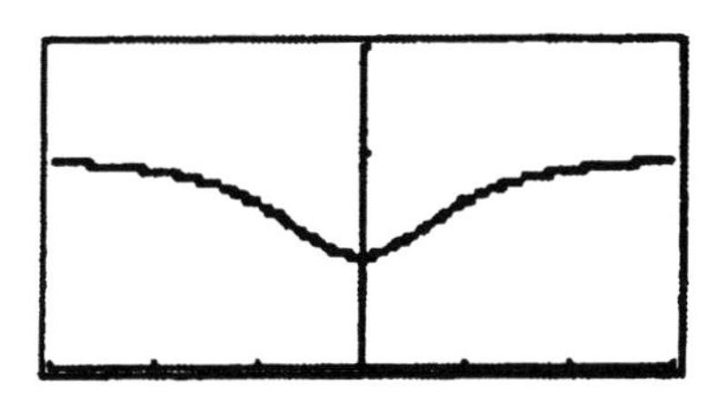

19. a.

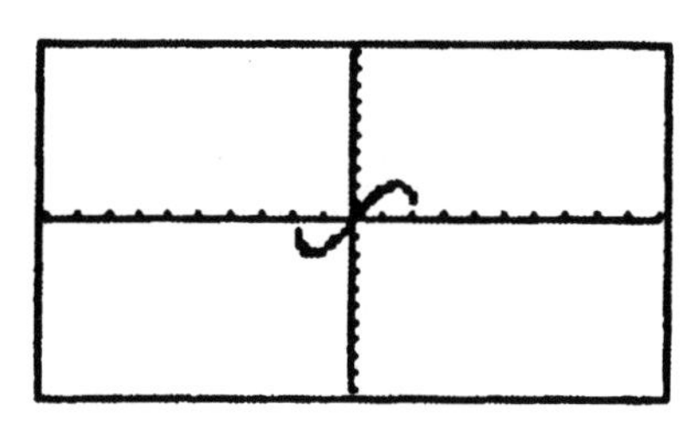

b.

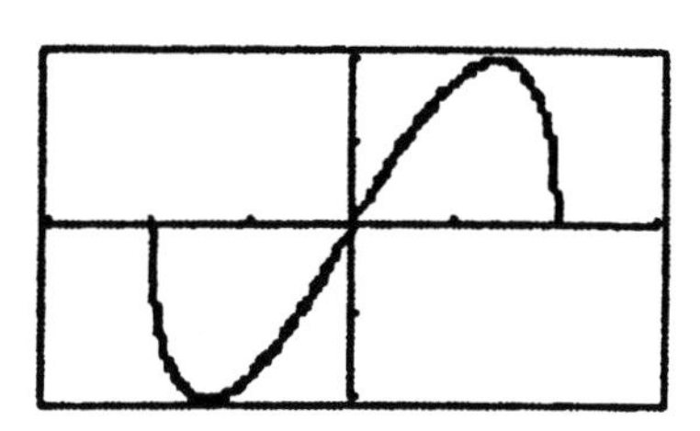

21.

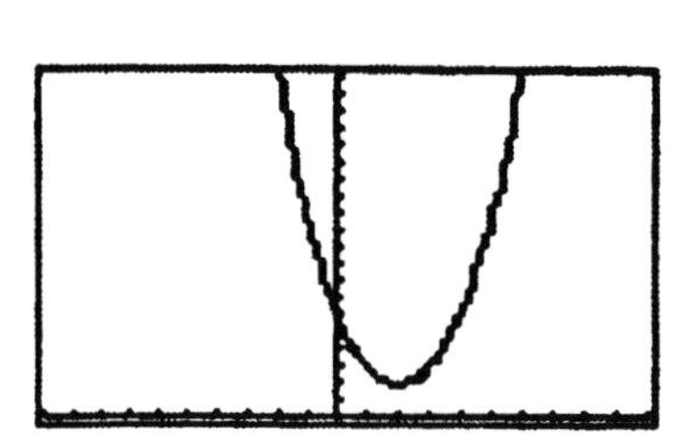

23.

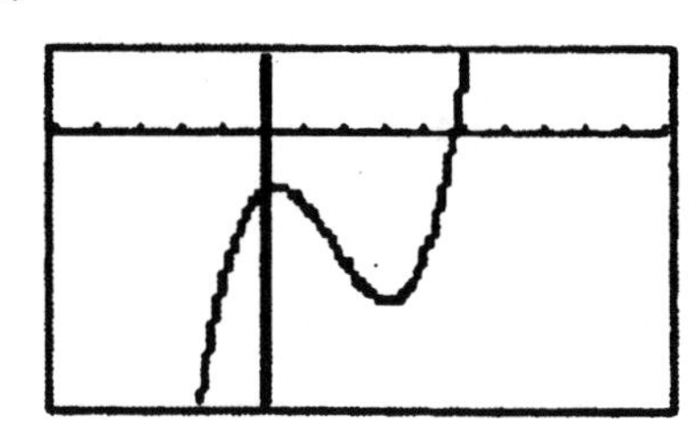

25.

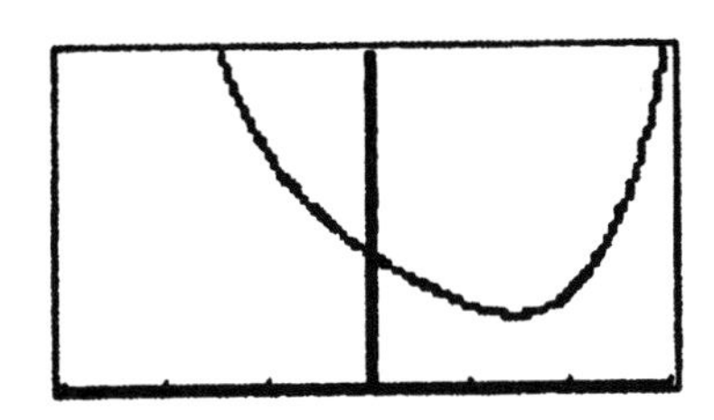

27.

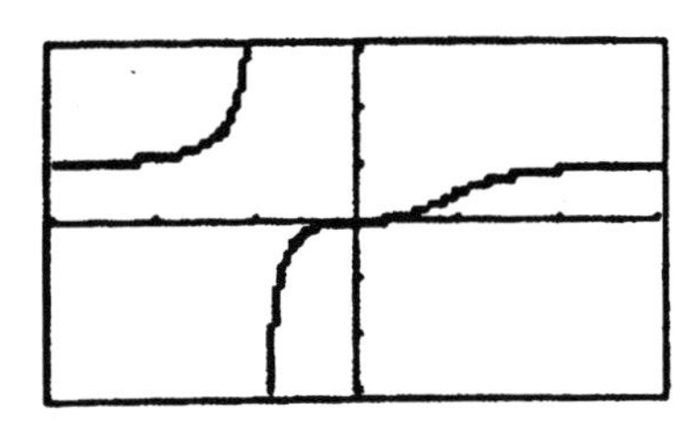

29.

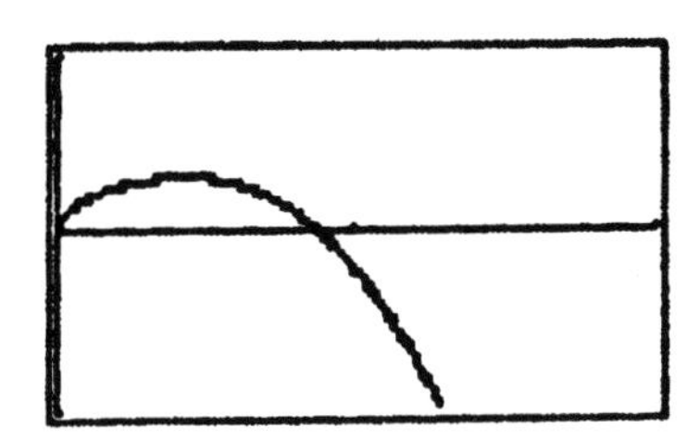

31. 18; $f(-1) = -3(-1)^3 + 5(-1)^2 - 2(-1) + 8 = 3 + 5 + 2 + 8 = 18.$

33. 2; $f(1) = \dfrac{(1)^4 - 3(1)^2}{1-2} = \dfrac{1-3}{-1} = 2.$

35. $f(2.145) \approx 18.5505$

37. $f(1.28) \approx 17.3850$

39. $f(2.41) \approx 4.1616$

41. $f(0.62) \approx 1.7214$

43. a.

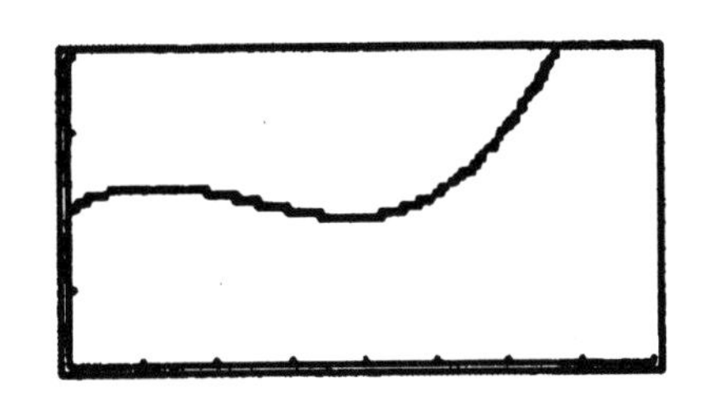

b. $f(2) \approx 2.1762$, or approximately 2.2%/yr

$f(4) \approx 1.9095$, or approximately 1.91 %/yr.

45. a.

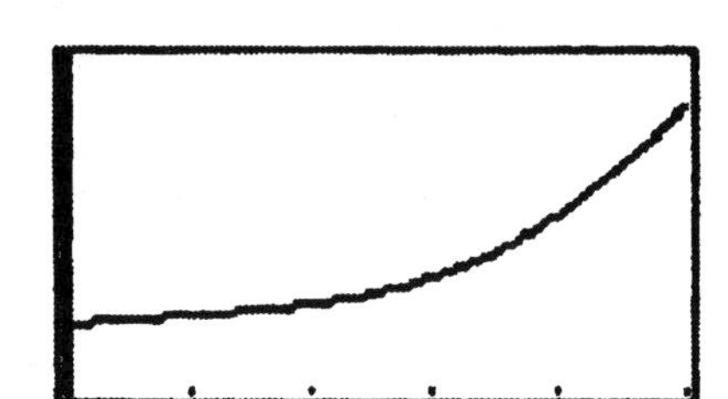

b. $f(6) = 44.7$;

$f(8) = 52.7$;

$f(11) = 129.2$.

EXERCISES 10.2, page 595

1. $(f + g)(x) = f(x) + g(x) = (x^3 + 5) + (x^2 - 2) = x^3 + x^2 + 3$.

3. $fg(x) = f(x)g(x) = (x^3 + 5)(x^2 - 2) = x^5 - 2x^3 + 5x^2 - 10$.

5. $\dfrac{f}{g}(x) = \dfrac{f(x)}{g(x)} = \dfrac{x^3+5}{x^2-2}$.

7. $\dfrac{fg}{h}(x) = \dfrac{f(x)g(x)}{h(x)} = \dfrac{(x^3+5)(x^2-2)}{2x+4} = \dfrac{x^5-2x^3+5x^2-10}{2x+4}$

9. $(f+g)(x) = x - 1 + \sqrt{x+1}$.

11. $(fg)(x) = (x-1)\sqrt{x+1}$

13. $\dfrac{g}{h}(x) = \dfrac{g(x)}{h(x)} = \dfrac{\sqrt{x+1}}{2x^3-1}$.

15. $\dfrac{fg}{h}(x) = \dfrac{(x-1)(\sqrt{x+1})}{2x^3-1}$

17. $\dfrac{f-h}{g}(x) = \dfrac{x-1-(2x^3-1)}{\sqrt{x+1}} = \dfrac{x-2x^3}{\sqrt{x+1}}$.

19. $(f+g)(x) = x^2 + 5 + \sqrt{x} - 2 = x^2 + \sqrt{x} + 3$.
    $(f-g)(x) = x^2 + 5 - (\sqrt{x} - 2) = x^2 - \sqrt{x} + 7$.

$(fg)(x) = (x^2+5)(\sqrt{x}-2)$.

$(\frac{f}{g})(x) = \frac{x^2+5}{\sqrt{x}-2}$.

21. $(f+g)(x) = \sqrt{x+3} + \frac{1}{x-1} = \frac{(x-1)\sqrt{x+3}+1}{x-1}$.

$(f-g)(x) = \sqrt{x+3} - \frac{1}{x-1} = \frac{(x-1)\sqrt{x+3}-1}{x-1}$.

$(fg)(x) = \sqrt{x+3}\left(\frac{1}{x-1}\right) = \frac{\sqrt{x+3}}{x-1}$.

$(\frac{f}{g}) = \sqrt{x+3}(x-1)$.

23. $(f+g)(x) = \frac{x+1}{x-1} + \frac{x+2}{x-2} = \frac{(x+1)(x-2)+(x+2)(x-1)}{(x-1)(x-2)}$

$= \frac{x^2-x-2+x^2+x-2}{(x-1)(x-2)} = \frac{2x^2-4}{(x-1)(x-2)} = \frac{2(x^2-2)}{(x-1)(x-2)}$.

$(f-g)(x) = \frac{x+1}{x-1} - \frac{x+2}{x-2} = \frac{(x+1)(x-2)-(x+2)(x-1)}{(x-1)(x-2)}$

$= \frac{x^2-x-2-x^2-x+2}{(x-1)(x-2)} = \frac{-2x}{(x-1)(x-2)}$.

$(fg)(x) = \frac{(x+1)(x+2)}{(x-1)(x-2)}$.

$(\frac{f}{g}) = \frac{(x+1)(x-2)}{(x-1)(x+2)}$.

25. $(f\circ g)(x) = f(g(x)) = f(x^2) = (x^2)^2 + x^2 + 1 = x^4 + x^2 + 1$.

$(g\circ f)(x) = g(f(x)) = g(x^2+x+1) = (x^2+x+1)^2$.

27.

$(g\circ f)(x) = g(f(x)) = g(\sqrt{x}+1) = (\sqrt{x}+1)^2 - 1 = x + 2\sqrt{x} + 1 - 1 = x + 2\sqrt{x}$.

29. $(f\circ g)(x) = f(g(x)) = f\left(\frac{1}{x}\right) = \frac{1}{x} \div \left(\frac{1}{x^2}+1\right) = \frac{1}{x}\cdot\frac{x^2}{x^2+1} = \frac{x}{x^2+1}$.

$$(g \circ f)(x) = g(f(x)) = g\left(\frac{x}{x^2+1}\right) = \frac{x^2+1}{x}.$$

31. $h(2) = g[f(2)]$. But $f(2) = 4 + 2 + 1 = 7$, so $h(2) = g(7) = 49$.

33. $h(2) = g[f(2)]$. But $f(2) = \dfrac{1}{2(2)+1} = \dfrac{1}{5}$, so $h(2) = g(\frac{1}{5}) = \dfrac{1}{\sqrt{5}} = \dfrac{\sqrt{5}}{5}$.

35. $f(x) = 2x^3 + x^2 + 1$, $g(x) = x^5$.

37. $f(x) = x^2 - 1$, $g(x) = \sqrt{x}$.

39. $f(x) = x^2 - 1$, $g(x) = \dfrac{1}{x}$.

41. $f(x) = 3x^2 + 2$, $g(x) = \dfrac{1}{x^{3/2}}$.

43. $f(a+h) - f(a) = [3(a+h)+4] - (3a+4)$
$= 3a + 3h + 4 - 3a - 4 = 3h.$

45. $f(a+h) - f(a) = 4 - (a+h)^2 - (4 - a^2)$
$= 4 - a^2 - 2ah - h^2 - 4 + a^2$
$= -2ah - h^2 = -h(2a + h).$

47. $$\frac{f(a+h)-f(a)}{h} = \frac{[(a+h)^2+1]-(a^2+1)}{h} = \frac{a^2+2ah+h^2+1-a^2-1}{h}$$
$$= \frac{h(2a+h)}{h} = 2a+h.$$

49. $C(x) = 0.6x + 12{,}100$.

51. a. $P(x) = R(x) - C(x)$
$= -0.1x^2 + 500x - (0.000003x^3 - 0.03x^2 + 200x + 100{,}000)$
$= -0.000003x^3 - 0.07x^2 + 300x - 100{,}000.$

b. $P(1500) = -0.000003(1500)^3 - 0.07(1500)^2 + 300(1500) - 100{,}000$
$= 182{,}375$ or \$182,375.

53. a. $$N(r(t)) = \frac{7}{1+0.02\left(\frac{10t+150}{t+10}\right)^2}.$$

b. $$N(r(0)) = \frac{7}{1+0.02\left(\frac{10(0)+150}{0+10}\right)^2}$$

$$= \frac{7}{1+0.02\left(\frac{150}{10}\right)^2} = \frac{7}{5.5} \approx 1.27,$$

or 1.27 million units.

$$N(r(12)) = \frac{7}{1+0.02\left(\frac{120+150}{12+10}\right)^2} = \frac{7}{1+0.02\left(\frac{270}{22}\right)^2} = \frac{7}{4.01} \approx 1.74,$$

or 1.74 million units.

$$N(r(18)) = \frac{7}{1+0.02\left(\frac{180+150}{18+10}\right)^2} = \frac{7}{1+0.02\left(\frac{330}{28}\right)^2} = \frac{7}{3.78} \approx 1.85,$$

or 1.85 million units.

55. $N(t) = 1.42(x(t)) = \dfrac{7(t+10)^2}{(t+10)^2+2(t+15)^2} = \dfrac{9.94(t+10)^2}{(t+10)^2+2(t+15)^2}.$

The number of jobs created 6 months from now will be

$$N(6) = \frac{9.94(16)^2}{(16)^2+2(21)^2} = 2.24, \quad \text{or 2.24 million jobs.}$$

The number of jobs created 12 months from now will be

$$N(12) = \frac{9.94(22)^2}{(22)^2+2(27)^2} = 2.48, \quad \text{or 2.48 million jobs.}$$

EXERCISES 10.3, page 605

1. f is a polynomial function in x of degree 6.

3. Expanding $G(x) = 2(x^2 - 3)^3$, we have

$$G(x) = 2x^6 - 18x^4 + 54x^2 - 54,$$

and we conclude that G is a polynomial function in x of degree 6.

5 f is neither a polynomial nor a rational function.

7. The individual's disposable income is

$$D = (1 - 0.28)40{,}000 = 28{,}800, \quad \text{or } \$28{,}800.$$

9. The child should receive

$$D(4) = \left(\frac{4+1}{24}\right)(500) = 104.17, \quad \text{or } 104 \text{ mg.}$$

11. When 1000 units are produced,

$$R(1000) = -0.1(1000)^2 + 500(1000) = 400{,}000, \quad \text{or } \$400{,}000.$$

13. a. The given data implies that $R(40) = 50$, that is,

$$\frac{100(40)}{b+40} = 50$$

$$50(b+40) = 4000,$$

or

$$b = 40.$$

Therefore, the required response function is

$$R(x) = \frac{100x}{40+x}.$$

b. The response will be

$$R(60) = \frac{100(60)}{40+60} = 60, \text{ or approximately 60 percent.}$$

15. Using the formula given in Problem 14, we have

$$V(2) = 100{,}000 - \frac{(100{,}000 - 30{,}000)}{5}(2) = 100{,}000 - \frac{70{,}000}{5}(2)$$

$$= 72{,}000, \text{ or } \$72{,}000.$$

17. $h(t) = f(t) - g(t) = \dfrac{110}{\frac{1}{2}t+1} - 26(\frac{1}{4}t^2 - 1)^2 - 52.$

$$h(0) = f(0) - g(0) = \frac{110}{\frac{1}{2}(0)+1} - 26\left[\tfrac{1}{4}(0)^2 - 1\right]^2 - 52 = 110 - 26 - 52 = 32, \text{ or } \$32.$$

$$h(1) = f(1) - g(1) = \frac{110}{\frac{1}{2}(1)+1} - 26\left[\tfrac{1}{4}(1)^2 - 1\right]^2 - 52 = 6.71, \text{ or } \$6.71.$$

$$h(2) = f(2) - g(2) = \frac{110}{\frac{1}{2}(2)+1} - 26\left[\tfrac{1}{4}(2)^2 - 1\right]^2 - 52 = 3, \text{ or } \$3.$$

We conclude that the price gap was narrowing.

19. a.

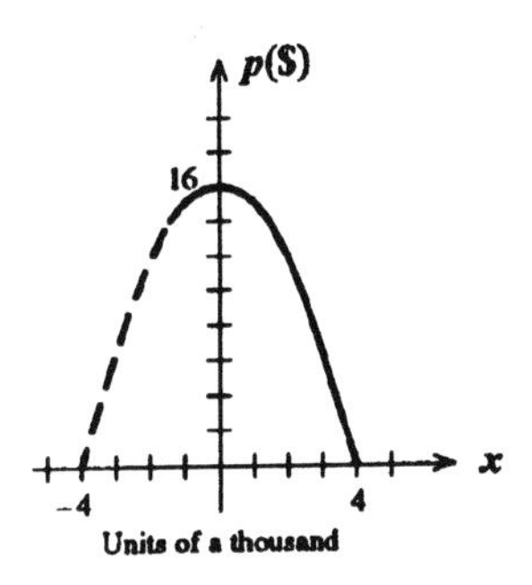

b. If $p = 7$, we have $7 = -x^2 + 16$, or $x^2 = 9$, so that $x = \pm 3$. Therefore, the quantity demanded when the unit price is \$7 is 3000 units.

21. a.

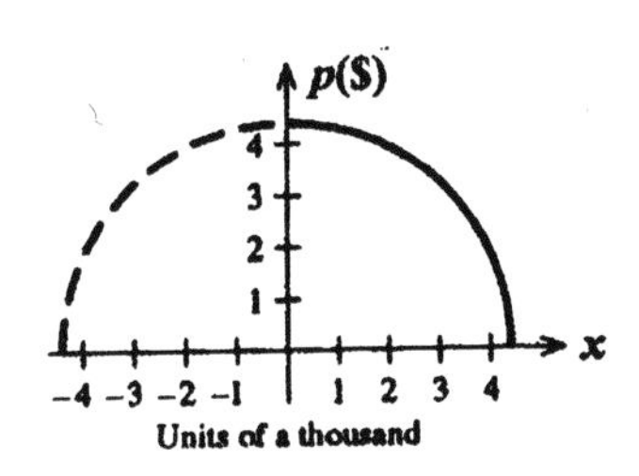

b. If $p = 3$, then $3 = \sqrt{18 - x^2}$, and $9 = 18 - x^2$, so that $x^2 = 9$ and $x = \pm 3$. Therefore, the quantity demanded when the unit price is \$3 is 3000 units.

23. Substituting $x = 6$ and $p = 8$ into the given equation gives

$$8 = \sqrt{-36a + b}, \quad \text{or } -36a + b = 64.$$

Next, substituting $x = 8$ and $p = 6$ into the equation gives

$$6 = \sqrt{-64a + b}, \quad \text{or } -64a + b = 36.$$

Solving the system

$$-36a + b = 64$$

$$-64a + b = 36$$

for a and b, we find $a = 1$ and $b = 100$. Therefore the demand equation is

$$p = \sqrt{-x^2 + 100}.$$

When the unit price is set at $7.50, we have

$$7.5 = \sqrt{-x^2 + 100}, \text{ or } 56.25 = -x^2 + 100$$

from which we deduce that $x = \pm 6.614$. Therefore, the quantity demanded is 6614 units.

25. a.

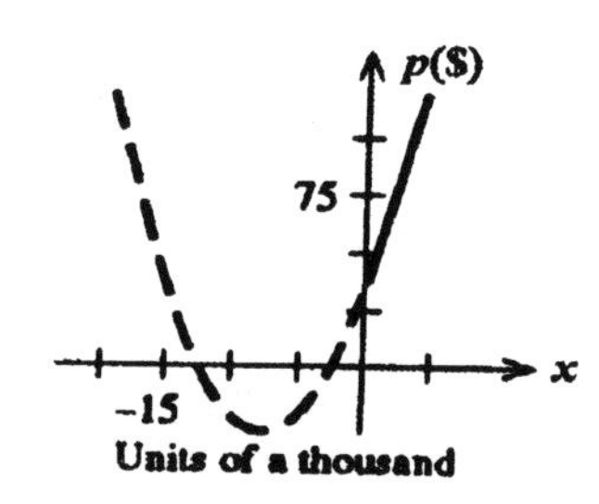

b. If $x = 2$, then $p = 2^2 + 16(2) + 40 = 76$, or $76.

27. a.

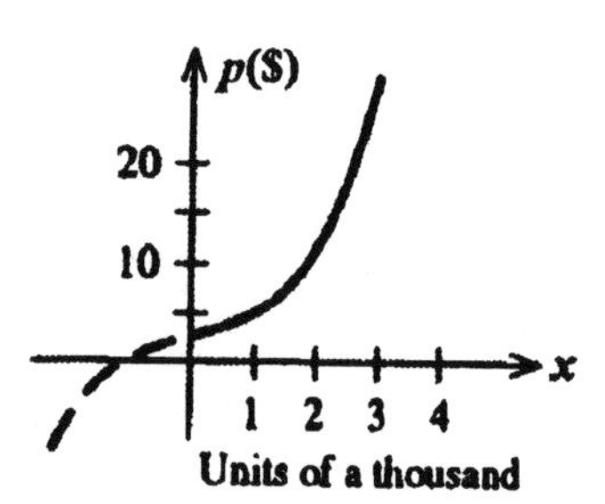

b. $p = 2^3 + 2(2) + 3 = 15$, or $15.

29. Substituting $x = 10{,}000$ and $p = 20$ into the given equation yields

$$20 = a\sqrt{10.000} + b = 100a + b.$$

Next, substituting $x = 62{,}500$ and $p = 35$ into the equation yields

$$35 = a\sqrt{62{,}500} + b = 250a + b.$$

Subtracting the first equation from the second yields

$$15 = 150a, \text{ or } a = \tfrac{1}{10}.$$

Substituting this value of a into the first equation gives $b = 10$. Therefore, the required equation is $p = \frac{1}{10}\sqrt{x} + 10$. The graph of the supply function follows.

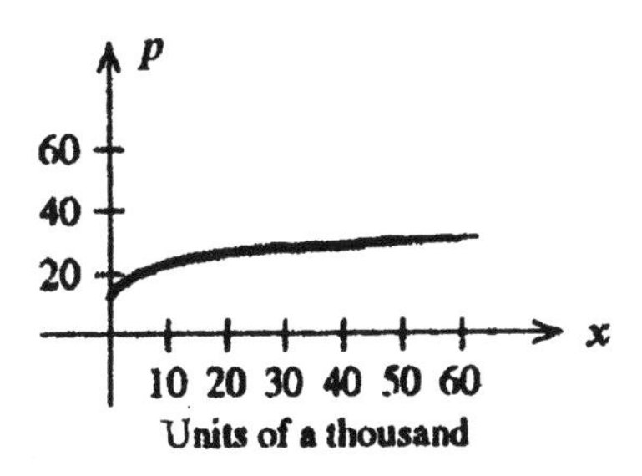

Substituting $x = 40{,}000$ into the supply equation yields

$$p = \tfrac{1}{10}\sqrt{40{,}000} + 10 = 30, \quad \text{or } \$30.$$

31. We solve the system of equations

$$p = -x^2 - 2x + 100 \quad \text{and} \quad p = 8x + 25.$$

Thus, $-x^2 - 2x + 100 = 8x + 25$, or $x^2 + 10x - 75 = 0$.

Factoring this equation, we have

$$(x + 15)(x - 5) = 0,$$

or $x = -15$ and $x = 5$. Rejecting the negative root, we have $x = 5$ and the corresponding value of p is

$$p = 8(5) + 25 = 65.$$

We conclude that the equilibrium quantity is 5000 and the equilibrium price is \$65.

33. We solve the system

$$p = 60 - 2x^2$$

and $$p = x^2 + 9x + 30.$$

Equating these two equations, we have

$$x^2 + 9x + 30 = 60 - 2x^2$$
$$3x^2 + 9x - 30 = 0$$
$$x^2 + 3x - 10 = 0$$
$$(x + 5)(x - 2) = 0$$

and $x = -5$ or $x = 2$. We take $x = 2$. The corresponding value of p is 5. Therefore, the equilibrium quantity is 2000 and the equilibrium price is \$52.

35. Equating the two equations, we have

$$144 - x^2 = 48 + \tfrac{1}{2}x^2$$

$$288 - 2x^2 = 96 + x^2$$
$$3x^2 = 192$$
$$x^2 = 64,$$

or $x = \pm 8$. We take $x = 8$, and the corresponding value of p is 144 - 8^2 = 80. We conclude that the equilibrium quantity is 8000 tires and the equilibrium price is $80.

EXERCISES 10.4, page 626

1. $\lim_{x \to 2} f(x) = 3.$

3. $\lim_{x \to 3} f(x) = 3.$

5. $\lim_{x \to -2} f(x) = 3.$

7. The limit does not exist. If we consider any value of x to the right of $x = -2$, $f(x) \leq 2$. If we consider values of x to the left of $x = -2$, $f(x) \geq -2$. Since $f(x)$ does not approach any one number as x approaches $x = -2$, we conclude that the limit does not exist.

9.

x	1.9	1.99	1.999	2.001	2.01	2.1
$f(x)$	4.61	4.9601	4.9960	5.004	5.0401	5.41

$\lim_{x \to 2} (x^2 + 1) = 5.$

11.

x	-0.1	-0.01	-0.001	0.001	0.01	0.1
$f(x)$	-1	-1	-1	1	1	1

The limit does not exist.

13.

x	0.9	0.99	0.999	1.001	1.01	1.1
$f(x)$	100	10,000	1,000,000	1,000,000	10,000	100

The limit does not exist.

15.

x	0.9	0.99	0.999	1.001	1.01	1.1
$f(x)$	2.9	2.99	2.999	3.001	3.01	3.1

$\lim_{x\to 1} \frac{x^2+x-2}{x-1} = 3.$

17.

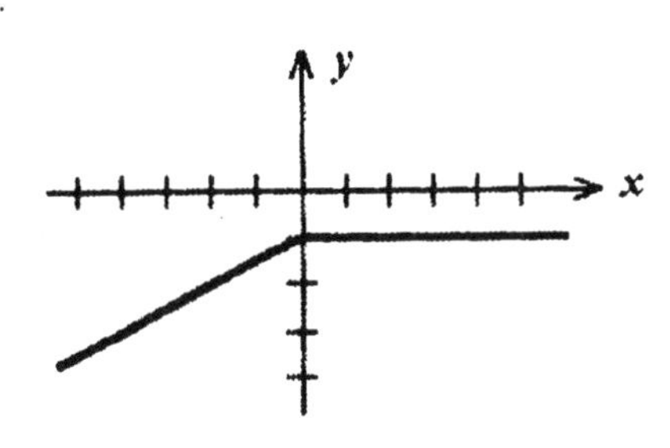

$\lim_{x\to 0} f(x) = -1$

19.

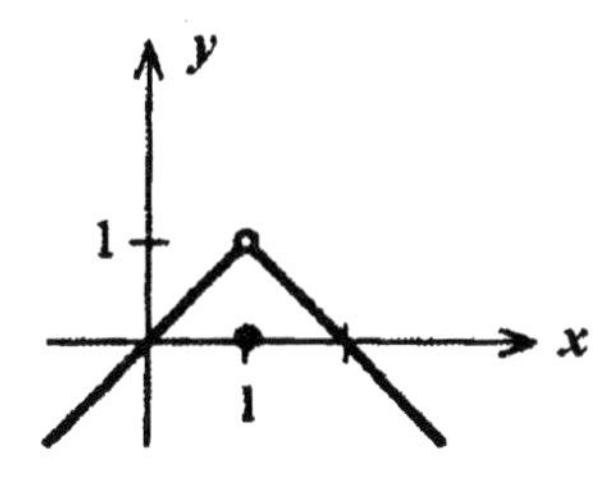

$\lim_{x\to 1} f(x) = 1$

21.

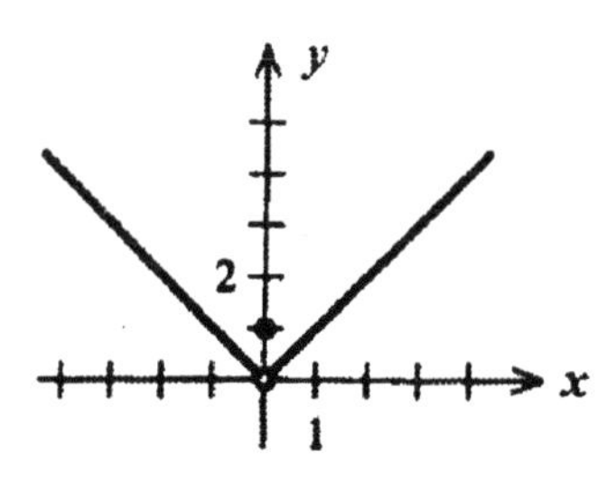

$\lim_{x\to 0} f(x) = 0$

23. $\lim_{x\to 2} 3 = 3$

25. $\lim_{x\to 3} x = 3$

27. $\lim_{x\to 1}(1-2x^2) = 1-2(1)^2 = -1$

29. $\lim_{x\to 1}(2x^3-3x^2+x+2) = 2(1)^3-3(1)^2+1+2 = 2.$

31. $\lim_{s\to 0}(2s^2-1)(2s+4) = (-1)(4) = -4.$

33. $\lim\limits_{x\to 2}\dfrac{2x+1}{x+2}=\dfrac{2(2)+1}{2+2}=\dfrac{5}{4}.$

35. $\lim\limits_{x\to 2}\sqrt{x+2}=\sqrt{2+2}=2.$

37. $\lim\limits_{x\to -3}\sqrt{2x^4+x^2}=\sqrt{2(-3)^4+(-3)^2}=\sqrt{162+9}=\sqrt{171}=3\sqrt{19}.$

39. $\lim\limits_{x\to -1}\dfrac{\sqrt{x^2+8}}{2x+4}=\dfrac{\sqrt{(-1)^2+8}}{2(-1)+4}=\dfrac{\sqrt{9}}{2}=\dfrac{3}{2}.$

41. $\lim\limits_{x\to a}[f(x)-g(x)]=\lim\limits_{x\to a}f(x)-\lim\limits_{x\to a}g(x)=3-4=-1.$

43. $\lim\limits_{x\to a}[2f(x)-3g(x)]=\lim\limits_{x\to a}2f(x)-\lim\limits_{x\to a}3g(x)=2(3)-3(4)=-6.$

45. $\lim\limits_{x\to a}\sqrt{g(x)}=\lim\limits_{x\to a}\sqrt{4}=2.$

47. $\lim\limits_{x\to a}\dfrac{2f(x)-g(x)}{f(x)g(x)}=\dfrac{2(3)-(4)}{(3)(4)}=\dfrac{2}{12}=\dfrac{1}{6}.$

49. $\lim\limits_{x\to 1}\dfrac{x^2-1}{x-1}=\lim\limits_{x\to 1}\dfrac{(x-1)(x+1)}{x-1}=\lim\limits_{x\to 1}(x+1)=1+1=2.$

51. $\lim\limits_{x\to 0}\dfrac{x^2-x}{x}=\lim\limits_{x\to 0}\dfrac{x(x-1)}{x}=\lim\limits_{x\to 0}(x-1)=0-1=-1.$

53. $\lim\limits_{x\to -5}\dfrac{x^2-25}{x+5}=\lim\limits_{x\to -5}\dfrac{(x+5)(x-5)}{x+5}=\lim\limits_{x\to -5}(x-5)=-10.$

55. $\lim\limits_{x\to 1}\dfrac{x}{x-1}$ does not exist.

57. $\lim\limits_{x\to -2}\dfrac{x^2-x-6}{x^2+x-2}=\lim\limits_{x\to -2}\dfrac{(x-3)(x+2)}{(x+2)(x-1)}=\lim\limits_{x\to -2}\dfrac{x-3}{x-1}=\dfrac{-2-3}{-2-1}=\dfrac{5}{3}.$

59. $\lim_{x\to 1}\frac{\sqrt{x}-1}{x-1}=\lim_{x\to 1}\frac{\sqrt{x}-1}{x-1}\cdot\frac{\sqrt{x}+1}{\sqrt{x}+1}=\lim_{x\to 1}\frac{x-1}{(x-1)(\sqrt{x}-1)}=\lim_{x\to 1}\frac{1}{(\sqrt{x}-1)}=\frac{1}{2}.$

61. $\lim_{x\to 1}\frac{x-1}{x^3+x^2-2x}=\lim_{x\to 1}\frac{x-1}{x(x-1)(x+2)}=\lim_{x\to 1}\frac{1}{x(x+2)}=\frac{1}{3}.$

63. $\lim_{x\to\infty} f(x)=-\infty$ (does not exist) and $\lim_{x\to-\infty} f(x)=\infty$ (does not exist).

65. $\lim_{x\to\infty} f(x)=0$ and $\lim_{x\to-\infty} f(x)=0.$

67. $\lim_{x\to\infty} f(x)=-\infty$ (does not exist) and $\lim_{x\to-\infty} f(x)=-\infty$ (does not exist).

69.

x	1	10	100	1000
$f(x)$	0.5	0.009901	0.0001	0.000001

x	-1	-10	-100	-1000
$f(x)$	0.5	0.009901	0.0001	0.000001

$\lim_{x\to\infty} f(x)=0$ and $\lim_{x\to-\infty} f(x)=0$

71.

x	1	5	10	100	1000
$f(x)$	12	360	2910	2.99×10^6	2.999×10^9

x	-1	-5	-10	-100	-1000
$f(x)$	6	-390	-3090	-3.01×10^6	-3.0×10^9

$\lim_{x\to\infty} f(x)=\infty$ (does not exist) and $\lim_{x\to-\infty} f(x)=-\infty$ (does not exist).

73. $\lim_{x\to\infty}\frac{3x+2}{x-5}=\lim_{x\to\infty}\frac{3+\frac{2}{x}}{1-\frac{5}{x}}=\frac{3}{1}=3.$

75. $\lim_{x\to-\infty}\frac{3x^3+x^2+1}{x^3+1}=\lim_{x\to-\infty}\frac{3+\frac{1}{x}+\frac{1}{x^3}}{1+\frac{1}{x^3}}=3.$

77. $\lim_{x\to-\infty}\frac{x^4+1}{x^3-1}=\lim_{x\to-\infty}\frac{x+\frac{1}{x^3}}{1-\frac{1}{x^3}}=-\infty$; that is, the limit does not exist.

79. $\lim_{x\to\infty}\frac{x^5-x^3+x-1}{x^6+2x^2+1}=\lim_{x\to\infty}\frac{\frac{1}{x}-\frac{1}{x^3}+\frac{1}{x^5}-\frac{1}{x^6}}{1+\frac{2}{x^4}+\frac{1}{x^6}}=0.$

81. a. The cost of removing 50 percent of the pollutant is

$$C(50)=\frac{0.5(50)}{100-50}=0.5\text{, or \$500,000.}$$

Similarly, we find that the cost of removing 60, 70, 80, 90, and 95 percent of the pollutants is \$750,000; \$1,166,667; \$2,000,000, \$4,500,000, and \$9,500,000, respectively.

b. $\lim_{x\to 100}\frac{0.5x}{100-x}=\infty,$

which means that the cost of removing the pollutant increases astronomically if we wish to remove almost all of the pollutant.

83. $\lim_{x\to\infty}\overline{C}(x)=\lim_{x\to\infty}2.2+\frac{2500}{x}=2.2,$ or \$2.20 per record.

In the long-run, the average cost of producing x video discs will approach \$2.20/disc.

85. a. $T(1)=\frac{120}{1+4}=24,$ or \$24 million.

$T(2)=\frac{120(4)}{8}=60,$ \$60 million.

$T(3)=\frac{120(9)}{13}=83.1,$ or \$83.1 million.

b. In the long run, the movie will gross

$$\lim_{x\to\infty}\frac{120x^2}{x^2+4}=\lim_{x\to\infty}\frac{120}{1+\frac{4}{x^2}}=120, \text{ or \$120 million.}$$

87. a. The average cost of driving 5000 miles per year is

$$C(5)=\frac{2010}{5^{2.2}}+17.80=76.07,$$

or 76.1 cents per mile. Similarly, we see that the average cost of driving 10,000 miles per year; 15,000 miles per year; 20,000 miles per year; and 25,000 miles per year is 30.5, 23; 20.6, and 19.5 cents per mile, respectively.

b.

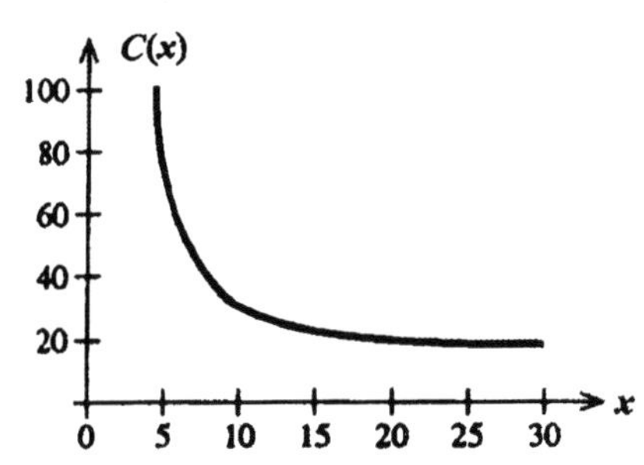

c It approaches 17.80 cents per mile.

89. Consider the functions $f(x)=1/x$ and $g(x)=-1/x$. Observe that

$$\lim_{x\to0}f(x) \quad \text{and} \quad \lim_{x\to0}g(x)$$

do not exist, but $\lim_{x\to0}[f(x)+g(x)]=\lim_{x\to0}0=0.$

This example does not contradict Theorem 1 because the hypothesis of Theorem 1 says that if $\lim_{x\to0}f(x)$ and $\lim_{x\to0}g(x)$ both exist, then the limit of the sum of *f* and *g* also exists. It does not say that if the former do not exist, then the latter might not exist.

USING TECHNOLOGY EXERCISES 10.4, page 632

1. 5 3. 3 5. $\frac{2}{3}$ 7. $\frac{1}{2}$ 9. e^2, or 7.38906

11. From the graph we see that $f(x)$ does not approach any finite number as x approaches 3.

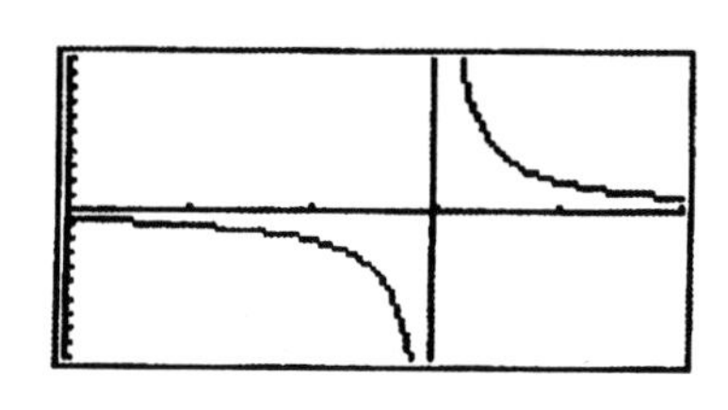

13. a.

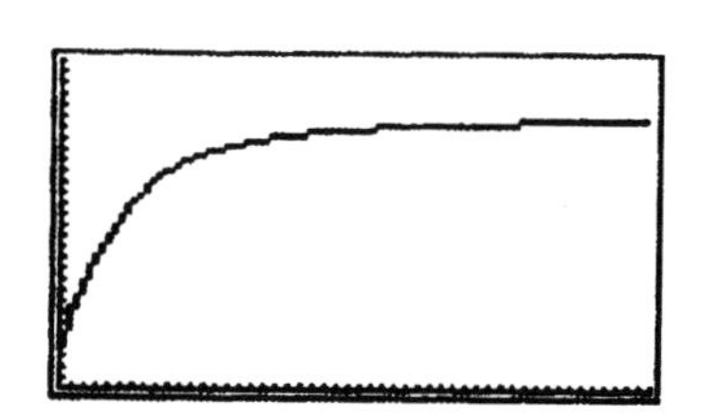

b. $\lim_{t\to\infty}\frac{25t^2+125t+200}{t^2+5t+40}=25$, so in the long run the population will approach 25,000.

EXERCISES 10.5, page 644

1. $\lim_{x\to 2^-} f(x)=3$, $\lim_{x\to 2^+} f(x)=2$, $\lim_{x\to 2} f(x)$ does not exist.

3. $\lim_{x\to -1^-} f(x)=\infty$, $\lim_{x\to -1^+} f(x)=2$. Therefore $\lim_{x\to -1} f(x)$ does not exist.

5. $\lim_{x\to 1^-} f(x)=0$, $\lim_{x\to 1^+} f(x)=2$, $\lim_{x\to 1} f(x)$ does not exist.

7. $\lim_{x\to 0^-} f(x)=-2$, $\lim_{x\to 0^+} f(x)=2$, $\lim_{x\to 0} f(x)$ does not exist.

9. True 11. True 13. False 15. True

17. False 19. True

21. $\lim_{x\to 1^+}(2x+4)=6$.

23. $\lim_{x\to 2^-}\frac{x-3}{x+2}=\frac{2-3}{2+2}=-\frac{1}{4}$.

25. $\lim_{x\to 0^+}\frac{1}{x}$ does not exist because $1/x\to\infty$ as $x\to 0$ from the right..

27. $\lim_{x\to 0^+} \dfrac{x-1}{x^2+1} = \dfrac{-1}{1} = -1.$.

29. $\lim_{x\to 0^+} \sqrt{x} = \sqrt{\lim_{x\to 0^+} x} = 0.$

31. $\lim_{x\to -2^+} (2x+\sqrt{2+x}) = \lim_{x\to -2^+} 2x + \lim_{x\to -2^+} \sqrt{2+x} = -4+0 = -4.$

33. $\lim_{x\to 1^-} \dfrac{1+x}{1-x} = \infty$, that is, the limit does not exist.

35. $\lim_{x\to 2^-} \dfrac{x^2-4}{x-2} = \lim_{x\to 2^-} \dfrac{(x+2)(x-2)}{x-2} = \lim_{x\to 2^-} (x+2) = 4.$

37. $\lim_{x\to 3^+} \dfrac{x^2-9}{x+3} = \dfrac{9-9}{3+3} = 0.$

39. $\lim_{x\to 0^+} f(x) = \lim_{x\to 0^+} x^2 = 0$, $\lim_{x\to 0^-} f(x) = \lim_{x\to 0^-} 2x = 0$

41. $\lim_{x\to 1^+} f(x) = \lim_{x\to 1^+} \sqrt{x+3} = \sqrt{4} = 2.$

$\lim_{x\to 1^-} f(x) = \lim_{x\to 1^-} (2+\sqrt{x}) = 2+\sqrt{x} = 2+\sqrt{1} = 3.$

43. The function is discontinuous at $x = 0$. Conditions 2 and 3 are violated.

45. The function is continuous everywhere.

47. The function is discontinuous at $x = 0$. Condition 3 is violated.

49. The function is discontinuous at $x = 0$. Condition 3 is violated.

51. f is continuous for all values of x.

53. f is continuous for all values of x. Note that $x^2 + 1 \geq 1 > 0$.

55. f is discontinuous at $x = 1/2$, where the denominator is 0.

57. Observe that $x^2 + x - 2 = (x + 2)(x - 1) = 0$ if $x = -2$ or $x = 1$. So, f is discontinuous at these values of x.

59. f is continuous everywhere since all three conditions are satisfied.

61. f is continuous everywhere because all three conditions are satisfied.

63. f is continuous everywhere since all three conditions are satisfied. Observe that

$$\lim_{x\to 1} f(x) = \lim_{x\to 1}\frac{x^2-1}{x-1} = \lim_{x\to 1}\frac{(x-1)(x+1)}{x-1} = \lim_{x\to 1}(x+1) = 2 = f(1).$$

65. f is continuous everywhere since all three conditions are satisfied.

67. Since the denominator $x^2 - 1 = (x - 1)(x + 1) = 0$ if $x = -1$ or 1, we see that f is discontinuous at these points.

69. Since $x^2 - 3x + 2 = (x - 2)(x - 1) = 0$ if $x = 1$ or 2, we see that the denominator is zero at these points and so f is discontinuous at these points.

71. The function f is discontinuous at $x = 1, 2, 3, \ldots, 11$ because the limit of f does not exist at these points.

73. Having made steady progress up to $x = x_1$, Michael's progress came to a standstill. Then at $x = x_2$ a sudden break-through occurs and he then continues to successfully complete the solution to the problem.

75. Conditions 2 and 3 are not satisfied at each of these points.

77. The graph of f follows.

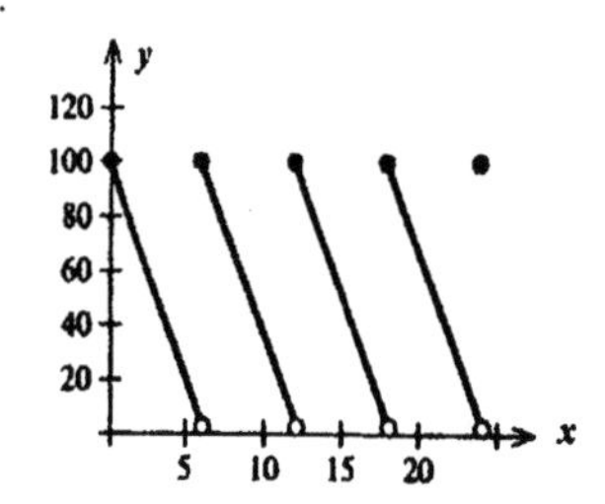

f is discontinuous at $x = 6, 12, 18, 24$.

79.

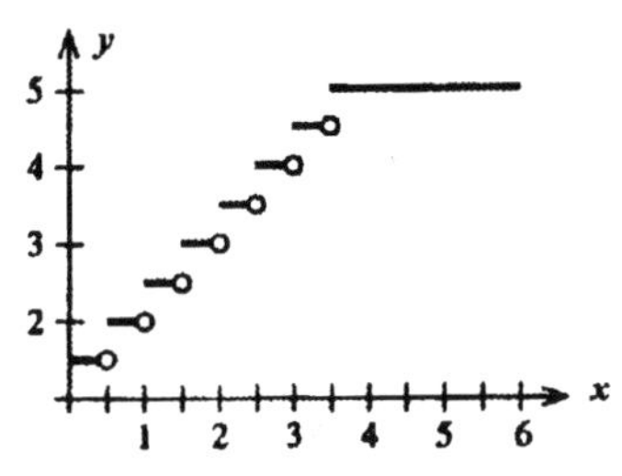

f is discontinuous at $x = \frac{1}{2}, 1, 1\frac{1}{2}, \ldots, 4$.

81. a. $\lim\limits_{v \to u^+} \dfrac{aLv^3}{v-u} = \infty$ and this shows that, when the speed of the fish is very close to that of the current, the energy expended by the fish will be enormous.

b. $\lim\limits_{v \to \infty} \dfrac{aLv^3}{v-u} = \infty$ and this says that if the speed of the fish increases greatly, so does the amount of energy required to swim a distance of L ft.

83. Since

$$\lim_{x \to -2} \frac{x^2-4}{x+2} = \lim_{x \to -2} \frac{(x-2)(x+2)}{x+2} = \lim_{x \to -2} (x-2) = -4,$$

we define $f(-2) = k = -4$, that is, take $k = -4$.

85. a. No. Consider the function $f(x) = 0$ for all x in $(-\infty, \infty)$ and

$$g(x) = \begin{cases} 1 & \text{if } x < 0 \\ -1 & \text{if } x \geq 0 \end{cases}.$$

b. No. Consider the functions f and g of Exercise 84b.

87. f is a polynomial and is therefore continuous on $[-1,1]$.

$f(-1) = (-1)^3 - 2(-1)^2 + 3(-1) + 2 = -1 - 2 - 3 + 2 = -4$.

$f(1) = 1 - 2 + 3 + 2 = 4$.

Since $f(-1)$ and $f(1)$ have opposite signs, we see that f has at least one zero in $(-1,1)$.

89. f is continuous on $[14,16]$ and

$f(14) = 2(14)^{5/3} - 5(14)^{4/3} \approx -6.06$

$f(16) = 2(16)^{5/3} - 5(16)^{4/3} \approx 1.60$,

and so f has at least one zero in $(14,16)$.

91. $f(0) = 6$ and $f(3) = 3$ and f is continuous on [0,3]. So the Intermediate Value Theorem guarantees that there is at least one value of x for which $f(x) = 2$. Solving

$$f(x) = x^2 - 4x + 6 = 2$$

we find, $x^2 - 4x + 4 = (x - 2)^2 = 0$, or $x = \pm 2$. Since -2 does not lie in [0,3], we see that $x = 2$.

93.

Step	Root of f(x) = 0 lies in
1	(1,2)
2	(1,1.5)
3	(1.25,1.5)
4	(1.25,1.375)
5	(1.3125,1.375)
6	(1.3125,1.34375)
7	(1.328125,1.34375)
8	(1.3359375,1.34375)
9	(1.33984375,1.34375)

We see that the required root is approximately 1.34.

95. False. Consider the function $f(x) = x^2 - 1$ on the interval [-2,2].Here, $f(-2) = f(2) = 3$, but f has zeros at $x = -1$ and $x = 1$.

97. a. f is a rational function whose denominator is never zero, and so it is continuous for all values of x.
b. Since the numerator, x^2, is nonnegative and the denominator is $x^2 + 1 \geq 1$ for all values of x, we see that $f(x)$ is nonnegative for all values of x.
c. $f(0) = \dfrac{0}{0+1} = \dfrac{0}{1} = 0$ and so f has a zero at $x = 0$. This does not contradict Theorem 4.

99. Consider the function f defined by

$$f(x) = \begin{cases} -1 & \text{if } -1 \leq x < 0 \\ 1 & \text{if } \quad 0 \leq x < 1 \end{cases}.$$

Then $f(-1) = -1$ and $f(1) = 1$. But, if we take the number 1/2 which lies between $y = -1$ and $y = 1$, there is no value of x such that $f(x) = 1/2$.

USING TECHNOLOGY EXERCISES 10.5, page 652

1. $x = 0,1$

3. $x = 2$

5. $x = 0, \frac{1}{2}$

7. $x = -\frac{1}{2}, 2$

9. $x = -2, 1$

11.

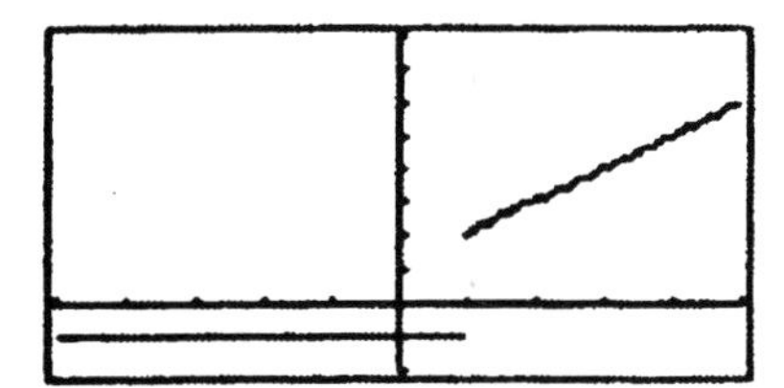

13.

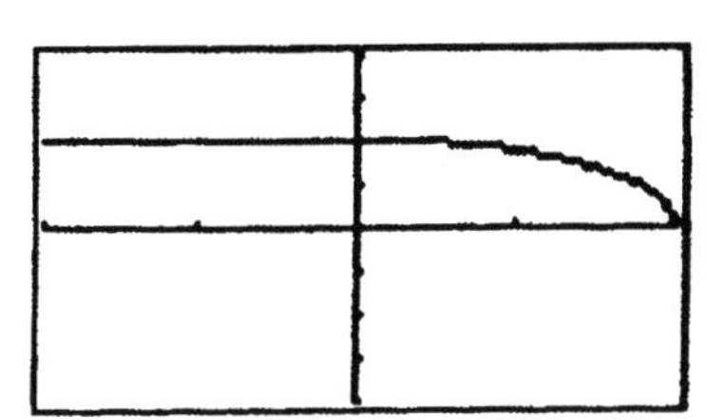

15.

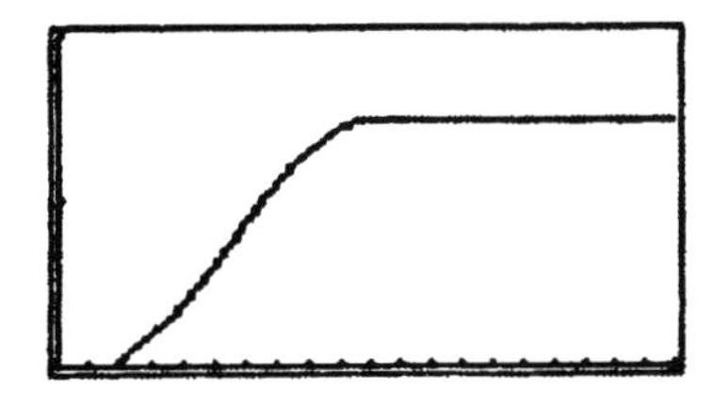

EXERCISES 10.6, page 669

1. The rate of change of the average infant's weight when $t = 3$ is (7.5)/5, or 1.5 lb/month. The rate of change of the average infant's weight when $t = 18$ is (3.5)/6, or approximately 0.6 lb/month. The average rate of change over the infant's first year of life is (22.5 – 7.5)/(12), or 1.25 lb/month.

3. The rate of change of the percentage of households watching television at 4 P.M. is (12.3)/4, or approximately 3.1 percent per hour. The rate at 11 P.M. is (–42.3)/2 = –21.15; that is, it is dropping off at the rate of 21.15 percent per hour.

5. a. Car A is travelling faster than Car B at t_1 because the slope of the tangent line to the graph of f is greater than the slope of the tangent line to the graph of g at t_1.

   b. Their speed is the same because the slope of the tangent lines are the same at t_2.

   c. Car B is travelling faster than Car A.

d. They have both covered the same distance and are once again side by side at t_3.

7. a. P_2 is decreasing faster at t_1 because the slope of the tangent line to the graph of g at t_1 is greater than the slope of the tangent line to the graph of f at t_1.
b. P_1 is decreasing faster than P_2 at t_2.
c. Bactericide B is more effective in the short run, but bactericide A is more effective in the long run.

9. **Step 1** $f(x+h) = 13$
Step 2 $f(x+h) - f(x) = 13 - 13 = 0$
Step 3 $\dfrac{f(x+h)-f(x)}{h} = \dfrac{0}{h} = 0$
Step 4 $f'(x) = \lim\limits_{h\to 0} \dfrac{f(x+h)-f(x)}{h} = \lim\limits_{h\to 0} 0 = 0$

11. **Step 1** $f(x+h) = 2(x+h) + 7$
Step 2 $f(x+h) - f(x) = 2(x+h) + 7 - (2x+7) = 2h$
Step 3 $\dfrac{f(x+h)-f(x)}{h} = \dfrac{2h}{h} = 2$
Step 4 $f'(x) = \lim\limits_{h\to 0} \dfrac{f(x+h)-f(x)}{h} = \lim\limits_{h\to 0} 2 = 2$

13. **Step 1** $f(x+h) = 3(x+h)^2 = 3x^2 + 6xh + 3h^2$
Step 2 $f(x+h) - f(x) = (3x^2 + 6xh + 3h^2) - 3x^2 = 6xh + 3h^2 = h(6x+3h)$
Step 3 $\dfrac{f(x+h)-f(x)}{h} = \dfrac{h(6x+3h)}{h} = 6x + 3h$
Step 4 $f'(x) = \lim\limits_{h\to 0} \dfrac{f(x+h)-f(x)}{h} = \lim\limits_{h\to 0} (6x+3h) = 6x.$

15. **Step 1** $f(x+h) = -(x+h)^2 + 3(x+h) = -x^2 - 2xh - h^2 + 3x + 3h$
Step 2 $f(x+h) - f(x) = (-x^2 - 2xh - h^2 + 3x + 3h) - (-x^2 + 3x)$
$= -2xh - h^2 + 3h = h(-2x - h + 3)$
Step 3 $\dfrac{f(x+h)-f(x)}{h} = \dfrac{h(-2x-h+3)}{h} = -2x - h + 3$
Step 4 $f'(x) = \lim\limits_{h\to 0} \dfrac{f(x+h)-f(x)}{h} = \lim\limits_{h\to 0} (-2x - h + 3) = -2x + 3.$

17. Using the four-step process

Step 1 $f(x+h)=2(x+h)+7=2x+2h+7$

Step 2 $f(x+h)-f(x)=2x+2h+7-2x-7=2h$

Step 3 $\dfrac{f(x+h)-f(x)}{h}=\dfrac{2h}{h}=2$

Step 4 $f'(x)=\lim\limits_{h\to 0}\dfrac{f(x+h)-f(x)}{h}=\lim\limits_{h\to 0}2=2$

we find that $f'(x)=2$. In particular, the slope at $x=2$ is also 2. Therefore, a required equation is

$$y-11=2(x-2) \qquad \text{or} \qquad y=2x+7.$$

19. We first compute $f'(x)=6x$ (see Problem 13). Since the slope of the tangent line is $f'(1)=6$, we use the point-slope form of the equation of a line and find that a required equation is

$$y-3=6(x-1), \text{ or } y=6x-3.$$

21. We first compute $f'(x)$ using the four-step process.

Step 1 $f(x+h)=-\dfrac{1}{x+h}$

Step 2 $f(x+h)-f(x)=-\dfrac{1}{x+h}+\dfrac{1}{x}=\dfrac{-x+(x+h)}{x(x+h)}=\dfrac{h}{x(x+h)}]$

Step 3 $\dfrac{f(x+h)-f(x)}{h}=\dfrac{\frac{h}{x(x+h)}}{h}=\dfrac{1}{x(x+h)}$

Step 4 $f'(x)=\lim\limits_{h\to 0}\dfrac{f(x+h)-f(x)}{h}=\lim\limits_{h\to 0}\dfrac{1}{x(x+h)}=\dfrac{1}{x^2}.$

The slope of the tangent line is $f'(3)=1/9$. Therefore, a required equation is

$$y-(-\tfrac{1}{3})=\tfrac{1}{9}(x-3) \quad \text{or} \quad y=\tfrac{1}{9}x-\tfrac{2}{3}.$$

23. a. We use the four-step process.

Step 1 $f(x+h)=2(x+h)^2+1=2x^2+4xh+2h^2+1$

Step 2 $f(x+h)-f(x)=(2x^2+4xh+2h^2+1)-(2x^2+1)=4xh+2h^2$
$=h(4x+2h)$

Step 3 $\dfrac{f(x+h)-f(x)}{h}=\dfrac{h(4x+2h)}{h}=4x+2h$

Step 4 $f'(x)=\lim\limits_{h\to 0}\dfrac{f(x+h)-f(x)}{h}=\lim\limits_{h\to 0}(4x+2h)=4x$

b. The slope of the tangent line is $f'(1) = 4(1) = 4$. Therefore, an equation is $y - 3 = 4(x - 1)$ or $y = 4x - 1$.

c.

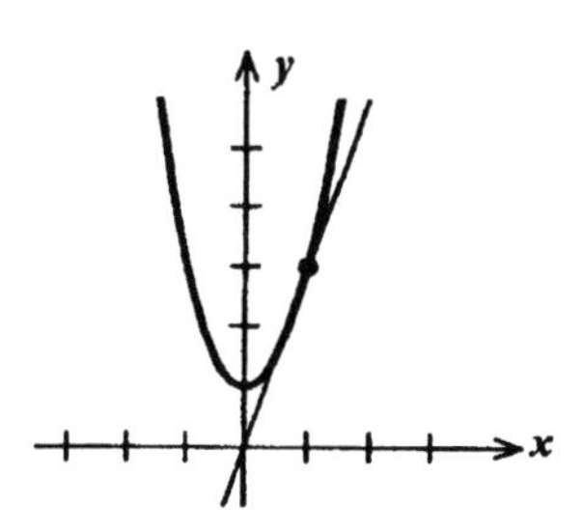

25. a. We use the four-step process:

Step 1 $f(x+h) = (x+h)^2 - 2(x+h) + 1 = x^2 + 2xh + h^2 - 2x - 2h + 1$

Step 2 $f(x+h) - f(x) = (x^2 + 2xh + h^2 - 2x - 2h + 1) - (x^2 - 2x + 1)]$

$$= 2xh + h^2 - 2h = h(2x + h - 2)$$

Step 3 $\dfrac{f(x+h) - f(x)}{h} = \dfrac{h(2x+h-2)}{h} = 2x + h - 2$

Step 4 $f'(x) = \lim\limits_{h\to 0} \dfrac{f(x+h) - f(x)}{h} = \lim\limits_{h\to 0} (2x + h - 2) = 2x - 2.$

b. At a point on the graph of f where the tangent line to the curve is horizontal, $f'(x) = 0$. Then $2x - 2 = 0$, or $x = 1$. Since $f(1) = 1 - 2 + 1 = 0$, we see that the required point is (1,0).

c.

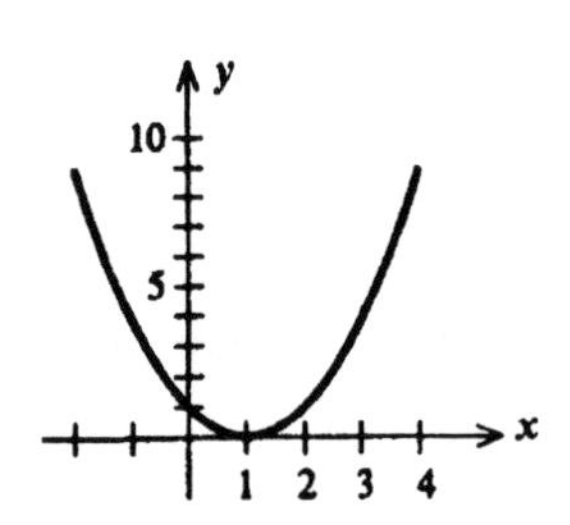

d. It is changing at the rate of 0 units per unit change in x.

27. a. $$\frac{f(3) - f(2)}{3 - 2} = \frac{(3^2 + 3) - (2^2 + 2)}{1} = 6$$

$$\frac{f(2.5) - f(2)}{2.5 - 2} = \frac{(2.5^2 + 2.5) - (2^2 + 2)}{0.5} = 5.5$$

$$\frac{f(2.1)-f(2)}{2.1-2}=\frac{(2.1^2+2.1)-(2^2+2)}{0.1}=5.1 \ .$$

b. We first compute $f'(x)$ using the four-step process.

Step 1 $f(x+h)=(x+h)^2+(x+h)=x^2+2xh+h^2+x+h$

Step 2 $f(x+h)-f(x)=(x^2+2xh+h^2+x+h)-(x^2+x)]$

$$=2xh+h^2+h=h(2x+h+1)$$

Step 3 $\dfrac{f(x+h)-f(x)}{h}=\dfrac{h(2x+h+1)}{h}=2x+h+1$

Step 4 $f'(x)=\lim\limits_{h\to 0}\dfrac{f(x+h)-f(x)}{h}=\lim\limits_{h\to 0}(2x+h+1)=2x+1.$

The instantaneous rate of change of y at $x = 2$ is $f'(2)=5$ or 5 units per unit change in x.

c. The results in (a) suggest that the average rates of change of f at $x = 2$ approach 5 as the interval $[2, 2+h]$ gets smaller and smaller ($h = 1$, 0.5, and 0.1). This number is the instantaneous rate of change of f at $x = 2$ as computed in (b).

29. a. The average velocity of the car over [20,21] is

$$\frac{f(21)-f(20)}{21-20}=\frac{[2(21)^2+48(21)]-[2(20)^2+48(20)]}{1}=130 \text{ ft/sec}$$

Its average velocity over [20,20.1] is

$$\frac{f(20.1)-f(20)}{20.1-20}=\frac{[2(20.1)^2+48(20.1)]-[2(20)^2+48(20)]}{0.1}=128.2 \text{ ft/sec}$$

Its average velocity over [20,20.01]

$$\frac{f(20.01)-f(20)}{20.01-20}=\frac{[2(20.01)^2+48(20.01)]-[2(20)^2+48(20)]}{0.01}=128.02 \text{ ft/sec}$$

b. We first compute $f'(t)$ using the four-step process.

Step 1 $f(t+h)=2(t+h)^2+48(t+h)=2t^2+4th+2h^2+48t+48h$

Step 2 $f(t+h)-f(t)=(2t^2+4th+2h^2+48t+48h)-(2t^2+48t)]$

$$=4th+2h^2+48h=h(4t+2h+48).$$

Step 3 $\dfrac{f(t+h)-f(t)}{h}=\dfrac{h(4t+2h+48)}{h}=4t+2h+48$

Step 4 $f'(t)=\lim\limits_{h\to 0}(4t+2h+48)=4t+48$

The instantaneous velocity of the car at $t = 20$ is

$f'(20) = 4(20) + 48$, or 128 ft/sec.

c. Our results shows that the average velocities do approach the instantaneous velocity as the intervals over which they are computed decreases.

31. a. We solve the equation $16t^2 = 400$ obtaining $t = 5$ which is the time it takes the screw driver to reach the ground.

b. The average velocity over the time [0,5] is

$$\frac{f(5)-f(0)}{5-0} = \frac{16(25)-0}{5} = 80, \text{ or } 80 \text{ ft/sec.} \quad [\text{Let } s = f(t) = 16t^2.]$$

c. The velocity of the screwdriver at time t is

$$v(t) = \lim_{h\to 0}\frac{f(t+h)-f(t)}{h} = \lim_{h\to 0}\frac{16(t+h)^2-16t^2}{h}$$

$$= \lim_{h\to 0}\frac{16t^2+32th+16h^2-16t^2}{h} = \lim_{h\to 0}\frac{(32t+16h)h}{h}.$$

In particular, the velocity of the screwdriver when it hits the ground (at $t = 5$) is

$v(5) = 32(5) = 160$, or 160 ft/sec.

33. a. The average rate of change of V is

$$\frac{f(3)-f(2)}{3-2} = \frac{\frac{1}{3}-\frac{1}{2}}{1} = -\frac{1}{6}, \qquad [\text{Write } V = f(p) = \frac{1}{p}.]$$

or a decrease of $\frac{1}{6}$ liters/atmosphere.

b.

$$V'(t) = \lim_{h\to 0}\frac{f(p+h)-f(p)}{h} = \lim_{h\to 0}\frac{\frac{1}{p+h}-\frac{1}{p}}{h}$$

$$= \lim_{h\to 0}\frac{p-(p+h)}{hp(p+h)} = \lim_{h\to 0}-\frac{1}{p(p+h)} = -\frac{1}{p^2}.$$

In particular, the rate of change of V when $p = 2$ is

$$V'(2) = -\frac{1}{2^2}, \text{ or a decrease of } \tfrac{1}{4} \text{ liters/atmosphere}$$

35. a. Using the four-step process, we find that

$$P'(x) = \lim_{h\to 0}(-\tfrac{2}{3}x - \tfrac{1}{3}h + 7) = -\tfrac{2}{3}x + 7.$$

b. $P'(10) = -\frac{2}{3}(10) + 7 \approx 0.333$, or \$333 per quarter.

$P'(30) = -\frac{2}{3}(30) + 7 \approx -13$, or a decrease of \$13,000 per quarter.

37. We first compute $N'(t)$ using the four–step process.

Step 1 $N(t+h) = (t+h)^2 + 2(t+h) + 50$
$= t^2 + 2th + h^2 + 2t + 2h + 50$

Step 2 $N(t+h) - N(t)$
$= (t^2 + 2th + h^2 + 2t + 2h + 50) - (t^2 + 2t + 50)$
$= 2th + h^2 + 2h = h(2t + h + 2).$

Step 3 $\dfrac{N(t+h) - N(t)}{h} = 2t + h + 2.$

Step 4 $N'(t) = \lim_{h \to 0}(2t + h + 2) = 2t + 2.$

The rate of change of the country's GNP two years from now will be $N'(2) = 6$, or \$6 billion/yr. The rate of change four years from now will be $N'(4) = 10$, or \$10 billion/yr.

39. $\dfrac{f(a+h) - f(a)}{h}$ gives the average rate of change of the seal population over the time interval $[a, a+h]$. $\lim_{h \to 0} \dfrac{f(a+h) - f(a)}{h}$ gives the instantaneous rate of change of the seal population at $x = a$.

41. $\dfrac{f(a+h) - f(a)}{h}$ gives the average rate of change of the country's industrial production over the time interval $[a, a+h]$.
$\lim_{h \to 0} \dfrac{f(a+h) - f(a)}{h}$ gives the instantaneous rate of change of the country's industrial production at $x = a$.

43. $\dfrac{f(a+h) - f(a)}{h}$ gives the average rate of change of the atmospheric pressure over the altitudes $[a, a+h]$.
$\lim_{h \to 0} \dfrac{f(a+h) - f(a)}{h}$ gives the instantaneous rate of change of the atmospheric pressure at $x = a$.

45. a. f has a limit at $x = a$.
b. f is not continuous at $x = a$ because $f(a)$ is not defined.
c. f is not differentiable at $x = a$ because it is not continuous there.

47. a. f has a limit at $x = a$.   b. f is continuous at $x = a$.
c. f is not differentiable at $x = a$ because f has a kink at the point $x = a$.

49. a. f does not have a limit at $x = a$ because it is unbounded in the neighborhood of a.
b. f is not continuous at $x = a$.
c. f is not differentiable at $x = a$ because it is not continuous there.

51. Our computations yield the following results:
5.06060, 5.06006, 5.060006, 5.0600006, 5.06000006;
The rate of change of the total cost function when the level of production is 100 cases a day is approximately \$5.06.

53. f does not have a derivative at $x = 1$ because it is not continuous there.

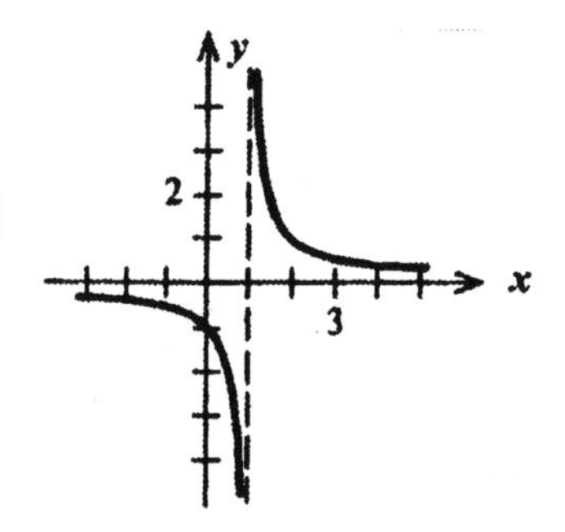

55. f is continuous at $x = 0$, but $f'(0)$ does not exist because the graph of f has a vertical tangent line at $x = 0$. The graph of f follows.

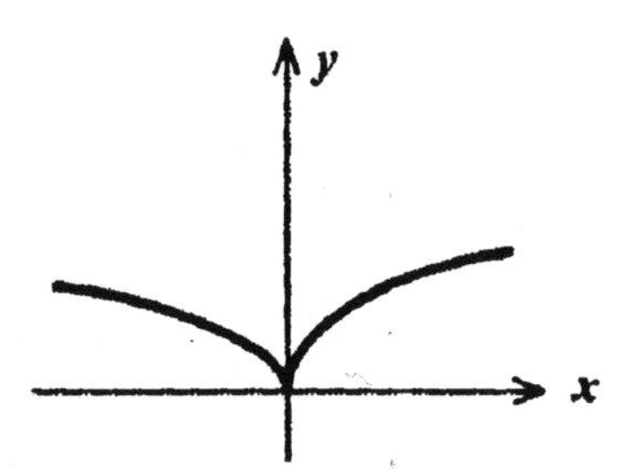

57. From
$$f(x) - f(a) = \left[\frac{f(x) - f(a)}{x - a}\right](x - a)$$
we see that

$$\lim_{x\to a}[f(x)-f(a)] = \lim_{x\to a}\left[\frac{f(x)-f(a)}{x-a}\right]\lim_{x\to a}(x-a)$$

$$= f'(a)\cdot 0 = 0$$

and so $\lim_{x\to a} f(x) = f(a)$. This shows that f is continuous at $x = a$.

USING TECHNOLOGY EXERCISES 10.6, page 677

1. a. $y = 4x - 3$

   b.

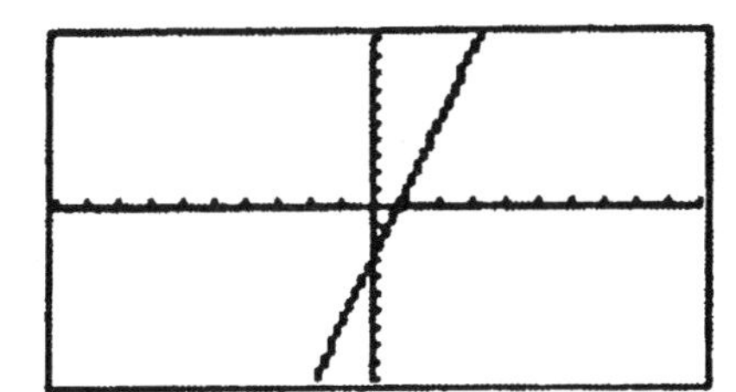

3. a. $y = -7x - 8$

   b.

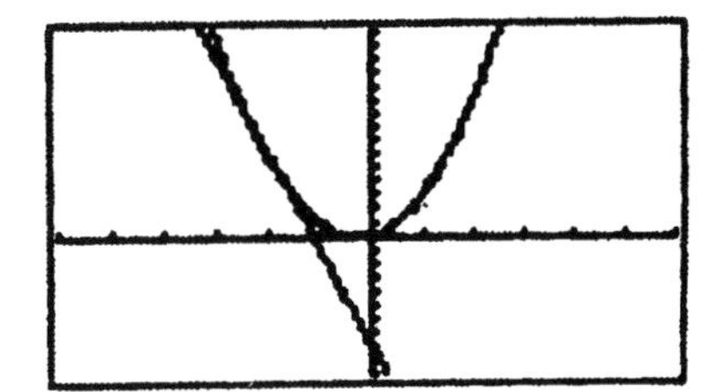

5. a. $y = 9x - 11$

   b.

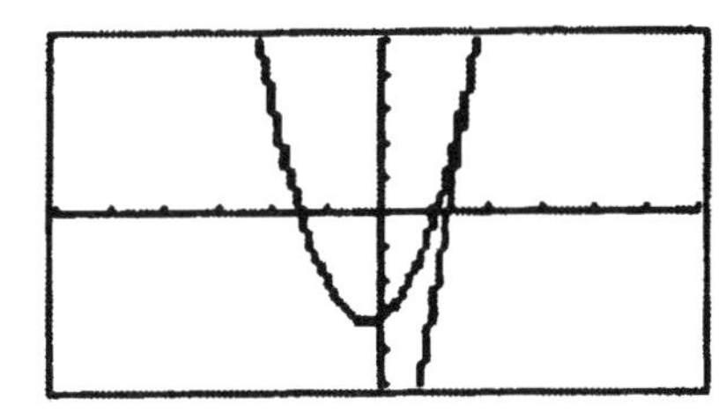

7. a. $y = 2$

   b.

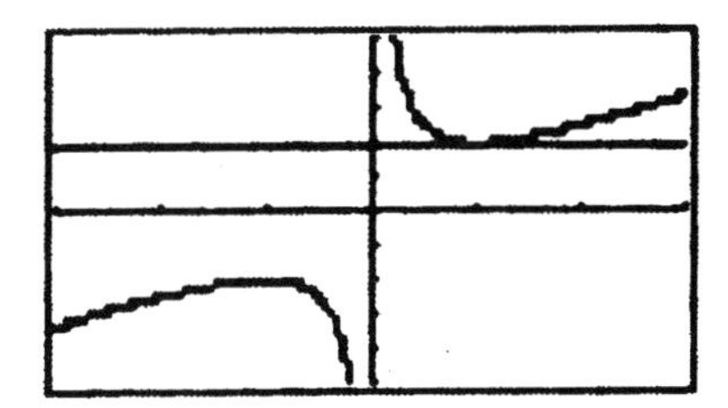

9. a. $y = \frac{1}{4}x + 1$

   b.

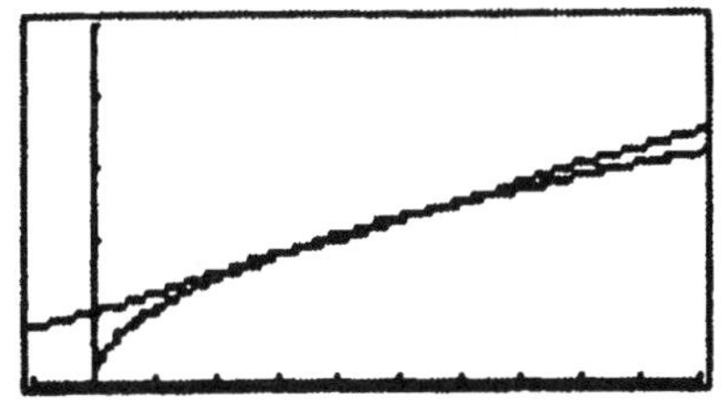

11. a. 4 b. $y = 4x - 1$

    c.

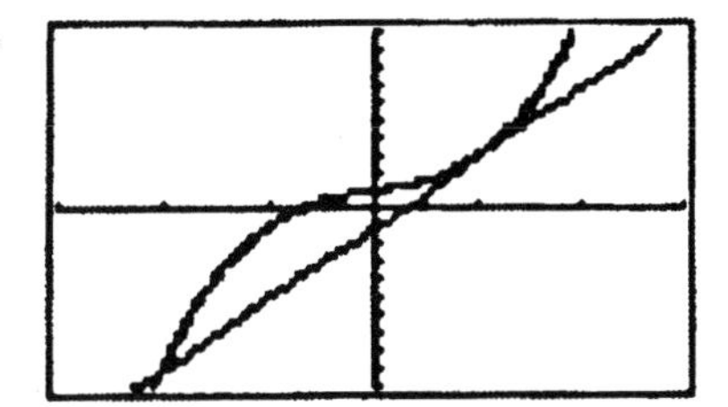

13. a. 20 b. $y = 20x - 35$

c.

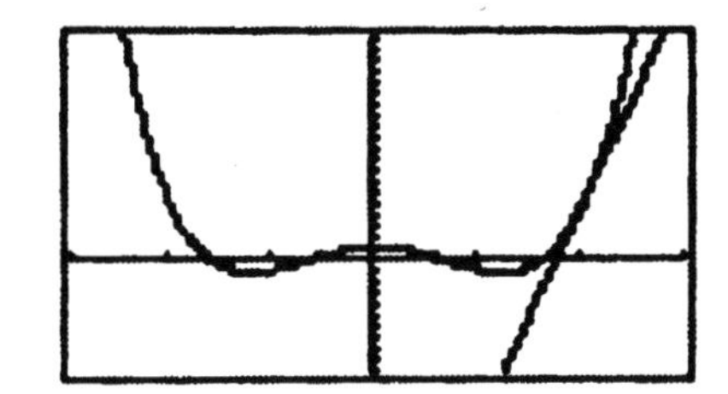

15. a.  $\frac{3}{4}$ b. $y = \frac{3}{4}x - 1$

c.

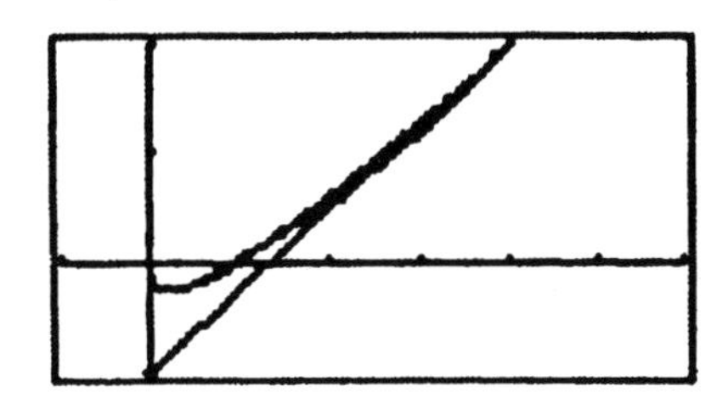

17. a. $-\frac{1}{4}$ b. $y = -\frac{1}{4}x + \frac{3}{4}$

c.

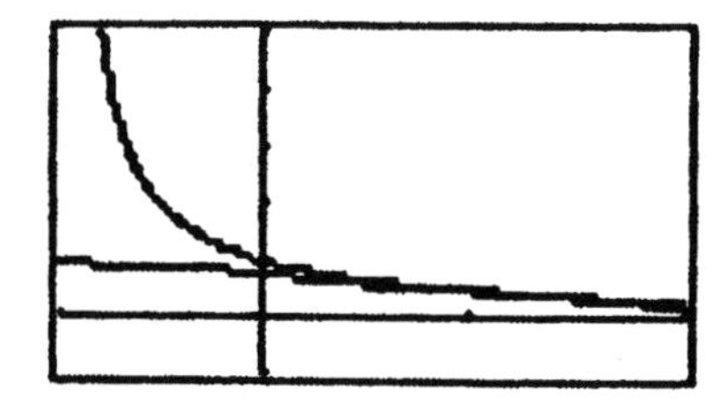

19. a. 4.02 b. $y = 4.02x - 3.57$

c.

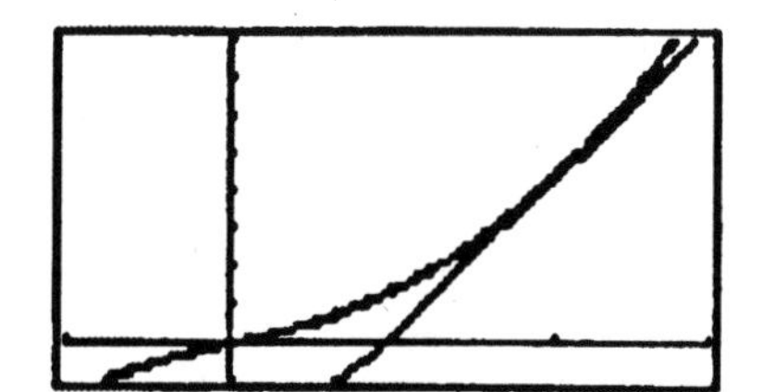

CHAPTER 10 REVIEW, page 679

1. a. $9 - x \geq 0$ gives $x \leq 9$ and the domain is $(-\infty, 9]$.
   b. $2x^2 - x - 3 = (2x - 3)(x + 1)$, and $x = 3/2$ or $x = -1$.
   Since the denominator of the given expression is zero at these points, we see that the domain of f cannot include these points and so the domain of f is
   $(-\infty, -1) \cup (-1, \frac{3}{2}) \cup (\frac{3}{2}, \infty)$.

3. a.

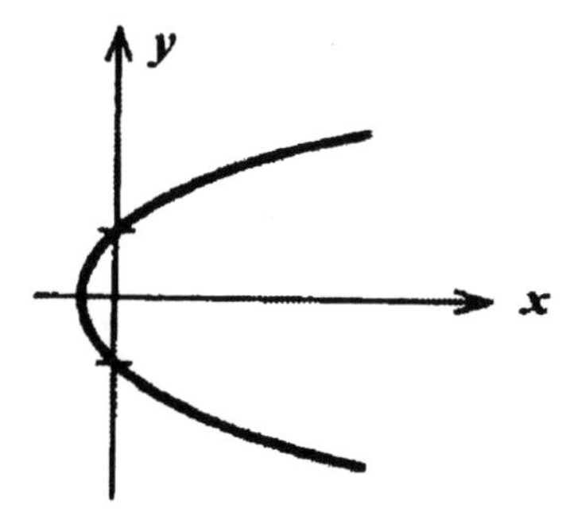

b. For each value of $x > 0$, there are two values of y. We conclude that y is not a function of x. Equivalently, the function fails the vertical line test.
c. Yes. For each value of y, there is only 1 value of x.

5. a. $f(x)g(x) = \dfrac{2x+3}{x}$ b. $\dfrac{f(x)}{g(x)} = \dfrac{1}{x(2x+3)}$

c. $f(g(x)) = \dfrac{1}{2x+3}$. d. $g(f(x)) = 1\left(\dfrac{1}{x}\right) + 3 = \dfrac{2}{x} + 3.$

7. $\lim\limits_{x \to 1}(x^2 + 1) = \lim\limits_{x \to 1}[(1)^2 + 1] = 1 + 1 = 2.$

9. $\lim\limits_{x \to 3} \dfrac{x-3}{x-4} = \dfrac{3-3}{3-4} = 0.$

11. $\lim\limits_{x \to -2} \dfrac{x^2 - 2x - 3}{x^2 + 5x + 6}$ does not exist. (The denominator is 0 at $x = -2$.)

13. $\lim\limits_{x \to 3} \dfrac{4x-3}{\sqrt{x+1}} = \dfrac{12-3}{\sqrt{4}} = \dfrac{9}{2}.$

15. $\lim\limits_{x \to 1^-} \dfrac{\sqrt{x}-1}{x-1} = \lim\limits_{x \to 1^-} \dfrac{(\sqrt{x}-1)(\sqrt{x}+1)}{(x-1)(\sqrt{x}+1)} = \lim\limits_{x \to 1^-} \dfrac{x-1}{(x-1)(\sqrt{x}+1)}$

$= \lim\limits_{x \to 1^-} \dfrac{1}{\sqrt{x}+1} = \dfrac{1}{2}.$

17. $\lim\limits_{x \to -\infty} \dfrac{x+1}{x} = \lim\limits_{x \to -\infty}\left(1 + \dfrac{1}{x}\right) = 1.$

19. $\lim\limits_{x \to -\infty} \dfrac{x^2}{x+1} = \lim\limits_{x \to -\infty} x \cdot \dfrac{1}{1 + \dfrac{1}{x}} = -\infty$, so the limit does not exist.

21. $\lim_{x\to 2^+} f(x) = \lim_{x\to 2^+}(x+2) = 4;$

$\lim_{x\to 2^-} f(x) = \lim_{x\to 2^-}(4-x) = 2.$

Therefore, $\lim_{x\to 2} f(x)$ does not exist.

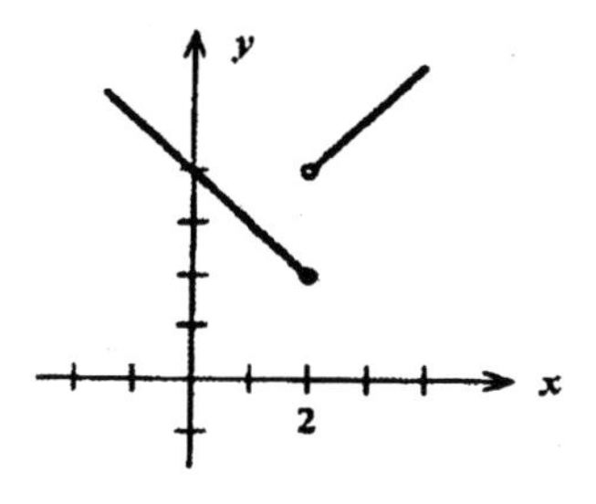

23. Since the denominator

$$4x^2 - 2x - 2 = 2(2x^2 - x - 1) = 2(2x+1)(x-1) = 0$$

if $x = -1/2$ or 1, we see that f is discontinuous at these points.

25. The function is discontinuous at $x = 0$.

27. Using the four-step process, we find

Step 1 $f(x+h) = 3(x+h) + 5 = 3x + 3h + 5$

Step 2 $f(x+h) - f(x) = 3x + 3h + 5 - 3x - 5 = 3h$

Step 3 $\dfrac{f(x+h)-f(x)}{h} = \dfrac{3h}{h} = 3.$

Step 4 $f'(x) = \lim_{h\to 0}\dfrac{f(x+h)-f(x)}{h} = \lim_{h\to 0}(3) = 3.$

29. We use the four-step process to obtain

Step 1 $f(x+h) = \tfrac{3}{2}(x+h)+5 = \tfrac{3}{2}x + \tfrac{3}{2}h + 5.$

Step 2 $f(x+h) - f(x) = \tfrac{3}{2}x + \tfrac{3}{2}h + 5 - \tfrac{3}{2}x - 5 = \tfrac{3}{2}h.$

Step 3 $\dfrac{f(x+h)-f(x)}{h} = \dfrac{3}{2}.$

Step 4 $f'(x) = \lim_{h\to 0}\dfrac{f(x+h)-f(x)}{h} = \lim_{h\to 0}\dfrac{3}{2} = \dfrac{3}{2}.$

Therefore, the slope of the tangent line to the graph of the function f at the point (-2,2) is 3/2. To find the equation of the tangent line to the curve at the point (-2,2), we use the point–slope form of the equation of a line obtaining

$$y - 2 = \tfrac{3}{2}[x - (-2)] \quad \text{or} \quad y = \tfrac{3}{2}x + 5.$$

31. a. f is continuous at $x = a$ because the three conditions for continuity are satisfied at $x = a$; that is,

   i. $f(x)$ is defined

   ii. $\lim_{x \to a} f(x)$ exists

   iii. $\lim_{x \to a} f(x) = f(a)$

   b. f is not differentiable at $x = a$ because the graph of f has a kink at $x = a$.

33. a. The line passes through (0, 2.4) and (5, 7.4) and has slope $m = \dfrac{7.4 - 2.4}{5 - 0} = 1.$

   Letting y denote the sales, we see that an equation of the line is
   $$y - 2.4 = 1(t - 0), \text{ or } y = t + 2.4.$$
   We can also write this in the form $S(t) = t + 2.4$.

   b. The sales in 1997 were $S(3) = 3 + 2.4 = 5.4$, or \$5.4 million.

35. Substituting the first equation into the second yields
   $$3x - 2(\tfrac{3}{4}x + 6) + 3 = 0 \quad \text{or} \quad \tfrac{3}{2}x - 12 + 3 = 0$$
   or $x = 6$. Substituting this value of x into the first equation then gives $y = 21/2$, so the point of intersection is $(6, \frac{21}{2})$.

37. We solve the system
   $$3x + p - 40 = 0$$
   $$2x - p + 10 = 0.$$
   Adding, we obtain $5x - 30 = 0$, or $x = 6$. So,
   $$p = 2x + 10 = 12 + 10 = 22.$$
   Therefore, the equilibrium quantity is 6000 and the equilibrium price is \$22.

39. $R(30) = -\frac{1}{2}(30)^2 + 30(30) = 450$, or \$45,000.

41. $T = f(n) = 4n\sqrt{n - 4}$.
   $$f(4) = 0,\ f(5) = 20\sqrt{1} = 20,\ f(6) = 24\sqrt{2} \approx 33.9,\ f(7) = 28\sqrt{3} \approx 48.5,$$
   $$f(8) = 32\sqrt{4} = 64,\ f(9) = 36\sqrt{5} \approx 80.5,\ f(10) = 40\sqrt{6} \approx 98,$$
   $$f(11) = 44\sqrt{7} \approx 116 \quad \text{and} \quad f(12) = 48\sqrt{8} \approx 135.8.$$

The graph of f follows:

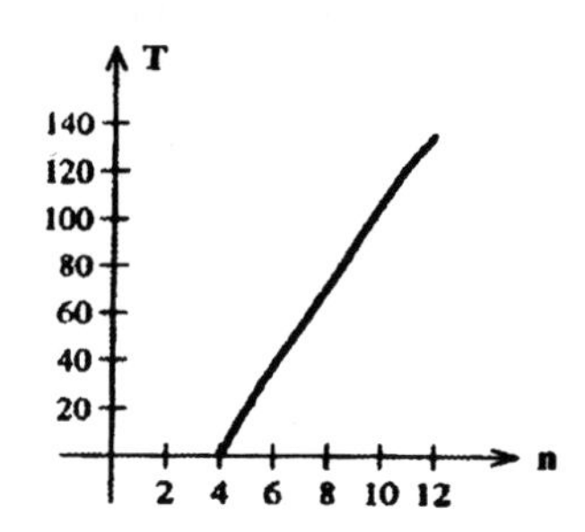

43.

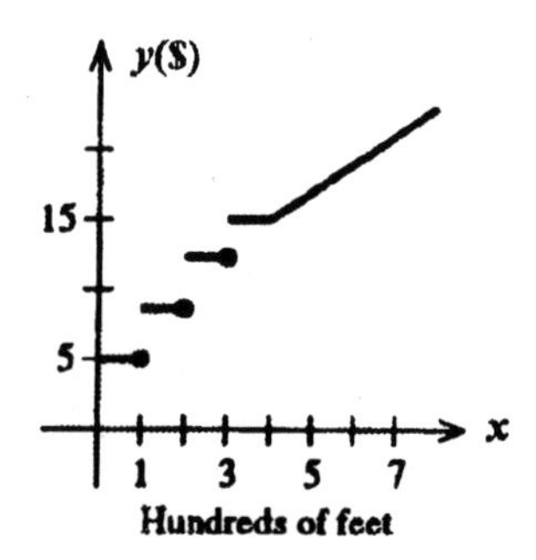

The function is discontinuous at $x = 100$, 200, and 300.

CHAPTER 11

EXERCISES 11.1, page 691

1. $f'(x) = \dfrac{d}{dx}(-3) = 0.$

3. $f'(x) = \dfrac{d}{dx}(x^5) = 5x^4.$

5. $f'(x) = \dfrac{d}{dx}(x^{2.1}) = 2.1x^{1.1}.$

7. $f'(x) = \dfrac{d}{dx}(3x^2) = 6x.$

9. $f'(r) = \dfrac{d}{dr}(\pi r^2) = 2\pi r.$

11. $f'(x) = \dfrac{d}{dx}(9x^{1/3}) = \dfrac{1}{3}(9)x^{(1/3-1)} = 3x^{-2/3}.$

13. $f'(x) = \dfrac{d}{dx}(3\sqrt{x}) = \dfrac{d}{dx}(3x^{1/2}) = \dfrac{1}{2}(3)x^{-1/2} = \dfrac{3}{2}x^{-1/2} = \dfrac{3}{2\sqrt{x}}.$

15. $f'(x) = \dfrac{d}{dx}(7x^{-12}) = (-12)(7)x^{(-12-1)} = -84x^{-13}.$

17. $f'(x) = \dfrac{d}{dx}(5x^2 - 3x + 7) = 10x - 3.$

19. $f'(x) = \dfrac{d}{dx}(-x^3 + 2x^2 - 6) = -3x^2 + 4x.$

21. $f'(x) = \dfrac{d}{dx}(0.03x^2 - 0.4x + 10) = 0.06x - 0.4.$

23. If $f(x) = \dfrac{x^3 - 4x^2 + 3}{x} = x^2 - 4x + \dfrac{3}{x},$

    then $f'(x) = \dfrac{d}{dx}(x^2 - 4x + 3x^{-1}) = 2x - 4 - \dfrac{3}{x^2}.$

25. $f'(x)=\dfrac{d}{dx}\left(4x^4-3x^{5/2}+2\right)=16x^3-\tfrac{15}{2}x^{3/2}$.

27. $f'(x)=\dfrac{d}{dx}\left(3x^{-1}+4x^{-2}\right)=-3x^{-2}-8x^{-3}$.

29. $f'(t)=\dfrac{d}{dt}\left(4t^{-4}-3t^{-3}+2t^{-1}\right)=-16t^{-5}+9t^{-4}-2t^{-2}$.

31. $f'(x)=\dfrac{d}{dx}\left(2x-5x^{1/2}\right)=2-\dfrac{5}{2}x^{-1/2}=2-\dfrac{5}{2\sqrt{x}}$.

33. $f'(x)=\dfrac{d}{dx}\left(2x^{-2}-3x^{-1/3}\right)=-4x^{-3}+x^{-4/3}=-\dfrac{4}{x^3}+\dfrac{1}{x^{4/3}}$.

35. $f'(x)=\dfrac{d}{dx}\left(2x^3-4x\right)=6x^2-4$.

a. $f'(-2)=6(-2)^2-4=20$.

b. $f'(0)=6(0)-4=-4$.

c. $f'(2)=6(2)^2-4=20$.

37. The given limit is $f'(1)$ where $f(x)=x^3$. Since $f'(x)=3x^2$, we have

$$\lim_{h\to 0}\frac{(1+h)^3-1}{h}=f'(1)=3.$$

39. Let $f(x)=3x^2-x$. Then

$$\lim_{h\to 0}\frac{3(2+h)^2-(2+h)-10}{h}=\lim_{h\to 0}\frac{f(2+h)-f(2)}{h}$$

because $f(2+h)-f(2)=3(2+h)^2-(2+h)-[3(4)-2]$
$=3(2+h)^2-(2+h)-10$.

But the last limit is $f'(2)$. Since $f'(x)=6x-1$, we have $f'(2)=11$.

Therefore, $\displaystyle\lim_{h\to 0}\frac{3(2+h)^2-(2+h)-10}{h}=11$.

41. The slope of the tangent line at any point $(x, f(x))$ on the graph of f is

$$f'(x) = 4x - 3.$$

In particular, the slope of the tangent line at the point (2,6) is

$$f'(2) = 4(2) - 3 = 5.$$

An equation of the required tangent line is

$$y - 6 = 5(x - 2) \quad \text{or} \quad y = 5x - 4.$$

43. $f'(x) = 4x^3 - 9x^2 + 4x - 1$. The slope is $f'(1) = 4 - 9 + 4 - 1 = -2$. An equation of the tangent line is $y - 0 = -2(x - 1)$ or $y = -2x + 2$.

45. a. $f'(x) = 3x^2$. At a point where the tangent line is horizontal, $f'(x) = 0$, or $3x^2 = 0$ giving $x = 0$. Therefore, the point is (0,0).

b

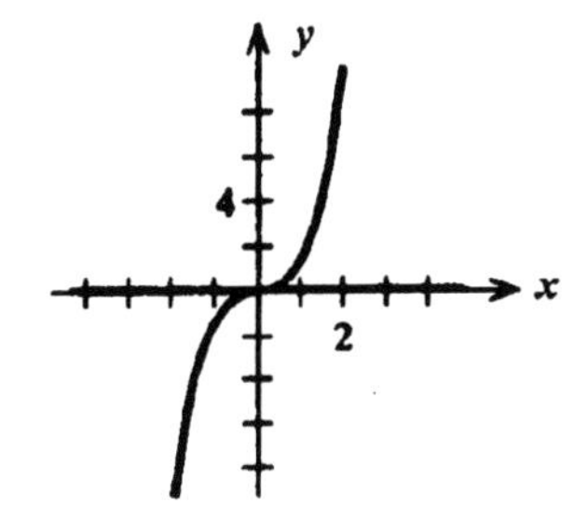

47. a. The slope of the tangent line at any point $(x, f(x))$ on the graph of f is

$$f'(x) = 3x^2.$$

At the point(s) where the slope is 12, we have

$$3x^2 = 12, \text{ or } x = \pm 2.$$

The required points are (-2,-7) and (2,9).

b. The tangent line at (-2,-7) has equation

$$y - (-7) = 12[x - (-2)], \quad \text{or} \quad y = 12x + 17,$$

and the tangent line at (2,9) has equation

$$y - 9 = 12(x - 2), \quad \text{or} \quad y = 12x - 15.$$

c.

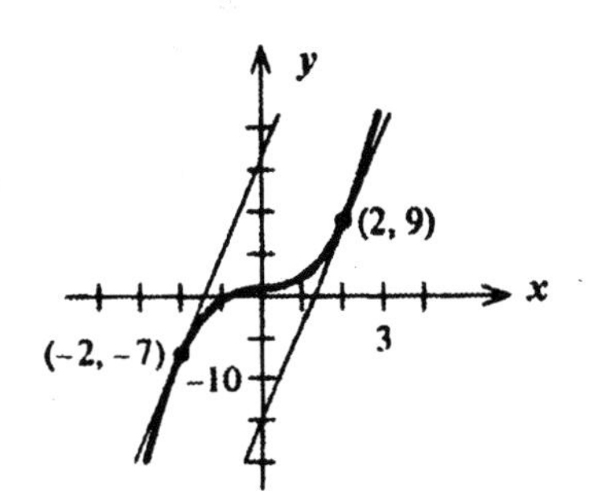

49. If $f(x) = \frac{1}{4}x^4 - \frac{1}{3}x^3 - x^2$, then $f'(x) = x^3 - x^2 - 2x$.

a. $f'(x) = x^3 - x^2 - 2x = -2x$

$$x^3 - x^2 = 0$$
$$x^2(x - 1) = 0$$

and $x = 0$ or $x = 1$.

$$f(1) = \tfrac{1}{4}(1)^4 - \tfrac{1}{3}(1)^3 - (1)^2 = -\tfrac{13}{12}.$$
$$f(0) = \tfrac{1}{4}(0)^4 - \tfrac{1}{3}(0)^3 - (0)^2 = 0.$$

We conclude that the corresponding points on the graph are $(1, -\frac{13}{12})$ and $(0,0)$.

b. $f'(x) = x^3 - x^2 - 2x = 0$

$$x(x^2 - x - 2) = 0$$
$$x(x - 2)(x + 1) = 0$$

and $x = 0, 2$, or -1.

$$f(0) = 0$$
$$f(2) = \frac{1}{4}(2)^4 - \frac{1}{3}(2)^3 - (2)^2 = 4 - \frac{8}{3} - 4 = -\frac{8}{3}.$$
$$f(-1) = \frac{1}{4}(-1)^4 - \frac{1}{3}(-1)^3 - (-1)^2 = \frac{1}{4} + \frac{1}{3} - 1 = -\frac{5}{12}.$$

We conclude that the corresponding points are $(0,0)$, $(2, -\frac{8}{3})$ and $(-1, -\frac{5}{12})$.

c. $f'(x) = x^3 - x^2 - 2x = 10x$

$$x^3 - x^2 - 12x = 0$$
$$x(x^2 - x - 12) = 0$$
$$x(x - 4)(x + 3) = 0$$

and $x = 0, 4$, or -3.

$$f(0) = 0$$
$$f(4) = \tfrac{1}{4}(4)^4 - \tfrac{1}{3}(4)^3 - (4)^2 = 48 - \tfrac{64}{3} = \tfrac{80}{3}.$$
$$f(-3) = \tfrac{1}{4}(-3)^4 - \tfrac{1}{3}(-3)^3 - (-3)^2 = \tfrac{81}{4} + 9 - 9 = \tfrac{81}{4}.$$

We conclude that the corresponding points are $(0,0)$, $(4, \frac{80}{3})$ and $(-3, \frac{81}{4})$.

51. $V'(r) = 4\pi r^2$.

a. $V'(\frac{2}{3}) = 4\pi(\frac{4}{9}) = \frac{16}{9}\pi$ cm^3/cm.

b. $V'(\frac{5}{4}) = 4\pi(\frac{25}{16}) = \frac{25}{4}\pi$. cm^3/cm.

53. $$\frac{dA}{dx} = 26.5\frac{d}{dx}(x^{-0.45}) = 26.5(-0.45)x^{-1.45} = -\frac{11.925}{x^{1.45}}.$$

Therefore, $\left.\frac{dA}{dx}\right|_{x=0.25} = -\frac{11.925}{(0.25)^{1.45}} \approx -89.01$ and $\left.\frac{dA}{dx}\right|_{x=2} = -\frac{11.925}{(2)^{1.45}} \approx -4.36$

Our computations reveal that if you make 0.25 stops per mile, your average speed will decrease at the rate of approximately 89.01 mph per stop per mile. If you make 2 stops per mile, your average speed will decrease at the rate of approximately 4.36 mph per stop per mile.

55. $I'(t) = -0.6t^2 + 6t$.

a. In 1993, it was changing at a rate of $I'(5) = -0.6(25) + 6(5)$, or 15 points/yr. In 1995, it was $I'(7) = -0.6(49) + 6(7)$, or 12.6 pts/yr. In 1998, it was $I'(10) = -0.6(100) + 6(10)$, or 0 pts/yr.

b. The average rate of increase of the CPI over the period from 1993 to 1998 was

$$\frac{I(10)-I(5)}{5} = \frac{[-0.2(1000)+3(100)+100]-[-0.2(125)+3(25)+100]}{5}$$

$$= \frac{200-150}{5} = 10, \text{ or } 10 \text{ pts/yr.}$$

57. The rate at which the population will be increasing at any time t is

$$P'(t) = 45t^{1/2} + 20.$$

Nine months from now the population will be increasing at

$$P'(9) = 45(3) + 20, \text{ or } 155 \text{ people/month.}$$

Sixteen months from now the population will be increasing at

$$P'(16) = 45(4) + 20, \text{ or } 200 \text{ people/month.}$$

59. $N'(t) = 6t^2 + 6t - 4$.

$N'(2) = 6(4) + 6(2) - 4 = 32$, or 32 turtles/yr.

$N'(8) = 6(64) + 6(8) - 4 = 428$, or 428 turtles/yr.

The population ten years after implementation of the conservation measures will be $N(10) = 2(1000) + 3(100) - 4(10) + 1000$, or 3260 turtles.

61. a. $v = f'(t) = 120 - 30t$ b. $v(0) = 120$ ft/sec

c. Setting $v = 0$ gives $120 - 30t = 0$, or $t = 4$. Therefore, the stopping distance is

$$f(4) = 120(4) - 15(16) \text{ or } 240 \text{ ft.}$$

63. a. The number of temporary workers at the beginning of 1994 ($t = 3$) was

$$N(3) = 0.025(9) + 0.255(3) + 1.505 = 2.495 \text{ million.}$$

b. $N'(t) = 0.05t + 0.255$.

So, at the beginning of 1994 ($t = 3$), the number of temporary workers was growing at the rate of

$$N'(3) = 0.05(3) + 0.255 = 0.405, \text{ or } 405{,}000 \text{ per year.}$$

65. a. $f'(t) = 20 - 40\left(\frac{1}{2}\right)t^{-1/2} = 20\left(1 - \frac{1}{\sqrt{t}}\right).$

b. $f(0) = 20(0) - 40\sqrt{0} + 50 = 50$

$f(1) = 20(1) - 40\sqrt{1} + 50 = 30$

$f(2) = 20(2) - 40\sqrt{2} + 50 \approx 33.43.$

The average velocity at 6, 7, and 8 A.M. is 50 mph, 30 mph, and 33.43 mph, respectively.

c. $f'(\frac{1}{2}) = 20 - 20(\frac{1}{2})^{-1/2} \approx -8.28.$

$f'(1) = 20 - 20(1)^{-1/2} \approx 0.$

$f'(2) = 20 - 20(2)^{-1/2} \approx 5.86.$

At 6:30 A.M. the average velocity is decreasing at the rate of 8.28 mph/hr; at 7 A.M., it is unchanged, and at 8 A.M., it is increasing at the rate of 5.86 mph.

67. $$\frac{d}{dx}(x^3) = \lim_{h\to 0}\frac{(x+h)^3 - x^3}{h} = \lim_{h\to 0}\frac{x^3 + 3x^2h + 3xh^2 + h^3 - x^3}{h}$$

$$= \lim_{h\to 0}\frac{h(3x^2 + 3xh + h^2)}{h} = \lim_{h\to 0}(3x^2 + 3xh + h^2) = 3x^2.$$

USING TECHNOLOGY EXERCISES 11.1, page 693

1. 1

3. 0.4226

5. 0.1613

7. a.

b. 3.4295 parts/million; 105.4332 parts/million

9. a.

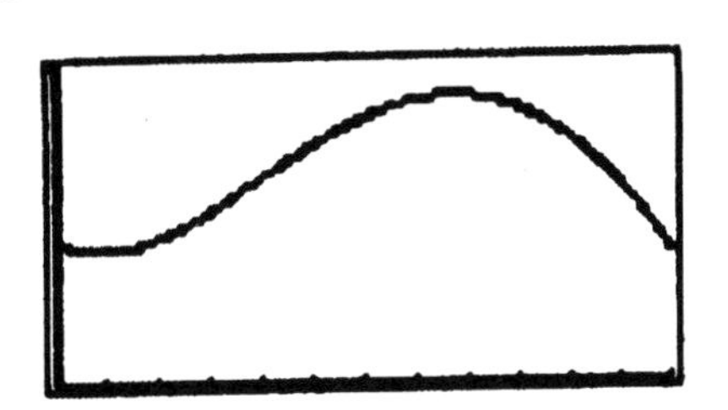

b. decreasing at the rate of 9 days/yr

increasing at the rate of 13 days/yr

EXERCISES 11.2, page 705

1. $f(x) = 2x(x^2 + 1).$

$$f'(x) = 2x\frac{d}{dx}(x^2+1) + (x^2+1)\frac{d}{dx}(2x)$$
$$= 2x(2x) + (x^2+1)(2) = 6x^2 + 2.$$

3. $f(t) = (t - 1)(2t + 1)$

$$f'(t) = (t-1)\frac{d}{dt}(2t+1) + (2t+1)\frac{d}{dt}(t-1) = (t-1)(2) + (2t+1)(1)$$
$$= 4t - 1.$$

5. $f(x) = (3x + 1)(x^2 - 2)$

$$f'(x) = (3x+1)\frac{d}{dx}(x^2-2) + (x^2-2)\frac{d}{dx}(3x+1)$$
$$= (3x+1)(2x) + (x^2-2)(3) = 9x^2 + 2x - 6.$$

7. $f(x) = (x^3 - 1)(x + 1).$

$$f'(x) = (x^3-1)\frac{d}{dx}(x+1) + (x+1)\frac{d}{dx}(x^3-1)$$
$$= (x^3-1)(1) + (x+1)(3x^2) = 4x^3 + 3x^2 - 1.$$

9. $f(w) = (w^3 - w^2 + w - 1)(w^2 + 2).$

$$f'(w) = (w^3 - w^2 + w - 1)\frac{d}{dw}(w^2+2) + (w^2+2)\frac{d}{dw}(w^3 - w^2 + w - 1)$$
$$= (w^3 - w^2 + w - 1)(2w) + (w^2+2)(3w^2 - 2w + 1)$$
$$= 2w^4 - 2w^3 + 2w^2 - 2w + 3w^4 - 2w^3 + w^2 + 6w^2 - 4w + 2$$
$$= 5w^4 - 4w^3 + 9w^2 - 6w + 2.$$

11. $f(x) = (5x^2+1)(2\sqrt{x}-1)$

$$f'(x) = (5x^2+1)\frac{d}{dx}(2x^{1/2}-1) + (2x^{1/2}-1)\frac{d}{dx}(5x^2+1)$$
$$= (5x^2+1)(x^{-1/2}) + (2x^{1/2}-1)(10x)$$
$$= 5x^{3/2} + x^{-1/2} + 20x^{3/2} - 10x$$
$$= \frac{25x^2 - 10x\sqrt{x} + 1}{\sqrt{x}}.$$

13. $f(x) = (x^2-5x+2)(x-\frac{2}{x})$

$$f'(x) = (x^2-5x+2)\frac{d}{dx}(x-\frac{2}{x}) = (x-\frac{2}{x})\frac{d}{dx}(x^2-5x+2)$$
$$= \frac{(x^2-5x+2)(x^2+2)}{x^2} + \frac{(x^2-2)(2x-5)}{x}$$
$$= \frac{(x^2-5x+2)(x^2+2) + x(x^2-2)(2x-5)}{x^2}$$
$$= \frac{x^4+2x^2-5x^3-10x+2x^2+4+2x^4-5x^3-4x^2+10x}{x^2}$$
$$= \frac{3x^4-10x^3+4}{x^2}.$$

15. $f(x) = \frac{1}{x-2}$. $f'(x) = \frac{(x-2)\frac{d}{dx}(1) - (1)\frac{d}{dx}(x-2)}{(x-2)^2} = \frac{0-1(1)}{(x-2)^2} = -\frac{1}{(x-2)^2}.$

17. $f(x) = \frac{x-1}{2x+1}$.

$$f'(x) = \frac{(2x+1)\frac{d}{dx}(x-1) - (x-1)\frac{d}{dx}(2x+1)}{(2x+1)^2}$$
$$= \frac{2x+1-(x-1)(2)}{(2x+1)^2} = \frac{3}{(2x+1)^2}.$$

19. $f(x) = \frac{1}{x^2+1}$.

$$f'(x) = \frac{(x^2+1)\frac{d}{dx}(1) - (1)\frac{d}{dx}(x^2+1)}{(x^2+1)^2}$$

$$= \frac{(x^2+1)(0) - 1(2x)}{(x^2+1)^2} = -\frac{2x}{(x^2+1)^2}.$$

21. $f(s) = \frac{s^2-4}{s+1}$.

$$f'(s) = \frac{(s+1)\frac{d}{ds}(s^2-4) - (s^2-4)\frac{d}{ds}(s+1)}{(s+1)^2} = \frac{s^2+2s+4}{(s+1)^2}.$$

23. $f(x) = \frac{\sqrt{x}}{x^2+1}$.

$$f'(x) = \frac{(x^2+1)\frac{d}{dx}(x^{1/2}) - (x^{1/2})\frac{d}{dx}(x^2+1)}{(x^2+1)^2}$$

$$= \frac{(\frac{1}{2}x^{-1/2})[(x^2+1) - 4x^2]}{(x^2+1)^2} = \frac{1-3x^2}{2\sqrt{x}(x^2+1)^2}.$$

25. $f(x) = \frac{x^2+2}{x^2+x+1}$.

$$f'(x) = \frac{(x^2+x+1)\frac{d}{dx}(x^2+2) - (x^2+2)\frac{d}{dx}(x^2+x+1)}{(x^2+x+1)^2}$$

$$= \frac{(x^2+x+1)(2x) - (x^2+2)(2x+1)}{(x^2+x+1)^2}$$

$$= \frac{2x^3+2x^2+2x-2x^3-x^2-4x-2}{(x^2+x+1)^2} = \frac{x^2-2x-2}{(x^2+x+1)^2}.$$

27. $f(x) = \frac{(x+1)(x^2+1)}{x-2} = \frac{(x^3+x^2+x+1)}{x-2}$.

$$f'(x) = \frac{(x-2)\frac{d}{dx}(x^3+x^2+x+1)-(x^3+x^2+x+1)\frac{d}{dx}(x-2)}{(x-2)^2}$$

$$= \frac{(x-2)(3x^2+2x+1)-(x^3+x^2+x+1)}{(x-2)^2}$$

$$= \frac{3x^3+2x^2+x-6x^2-4x-2-x^3-x^2-x-1}{(x-2)^2}$$

$$= \frac{2x^3-5x^2-4x-3}{(x-2)^2}.$$

29. $f(x) = \frac{x}{x^2-4} - \frac{x-1}{x^2+4} = \frac{x(x^2+4)-(x-1)(x^2-4)}{(x^2-4)(x^2+4)} = \frac{x^2+8x-4}{(x^2-4)(x^2+4)}.$

$$f'(x) = \frac{(x^2-4)(x^2+4)\frac{d}{dx}(x^2+8x-4)-(x^2+8x-4)\frac{d}{dx}(x^4-16)}{(x^2-4)^2(x^2+4)^2}$$

$$= \frac{(x^2-4)(x^2+4)(2x+8)-(x^2+8x-4)(4x^3)}{(x^2-4)^2(x^2+4)^2}$$

$$= \frac{2x^5+8x^4-32x-128-4x^5-32x^5-32x^4+16x^3}{(x^2-4)^2(x^2+4)^2}$$

$$= \frac{-2x^5-24x^4+16x^3-32x-128}{(x^2-4)^2(x^2+4)^2}.$$

31. $h'(x) = f(x)g'(x) + f'(x)g(x)$, by the Product Rule. Therefore,
$h'(1) = f(1)g'(1) + f'(1)g(1) = (2)(3) + (-1)(-2) = 8.$

33. Using the Quotient Rule followed by the Product Rule, we have

$$h'(x) = \frac{[x+g(x)]\frac{d}{dx}[xf(x)] - xf(x)\frac{d}{dx}[x+g(x)]}{[x+g(x)]^2}$$

$$= \frac{[x+g(x)][xf'(x)+f(x)] - xf(x)[1+g'(x)]}{[x+g(x)]^2}$$

Therefore,

$$h'(1) = \frac{[1+g(1)][f'(1)+f(1)] - f(1)[1+g'(1)]}{[1+g(1)]^2}$$

$$= \frac{(1-2)(-1+2)-2(1+3)}{(1-2)^2} = \frac{-1-8}{1} = -9.$$

35. $f(x) = (2x-1)(x^2+3)$

$$f'(x) = (2x-1)\frac{d}{dx}(x^2+3) + (x^2+3)\frac{d}{dx}(2x-1)$$

$$= (2x-1)(2x) + (x^2+3)(2) = 6x^2 - 2x + 6 = 2(3x^2 - x + 3).$$

At $x = 1, f'(1) = 2[3(1)^2 - (1) + 3] = 2(5) = 10.$

37. $f(x) = \dfrac{x}{x^4 - 2x^2 - 1}.$

$$f'(x) = \frac{(x^4-2x^2-1)\frac{d}{dx}(x) - x\frac{d}{dx}(x^4-2x^2-1)}{(x^4-2x^2-1)^2}$$

$$= \frac{(x^4-2x^2-1)(1) - x(4x^3-4x)}{(x^4-2x^2-1)^2} = \frac{-3x^4+2x^2-1}{(x^4-2x^2-1)^2}.$$

Therefore,

$$f'(1) = \frac{-3+2-1}{(1-2-1)^2} = -\frac{2}{4} = -\frac{1}{2}.$$

39. $f(x) = (x^3+1)(x^2-2).$

$$f'(x) = (x^3+1)\frac{d}{dx}(x^2-2) + (x^2-2)\frac{d}{dx}(x^3+1)$$

$$= (x^3+1)(2x) + (x^2-2)(3x^2).$$

The slope of the tangent line at (2,18) is

$$f'(2) = (8 + 1)(4) + (4 - 2)(12) = 60.$$

An equation of the tangent line is

$$y - 18 = 60(x - 2), \text{ or } y = 60x - 102.$$

41. $f(x) = \dfrac{x+1}{x^2+1}.$

$$f'(x) = \frac{(x^2+1)\frac{d}{dx}(x+1) - (x+1)\frac{d}{dx}(x^2+1)}{(x^2+1)^2}$$

$$= \frac{(x^2+1)(1) - (x+1)(2x)}{(x^2+1)^2} = \frac{-x^2 - 2x + 1}{(x^2+1)^2}.$$

At $x = 1$, $f'(1) = \frac{-1-2+1}{4} = -\frac{1}{2}$.

Therefore, the slope of the tangent line at $x = 1$ is -1/2. Then an equation of the tangent line is

$$y - 1 = -\tfrac{1}{2}(x-1) \quad \text{or} \quad y = -\tfrac{1}{2}x + \tfrac{3}{2}.$$

43. $f(x) = (x^3+1)(3x^2 - 4x + 2)$

$$f'(x) = (x^3+1)\frac{d}{dx}(3x^2-4x+2) + (3x^2-4x+2)\frac{d}{dx}(x^3+1)$$

$$= (x^3+1)(6x-4) + (3x^2-4x+2)(3x^2)$$

$$= 6x^4 + 6x - 4x^3 - 4 + 9x^4 - 12x^3 + 6x^2$$

$$= 15x^4 - 16x^3 + 6x^2 + 6x - 4.$$

At $x = 1$, $f'(1) = 15(1)^4 - 16(1)^3 + 6(1) + 6(1) - 4 = 7$. The slope of the tangent line at the point $x = 1$ is 7. The equation of the tangent line is

$$y - 2 = 7(x - 1), \quad \text{or} \quad y = 7x - 5.$$

45. $f(x) = (x^2+1)(2-x)$

$$f'(x) = (x^2+1)\frac{d}{dx}(2-x) + (2-x)\frac{d}{dx}(x^2+1)$$

$$= (x^2+1)(-1) + (2-x)(2x) = -3x^2 + 4x - 1.$$

At a point where the tangent line is horizontal, we have

$$f'(x) = -3x^2 + 4x - 1 = 0$$

or $\quad 3x^2 - 4x + 1 = (3x-1)(x-1) = 0,$

giving $x = 1/3$ or $x = 1$. Since $f(\tfrac{1}{3}) = (\tfrac{1}{9}+1)(2-\tfrac{1}{3}) = \tfrac{50}{27}$, and $f(1) = 2(2 - 1) = 2$, we see that the required points are $(\tfrac{1}{3}, \tfrac{50}{27})$ and $(1,2)$.

47. $f(x) = (x^2+6)(x-5)$

$$f'(x) = (x^2+6)\frac{d}{dx}(x-5) + (x-5)\frac{d}{dx}(x^2+6)$$

$$= (x^2+6)(1) + (x-5)(2x) = x^2 + 6 + 2x^2 - 10x = 3x^2 - 10x + 6.$$

At a point where the slope of the tangent line is -2, we have

$f'(x) = 3x^2 - 10x + 6 = -2.$

This gives

$$3x^2 - 10x + 8 = (3x - 4)(x - 2) = 0.$$

So $x = \frac{4}{3}$ or $x = 2$.

Since $f(\frac{4}{3}) = (\frac{16}{9} + 6)(\frac{4}{3} - 5) = -\frac{770}{27}$ and $f(2) = (4 + 6)(2 - 5) = -30$,
the required points are $(\frac{4}{3}, -\frac{770}{27})$ and $(2, -30)$.

49. $$y' = \frac{(1+x^2)\frac{d}{dx}(1) - (1)\frac{d}{dx}(1+x^2)}{(1+x^2)^2} = \frac{-2x}{(1+x^2)^2}.$$

So, the slope of the tangent line at $(1, \frac{1}{2})$ is

$$y'\big|_{x=1} = \frac{-2x}{(1+x^2)^2}\bigg|_{x=1} = \frac{-2}{4} = -\frac{1}{2}$$

and the equation of the tangent line is

$$y - \tfrac{1}{2} = -\tfrac{1}{2}(x - 1), \text{ or } y = -\tfrac{1}{2}x + 1.$$

Next, the slope of the required normal line is 2 and its equation is

$$y - \tfrac{1}{2} = 2(x - 1), \quad \text{or} \quad y = 2x - \tfrac{3}{2}.$$

51. $$C(x) = \frac{0.5x}{100 - x}.$$

$$C'(x) = \frac{(100 - x)(0.5) - 0.5x(-1)}{(100 - x)^2} = \frac{50}{(100 - x)^2}.$$

$$C'(80) = \frac{50}{20^2} = 0.125, \; C'(90) = \frac{50}{10^2} = 0.5,$$

$$C'(95) = \frac{50}{5^2} = 2; \; C'(99) = \frac{50}{1} = 50.$$

The rates of change of the cost in removing 80%, 90%, and 99% of the toxic waste are 0.125, 0.5, 2, and 50 million dollars per 1% more of the waste to be removed, respectively.

53. $$N(t) = \frac{10{,}000}{1 + t^2} + 2000$$

$$N'(t) = \frac{d}{dt}[10{,}000(1 + t^2)^{-1} + 2000] = -\frac{10{,}000}{(1 + t^2)^2}(2t) = -\frac{20{,}000t}{(1 + t^2)^2}.$$

The rate of change after 1 minute and after 2 minutes is

$$N'(1) = -\frac{20{,}000}{(1+1^2)^2} = -5000; \; N'(2) = -\frac{20{,}000(2)}{(1+2^2)^2} = -1600.$$

The population of bacteria after one minute is

$$N(1) = \frac{10{,}000}{1+1} + 2000 = 7000.$$

The population after two minutes is

$$N(2) = \frac{10{,}000}{1+4} + 2000 = 4000.$$

55. a. $$N(t) = \frac{60t+180}{t+6}.$$

$$N'(t) = \frac{(t+6)\frac{d}{dt}(60t+180) - (60t+180)\frac{d}{dt}(t+6)}{(t+6)^2}$$

$$= \frac{(t+6)(60) - (60t+180)(1)}{(t+6)^2} = \frac{180}{(t+6)^2}.$$

b. $$N'(1) = \frac{180}{(1+6)^2} = 3.7, \; N'(3) = \frac{180}{(3+6)^2} = 2.2, \; N'(4) = \frac{180}{(4+6)^2} = 1.8,$$

$$N'(7) = \frac{180}{(7+6)^2} = 1.1$$

We conclude that the rate at which the average student is increasing his or her speed one week, three weeks, four weeks, and seven weeks into the course is 3.7, 2.2, 1.8, and 1.1 words per minute, respectively.

c.

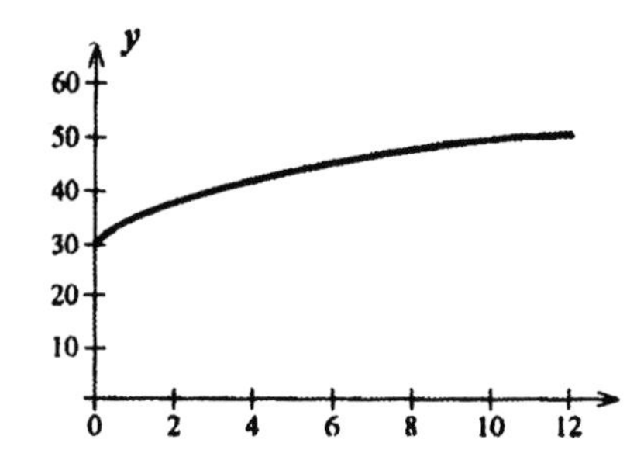

d. $$N(12) = \frac{60(12)+180}{12+6} = 50, \text{ or 50 words/minute.}$$

57. $f'(t) = \dfrac{(t+2)(0.055) - (0.055t + 0.26)(1)}{(t+2)^2} = -\dfrac{0.15}{(t+2)^2}$.

At the beginning, the formaldehyde level is changing at the rate of

$$f'(0) = -\frac{0.15}{4} = -0.0375;$$

that is, it is dropping at the rate of 0.0375 parts per million per year. Next,

$$f'(3) = -\frac{0.15}{6^2} = -0.006,$$

and so the level is dropping at the rate of 0.006 parts per million per year at the beginning of the fourth year ($t = 3$).

59. Let $f(x) = u(x)v(x)$ and $g(x) = w(x)$. Then $h(x) = f(x)g(x)$. Therefore,

$$h'(x) = f'(x)g(x) + f(x)g'(x).$$

But $\quad f'(x) = u(x)v'(x) + u'(x)v(x).$

Therefore,

$$\begin{aligned} h'(x) &= [u(x)v'(x) + u'(x)v(x)]g(x) + u(x)v(x)w'(x) \\ &= u(x)v(x)w'(x) + u(x)v'(x)w(x) + u'(x)v(x)w(x). \end{aligned}$$

EXERCISES 11.3, page 720

1. $f(x) = (2x-1)^4$.

$$f'(x) = 4(2x-1)^3 \frac{d}{dx}(2x-1) = 4(2x-1)^3(2) = 8(2x-1)^3.$$

3. $f(x) = (x^2+2)^5$.

$$f'(x) = 5(x^2+2)^4(2x) = 10x(x^2+4)^4.$$

5. $f(x) = (2x-x^2)^3$.

$$\begin{aligned} f'(x) &= 3(2x-x^2)^2 \frac{d}{dx}(2x-x^2) = 3(2x-x^2)^2(2-2x) \\ &= 6x^2(1-x)(2-x)^2. \end{aligned}$$

7. $f(x) = (2x+1)^{-2}$.

$$f'(x) = -2(2x+1)^{-3} \frac{d}{dx}(2x+1) = -2(2x+1)^{-3}(2) = -4(2x+1)^{-3}.$$

9. $f(x) = (x^2 - 4)^{3/2}$.

$$f'(x) = \tfrac{3}{2}(x^2 - 4)^{1/2} \frac{d}{dx}(x^2 - 4) = \tfrac{3}{2}(x^2 - 4)^{1/2}(2x) = 3x(x^2 - 4)^{1/2}.$$

11. $f(x) = \sqrt{3x - 2} = (3x - 2)^{1/2}$.

$$f'(x) = \frac{1}{2}(3x - 2)^{-1/2}(3) = \frac{3}{2}(3x - 2)^{-1/2} = \frac{3}{2\sqrt{3x - 2}}.$$

13. $f(x) = \sqrt[3]{1 - x^2}$.

$$f'(x) = \frac{d}{dx}(1 - x^2)^{1/3} = \frac{1}{3}(1 - x^2)^{-2/3}\frac{d}{dx}(1 - x^2)$$

$$= \frac{1}{3}(1 - x^2)^{-2/3}(-2x) = -\frac{2}{3}x(1 - x^2)^{-2/3}.$$

15. $f(x) = \dfrac{1}{(2x + 3)^3} = (2x + 3)^{-3}$.

$$f'(x) = -3(2x + 3)^{-4}(2) = -6(2x + 3)^{-4} = -\frac{6}{(2x + 3)^4}.$$

17. $f(t) = \dfrac{1}{\sqrt{2t - 3}}$.

$$f'(t) = \frac{d}{dt}(2t - 3)^{-1/2} = -\frac{1}{2}(2t - 3)^{-3/2}(2) = -(2t - 3)^{-3/2}.$$

19. $y = \dfrac{1}{(4x^4 + x)^{3/2}}$.

$$\frac{dy}{dx} = \frac{d}{dx}(4x^4 + x)^{-3/2} = -\frac{3}{2}(4x^4 + x)^{-5/2}(16x^3 + 1) = -\frac{3}{2}(16x^3 + 1)(4x^4 + x)^{-5/2}.$$

21. $f(x) = (3x^2 + 2x + 1)^{-2}$.

$$f'(x) = -2(3x^2 + 2x + 1)^{-3}\frac{d}{dx}(3x^2 + 2x + 1)$$

$$= -2(3x^2 + 2x + 1)^{-3}(6x + 2) = -4(3x + 1)(3x^2 + 2x + 1)^{-3}.$$

23. $f(x) = (x^2 + 1)^3 - (x^3 + 1)^2$.

$$f'(x) = 3(x^2+1)^2 \frac{d}{dx}(x^2+1) - 2(x^3+1)\frac{d}{dx}(x^3+1)$$
$$= 3(x^2+1)^2(2x) - 2(x^3+1)(3x^2)$$
$$= 6x[(x^2+1)^2 - x(x^3+1)] = 6x(2x^2 - x + 1).$$

25. $f(t) = (t^{-1} - t^{-2})^3.$

$$f'(t) = 3(t^{-1} - t^{-2})^2 \frac{d}{dt}(t^{-1} - t^{-2}) = 3(t^{-1} - t^{-2})^2(-t^{-2} + 2t^{-3}).$$

27. $f(x) = \sqrt{x+1} + \sqrt{x-1} = (x+1)^{1/2} + (x-1)^{1/2}.$

$$f'(x) = \tfrac{1}{2}(x+1)^{-1/2}(1) + \tfrac{1}{2}(x-1)^{-1/2}(1) = \tfrac{1}{2}[(x+1)^{-1/2} + (x-1)^{-1/2}].$$

29. $f(x) = 2x^2(3-4x)^4.$

$$f'(x) = 2x^2(4)(3-4x)^3(-4) + (3-4x)^4(4x) = 4x(3-4x)^3(-8x + 3 - 4x)$$
$$= 4x(3-4x)^3(-12x+3) = (-12x)(4x-1)(3-4x)^3.$$

31. $f(x) = (x-1)^2(2x+1)^4.$

$$f'(x) = (x-1)^2 \frac{d}{dx}(2x+1)^4 + (2x+1)^4 \frac{d}{dx}(x-1)^2 \quad \text{[Product Rule]}$$
$$= (x-1)^2(4)(2x+1)^3 \frac{d}{dx}(2x+1) + (2x+1)^4(2)(x-1)\frac{d}{dx}(x-1)$$
$$= 8(x-1)^2(2x+1)^3 + 2(x-1)(2x+1)^4$$
$$= 2(x-1)(2x+1)^3(4x - 4 + 2x + 1) = 6(x-1)(2x-1)(2x+1)^3.$$

33. $f(x) = \left(\dfrac{x+3}{x-2}\right)^3.$

$$f'(x) = 3\left(\frac{x+3}{x-2}\right)^2 \frac{d}{dx}\left(\frac{x-3}{x-2}\right) = 3\left(\frac{x+3}{x-2}\right)^2\left[\frac{(x-2)(1) - (x+3)(1)}{(x-2)^2}\right]$$
$$= 3\left(\frac{x+3}{x-2}\right)^2\left[-\frac{5}{(x-2)^2}\right] = -\frac{15(x+3)^2}{(x-2)^4}.$$

35. $s(t) = \left(\dfrac{t}{2t+1}\right)^{3/2}.$

$$s'(t) = \frac{3}{2}\left(\frac{t}{2t+1}\right)^{1/2} \frac{d}{dt}\left(\frac{t}{2t+1}\right) = \frac{3}{2}\left(\frac{t}{2t+1}\right)^{1/2}\left[\frac{(2t+1)(1)-t(2)}{(2t+1)^2}\right]$$

$$= \frac{3}{2}\left(\frac{t}{2t+1}\right)^{1/2}\left[\frac{1}{(2t+1)^2}\right] = \frac{3t^{1/2}}{2(2t+1)^{5/2}}.$$

37. $g(u) = \left(\dfrac{u+1}{3u+2}\right)^{1/2}.$

$$g'(u) = \frac{1}{2}\left(\frac{u+1}{3u+2}\right)^{-1/2} \frac{d}{du}\left(\frac{u+1}{3u+2}\right)$$

$$= \frac{1}{2}\left(\frac{u+1}{3u+2}\right)^{-1/2}\left[\frac{(3u+2)(1)-(u+1)(3)}{(3u+2)^2}\right]$$

$$= -\frac{1}{2\sqrt{u+1})(3u+2)^{3/2}}.$$

39. $f(x) = \dfrac{x^2}{(x^2-1)^4}.$

$$f'(x) = \frac{(x^2-1)\frac{d}{dx}(x^2) - (x^2)\frac{d}{dx}(x^2-1)^4}{\left[(x^2-1)^4\right]^2}$$

$$= \frac{(x^2-1)^4(2x) - x^2(4)(x^2-1)^3(2x)}{(x^2-1)^8}$$

$$= \frac{(x^2-1)^3(2x)(x^2-1-4x^2)}{(x^2-1)^8} = \frac{(-2x)(3x^2+1)}{(x^2-1)^5}.$$

41. $h(x) = \dfrac{(3x^2+1)^3}{(x^2-1)^4}.$

$$h'(x) = \frac{(x^2-1)^4(3)(3x^2+1)^2(6x) - (3x^2+1)^3(4)(x^2-1)^3(2x)}{(x^2-1)^8}$$

$$= \frac{2x(x^2-1)^3(3x^2+1)^2[9(x^2-1) - 4(3x^2+1)]}{(x^2-1)^8}$$

$$= -\frac{2x(3x^2+13)(3x^2+1)^2}{(x^2-1)^5}.$$

43. $f(x) = \dfrac{\sqrt{2x+1}}{x^2-1}$.

$$f'(x) = \frac{(x^2-1)(\frac{1}{2})(2x+1)^{-1/2}(2) - (2x+1)^{1/2}(2x)}{(x^2-1)^2}$$

$$= \frac{(2x+1)^{-1/2}[(x^2-1) - (2x+1)(2x)]}{(x^2-1)^2} = -\frac{3x^2+2x+1}{\sqrt{2x+1}(x^2-1)^2}.$$

45. $g(t) = \dfrac{(t+1)^{1/2}}{(t^2+1)^{1/2}}$.

$$g'(t) = \frac{(t^2+1)^{1/2}\frac{d}{dt}(t+1)^{1/2} - (t+1)^{1/2}\frac{d}{dt}(t^2+1)^{1/2}}{t^2+1}$$

$$= \frac{(t^2+1)^{1/2}(\frac{1}{2})(t+1)^{-1/2}(1) - (t+1)^{1/2}(\frac{1}{2})(t^2+1)^{-1/2}(2t)}{t^2+1}$$

$$= \frac{\frac{1}{2}(t+1)^{-1/2}(t^2+1)^{-1/2}[(t^2+1) - 2t(t+1)]}{t^2+1}$$

$$= -\frac{t^2+2t-1}{2\sqrt{t+1}(t^2+1)^{3/2}}.$$

47. $y = g(u) = u^{4/3}$ and $\dfrac{dy}{du} = \dfrac{4}{3}u^{1/3}$, $u = f(x) = 3x^2-1$, and $\dfrac{du}{dx} = 6x$.

$$\frac{dy}{dx} = \frac{dy}{du}\cdot\frac{du}{dx} = \tfrac{4}{3}u^{1/3}(6x)$$

So $= \tfrac{4}{3}(3x^2-1)^{1/3}6x$

$= 8x(3x^2-1)^{1/3}$.

49. $\dfrac{dy}{du} = -\dfrac{2}{3}u^{-5/3} = -\dfrac{2}{3u^{5/3}}$, $\dfrac{du}{dx} = 6x^2-1$.

$$\frac{dy}{dx} = \frac{dy}{du}\cdot\frac{du}{dx} = -\frac{2(6x^2-1)}{3u^{5/3}} = -\frac{2(6x^2-1)}{3(2x^3-x+1)^{5/3}}.$$

51. $\dfrac{dy}{du} = \tfrac{1}{2}u^{-1/2} - \tfrac{1}{2}u^{-3/2}$, $\dfrac{du}{dx} = 3x^2-1$.

$$\frac{dy}{dx}=\frac{dy}{du}\cdot\frac{du}{dx}=\left[\frac{1}{2\sqrt{x^3-x}}-\frac{1}{2(x^3-x)^{3/2}}\right](3x^2-1)$$
$$=\frac{(3x^2-1)(x^3-x-1)}{2(x^3-x)^{3/2}}.$$

53. $f(x)=(1-x)(x^2-1)^2$.
$$f'(x)=(1-x)2(x^2-1)(2x)+(-1)(x^2-1)^2$$
$$=(x^2-1)(4x-4x^2-x^2+1)$$
$$=(x^2-1)(-5x^2+4x+1).$$
Therefore, the slope of the tangent line at (2,–9) is
$$f'(2)=[(2)^2-1][-5(2)^2+4(2)+1]=-33.$$
Then the required equation is
$$y+9=-33(x-2),\text{ or }\quad y=-33x+57.$$

55. $f'(x)=\sqrt{2x^2+7}+x(\tfrac{1}{2})(2x^2+7)^{-1/2}(4x)$.
The slope of the tangent line is $f'(3)=\sqrt{25}+(\tfrac{3}{2})(25)^{-1/2}(12)=\tfrac{43}{5}$.
An equation of the tangent line is $y-15=\tfrac{43}{5}(x-3)$ or $y=\tfrac{43}{5}x-\tfrac{54}{5}$.

57. $N(x)=(60+2x)^{2/3}$.
$$N'(x)=\tfrac{2}{3}(60+2x)^{-1/3}\frac{d}{dx}(60+2x)=\tfrac{4}{3}(60+2x)^{-1/3}.$$
The rate of increase at the end of the second week is
$$N'(2)=\tfrac{4}{3}(64)^{-1/3}=\tfrac{1}{3},\text{ or }\tfrac{1}{3}\text{ million/week}$$
At the end of the 12th week, $N'(12)=\tfrac{4}{3}(84)^{-1/3}\approx 0.3$ million/wk. The number of viewers in the 2nd and 24th week are $N(2)=(60+4)^{2/3}=16$ million and $N(24)=(60+48)^{2/3}=22.7$ million, respectively.

59. $C(t)=0.01(0.2t^2+4t+64)^{2/3}$.

a. $$C'(t)=0.01(\tfrac{2}{3})(0.2t^2+4t+64)^{-1/3}\frac{d}{dt}(0.2t^2+4t+64)$$
$$=(0.01)(0.667)(0.4t+4)(0.2t^2+4t+4)^{-1/3}$$
$$=0.027(0.1t+1)(0.2t^2+4t+64)^{-1/3}.$$

b. $C'(5)=0.007[0.4(5)+4][0.2(25)+4(5)+64]^{-1/3}\approx 0.009$, or 0.009 parts per million per year.

61. a. $A(t) = 0.03t^3(t-7)^4 + 60.2$

$$A'(t) = 0.03[3t^2(t-7)^4 + t^3(4)(t-7)^3)] = 0.03t^2(t-7)^3[3(t-7) + 4t]$$
$$= 0.21t^2(t-3)(t-7)^3.$$

b. $A'(1) = 0.21(-2)(-6)^3 = 90.72$; A'(3) = 0. $A'(4) = 0.21(16)(1)(-3)^3 = -90.72$. The amount of pollutant is increasing at the rate of 90.72 units/hr at 8 A.M. Its rate of change is 0 units/hr at 10 A.M.;and its rate of change is –90.72 units/hr at 11 A.M.

63. $$P(t) = \frac{300\sqrt{\frac{1}{2}t^2 + 2t + 25}}{t+25} = \frac{300(\frac{1}{2}t^2 + 2t + 25)^{1/2}}{t+25}.$$

$$P'(t) = 300\left[\frac{(t+25)\frac{1}{2}(\frac{1}{2}t^2 + 2t + 25)^{-1/2}(t+2) - (\frac{1}{2}t^2 + 2t + 25)^{1/2}(1)}{(t+25)^2}\right]$$

$$= 300\left[\frac{(\frac{1}{2}t^2 + 2t + 25)^{-1/2}[(t+25)(t+2) - 2(\frac{1}{2}t^2 + 2t + 25)^{1/2}}{(t+25)^2}\right]$$

$$= \frac{3450t}{(t+25)^2\sqrt{\frac{1}{2}t^2 + 2t + 25}}.$$

Ten seconds into the run, the athlete's pulse rate is increasing at

$$P'(10) = \frac{3450(10)}{(35)^2\sqrt{50 + 20 + 25}} \approx 2.9,$$

or approximately 2.9 beats per minute per minute. Sixty seconds into the run, it is increasing at

$$P'(60) = \frac{3450(60)}{(85)^2\sqrt{1800 + 120 + 25}} \approx 0.65,$$

or approximately 0.7 beats per minute per minute. Two minutes into the run, it is increasing at

$$P'(120) = \frac{3450(120)}{(145)^2\sqrt{7200 + 240 + 25}} \approx 0.23,$$

or approximately 0.2 beats per minute per minute. The pulse rate two minutes into the run is given by

$$P(120) = \frac{300\sqrt{7200 + 240 + 25}}{120 + 25} \approx 178.8,$$

or approximately 179 beats per minute.

65. The area is given by $A = \pi r^2$. The rate at which the area is increasing is given by

dA/dt, that is, $\dfrac{dA}{dt} = \dfrac{d}{dt}(\pi r^2) = \dfrac{d}{dt}(\pi r^2)\dfrac{dr}{dt} = 2\pi r\dfrac{dr}{dt}$.

If $r = 40$ and $dr/dt = 2$, then $\dfrac{dA}{dt} = 2\pi(40)(2) = 160\pi$, that is, it is increasing at the rate of 160π or approximately 503 sq ft/sec.

67. $\dfrac{dS}{dt} = g'(x)f'(t) = (-0.0015x)(12.5t + 19.75)$.

When $t = 4$, we have $x = f(4) = 6.25(16) + 19.75(4) + 74.75 = 253.75$

and $\left.\dfrac{dS}{dt}\right|_{t=4} = (-0.0015)(253.75)[12.5(4) + 19.75] \approx -26.55$;

that is, the average speed will be dropping at the rate of approximately 27 mph per decade. The average speed of traffic flow at that time will be

$$S = g(f(4)) = -0.00075(253.75^2) + 67.5 = 19.2,$$

or approximately 19 mph.

69. $N(x) = 1.42x$ and $x(t) = \dfrac{7t^2 + 140t + 700}{3t^2 + 80t + 550}$. The number of construction jobs as a function of time is $n(t) = N[x(t)]$. Using the Chain Rule,

$$n'(t) = \frac{dN}{dx}\cdot\frac{dx}{dt} = 1.42\frac{dx}{dt}$$

$$= (1.42)\left[\frac{(3t^2 + 80t + 550)(14t + 140) - (7t^2 + 140t + 700)(6t + 80)}{(3t^2 + 80t + 550)^2}\right]$$

$$= \frac{1.42(140t^2 + 3500t + 21000)}{(3t^2 + 80t + 550)^2}.$$

$$n'(1) = \frac{1.42(140 + 3500 + 21000)}{(3 + 80 + 550)^2} \approx 0.0873216,$$

or approximately 87,322 jobs per year.

71. We want $\dfrac{dR}{dt} = \dfrac{dR}{dp}\cdot\dfrac{dp}{dt}$. Now

$$\frac{dR}{dp} = 1000\left[\frac{(p+2)(1) - (p+4)(1)}{(p+2)^2}\right] = -\frac{2400}{(p+2)^2}$$

$$\frac{dp}{dt} = 50\left[\frac{(t^2+4t+8)(2t+2)-(t^2+2t+4)(2t+4)}{(t^2+4t+8)^2}\right]$$
$$= \frac{100t(t+4)}{(t^2+4t+8)^2}.$$

When $t = 2$,

$$\frac{dR}{dt} = -\frac{2000}{(p+2)^2}\cdot\frac{100t(t+4)}{(t^2+4t+8)^2}\Bigg|_{t=2} = -\frac{2000}{(32)^2}\cdot\frac{100(2)(6)}{(4+8+8)^2}$$
$$\approx -5.86,$$

that is, the passage will decrease at the rate of approximately \$5.86 per passenger per year.

73. Let $f(x) = x^r = x^{m/n} = (x^m)^{1/n}$. Then $[f(x)]^n = x^m$.

$$n[f(x)]^{n-1} f'(x) = \frac{m}{n}[f(x)]^{-n+1} x^{m-1} = \frac{m}{n}(x^{m/n})^{-n+1} x^{m-1}$$

Therefore,
$$= \frac{m}{n} x^{[m(-n+1)/n]+m-1} = \frac{m}{n} x^{(m-n)/n} = \frac{m}{n} x^{(m/n)-1}$$
$$= rx^{r-1}.$$

USING TECHNOLOGY EXERCISES 11.3 page 719

1. 0.5774   3. 0.9390   5. –4.9498

7. a. 10,146,200/decade   b. 7,810,520/decade

EXERCISES 11.4, page 736

1. a. $C(x)$ is always increasing because as x, the number of units produced increases, the greater the amount of money that must be spent on production.
   b. This occurs at x = 4, or a production level of 4000. You can see this by looking at the slopes of the tangent lines for x less than, equal to, and a little larger then $x = 4$.

3. a. $C(101) - C(100) = [0.0002(101)^3 - 0.06(101)^2 + 120(101) + 5000]$
$- [0.0002(100)^3 - 0.06(100)^2 + 120(100) + 5000]$
$\approx$ 114, or approximately \$114.
   Similarly, we find $C(201) - C(200) \approx$ \$120.16; $C(301) - C(300) \approx$ \$138.12.
   b. We compute $C'(x) = 0.0006x^2 - 0.12x + 120$. So the required quantities are

$C'(100) = 0.0006(100)^2 - 0.12(100) + 120 = 114$, or \$114,
$C'(200) = 0.0006(200)^2 - 0.12(200) + 120 = 120$, or \$120,
and $C'(300) = 138$, or \$138.

5. a. $\overline{C}(x) = \frac{C(x)}{x} = \frac{5000}{x} + 2.$ b. $\overline{C}'(x) = -\frac{5000}{x^2}.$

c. Since the marginal average cost function is negative for $x > 0$, the rate of change of the average cost function is negative for all $x > 0$.

7. $\overline{C}(x) = \frac{C(x)}{x} = \frac{0.0002x^3 - 0.06x^2 + 120x + 5000}{x}$

$$= 0.0002x^2 - 0.06x + 120 + \frac{5000}{x}.$$

$\overline{C}'(x) = 0.0004x - 0.06 - \frac{5000}{x^2}.$

9. a. $R(x) = px = x(-0.04x + 800) = -0.04x^2 + 800x$
b. $R'(x) = -0.08x + 800$
c. $R'(5000) = -0.08(5000) + 800 = 400.$
This says that when the level of production is 5000 units the production of the next speaker system will bring an additional revenue of \$400.

11. a. $P(x) = R(x) - C(x) = (-0.04x^2 + 800x) - (200x + 300{,}000)$
$= -0.04x^2 + 600x - 300{,}000.$
b. $P'(x) = -0.08x + 600$
c. $P'(5000) = -0.08(5000) + 600 = 200$
$P'(8000) = -0.08(8000) + 600 = -40.$
d.

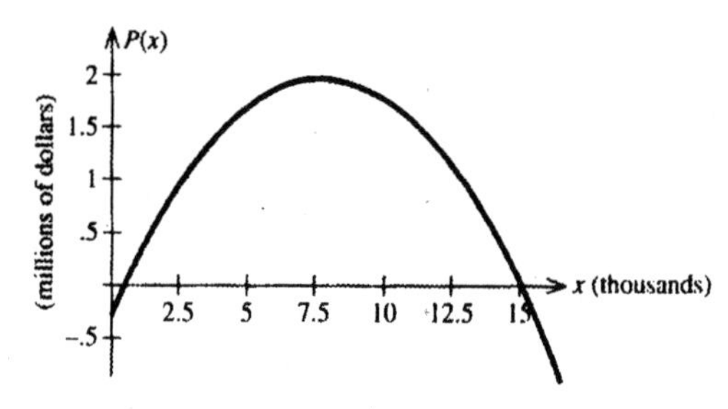

The profit realized by the company increases as production increases, peaking at a level of production of 7500 units. Beyond this level of production, the profit begins to fall.

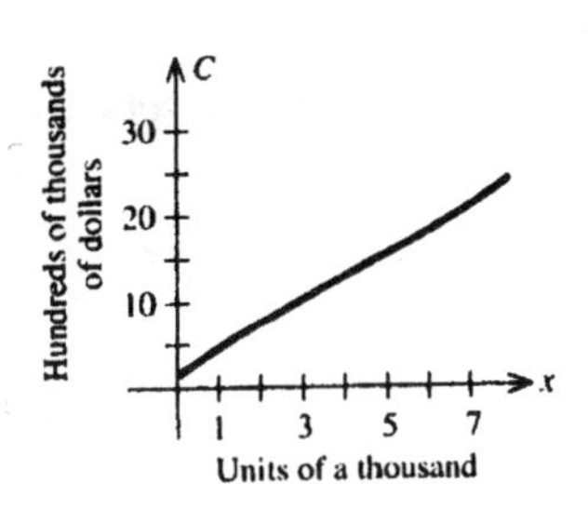

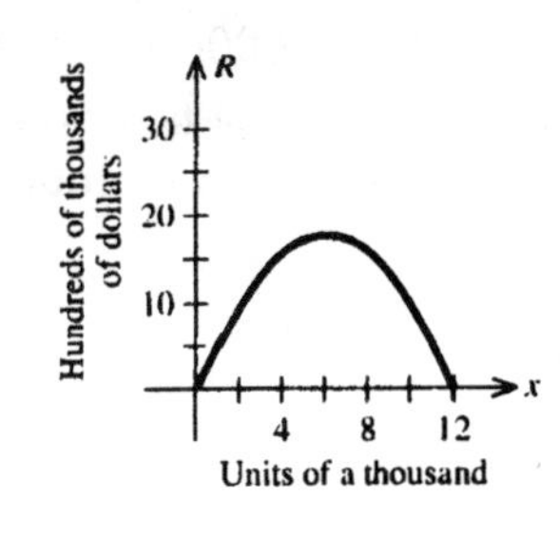

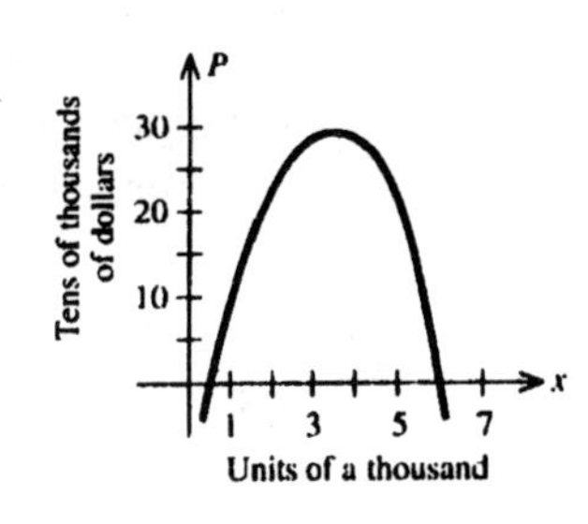

13. a. $R(x) = xp(x) = -0.006x^2 + 180x$

$P(x) = R(x) - C(x)$

$= -0.006x^2 + 180x - (0.000002x^3 - -0.02x^2 + 120x + 60{,}000)$

$= -0.000002x^3 + 0.014x^2 + 60x - 60{,}000.$

b. $C'(x) = 0.000006x^2 - 0.04x + 120$

$R'(x) = -0.012x + 180$

$P'(x) = -0.000006x^2 + 0.028x + 60.$

c. $C'(2000) = 0.000006(2000)^2 - 0.04(2000) + 120 = 64$

$R'(2000) = -0.012(2000) + 180 = 156$

$P'(2000) = -0.000006(2000)^2 + 0.028(2000) + 60 = 92.$

d.

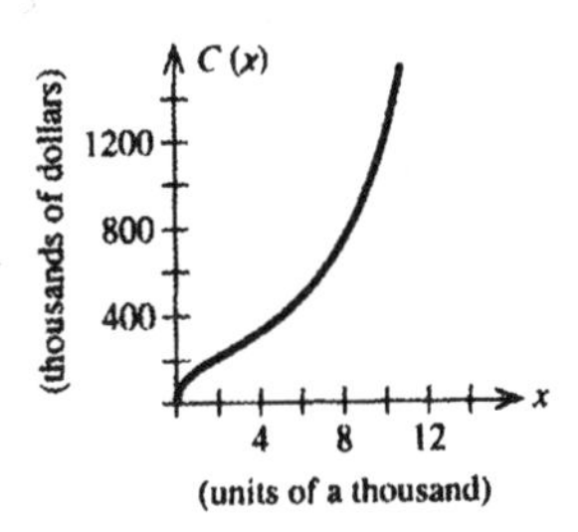

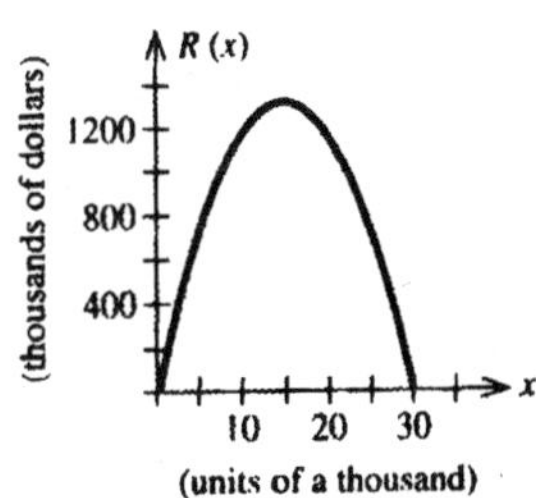

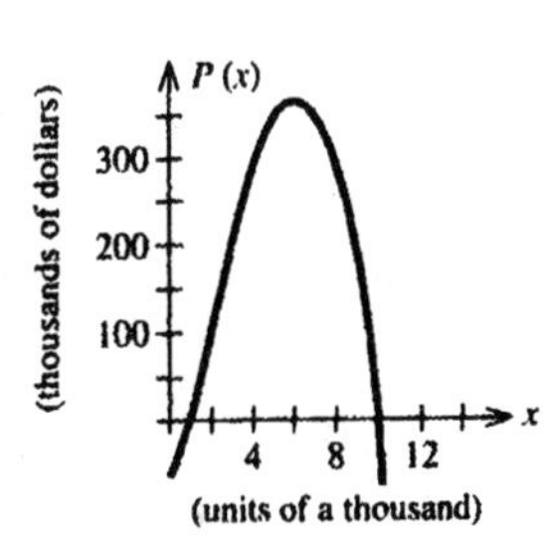

15. $C(x) = 0.000002x^3 - 0.02x + 120x + 60{,}000$

$$\overline{C}(x) = 0.000002x^2 - 0.02x + 120 + \frac{60{,}000}{x}.$$

a. The marginal average cost function is given by

$$\overline{C}'(x) = 0.000004x - 0.02 - \frac{60{,}000}{x^2}$$

b. $$\overline{C}'(5000) = 0.000004(5000) - 0.02 - \frac{60{,}000}{(5000)^2}$$

$$= 0.02 - 0.02 - 0.0024 = -0.0024.$$

$$\overline{C}'(10000) = 0.000004(10000) - 0.02 - \frac{60{,}000}{(10000)^2}$$

$= 0.04 - 0.02 - 0.0006 = 0.0194.$

We conclude that the average cost is decreasing when 5000 TV sets are produced and increasing when 10,000 units are produced.

17. $\dfrac{dC}{dx} = \dfrac{d}{dx}(0.712x + 95.05) = 0.712.$

19. $\dfrac{dS}{dx} = \dfrac{d}{dx}[x - C(x)] = 1 - \dfrac{dC}{dx}.$

21. Here $C(x) = 0.873x^{1.1} + 20.34$. So $C'(x) = 0.9603x^{0.1}$ and

$$\frac{dS}{dx} = 1 - \frac{dC}{dx} = 1 - 0.9603x^{0.1}.$$

When $x = 10$, we have

$$\frac{dS}{dx} = 1 - 0.9603(10)^{0.1} = -0.209,$$

or –$0.209 billion per billion dollars.

23. Here $x = f(p) = -\frac{5}{4}p + 20$ and so $f'(p) = -\frac{5}{4}$. Therefore,

$$E(p) = -\frac{pf'(p)}{f(p)} = -\frac{p(-\frac{5}{4})}{-\frac{5}{4}p + 20} = \frac{5p}{80 - 5p}.$$

$$E(10) = \frac{5(10)}{80 - 5(10)} = \frac{50}{30} = \frac{5}{3} > 1,$$

and so the demand is elastic.

25. Solving the demand equation for x, we have

$$0.4x = -p + 20 \quad \text{or} \quad x = f(p) = -\tfrac{5}{2}p + 50.$$

Then $f'(p) = -\frac{5}{2}$, and so

$$E(p) = -\frac{pf'(p)}{f(p)} = -\frac{p(-\frac{5}{2})}{-\frac{5}{2}p + 50} = \frac{5p}{100 - 5p}.$$

$E(10) = \dfrac{50}{50} = 1$, and so the demand is unitary.

27. $x^2 = 169 - p$ and $f(p) = (169 - p)^{1/2}$.

Next, $f'(p) = \frac{1}{2}(169 - p)^{-1/2}(-1) = -\frac{1}{2}(169 - p)^{-1/2}.$

Then the elasticity of demand is given by

$$E(p) = -\frac{pf'(p)}{f(p)} = -\frac{p(-\frac{1}{2})(169-p)^{-1/2}}{(169-p)^{1/2}} = \frac{\frac{1}{2}p}{169-p}.$$

Therefore, when $p = 29$,

$$E(p) = \frac{\frac{1}{2}(29)}{169-29} = \frac{14.5}{140} = 0.104.$$

Since $E(p) < 1$, we conclude that demand is inelastic at this price.

29. $f(p) = \frac{1}{5}(225-p^2);\ f'(p) = \frac{1}{5}(-2p) = -\frac{2}{5}p.$
Then the elasticity of demand is given by

$$E(p) = -\frac{pf'(p)}{f(p)} = -\frac{p(-\frac{2}{5}p)}{\frac{1}{5}(225-p^2)} = \frac{2p^2}{225-p^2}.$$

a. When $p = 8$, $\quad E(8) = \dfrac{2(64)}{225-64} = 0.8 < 1$

and the demand is inelastic. When $p = 10$, $\quad E(10) = \dfrac{2(100)}{225-100} = 1.6 > 1$

and the demand is elastic when $p = 10$.

b. The demand is unitary when $E = 1$. Solving $\dfrac{2p^2}{225-p^2} = 1$,

we find $2p^2 = 225 - p^2$, $3p^2 = 225$, and $p = 8.66$. So the demand is unitary when $p = 8.66$.

c. Since demand is elastic when $p = 10$, lowering the unit price will cause the revenue to increase.

d. Since the demand is inelastic at $p = 8$, a slight increase in the unit price will cause the revenue to increase.

31. $f(p) = \frac{2}{3}(36-p^2)^{1/2}$
$f'(p) = \frac{2}{3}(\frac{1}{2})(36-p^2)^{-1/2}(-2p) = -\frac{2}{3}p(36-p^2)^{-1/2}.$

Then the elasticity of demand is given by

$$E(p) = -\frac{pf'(p)}{f(p)} = -\frac{-\frac{2}{3}p(36-p^2)^{-1/2}p}{\frac{2}{3}(36-p^2)^{1/2}} = \frac{p^2}{36-p^2}.$$

When $p = 2$,

$$E(2) = \frac{4}{36-4} = \frac{1}{8} < 1,$$

and we conclude that the demand is inelastic.

b. Since the demand is inelastic, the revenue will increase when the rental price is increased.

33. $f(p)=10\left(\frac{50-p}{p}\right)^{1/2}=10\left(\frac{50}{p}-1\right)^{1/2}$.

$$f'(p)=10(\tfrac{1}{2})\left(\frac{50}{p}-1\right)^{-1/2}\left(-\frac{50}{p^2}\right)=-\frac{250}{p^2}\left(\frac{50}{p}-1\right)^{-1/2}.$$

Then the elasticity of demand is given by

$$E(p)=-\frac{pf'(p)}{f(p)}=-\frac{p\left(-\frac{250}{p^2}\right)\left(\frac{50}{p}-1\right)^{-1/2}}{10\left(\frac{50}{p}-1\right)^{1/2}}$$

$$=-\frac{\frac{250}{p}}{10\left(\frac{50}{p}-1\right)}=\frac{25}{p\left(\frac{50-p}{p}\right)}=\frac{25}{50-p}.$$

Setting $E=1$, gives

$$1=\frac{25}{50-p}\text{ and so }25=50-p\text{, and }p=25.$$

Thus, if $p>25$, then $E>1$, and the demand is elastic; if $p=25$, then $E=1$ and the demand is unitary; and if $p<25$, then $E<1$ and the demand is inelastic.

EXERCISES 11.5, page 745

1. $f(x)=4x^2-2x+1$; $f'(x)=8x-2$; $f''(x)=8$.

3. $f(x)=2x^3-3x^2+1$; $f'(x)=6x^2-6x$; $f''(x)=12x-6=6(2x-1)$.

5. $h(t)=t^4-2t^3+6t^2-3t+10$; $h'(t)=4t^3-6t^2+12t-3$
$h''(t)=12t^2-12t+12=12(t^2-t+1)$.

7. $f(x)=(x^2+2)^5$; $f'(x)=5(x^2+2)^4(2x)=10x(x^2+2)^4$ and

$$f'(x) = 10(x^2+2)^4 + 10x(x^2+2)^3(2x)$$
$$= 10(x^2+2)^3[(x^2+2)+8x^2] = 10(9x^2+2)(x^2+2)^3.$$

9. $g(t) = (2t^2-1)^2(3t^2);$
$$g''(t) = 2(2t^2-1)(4t)(3t^2) - (2t^2-1)^2(6t)$$
$$= 6t(2t^2-1)[4t^2+(2t^2-1)] = 6t(2t^2-1)(6t^2-1)$$
$$= 6t(12t^4-8t^2+1) = 72t^5-48t^3+6t.$$
$$g''(t) = 360t^4-144t^2+6 = 6(60t^4-24t^2+1)$$

11. $f(x) = (2x^2+2)^{7/2};\ f'(x) = \frac{7}{2}(2x^2+2)^{5/2}(4x) = 14x(2x^2+2)^{5/2};$
$$f''(x) = 14(2x^2+2)^{5/2} + 14x(\tfrac{5}{2})(2x^2+2)^{3/2}(4x)$$
$$= 14(2x^2+2)^{3/2}[(2x^2+2)+10x^2] = 28(6x^2+1)(2x^2+2)^{3/2}.$$

13. $f(x) = x(x^2+1)^2;$
$$f'(x) = (x^2+1)^2 + x(2)(x^2+1)(2x)$$
$$= (x^2+1)[(x^2+1)+4x^2] = (x^2+1)(5x^2+1);$$
$$f''(x) = 2x(5x^2+1)+(x^2+1)(10x) = 2x(5x^2+1+5x^2+5)$$
$$= 4x(5x^2+3).$$

15. $f(x) = \dfrac{x}{2x+1};\ f'(x) = \dfrac{(2x+1)(1)-x(2)}{(2x+1)^2} = \dfrac{1}{(2x+1)^2};$

$$f''(x) = \frac{d}{dx}(2x+1)^{-2} = -2(2x+1)^{-3}(2) = -\frac{4}{(2x+1)^3}.$$

17. $f(s) = \dfrac{s-1}{s+1};\ f'(s) = \dfrac{(s+1)(1)-(s-1)(1)}{(s+1)^2} = \dfrac{2}{(s+1)^2}.$
$$f''(s) = 2\frac{d}{ds}(s+1)^{-2} = -4(s+1)^{-3} = -\frac{4}{(s+1)^3}.$$

19. $f(u) = \sqrt{4-3u} = (4-3u)^{1/2}.$
$$f'(u) = \tfrac{1}{2}(4-3u)^{-1/2}(-3) = -\frac{3}{2\sqrt{4-3u}}.$$

$$f''(u) = -\frac{3}{2}\cdot\frac{d}{du}(4-3u)^{-1/2} = -\frac{3}{2}\left(-\frac{1}{2}\right)(4-3u)^{-3/2}(-3) = -\frac{9}{4(4-3u)^{3/2}}.$$

21. $f(x) = 3x^4 - 4x^3$; $f'(x) = 12x^3 - 12x^2$; $f''(x) = 36x^2 - 24x$; $f'''(x) = 72x - 24$.

23. $f(x) = \frac{1}{x}$; $f'(x) = \frac{d}{dx}(x^{-1}) = -x^{-2}$; $f''(x) = 2x^{-3}$; $f'''(x) = -6x^{-4} = -\frac{6}{x^4}$.

25. $g(s) = (3s-2)^{1/2}$; $g'(s) = \frac{1}{2}(3s-2)^{-1/2}(3) = \frac{3}{2(3s-2)^{1/2}}$;

$$g''(s) = \frac{3}{2}\left(-\frac{1}{2}\right)(3s-2)^{-3/2}(3) = -\frac{9}{4}(3s-2)^{-3/2} = -\frac{9}{4(3s-2)^{3/2}};$$

$$g'''(s) = \frac{27}{8}(3s-2)^{-5/2}(3) = \frac{81}{8}(3s-2)^{-5/2} = \frac{81}{8(3s-2)^{5/2}}.$$

27. $f(x) = (2x-3)^4$; $f'(x) = 4(2x-3)^3(2) = 8(2x-3)^3$

$f''(x) = 24(2x-3)^2(2) = 48(2x-3)^2$

$f'''(x) = 96(2x-3)(2) = 192(2x-3)$.

29. Its velocity at any time t is $v(t) = \frac{d}{dt}(16t^2) = 32t$. The hammer strikes the ground when $16t^2 = 256$ or $t = 4$ (we reject the negative root). Therefore, its velocity at the instant it strikes the ground is $v(4) = 32(4) = 128$ ft/sec. Its acceleration at time t is $a(t) = \frac{d}{dt}(32t) = 32$. In particular, its acceleration at $t = 4$ is 32 ft/sec^2.

31. $N(t) = -0.1t^3 + 1.5t^2 + 100$.

a. $N'(t) = -0.3t^2 + 3t = 0.3t(10-t)$. Since $N'(t) > 0$ for $t = 0, 1, 2, \ldots, 7$, it is evident that $N(t)$ (and therefore the crime rate) was increasing from 1988 through 1995.

b. $N''(t) = -0.6t + 3 = 0.6(5-t)$. Now $N''(4) = 0.6 > 0$, $N''(5) = 0$, $N''(6) = -0.6 < 0$ and $N''(7) = -1.2 < 0$. This shows that the rate of the rate of change is decreasing beyond $t = 5$ (1990) and this shows that the program was working.

33. $h(t) = \frac{1}{16}t^4 - t^3 + 4t^2$.

a. $h'(t) = \frac{1}{4}t^3 - 3t^2 + 8t$

b. $h'(0) = 0$ or zero feet per second.

$h'(4) = \frac{1}{4}(64) - 3(16) + 8(4) = 0,$ or zero feet per second.

$h'(8) = \frac{1}{4}(8)^3 - 3(64) + 8(8) = 0,$ or zero feet per second.

c. $h''(t) = \frac{3}{4}t^2 - 6t + 8$

d. $h''(0) = 8$ ft/sec^2

$h''(4) = \frac{3}{4}(16) - 6(4) + 8 = -4$ ft/sec^2.

$h''(8) = \frac{3}{4}(64) - 6(8) + 8 = 8$ ft/sec^2.

e. $h(0) = 0$ feet

$h(4) = \frac{1}{16}(4)^4 - (4)^3 + 4(4)^2 = 16$ feet.

$h(8) = \frac{1}{16}(8)^4 - (8)^3 + 4(8)^2 = 0$ feet.

35. $A(t) = 100 - 17.63t + 1.915t^2 - 0.1316t^3 + 0.00468t^4 - 0.00006t^5$

$A'(t) = -17.63 + 3.83t - 0.3948t^2 + 0.01872t^3 - 0.0003t^4$

$A''(t) = 3.83 - 0.7896t + 0.05616t^2 - 0.0012t^3$.

So, $A'(10) = -3.09$ and $A''(10) = 0.35$.

Our computations show that 10 minutes after the start of the test,the smoke remaining is decreasing at a rate of 3 percent per minute but the rate at which the rate of smoke is decreasing is increasing at the rate of 0.35 percent per minute per minute.

37. Consider the function $f(x) = x^{(2n+1)/2} = x^{n+(1/2)}$.

Then $f'(x) = (n + \frac{1}{2})x^{n-(1/2)}$

$$f''(x) = (n + \tfrac{1}{2})(n - \tfrac{1}{2})x^{n-(3/2)}$$

....

$$f^{(n)}(x) = (n + \tfrac{1}{2})(n - \tfrac{1}{2}) \cdots \tfrac{3}{2}x^{1/2}$$

$$f^{(n+1)}(x) = (n + \tfrac{1}{2})(n - \tfrac{1}{2}) \cdots \tfrac{1}{2}x^{-1/2}.$$

The first n derivatives exist at $x = 0$, but the $(n + 1)$st derivative fails to be defined there.

USING TECHNOLOGY EXERCISES 11.5, page 747

1. –18 3. 15.2762 5. –0.62549 7. 0.197304

9. $f''(6) = -68.46214$ and it tells us that at the beginning of 1988, the rate of the rate of the rate at which banks were failing was 68 banks per year per year per year.

EXERCISES 11.6, page 757

1. a. Solving for y in terms of x, we have $y=-\frac{1}{2}x+\frac{5}{2}$. Therefore, $y'=-\frac{1}{2}$.
   b. Next, differentiating $x + 2y = 5$ implicitly, we have $1+2y'=0$, or $y'=-\frac{1}{2}$.

3. a. $xy = 1, y=\dfrac{1}{x}$, and $\dfrac{dy}{dx}=-\dfrac{1}{x^2}$.

   b. $$x\frac{dy}{dx}+y=0$$
   $$x\frac{dy}{dx}=-y$$
   $$\frac{dy}{dx}=-\frac{y}{x}=\frac{-\frac{1}{x}}{x}=-\frac{1}{x^2}.$$

5. $x^3-x^2-xy=4.$

   a. $-xy=4-x^3+x^2$
   $$y=-\frac{4}{x}+x^2-x$$
   $$y'=\frac{4}{x^2}+2x-1.$$

   b. $x^3-x^2-xy=4$
   $$3x^2-2x-x\frac{dy}{dx}-y=0$$
   $$-x\frac{dy}{dx}=-3x^2+2x+y$$
   $$\frac{dy}{dx}=3x-2-\frac{y}{x}$$
   $$=3x-2-\frac{1}{x}(-\frac{4}{x}+x^2-x)$$
   $$=3x-2+\frac{4}{x^2}-x+1$$
   $$=\frac{4}{x^2}+2x-1.$$

7. a. $\dfrac{x}{y}-x^2=1$ is equivalent to $\dfrac{x}{y}=x^2+1$, or $y=\dfrac{x}{x^2+1}$. Therefore,

$$y' = \frac{(x^2+1) - x(2x)}{(x^2+1)^2} = \frac{1-x^2}{(x^2+1)^2}.$$

b. Next, differentiating the equation $x - x^2y = y$ implicitly, we obtain

$$1 - 2xy - x^2y' = y',\ y'(1+x^2) = 1 - 2xy, \text{ or } \ y' = \frac{1-2xy}{(1+x^2)}.$$

(This may also be written in the form $-2y^2 + \frac{y}{x}$.)

To show that this is equivalent to the results obtained earlier, use the value of y obtained before, to get

$$y' = \frac{1-2x\left(\frac{x}{x^2+1}\right)}{1+x^2} = \frac{x^2+1-2x^2}{(1+x^2)^2} = \frac{1-x^2}{(1+x^2)^2}.$$

9. $x^2 + y^2 = 16$. Differentiating both sides of the equation implicitly, we obtain

$$2x + 2yy' = 0 \text{ and so } y' = -\frac{x}{y}.$$

11. $x^2 - 2y^2 = 16$. Differentiating implicitly with respect to x, we have

$$2x - 4y\frac{dy}{dx} = 0 \text{ and } \frac{dy}{dx} = \frac{x}{2y}.$$

13. $x^2 - 2xy = 6$. Differentiating both sides of the equation implicitly, we obtain

$$2x - 2y - 2xy' = 0 \text{ and so } y' = \frac{x-y}{x} = 1 - \frac{y}{x}.$$

15. $x^2y^2 - xy = 8$. Differentiating both sides of the equation implicitly, we obtain

$$2xy^2 + 2x^2yy' - y - xy' = 0,\ 2xy^2 - y + y'(2x^2y - x)$$

and so $y' = \dfrac{y(1-2xy)}{x(2xy-1)} = -\dfrac{y}{x}.$

17. $x^{1/2} + y^{1/2} = 1$. Differentiating implicitly with respect to x, we have

$$\tfrac{1}{2}x^{-1/2} + \tfrac{1}{2}y^{-1/2}\frac{dy}{dx} = 0.$$

Therefore, $\dfrac{dy}{dx} = -\dfrac{x^{-1/2}}{y^{-1/2}} = -\dfrac{\sqrt{y}}{\sqrt{x}}.$

19. $\sqrt{x+y}=x$. Differentiating both sides of the equation implicitly, we obtain

$$\tfrac{1}{2}(x+y)^{-1/2}(1+y')=1,\ 1+y'=2(x+y)^{1/2},$$

or $y'=2\sqrt{x+y}-1.$

21. $\dfrac{1}{x^2}+\dfrac{1}{y^2}=1.$ Differentiating both sides of the equation implicitly, we obtain

$$-\frac{2}{x^3}-\frac{2}{y^3}y'=0, \text{ or } y'=-\frac{y^3}{x^3}.$$

23. $\sqrt{xy}=x+y$. Differentiating both sides of the equation implicitly, we obtain

$$\tfrac{1}{2}(xy)^{-1/2}(xy'+y)=1+y'$$

$$xy'+y=2\sqrt{xy}(1+y')$$

$$y'(x-2\sqrt{xy})=2\sqrt{xy}-y$$

or

$$y'=-\frac{(2\sqrt{xy}-y)}{(2\sqrt{xy}-x)}=\frac{2\sqrt{xy}-y}{x-2\sqrt{xy}}.$$

25. $\dfrac{x+y}{x-y}=3x$, or $x+y=3x^2-3xy$. Differentiating both sides of the equation implicitly, we obtain $1+y'=6x-3xy'-3y$ or $y'=\dfrac{6x-3y-1}{3x+1}.$

27. $xy^{3/2}=x^2+y^2$. Differentiating implicitly with respect to x, we obtain

$$y^{3/2}+x(\tfrac{3}{2})y^{1/2}\frac{dy}{dx}=2x+2y\frac{dy}{dx}$$

$$2y^{3/2}+3xy^{1/2}\frac{dy}{dx}=4x+4y\frac{dy}{dx} \qquad \text{(Multiplying by 2.)}$$

$$(3xy^{1/2}-4y)\frac{dy}{dx}=4x-2y^{3/2}$$

$$\frac{dy}{dx}=\frac{2(2x-y^{3/2})}{3xy^{1/2}-4y}.$$

29. $(x+y)^3+x^3+y^3=0$. Differentiating implicitly with respect to x, we obtain

$$3(x+y)^2\left(1+\frac{dy}{dx}\right)+3x^2+3y^2\frac{dy}{dx}=0$$

$$(x+y)^2+(x+y)^2\frac{dy}{dx}+x^2+y^2\frac{dy}{dx}=0$$

$$[(x+y)^2+y^2]\frac{dy}{dx}=-[(x+y)^2+x^2]$$

$$\frac{dy}{dx}=-\frac{2x^2+2xy+y^2}{x^2+2xy+2y^2}.$$

31. $4x^2+9y^2=36$. Differentiating the equation implicitly, we obtain

$$8x+18yy'=0.$$

At the point (0,2), we have $0+36y'=0$ and the slope of the tangent line is 0. Therefore, an equation of the tangent line is $y=2$.

33. $x^2y^3-y^2+xy-1=0$. Differentiating implicitly with respect to x, we have

$$2xy^3+3x^2y^2\frac{dy}{dx}-2y\frac{dy}{dx}+y+x\frac{dy}{dx}=0.$$

At (1,1),

$$2+3\frac{dy}{dx}-2\frac{dy}{dx}+1+\frac{dy}{dx}=0$$

$$2\frac{dy}{dx}=-3 \text{ and } \frac{dy}{dx}=-\frac{3}{2}.$$

Using the point-slope form of an equation of a line, we have

$$y-1=-\tfrac{3}{2}(x-1)$$

and the equation of the tangent line to the graph of the function f at (1,1) is

$$y=-\tfrac{3}{2}x+\tfrac{5}{2}.$$

35. $xy=1$. Differentiating implicitly, we have

$$xy'+y=0, \text{ or } y'=-\frac{y}{x}.$$

Differentiating implicitly once again, we have

$$xy''+y'+y'=0.$$

Therefore, $$y''=-\frac{2y'}{x}=\frac{2\left(\frac{y}{x}\right)}{x}=\frac{2y}{x^2}.$$

37. $y^2 - xy = 8$. Differentiating implicitly we have $2yy' - y - xy' = 0$

and $y' = \dfrac{y}{2y-x}$. Differentiating implicitly again, we have

$$2(y')^2 + 2yy'' - y' - y' - xy'' = 0, \quad \text{or} \quad y'' = \frac{2y' - 2(y')^2}{2y-x}.$$

Then $$y'' = \frac{2\left(\dfrac{y}{2y-x}\right)\left(1 - \dfrac{y}{2y-x}\right)}{2y-x} = \frac{2y(2y-x-y)}{(2y-x)^3} = \frac{2y(y-x)}{(2y-x)^3}.$$

39. a. Differentiating the given equation with respect to t, we obtain

$$\frac{dV}{dt} = \pi r^2 \frac{dh}{dt} + 2\pi rh\frac{dr}{dt} = \pi r\left(r\frac{dh}{dt} + 2h\frac{dr}{dt}\right).$$

b. Substituting $r = 2$, $h = 6$, $\dfrac{dr}{dt} = 0.1$ *and* $\dfrac{dh}{dt} = 0.3$ into the expression for $\dfrac{dV}{dt}$

we obtain $\dfrac{dV}{dt} = \pi(2)[2(0.3) + 2(6)(0.1)] = 3.6\pi$

and so the volume is increasing at the rate of 3.6π cu in/sec.

41. We are given $\dfrac{dp}{dt} = 2$ and are required to find $\dfrac{dx}{dt}$ when $x = 9$ and $p = 63$.

Differentiating the equation $p + x^2 = 144$ with respect to t, we obtain

$$\frac{dp}{dt} + 2x\frac{dx}{dt} = 0.$$

When $x = 9$, $p = 63$, and $\dfrac{dp}{dt} = 2$,

$$2 + 2(9)\frac{dx}{dt} = 0$$

and $$\frac{dx}{dt} = -\frac{1}{9} \approx -0.111,$$

or decreasing at the rate of 111 tires per week.

43. $100x^2 + 9p^2 = 3600$. Differentiating the given equation implicitly with respect to t, we have

$$200x\frac{dx}{dt} + 18p\frac{dp}{dt} = 0.$$

Next, when $p = 14$, the given equation yields

$$100x^2 + 9(14)^2 = 3600$$
$$100x^2 = 1836,$$

or $x = 4.2849$. When $p = 14$, $\frac{dp}{dt} = -0.15$, and $x = 4.2849$, we have

$$200(4.2849)\frac{dx}{dt} + 18(14)(-0.15) = 0$$
$$\frac{dx}{dt} = 0.0441.$$

So the quantity demanded is increasing at the rate of 44 ten–packs per week.

45. From the results of Problem 44, we have

$$1250p\frac{dp}{dt} - 2x\frac{dx}{dt} = 0.$$

When $p = 1.0770$, $x = 25$, and $\frac{dx}{dt} = -1$, we find that

$$1250(1.077)\frac{dp}{dt} - 2(25)(-1) = 0,$$

and $$\frac{dp}{dt} = -\frac{50}{1250(1.077)} = -0.037.$$

We conclude that the price is decreasing at the rate of 3.7 cents per carton.

47. $p = -0.01x^2 - 0.2x + 8$. Differentiating the given equation implicitly with respect to p, we have

$$1 = -0.02x\frac{dx}{dp} - 0.2\frac{dx}{dp} = [0.02x + 0.2]\frac{dx}{dp}$$

or $$\frac{dx}{dp} = -\frac{1}{0.02x + 0.2}.$$

When $x = 15$, $p = -0.01(15)^2 - 0.2(15) + 8 = 2.75$

and $$\frac{dx}{dp} = -\frac{1}{0.02(15) + 0.2} = -2.$$

Therefore, $$E(p) = -\frac{pf'(p)}{f(p)} = -\frac{(2.75)(-2)}{15} = 0.37 < 1,$$

and the demand is inelastic.

49. $A = \pi r^2$. Differentiating with respect to t, we obtain

$$\frac{dA}{dt} = 2\pi r\frac{dr}{dt}.$$

When the radius of the circle is 40 ft and increasing at the rate of 2 ft/sec,

$$\frac{dA}{dt} = 2\pi(40)(2) = 160\pi \text{ ft}^2\text{ / sec.}$$

51. Let D denote the distance between the two cars, x the distance traveled by the car heading east, and y the distance traveled by the car heading north as shown in the diagram at the right. Then

$$D^2 = x^2 + y^2.$$

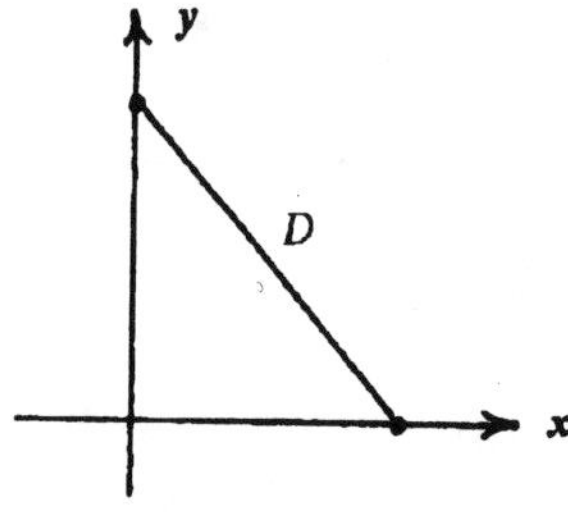

Differentiating with respect to t, we have

$$2D\frac{dD}{dt} = 2x\frac{dy}{dt} + 2y\frac{dy}{dt},$$

or $$\frac{dD}{dt} = \frac{x\frac{dx}{dt} + y\frac{dy}{dt}}{D}$$

When $t = 5$, $x = 30$, $y = 40$, $\frac{dx}{dt} = 2(5) + 1 = 11$, and $\frac{dy}{dt} = 2(5) + 3 = 13.$

Therefore, $\frac{dD}{dt} = \frac{(30)(11) + (40)(13)}{\sqrt{900 + 1600}} = 17$ ft/sec.

53. Referring to the diagram at the right, we see that

$$D^2 = 120^2 + x^2.$$

Differentiating this last equation with respect to t, we have

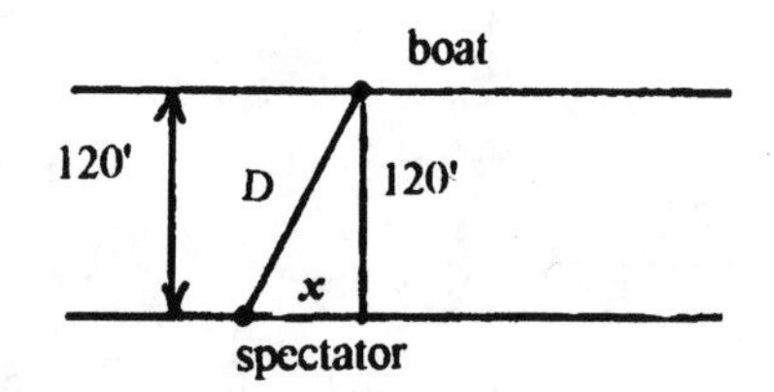

$$2\frac{dD}{dt} = 2x\frac{dx}{dt} \quad \text{and} \quad \frac{dD}{dt} = \frac{x\frac{dx}{dt}}{D}.$$

When $x = 50$, $D = \sqrt{120^2 + 50^2} = 130$

and $\frac{dD}{dt} = \frac{(20)(50)}{130} \approx 7.69$, or 7.69 ft/sec.

55. Referring to the diagram in the text, we see that $x^2 + y^2 = 400$. Differentiating this equation with respect to t, we have

$$2x\frac{dx}{dt} + 2y\frac{dy}{dt} = 0 \quad \text{and} \quad \frac{dy}{dt} = -\frac{x\frac{dx}{dt}}{y}.$$

When $x = 12$, $y^2 = 400 - 144 = 256$, and $y = 16$. Since $dx/dt = 2$,

$$\frac{dy}{dt} = -\frac{12(2)}{16} = -1.5$$

and the ladder is sliding down the wall at the rate of 1.5 ft/sec.

EXERCISES 11.7, page 768

1. $f(x) = 2x^2$ and $dy = 4x\,dx$.

3. $f(x) = x^3 - x$ and $dy = (3x^2 - 1)\,dx$.

5. $f(x) = \sqrt{x+1} = (x+1)^{1/2}$ and $dy = \frac{1}{2}(x+1)^{-1/2}\,dx = \frac{dx}{2\sqrt{x+1}}$.

7. $f(x) = 2x^{3/2} + x^{1/2}$ and

$$dy = (3x^{1/2} + \tfrac{1}{2}x^{-1/2})\,dx = \tfrac{1}{2}x^{-1/2}(6x+1)dx = \frac{6x+1}{2\sqrt{x}}dx.$$

9. $f(x) = x + \frac{2}{x}$ and $dy = \left(1 - \frac{2}{x^2}\right)dx = \frac{(x^2-2)}{x^2}dx$.

11. $f(x) = \frac{x-1}{x^2+1}$ and $dy = \frac{x^2+1-(x-1)2x}{(x^2+1)^2}dx = \frac{(-x^2+2x+1)}{(x^2+1)^2}dx$.

13. $f(x) = \sqrt{3x^2 - x} = (3x^2 - x)^{1/2}$ and

$$dy = \frac{1}{2}(3x^2 - x)^{-1/2}(6x-1)dx = \frac{6x-1}{2\sqrt{3x^2-x}}dx.$$

15. $f(x) = x^2 - 1$.

a. $dy = 2x\,dx$.

b. $dy \approx 2(1)(0.02) = 0.04$.

c. $\Delta y = [(1.02)^2 - 1] - [1 - 1] = 0.0404$.

17. $f(x) = \frac{1}{x}$.

a. $dy = -\frac{dx}{x^2}$.

b. $dy \approx -0.05$

c. $\Delta y = \dfrac{1}{-0.95} - \dfrac{1}{-1} = -0.05263.$

19. $y = \sqrt{x}$ and $dy = \dfrac{dx}{2\sqrt{x}}$. Therefore, $\sqrt{10} = 3 + \dfrac{1}{2 \cdot \sqrt{9}} = 3.167.$

21. $y = \sqrt{x}$ and $dy = \dfrac{dx}{2\sqrt{x}}$. Therefore, $\sqrt{49.5} = 7 + \dfrac{0.5}{2 \cdot 7} = 7.0358.$

23. $y = x^{1/3}$ and $dy = \frac{1}{3}x^{-2/3}\,dx$. Therefore, $\sqrt[3]{7.8} = 2 - \dfrac{0.2}{3 \cdot 4} = 1.983.$

25. $y = \sqrt{x}$ and $dy = \dfrac{dx}{2\sqrt{x}}$. Therefore, $\sqrt{0.089} = \frac{1}{10}\sqrt{8.9} = \frac{1}{10}\left[3 - \dfrac{0.1}{2 \cdot 3}\right] \approx 0.298.$

27. $y = f(x) = \sqrt{x} + \dfrac{1}{\sqrt{x}} = x^{1/2} + x^{-1/2}$. Therefore,

$$\frac{dy}{dx} = \frac{1}{2}x^{-1/2} - \frac{1}{2}x^{-3/2}$$

$$dy = \left(\frac{1}{2x^{1/2}} - \frac{1}{2x^{3/2}}\right)dx.$$

Letting $x = 4$ and $dx = 0.02$, we find

$$\sqrt{4.02} + \frac{1}{\sqrt{4.02}} - f(4) = f(4.02) - f(4) = \Delta y \approx dy$$

$$\sqrt{4.02} + \frac{1}{\sqrt{4.02}} = f(4) + dy\Big|_{\substack{x=4 \\ dx=0.02}}$$

$$\approx 2 + \frac{1}{2} + \left(\frac{1}{2 \cdot 2} - \frac{1}{2 \cdot 2\sqrt{2}}\right)(0.02)$$

$$\approx 2.50146.$$

29. The volume of the cube is given by $V = x^3$. Then $dV = 3x^2\,dx$ and when $x = 12$ and $dx = 0.02$, $dV = 3(144)\pm(0.02) = \pm\, 8.64$, and the possible error that might occur in calculating the volume is $\pm\, 8.64$ cm^3.

31. The volume of the tumor is given by $V = \frac{4}{3}\pi r^3$. Then $dV = 4\pi r^2\,dr$. When $r = 1.1$ and $dr = 0.005$, $dV = 4\pi(1.1)^2(\pm 0.005) = \pm 0.076$ cm^3.

33. $f(x) = 640x^{1/5}$ and $df = 128x^{-4/5}\,dx$. When $x = 243$ and $dx = 5$, we have

$$df = 128\left(\tfrac{5}{81}\right) \approx 7.9,$$

or approximately 7.9 billion dollars.

35. $P(x) = -\frac{1}{8}x^2 + 7x + 30$ and $dP = (-\frac{1}{4}x + 7)dx$. To estimate the increase in profits when the amount spent on advertising each quarter is increased from \$24,000 to \$26,000, we compute

$$dP = (-\tfrac{24}{4} + 7)(2) = 2, \qquad (dx = 2,\ x = 24)$$

or \$2000 dollars.

37. $s(x) = 0.3\sqrt{x} + 10$ and $s' = \dfrac{0.15}{\sqrt{x}}dx$. To estimate the change in price when the quantity supplied is increased from 10,000 units to 10,500 units, we compute

$$ds = \frac{(0.15)500}{100} = 0.75, \text{or 75 cents.}$$

39. $S = kW^{2/3}$ and $dS = \dfrac{0.2}{3W^{1/3}}dW$. To determine the percentage error in the calculation of the surface area of a horse that weighs 300 kg when the maximum error in measurement is 0.6 kg and $k = 0.1$, we compute

$$\frac{dS}{S} = \frac{0.2}{3W^{1/3}}dW \bullet \frac{1}{0.1W^{2/3}} = \frac{2}{3W}dW = \frac{2(0.6)}{3(300)} = 0.00133,$$

or 0.133 percent.

41. $p = \dfrac{55}{2x^2 + 1}$ and $dp = -\dfrac{220x}{(2x^2+1)^2}dx$. To find the error corresponding to a possible error of 15 percent in a forecast of 1.8 billion bushels, we compute

$$dp = -\frac{(220)(1.8) \pm (0.27)}{[2(1.8)^2 + 1]^2} \approx 1.91,$$

or approximately $\pm$ 1.91 percent.

43. a. $P = \dfrac{10{,}000r}{1 - \left(1 + \dfrac{r}{12}\right)^{-360}}$ and

$$dP = \frac{\left[1-\left(1+\frac{r}{12}\right)^{-360}\right]10{,}000 - 10{,}000r(360)\left(1+\frac{r}{12}\right)^{-361}\left(\frac{1}{12}\right)}{\left[1-\left(1+\frac{r}{12}\right)^{-360}\right]^2}dr$$

$$= \frac{10{,}000\left\{\left[1-\left(1+\frac{r}{12}\right)^{-360} - 30r\left(1+\frac{r}{12}\right)^{-361}\right]\right\}}{\left[1-\left(1+\frac{r}{12}\right)^{-360}\right]^2}dr$$

b. When $x = 0.09$, $dP = \dfrac{10{,}000(0.932113992 - 0.181927761)}{(0.932113992)^2}dr \approx 8634.378\,dr$.

When the interest rate increases from 9% per year to 9.2% per year

$dP = 8634.378(0.002) \approx 17.27$, or approximately \$17.27.

When the interest rate increases from 9% per year to 9.3% per year

$dP = 8634.378(0.003) \approx 25.90$, or approximately \$25.907.

When the interest rate increases from 9% per year to 9.4% per year

$dP = 8634.378(0.004) \approx 34.54$, or approximately \$34.54.

When the interest rate increases from 9% per year to 9.5% per year

$dP = 8634.378(0.005) \approx 43.17$, or approximately \$43.17.

USING TECHNOLOGY EXERCISES 11.7, page 771

1. $dy = f'(3)\,dx = 757.87(0.01) \approx 7.5787.$

3. $dy = f'(1)\,dx = 1.04067285926(0.03) \approx 0.031220185778.$

5. $dy = f'(4)(0.1) = -0.198761598(0.1) = -0.0198761598.$

7. If the interest rate changes from 10% to 10.3% per year, the monthly payment will increase by

   $dP = f'(0.1)(0.003) \approx 26.60279,$

   or approximately \$26.60 per month. If the interest rate changes from 10% to 10.4% per year, it will be \$35.47 per month. If the interest rate changes from 10% to 10.5% per year, it will be \$44.34 per month.

9. $dx = f'(40)(2) \approx -0.625$. That is, the quantity demanded will decrease by 625 watches per week.

CHAPTER 11 REVIEW, page 774

1. $f'(x) = \frac{d}{dx}(3x^5 - 2x^4 + 3x^2 - 2x + 1) = 15x^4 - 8x^3 + 6x - 2.$

3. $g'(x) = \frac{d}{dx}(-2x^{-3} + 3x^{-1} + 2) = 6x^{-4} - 3x^{-2}.$

5. $g'(t) = \frac{d}{dt}(2t^{-1/2} + 4t^{-3/2} + 2) = -t^{-3/2} - 6t^{-5/2}.$

7. $f'(t) = \frac{d}{dt}(t + 2t^{-1} + 3t^{-2}) = 1 - 2t^{-2} - 6t^{-3} = 1 - \frac{2}{t^2} - \frac{6}{t^3}.$

9. $h'(x) = \frac{d}{dx}(x^2 - 2x^{-3/2}) = 2x + 3x^{-5/2} = 2x + \frac{3}{x^{5/2}}.$

11. $g(t) = \frac{t^2}{2t^2 + 1}.$

$$g'(t) = \frac{(2t^2+1)\frac{d}{dt}(t^2) - t^2\frac{d}{dt}(2t^2+1)}{(2t^2+1)^2}$$

$$= \frac{(2t^2+1)(2t) - t^2(4t)}{(2t^2+1)^2} = \frac{2t}{(2t^2+1)^2}.$$

13. $f(x) = \frac{\sqrt{x} - 1}{\sqrt{x} + 1} = \frac{x^{1/2} - 1}{x^{1/2} + 1}.$

$$f'(x) = \frac{(x^{1/2}+1)(\frac{1}{2}x^{-1/2}) - (x^{1/2}-1)(\frac{1}{2}x^{-1/2})}{(x^{1/2}+1)^2}$$

$$= \frac{\frac{1}{2} + \frac{1}{2}x^{-1/2} - \frac{1}{2} + \frac{1}{2}x^{-1/2}}{(x^{1/2}+1)^2} = \frac{x^{-1/2}}{(x^{1/2}+1)^2} = \frac{1}{\sqrt{x}(\sqrt{x}+1)^2}.$$

15. $f(x) = \frac{x^2(x^2+1)}{x^2 - 1}.$

$$f'(x) = \frac{(x^2-1)\frac{d}{dx}(x^4+x^2)-(x^4+x^2)\frac{d}{dx}(x^2-1)}{(x^2-1)^2}$$

$$= \frac{(x^2-1)(4x^3+2x)-(x^4+x^2)(2x)}{(x^2-1)^2}$$

$$= \frac{4x^5+2x^3-4x^3-2x-2x^5-2x^3}{(x^2-1)^2}$$

$$= \frac{2x^5-4x^3-2x}{(x^2-1)^2} = \frac{2x(x^4-2x^2-1)}{(x^2-1)^2}.$$

17. $f(x) = (3x^3-2)^8; f'(x) = 8(3x^3-2)^7(9x^2) = 72x^2(3x^3-2)^7.$

19. $f'(t) = \frac{d}{dt}(2t^2+1)^{1/2} = \frac{1}{2}(2t^2+1)^{-1/2}\frac{d}{dt}(2t^2+1)$

$$= \frac{1}{2}(2t^2+1)^{-1/2}(4t) = \frac{2t}{\sqrt{2t^2+1}}.$$

21. $s(t) = (3t^2-2t+5)^{-2}$

$$s'(t) = -2(3t^2-2t+5)^{-3}(6t-2) = -4(3t^2-2t+5)^{-3}(3t-1)$$

$$= -\frac{4(3t-1)}{(3t^2-2t+5)^3}.$$

23. $h(x) = \left(x+\frac{1}{x}\right)^2 = (x+x^{-1})^2.$

$$h'(x) = 2(x+x^{-1})(1-x^{-2}) = 2\left(x+\frac{1}{x}\right)\left(1-\frac{1}{x^2}\right)$$

$$= 2\left(\frac{x^2+1}{x}\right)\left(\frac{x^2-1}{x^2}\right) = \frac{2(x^2+1)(x^2-1)}{x^3}.$$

25. $h'(t) = (t^2+t)\frac{d}{dt}(2t^2)+2t^2\frac{d}{dt}(t^2+t)^4$

$$= (t^2+t)^4(4t)+2t^2\cdot 4(t^2+t)^3(2t+1)$$

$$= 4t(t^2+t)^3[(t^2+t)+4t^2+2t] = 4t^2(5t+3)(t^2+t)^3.$$

27. $g(x) = x^{1/2}(x^2-1)^3.$

$$g'(x) = \frac{d}{dx}[x^{1/2}(x^2-1)^3] = x^{1/2} \cdot 3(x^2-1)^2(2x) + (x^2-1)^3 \cdot \tfrac{1}{2}x^{-1/2}$$

$$= \tfrac{1}{2}x^{-1/2}(x^2-1)^2[12x^2 + (x^2-1)]$$

$$= \frac{(13x^2-1)(x^2-1)^2}{2\sqrt{x}}.$$

29. $h(x) = \dfrac{(3x+2)^{1/2}}{4x-3}.$

$$h'(x) = \frac{(4x-3)\frac{1}{2}(3x+2)^{-1/2}(3) - (3x+2)^{1/2}(4)}{(4x-3)^2}$$

$$= \frac{\frac{1}{2}(3x+2)^{-1/2}[3(4x-3) - 8(3x+2)]}{(4x-3)^2} = -\frac{12x+25}{2\sqrt{3x+2}(4x-3)^2}.$$

31. $f(x) = 2x^4 - 3x^3 + 2x^2 + x + 4.$

$$f'(x) = \frac{d}{dx}(2x^4 - 3x^3 + 2x^2 + x + 4) = 8x^3 - 9x^2 + 4x + 1.$$

$$f''(x) = \frac{d}{dx}(8x^3 - 9x^2 + 4x + 1) = 24x^2 - 18x + 4 = 2(12x^2 - 9x + 2).$$

33. $h(t) = \dfrac{t}{t^2+4}.\ h'(t) = \dfrac{(t^2+4)(1) - 4(2t)}{(t^2+4)^2} = \dfrac{4-t^2}{(t^2+4)^2}.$

$$h''(t) = \frac{(t^2+4)^2(-2t) - (4-t^2)2(t^2+4)(2t)}{(t^2+4)^4}$$

$$= \frac{-2t(t^2+4)[(t^2+4) + 2(4-t^2)]}{(t^2+4)^4} = \frac{2t(t^2-12)}{(t^2+4)^3}.$$

35. $f'(x) = \dfrac{d}{dx}(2x^2+1)^{1/2} = \dfrac{1}{2}(2x^2+1)^{-1/2}(4x) = 2x(2x^2+1)^{-1/2}.$

$$f''(x) = 2(2x^2+1)^{1/2} + 2x \cdot (-\tfrac{1}{2})(2x^2+1)^{-3/2}(4x)$$

$$= 2(2x^2+1)^{-3/2}[(2x^2+1) - 2x^2] = \frac{2}{(2x^2+1)^{3/2}}.$$

37. $6x^2 - 3y^2 = 9$ so $12x - 6y\dfrac{dy}{dx} = 0$ and $-6y\dfrac{dy}{dx} = -12x.$

Therefore, $\dfrac{dy}{dx} = \dfrac{-12y}{-6y} = \dfrac{2x}{y}$.

39. $y^3 + 3x^2 = 3y$, so $3y^2y' + 6x = 3y'$, $3y^2y' - 3y' = -6x$,

and $y'(3y^2 - 3) = -6x$. Therefore, $y' = -\dfrac{6x}{3(y^2-1)} = -\dfrac{2x}{y^2-1}$.

41. $x^2 - 4xy - y^2 = 12$ so $2x - 4xy' - 4y - 2yy' = 0$ and $y'(-4x - 2y) = -2x + 4y$.

So $y' = \dfrac{-2(x-2y)}{-2(2x+y)} = \dfrac{x-2y}{2x+y}$.

43. $f(x) = 2x^3 - 3x^2 - 16x + 3$ and $f'(x) = 6x^2 - 6x - 16$.

a. To find the point(s) on the graph of f where the slope of the tangent line is equal to -4, we solve

$$6x^2 - 6x - 16 = -4,\ 6x^2 - 6x - 12 = 0,\ 6(x^2 - x - 2) = 0$$
$$6(x-2)(x+1) = 0$$

and $x = 2$ or $x = -1$. Then

$$f(2) = 2(2)^3 - 3(2)^2 - 16(2) + 3 = -25$$

and $f(-1) = 2(-1)^3 - 3(-1)^2 - 16(-1) + 3 = 14$

and the points are $(2,-25)$ and $(-1,14)$.

b. Using the point-slope form of the equation of a line, we find that

$$y - (-25) = -4(x - 2),\ y + 25 = -4x + 8, \text{ or } y = -4x - 17$$

and $\quad y - 14 = -4(x + 1)$, or $y = -4x + 10$

are the equations of the tangent lines at $(2,-25)$ and $(-1,14)$.

45. $y = (4 - x^2)^{1/2}$. $y' = \frac{1}{2}(4 - x^2)^{-1/2}(-2x) = -\dfrac{x}{\sqrt{4-x^2}}$.

The slope of the tangent line is obtained by letting $x = 1$, giving

$$m = -\frac{1}{\sqrt{3}} = -\frac{\sqrt{3}}{3}.$$

Therefore, an equation of the tangent line is

$$y - \sqrt{3} = -\frac{\sqrt{3}}{3}(x - 1), \quad \text{or} \quad y = -\frac{\sqrt{3}}{3}x + \frac{4\sqrt{3}}{3}.$$

47. $f(x) = (2x-1)^{-1}$; $f'(x) = -2(2x-1)^{-2}$, $f''(x) = 8(2x-1)^{-3} = \dfrac{8}{(2x-1)^3}$.

$$f'''(x) = -48(2x-1)^4 = -\frac{48}{(2x-1)^4}.$$

Since $(2x-1)^4 = 0$ when $x = 1/2$, we see that the domain of f''' is $(-\infty, \frac{1}{2}) \cup (\frac{1}{2}, \infty)$.

49. $C(x) = 2500 + 2.2x$.

a. The marginal cost is $C'(x) = 2.2$. The marginal cost when $x = 1000$ is $C'(1000) = 2.2$. The marginal cost when $x = 2000$ is $C'(2000) = 2.2$.

b. $$\overline{C}'(x) = \frac{C(x)}{x} = \frac{2500+2.2x}{x} = 2.2 + \frac{2500}{x}.$$

$$\overline{C}'(x) = -\frac{2500}{x^2}.$$

c. $$\lim_{x\to\infty} \overline{C}(x) = \lim_{x\to\infty}\left(2.2 + \frac{2500}{x}\right) = 2.2.$$

51. a. $R(x) = px = (2000 - 0.04x)x = 2000x - 0.04x^2$.

$$\begin{aligned} P(x) &= R(x) - C(x) \\ &= (2000x - 0.04x^2) - (0.000002x^3 - 0.02x^2 + 1000x + 120{,}000) \\ &= -0.000002x^3 - 0.02x^2 + 1000x - 120{,}000. \end{aligned}$$

$$\begin{aligned} \overline{C}(x) &= \frac{C(x)}{x} = \frac{0.000002x^3 - 0.02x^2 + 1000x + 120{,}000}{x} \\ &= 0.000002x^2 - 0.02x + 1000 + \frac{120{,}000}{x} \end{aligned}$$

b. $$\begin{aligned} C'(x) &= \frac{d}{dx}(0.000002x^3 - 0.02x^2 + 1000x + 120{,}000) \\ &= 0.000006x^2 - 0.04x + 1000. \end{aligned}$$

$$R'(x) = \frac{d}{dx}(2000x - 0.04x^2) = 2000 - 0.08x.$$

$$\begin{aligned} P'(x) &= \frac{d}{dx}(-0.000002x^3 - 0.02x^2 + 1000x - 120{,}000) \\ &= -0.000006x^2 - 0.04x + 1000 \end{aligned}$$

$$\begin{aligned} \overline{C}'(x) &= \frac{d}{dx}(0.000002x^2 - 0.02x + 1000 + 120{,}000x^{-1}) \\ &= 0.000004x - 0.02 - 120{,}000x^{-2}. \end{aligned}$$

c. $C'(3000) = 0.000006(3000)^2 - 0.04(3000) + 1000 = 934$

$R'(3000) = 2000 - 0.08(3000) = 1760.$

$P'(3000) = -0.000006(3000)^2 - 0.04(3000) + 1000 = 826.$

d. $\overline{C}'(5000) = 0.000004(5000) - 0.02 - 120{,}000(5000)^{-2} = -0.0048$

$\overline{C}'(8000) = 0.000004(8000) - 0.02 - 120{,}000(8000)^{-2} = 0.0101.$

At a level of production of 5000 machines, the average cost of each additional unit is decreasing at a rate of 0.48 cents. At a level of production of 8000 machines, the average cost of each additional unit is increasing at a rate of 1 cent per unit.

CHAPTER 12

EXERCISES 12.1, page 791

1. f is decreasing on $(-\infty,0)$ and increasing on $(0, \infty)$.

3. f is increasing on $(-\infty,-1) \cup (1,\infty)$, and decreasing on $(-1,1)$.

5. f is increasing on $(0,2)$ and decreasing on $(-\infty,0) \cup (2,\infty)$.

7. f is decreasing on $(-\infty,-1) \cup (1,\infty)$ and increasing on $(-1,1)$.

9. Increasing on (20.2), 20.6) $\cup$ (21.7,21.8), constant on (19.6,20.2) $\cup$ (20.6,21.1), and decreasing on (21.1,21.7) $\cup$ (21.8,22.7),

11. $f(x) = 3x + 5; f'(x) = 3 > 0$ for all x and so f is increasing on $(-\infty,\infty)$.

13. $f(x) = x^2 - 3x$.
    $f'(x) = 2x - 3$ is continuous everywhere and is equal to zero when $x = 3/2$. From the following sign diagram

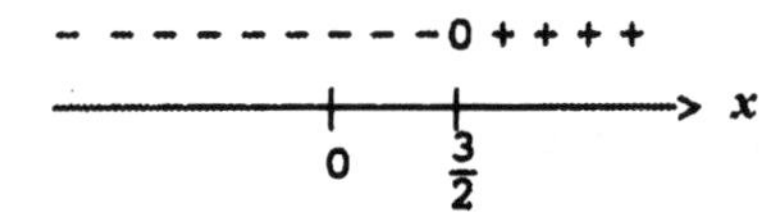

we see that f is decreasing on $(-\infty, \frac{3}{2})$ and increasing on $(\frac{3}{2}, \infty)$.

15. $g(x) = x - x^3$. $g'(x) = 1 - 3x^2$ is continuous everywhere and is equal to zero when $1 - 3x^2 = 0$, or $x = \pm\frac{\sqrt{3}}{3}$. From the following sign diagram

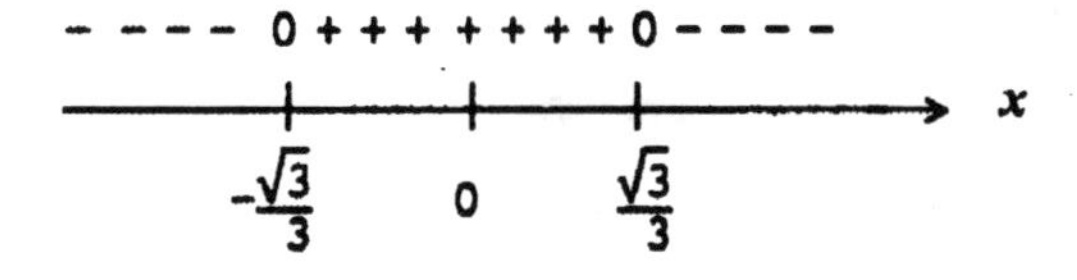

we see that f is decreasing on $(-\infty, -\frac{\sqrt{3}}{3}) \cup (\frac{\sqrt{3}}{3}, \infty)$ and increasing on $(-\frac{\sqrt{3}}{3}, \frac{\sqrt{3}}{3})$.

17. $g(x) = x^3 + 3x^2 + 1$; $g'(x) = 3x^2 + 6x = 3x(x + 2)$.
From the following sign diagram

+ + + + + + 0 - - 0 + + + + + x −2   0

we see that g is increasing on $(-\infty,-2) \cup (0,\infty)$ and decreasing on $(-2,0)$.

19. $f(x) = \frac{1}{3}x^3 - 3x^2 + 9x + 20$; $f'(x) = x^2 - 6x + 9 = (x - 3)^2 > 0$ for all x except $x = 3$, at which point $f'(3) = 0$. Therefore, f is increasing on $(-\infty,3) \cup (3,\infty)$.

21. $h(x) = x^4 - 4x^3 + 10$; $h'(x) = 4x^3 - 12x^2 = 4x^2(x - 3)$ if $x = 0$ or 3. From the sign

diagram of h',

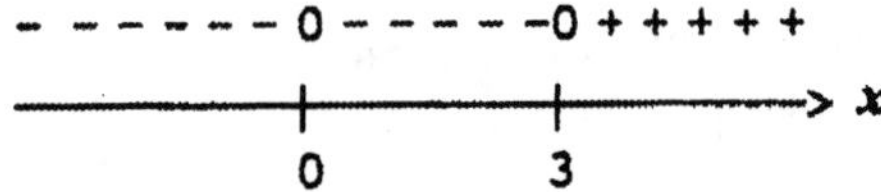

we see that h is increasing on $(3,\infty)$ and decreasing on $(-\infty,0) \cup (0,3)$.

23. $f(x) = \dfrac{1}{x-2} = (x-2)^{-1}$. $f'(x) = -1(x-2)^{-2}(1) = -\dfrac{1}{(x-2)^2}$ is discontinuous at $x = 2$ and is continuous everywhere else. From the sign diagram

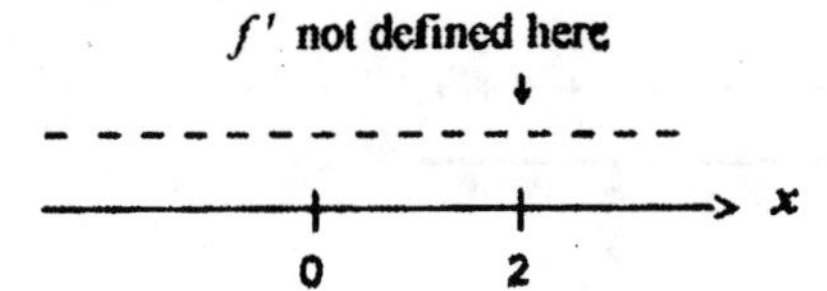

we see that f is decreasing on $(-\infty,2) \cup (2,\infty)$.

25. $h(t) = \dfrac{t}{t-1}$. $h'(t) = \dfrac{(t-1)(1) - t(1)}{(t-1)^2} = -\dfrac{1}{(t-1)^2}$.
From the following sign diagram,

h' not defined here

- - - - - - - - - - - - - - - - t 0   1

we see that $h'(t) < 0$ whenever it is defined. We conclude that h is decreasing on $(-\infty,1) \cup (1,\infty)$.

27. $f(x) = x^{3/5}$. $f'(x) = \frac{3}{5}x^{-2/5} = \frac{3}{5x^{2/5}}$. Observe that $f'(x)$ is not defined at $x = 0$, but is positive everywhere else and therefore increasing on $(-\infty,0) \cup (0,\infty)$.

29. $f(x) = \sqrt{x+1}$. $f'(x) = \frac{d}{dx}(x+1)^{1/2} = \frac{1}{2}(x+1)^{-1/2} = \frac{1}{2\sqrt{x+1}}$ and we see that $f'(x) > 0$ if $x > -1$. Therefore, f is increasing on $(-1, \infty)$.

31. $f(x) = \sqrt{16-x^2} = (16-x^2)^{1/2}$. $f'(x) = \frac{1}{2}(16-x^2)^{-1/2}(-2x) = -\frac{x}{\sqrt{16-x^2}}$.

Since the domain of f is $[-4,4]$, we consider the sign diagram for f' on this interval. Thus,

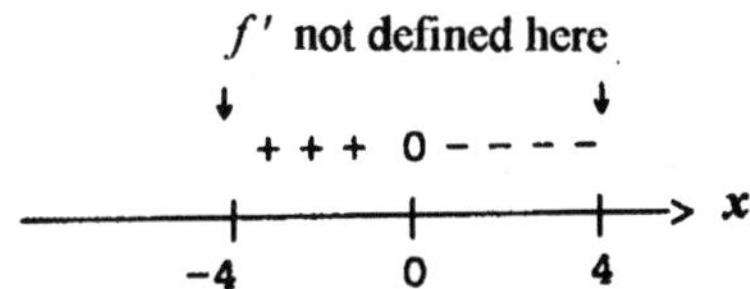

and we see that f is increasing on $(-4,0)$ and decreasing on $(0,4)$.

33. $f'(x) = \frac{d}{dx}(x - x^{-1}) = 1 + \frac{1}{x^2} = \frac{x^2+1}{x^2}$ and so $f'(x) > 0$ for all $x \neq 0$. Therefore, f is increasing on $(-\infty,0) \cup (0,\infty)$.

35. $f'(x) = \frac{d}{dx}(x-1)^{-2} = -2(x-1)^{-3} = -\frac{2}{(x-1)^3}$. From the sign diagram of f'

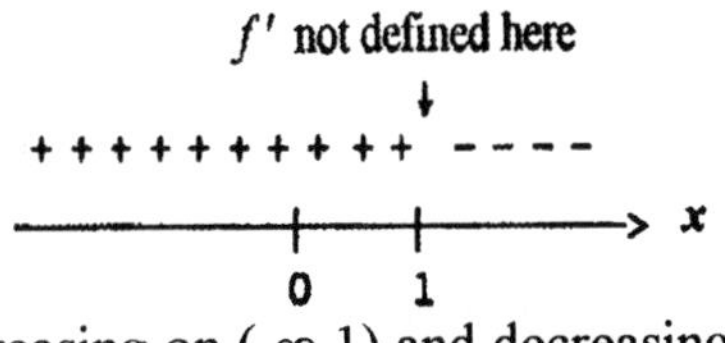

we see that f is increasing on $(-\infty,1)$ and decreasing on $(1,\infty)$.

37. f has a relative maximum of $f(0) = 1$ and relative minima of $f(-1) = 0$ and $f(1) = 0$.

39. f has a relative maximum of $f(-1) = 2$ and a relative minimum of $f(1) = -2$.

41. f has a relative maximum of $f(1) = 3$ and a relative minimum of $f(2) = 2$.

43. f has a relative minimum at $(0,2)$.

45. a

47. d

49. $f(x) = x^2 - 4x$. $f'(x) = 2x - 4 = 2(x - 2)$ has a critical point at $x = 2$. From the following sign diagram

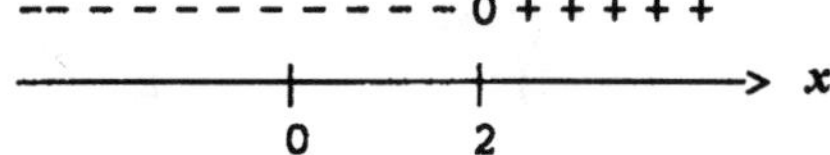

we see that $f(2) = -4$ is a relative minimum by the First Derivative Test.

51. $f(x) = \frac{1}{2}x^2 - 2x + 4$. $f'(x) = x - 2$ giving the critical point $x = 2$. The sign diagram for f' is

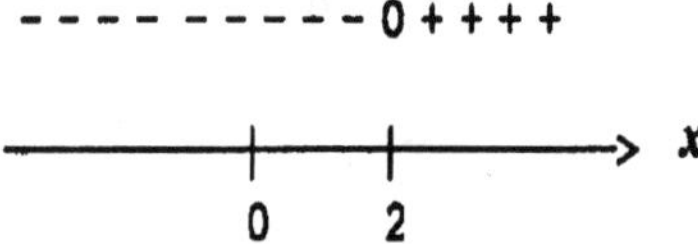

and we see that $f(2) = 2$ is a relative minimum.

53. $f(x) = x^{2/3} + 2$. $f'(x) = \dfrac{2}{3}x^{-1/3} = \dfrac{2}{3x^{1/3}}$ and is discontinuous at $x = 0$, a critical point. From the sign diagram

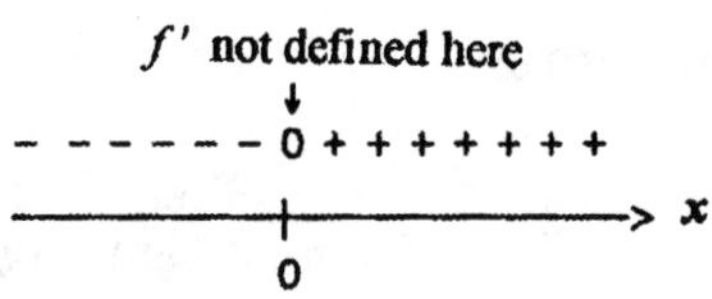

and the First Derivative Test we see that f has a relative minimum at (0,2).

55. $g(x) = x^3 - 3x^2 + 4$. $g'(x) = 3x^2 - 6x = 3x(x - 2) = 0$ if $x = 0$ or 2. From the sign

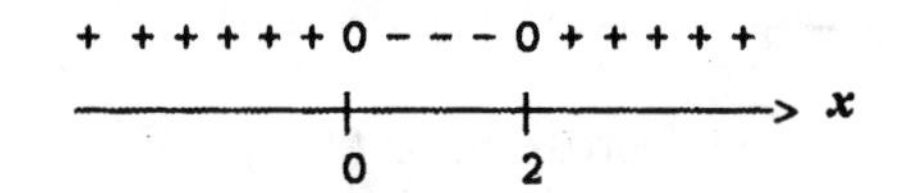

diagram, we see that the critical point $x = 0$ gives a relative maximum, whereas, $x = 2$ gives a relative minimum. The values are $g(0) = 4$ and $g(2) = 8 - 12 + 4 = 0$.

57. $F(x) = \frac{1}{3}x^3 - x^2 - 3x + 4$. Setting $F'(x) = x^2 - 2x - 3 = (x - 3)(x + 1) = 0$ gives $x = -1$ and $x = 3$ as critical points. From the sign diagram

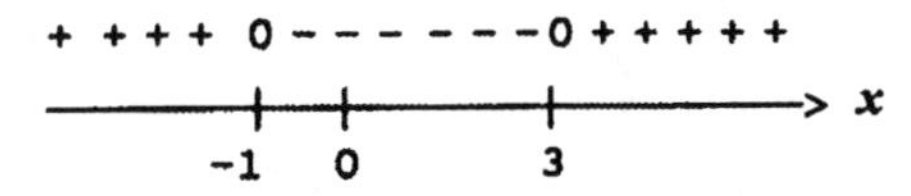

we see that $x = -1$ gives a relative maximum and $x = 3$ gives a relative minimum. The values are

$$F(-1) = -\tfrac{1}{3} - 1 + 3 + 4 = \tfrac{17}{3} \quad \text{and} \quad F(3) = 9 - 9 - 9 + 4 = -5,$$

respectively.

59. $g(x) = x^4 - 4x^3 + 8$. Setting $g'(x) = 4x^3 - 12x^2 = 4x^2(x - 3) = 0$ gives $x = 0$ and $x = 3$ as critical points. From the sign diagram

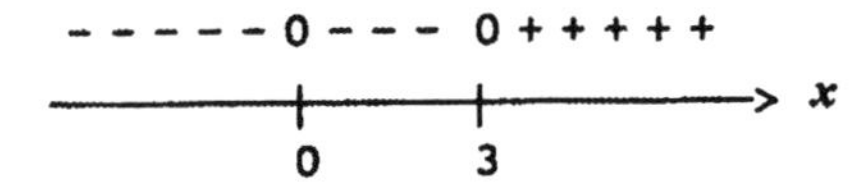

we see that $x = 3$ gives a relative minimum. Its value is $g(3) = 3^4 - 4(3)^3 + 8 = -19$.

61. $f(x) = 3x^4 - 2x^3 + 4$; $f'(x) = 12x^3 - 6x^2 = 6x^2(2x - 1) = 0$ if $x = 0$ or $1/2$. The sign diagram of f' is shown below.

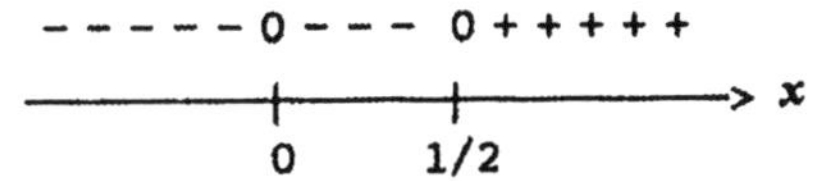

and shows that f has a relative minimum at $(\frac{1}{2}, \frac{63}{16})$.

63. $g'(x) = \dfrac{d}{dx}\left(1 + \dfrac{1}{x}\right) = -\dfrac{1}{x^2}$. Observe that g' is never zero for all values of x. Furthermore, g' is undefined at $x = 0$, but $x = 0$ is not in the domain of g. Therefore g has no critical points and so g has no relative extrema.

65. $f(x) = x + \dfrac{9}{x} + 2$. Setting $f'(x) = 1 - \dfrac{9}{x^2} = \dfrac{x^2 - 9}{x^2} = \dfrac{(x+3)(x-3)}{x^2} = 0$ gives $x = -3$ and $x = 3$ as critical points. From the sign diagram

f' not defined here

+ + + + 0 - - - 0 - - - 0 + + + + +

-3 0 3 x

we see that (-3,-4) is a relative maximum and (3,8) is a relative minimum.

67. $f(x)=\dfrac{x}{1+x^2}$. $f'(x)=\dfrac{(1+x^2)(1)-x(2x)}{(1+x^2)^2}=\dfrac{1-x^2}{(1+x^2)^2}=\dfrac{(1-x)(1+x)}{(1+x^2)^2}=0$ if $x=\pm 1$, and these are critical points of f. From the sign diagram of f'

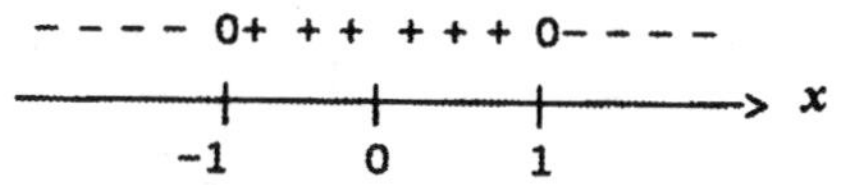

we see that f has a relative minimum at $(-1,-\frac{1}{2})$ and a relative maximum at $(1,\frac{1}{2})$.

69. $f(x)=\dfrac{x^2}{x^2-4}$. $f'(x)=\dfrac{(x^2-4)(2x)-x^2(2x)}{(x^2-4)^2}=-\dfrac{8x}{(x^2-4)^2}$ is continuous everywhere except at $x\pm 2$ and has a zero at $x=0$. Therefore, $x=0$ is the only critical point of f (the points $x=\pm 2$ do not lie in the domain of f). Using the following sign diagram of f'

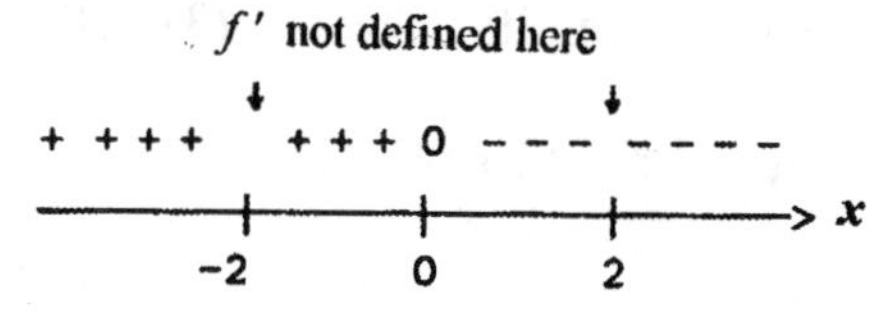

and the First Derivative Test, we conclude that $f(0)=0$ is a relative maximum of f.

71. $f(x)=(x-1)^{2/3}$. $f'(x)=\dfrac{2}{3}(x-1)^{-1/3}=\dfrac{2}{3(x-1)^{1/3}}$.
$f'(x)$ is discontinuous at $x=1$. The sign diagram for f' is

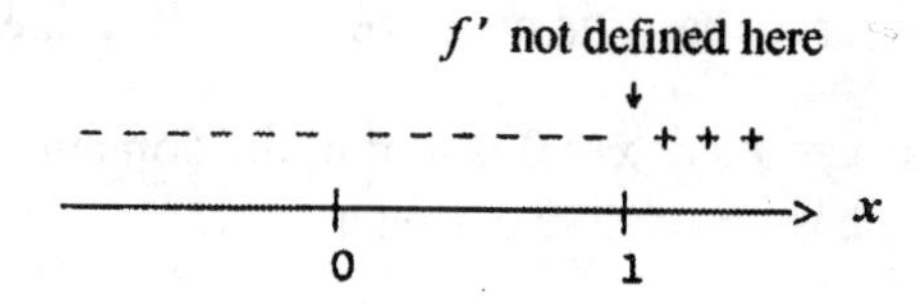

We conclude that $f(1)=0$ is a relative minimum.

73. $P(x)=-0.001x^2+8x-5000$. $P'(x)=-0.002x+8=0$ if $x=4000$. Observe that $P'(x)>0$ if $x<4000$ and $P'(x)<0$ if $x>4000$. So P is increasing on $(0,4000)$ and decreasing on $(4000,\infty)$.

75. $I(t) = \frac{1}{3}t^3 - \frac{5}{2}t^2 + 80$; $I'(t) = t^2 - 5t = t(t-5) = 0$ if $t = 0$ or 5. From the sign diagram,

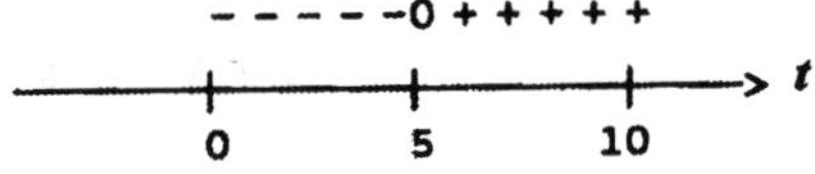

we see that I is decreasing on (0,5) and increasing on (5,10). After declining from 1984 through 1989, the index begins to increase after 1989.

77. $\overline{C}(x) = -0.0001x + 2 + \dfrac{2000}{x}$. $\overline{C}'(x) = -0.0001 - \dfrac{2000}{x^2} < 0$ for all values of x and so $\overline{C}$ is always decreasing.

79. $A(t) = -96.6t^4 + 403.6t^3 + 660.9t^2 + 250$

$A'(t) = -386.4t^3 + 1210.8t^2 + 1321.8t = t(386.4t^2 + 1210.8t + 1321.8)$.
Solving $A'(t) = 0$, we find $t = 0$ and

$$t = \frac{-1210.8 \pm \sqrt{(1210.8)^2 - 4(-386.4)(1321.8)}}{-2(386.4)} = \frac{-1210.8 \pm 1873.2}{-2(386.4)} \approx 4.$$

Since t lies in the interval [0,5], we see that the continuous function A' has zeros at $t = 0$ and $t = 4$. From the sign diagram

+ + + + 0 - -

0 4 5 → t

we see that f is increasing on (0,4) and decreasing on (4,5). We conclude that the cash in the Central Provident Trust Funds will be increasing from 1995 to 2015 and decreasing from 2015 to 2025.

81. $C(t) = \dfrac{t^2}{2t^3 + 1}$; $C'(t) = \dfrac{(2t^3+1)(2t) - t^2(6t^2)}{(2t^3+1)^2} = \dfrac{2t - 2t^4}{(2t^3+1)^2} = \dfrac{2t(1-t^3)}{(2t^3+1)^2}$.

From the sign diagram of C' on $(0,\infty)$,

+ + 0 - - - - - - -

0 1 4 → t

We see that the drug concentration is increasing on (0,1) and decreasing on (1,4).

83. $G(t) = N(t) - C(t) = (3.5t^2 + 26.7t + 436.2) - (24.3t + 365)$
$= 3.5t^2 + 2.4t + 71.2.$

$G'(t) = 7t + 2.4 > 0$ on $(0,10)$. This says that the gap is always increasing on the time interval $(0,10)$.

85. $f(x) = ax^2 + bx + c$; $f'(x) = 2ax + b = 2a\left(x + \frac{b}{2a}\right)$. Then f' is continuous everywhere and has a zero at $x = -\frac{b}{2a}$. The sign diagram of f' is

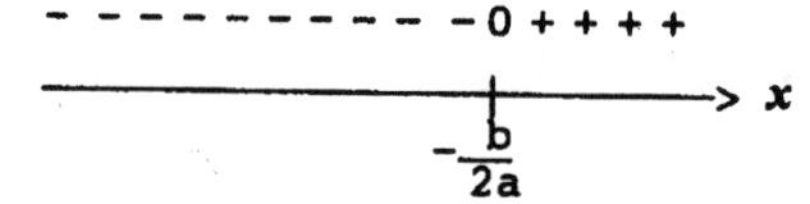

when $a > 0$, or

+ + + + + + + + + + + 0 - - - -

$-\frac{b}{2a}$ x

when $a < 0$. Therefore, if $a > 0$, f is decreasing on $(-\infty, -\frac{b}{2a})$ and increasing on $(-\frac{b}{2a}, \infty)$, and if $a < 0$, f is increasing on $(-\infty, -\frac{b}{2a})$ and decreasing on $(-\frac{b}{2a}, \infty)$.

87. a. $f'(x) = -2x$ if $x \neq 0$. $f'(-1) = 2$ and $f'(1) = -2$ so $f'(x)$ changes sign from positive to negative as we move across $x = 0$.
b. f does not have a relative maximum at $x = 0$ because $f(0) = 2$ but a neighborhood of $x = 0$, for example $(-\frac{1}{2}, \frac{1}{2})$, contains points with values larger than 2. This does not contradict the First Derivative Test because f is not continuous at $x = 0$.

89. $f(x) = ax^3 + bx^2 + cx + d$; $f'(x) = 3ax^2 + 2bx + c$. The roots of $f'(x) = 0$ are

$$x = \frac{-2b \pm \sqrt{4b^2 - 12ac}}{6a}$$

There is no relative extremum if and only if the roots are not real or are repeated. This occurs if and only if

$$4b^2 - 12ac = 4(b^2 - 3ac) \leq 0 \text{ or } b^2 - 3ac \leq 0.$$

91. $$f'(x) = \frac{(cx+d)a - (ax+b)c}{(cx+d)^2} = \frac{acx + ad - acx - bc}{(cx+d)^2} = \frac{ad - bc}{(cx+d)^2}.$$

So, if $ad - bc \neq 0$, then $f'(x) \neq 0$ for all x and f has no critical points. Therefore, f has no relative extremum in this case. If $ad - bc = 0$, then $f'(x) = 0$ for all x and so f is a constant function.

USING TECHNOLOGY EXERCISES 12.1, page 798

1. a. f is decreasing on $(-\infty,-0.2934)$ and increasing on $(-0.2934,\infty)$.
   b. Relative minimum: $f(-0.2934) = -2.5435$

3. a. f is increasing on $(-\infty,-1.6144) \cup (0.2390,\infty)$ and decreasing on $(-1.6144, 0.2390)$
   b. Relative maximum: $f(-1.6144) = 26.7991$; relative minimum: $f(0.2390) = 1.6733$

5. a. f is decreasing on $(-\infty,-1) \cup (0.33,\infty)$ and increasing on $(-1,0.33)$
   b. Relative maximum: $f(0.33) = 1.11$; relative minimum: $f(-1) = -0.63$

7. a. f is decreasing on $(-1,-0.71)$ and increasing on $(-0.71,1)$.
   b. f has a relative minimum at $(-0.71,-1.41)$.

9. a.

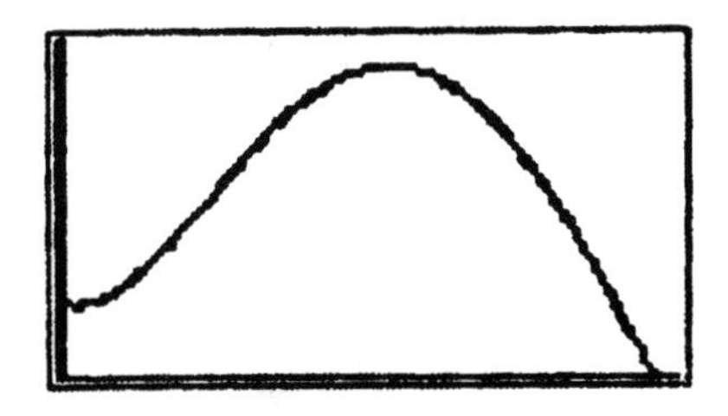

   b. f is decreasing on $(0,0.2398) \cup (6.8758,12)$ and increasing on $(0.2398,6.8758)$
   c. $(6.8758, 200.14)$; The rate at which the number of banks were failing reached a Gpeak of 200/yr during the latter part of 1988 ($t = 6.8758$).

11. a.

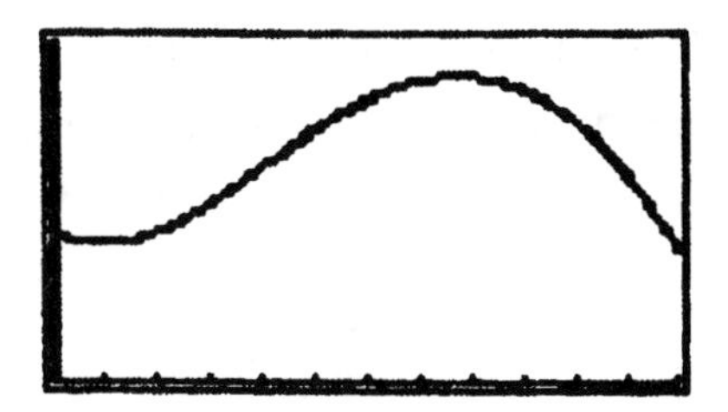

   b. f is decreasing on $(0, 0.8343) \cup (7.6726, 12)$ and increasing on $(0.8343, 7.6726)$.

The rate at which single-family homes in the greater Boston area were selling was decreasing during most of 1984, but started increasing in late 1984 and continued increasing until mid 1991 when it started decreasing again until 1996.

13. The PSI is increasing on the interval (0, 4.5) and decreasing on (4.5, 11). It is highest when $t = 4.5$ (11:30 A.M.) and has value 164.

EXERCISES 12.2, page 811

1. f is concave downward on $(-\infty,0)$ and concave upward on $(0,\infty)$. f has an inflection point at $(0,0)$.

3. f is concave downward on $(-\infty,0) \cup (0,\infty)$.

5. f is concave upward on $(-\infty,0) \cup (1,\infty)$ and concave downward on $(0,1)$. $(0,0)$ and $(1,-1)$ are inflection points of f.

7. f is concave downward on $(-\infty,-2) \cup (-2,2) \cup (2,\infty)$.

9. a

11. b

13. a. $D_1'(t) > 0$, $D_2'(t) > 0$, $D_1''(t) > 0$, and $D_2''(t) < 0$ on $(0,12)$.
    b. With or without the proposed promotional campaign, the deposits will increase, but with the promotion, the deposits will increase at an increasing rate whereas without the promotion, the deposits will increase at a decreasing rate.

15. The significance of the inflection point Q is that at the time t_0 corresponding to its t-coordinate, the restoration process is working at its peak.

17. $g(x) = x^4 + \frac{1}{2}x^2 + 6x + 10$; $g'(x) = 4x^3 + x + 6$ and $g''(x) = 12x^2 + 1$. We see that $g''(x) \geq 1$ for all values of x and so g is concave upward everywhere.

19. $f(x) = \dfrac{1}{x^4} = x^{-4}$; $f'(x) = -\dfrac{4}{x^5}$ and $f''(x) = \dfrac{20}{x^6} > 0$ for all values of x in $(-\infty,0) \cup (0,\infty)$ and so f is concave upward everywhere.

21. $g'(x) = \dfrac{d}{dx}(4-x^2)^{1/2} = -\frac{1}{2}(4-x^2)^{-1/2}(-2x) = x(4-x^2)^{-1/2}$.

$$g''(x) = (4 - x^2)^{-1/2} + x(-\tfrac{1}{2})(4 - x^2)^{-3/2}(-2x)$$

$$= (4 - x^2)^{-3/2}[(4 - x^2) + x^2] = \frac{4}{(4 - x^2)^{3/2}} > 0,$$

whenever it is defined and so g is concave upward wherever it is defined.

23. $f(x) = 2x^2 - 3x + 4$; $f'(x) = 4x - 3$ and $f''(x) = 4 > 0$ for all values of x. So f is concave upward on $(-\infty, \infty)$.

25. $f(x) = x^3 - 1$. $f'(x) = 3x^2$ and $f''(x) = 6x$. The sign diagram of f'' follows.

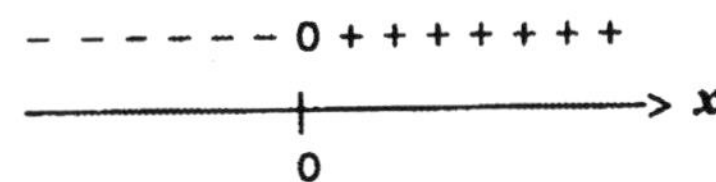

We see that f is concave downward on $(-\infty, 0)$ and concave upward on $(0, \infty)$.

27. $f(x) = x^4 - 6x^3 + 2x + 8$; $f'(x) = 4x^3 - 18x^2 + 2$ and $f''(x) = 12x^2 - 36x = 12x(x - 3)$. The sign diagram of f''

+ + + + + + 0 - - - - 0 + + +

0 3 → x

shows that f is concave upward on $(-\infty, 0) \cup (3, \infty)$ and is concave downward on $(0, 3)$.

29. $f(x) = x^{4/7}$. $f'(x) = \dfrac{4}{7}x^{-3/7}$ and $f''(x) = -\dfrac{12}{49}x^{-10/7} = -\dfrac{12}{49x^{10/7}}$.

Observe that $f''(x) < 0$ for all x different from zero. So f is concave downward on $(-\infty, 0) \cup (0, \infty)$.

31. $f(x) = (4 - x)^{1/2}$. $f'(x) = \dfrac{1}{2}(4 - x)^{-1/2}(-1) = -\dfrac{1}{2}(4 - x)^{-1/2}$;

$$f''(x) = \frac{1}{4}(4 - x)^{-3/2}(-1) = -\frac{1}{4(4 - x)^{3/2}} < 0.$$

whenever it is defined. So f is concave downward on $(-\infty, 4)$.

33. $f'(x) = \dfrac{d}{dx}(x - 2)^{-1} = -(x - 2)^{-2}$ and $f''(x) = 2(x - 2)^{-3} = \dfrac{2}{(x - 2)^3}$.

The sign diagram of f'' shows that f is concave downward on $(-\infty, 2)$ and concave

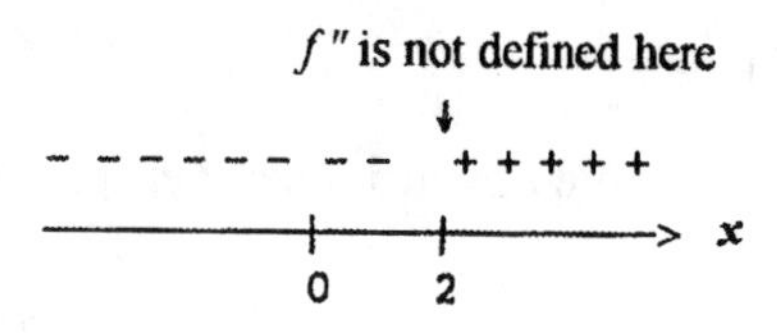

upward on $(2,\infty)$.

35. $f'(x) = \frac{d}{dx}(2+x^2)^{-1} = -(2+x^2)^{-2}(2x) = -2x(2+x^2)^{-2}$ and

$f''(x) = -2(2+x^2)^{-2} - 2x(-2)(2+x^2)^{-3}(2x)$

$$= 2(2+x^2)^{-3}[-(2+x^2)+4x^2] = \frac{2(3x^2-2)}{(2+x^2)^3} = 0 \text{ if } x = \pm\sqrt{2/3}.$$

From the sign diagram of f''

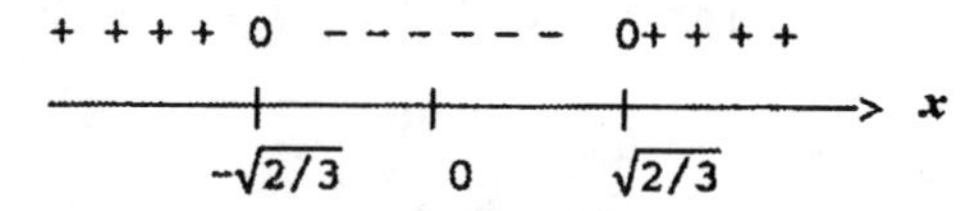

we see that f is concave upward on $(-\infty,-\sqrt{2/3}) \cup (\sqrt{2/3},\infty)$ and concave downward on $(-\sqrt{2/3},\sqrt{2/3})$.

37. $h(t) = \frac{t^2}{t-1}$; $h'(t) = \frac{(t-1)(2t) - t^2(1)}{(t-1)^2} = \frac{t^2-2t}{(t-1)^2}$;

$$h''(t) = \frac{(t-1)^2(2t-2) - (t^2-2t)2(t-1)}{(t-1)^2}$$

$$= \frac{(t-1)(2t^2-4t+2-2t^2+4t)}{(t-1)^2} = \frac{2}{(t-1)}.$$

The sign diagram of h'' is

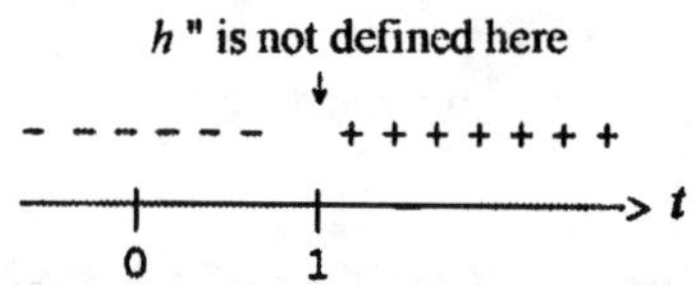

and tells us that h is concave downward on $(-\infty,1)$ and concave upward on $(1,\infty)$.

39. $g'(x) = 1 - 2x^{-3}$ and $g''(x) = 6x^{-4} = \frac{6}{x^4} > 0$ whenever $x \neq 0$. Therefore, g is concave upward on $(-\infty,0) \cup (0,\infty)$.

41. $g(t) = (2t - 4)^{1/3}$. $g'(t) = = \frac{1}{3}(2t-4)^{-2/3}(2) = \frac{2}{3}(2t-4)^{-2/3}$.

$$g''(t) = -\frac{4}{9}(2t-4)^{-5/3} = -\frac{4}{9(2t-4)^{5/3}}.$$

The sign diagram of g''

g" is not defined here

+ + + + + + + + + + + - - - -

0 2 x

tells us that g is concave upward on $(-\infty,2)$ and concave downward on $(2,\infty)$.

43. $f(x) = \frac{x^2}{x^2-1}$. $f'(x) = \frac{(x^2-1)(2x) - x^2(2x)}{(x^2-1)^2} = -\frac{2x}{(x^2-1)^2}$.

$$f''(x) = -\frac{(x^2-1)^2(2) - (-2x)(x^2-1)(2x)}{(x^2-1)^4} = -\frac{2[(x^2-1)(x^2-1) - 4x^2]}{(x^2-1)^4}$$

$$= \frac{2(3x^2+1)}{(x^2-1)^3}.$$

The sign diagram of f'' is

f" is not defined here

+ + + + - - - - - - - - + + + + +

-1 0 1 x

and we see that f is concave upward on $(-\infty,-1) \cup (1,\infty)$ and concave downward on $(-1,1)$.

45. $f(x) = x^3 - 2$. $f'(x) = 3x^2$ and $f''(x) = 6x$. $f''(x)$ is continuous everywhere and has a zero at $x = 0$. From the sign diagram of f''

- - - - - - - 0 + + + + + + +

0 x

we conclude that $(0,-2)$ is an inflection point of f.

47. $f(x) = 6x^3 - 18x^2 + 12x - 15$; $f'(x) = 18x^2 - 36x + 12$ and $f''(x) = 36x - 36 = 36(x - 1) = 0$ if $x = 1$. The sign diagram of f''

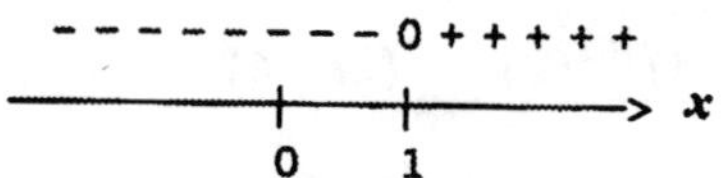

tells us that f has an inflection point at (1,-15).

49. $f(x) = 3x^4 - 4x^3 + 1$. $f'(x) = 12x^3 - 12x^2$ and $f''(x) = 36x^2 - 24x = 12x(3x - 2) = 0$ if $x = 0$ or $2/3$. These are candidates for inflection points. The sign diagram of f''

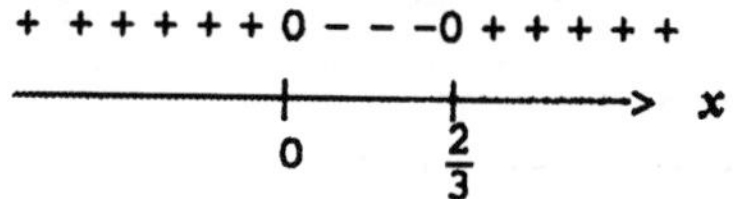

shows that (0,1) and $(\frac{2}{3}, \frac{11}{27})$ are inflection points of f.

51. $g(t) = t^{1/3}$, $g'(t) = \frac{1}{3}t^{-2/3}$ and $g''(t) = -\frac{2}{9}t^{-5/3} = -\dfrac{2}{9t^{5/3}}$. Observe that $t = 0$ is in the domain of g. Next, since $g''(t) > 0$ if $t < 0$ and $g''(t) < 0$, if $t > 0$, we see that (0,0) is an inflection point of g.

53. $f(x) = (x - 1)^3 + 2$. $f'(x) = 3(x - 1)^2$ and $f''(x) = 6(x - 1)$. Observe that $f''(x) < 0$ if $x < 1$ and $f''(x) > 0$ if $x > 1$ and so (1,2) is an inflection point of f.

55. $f(x) = \dfrac{2}{1+x^2} = 2(1+x^2)^{-1}$. $f'(x) = -2(1+x^2)^{-2}(2x) = -4x(1+x^2)^{-2}$.

$$f''(x) = -4(1+x^2)^{-2} - 4x(-2)(1+x^2)^{-3}(2x)$$

$$= 4(1+x^2)^{-3}[-(1+x^2)+4x^2] = \frac{4(3x^2-1)}{(1+x^2)^3},$$

is continuous everywhere and has zeros at $x = \pm\frac{\sqrt{3}}{3}$. From the sign diagram of f''

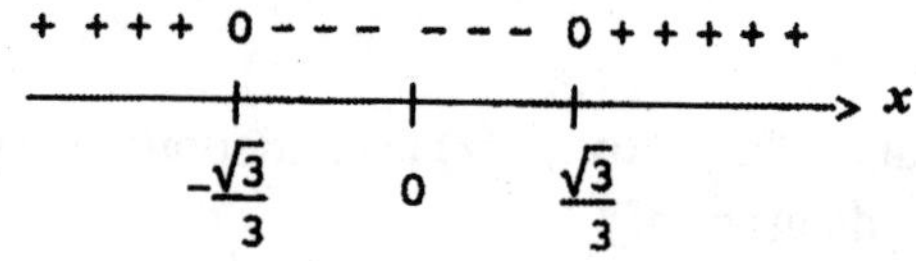

we conclude that $(-\frac{\sqrt{3}}{3}, \frac{3}{2})$ and $(\frac{\sqrt{3}}{3}, \frac{3}{2})$ are inflection points of f.

57. $f(x) = -x^2 + 2x + 4$ and $f'(x) = -2x + 2$. The critical point of f is $x = 1$. Since $f''(x) = -2$ and $f''(1) = -2 < 0$, we conclude that $f(1) = 5$ is a relative maximum of f.

59. $f(x) = 2x^3 + 1$; $f'(x) = 6x^2 = 0$ if $x = 0$ and this is a critical point of f. Next, $f''(x) = 12x$ and so $f''(0) = 0$. Thus, the Second Derivative Test fails. But the First

Derivative Test shows that (0,0) is not a relative extremum.

61. $f(x)=\frac{1}{3}x^3-2x^2-5x-10$. $f'(x)=x^2-4x-5=(x-5)(x+1)$ and this gives $x=-1$ and $x=5$ as critical points of f. Next, $f''(x)=2x-4$. Since $f''(-1)=-6<0$, we see

that $(-1,-\frac{22}{3})$ is a relative maximum. Next, $f''(5)=6>0$ and this shows that $(5,-\frac{130}{3})$ is a relative minimum.

63. $g(t)=t+\dfrac{9}{t}$. $g'(t)=1-\dfrac{9}{t^2}=\dfrac{t^2-9}{t^2}=\dfrac{(t+3)(t-3)}{t^2}$ and this shows that $t=\pm 3$ are

critical points of g. Now, $g''(t)=18t^{-3}=\dfrac{18}{t^3}$. Since $g''(-3)=-\dfrac{18}{27}<0$ the Second Derivative Test implies that g has a relative maximum at (-3,-6). Also, $g''(3)=\dfrac{18}{27}>0$ and so g has a relative minimum at (3,6).

65. $f(x)=\dfrac{x}{1-x}$. $f'(x)=\dfrac{(1-x)(1)-x(-1)}{(1-x)^2}=\dfrac{1}{(1-x)^2}$ is never zero.
So there are no critical points and f has no relative extrema.

67. $f(t)=t^2-\dfrac{16}{t}$. $f'(t)=2t+\dfrac{16}{t^2}=\dfrac{2t^3+16}{t^2}=\dfrac{2(t^3+8)}{t^2}$. Setting
$f'(t)=0$ gives $t=-2$ as a critical point. Next, we compute
$f''(t)=\dfrac{d}{dt}(2t+16t^{-2})=2-32t^{-3}=2-\dfrac{32}{t^3}$. Since $f''(-2)=2-\dfrac{32}{(-8)}=6>0$, we see that (-2,12) is a relative minimum.

69. $g(s)=\dfrac{s}{1+s^2}$; $g'(s)=\dfrac{(1+s^2)(1)-s(2s)}{(1+s^2)^2}=\dfrac{1-s^2}{(1+s^2)^2}=0$ gives $s=-1$ and $s=1$

as critical points of g. Next, we compute

$$g''(s)=\frac{(1+s^2)^2(-2s)-(1-s^2)2(1+s^2)(2s)}{(1+s^2)^4}$$
$$=\frac{2s(1+s^2)(-1-s^2-2+2s^2)}{(1+s^2)^4}=\frac{2s(s^2-3)}{(1+s^2)^3}.$$

Now, $g''(-1)=\frac{1}{2}>0$ and so $g(-1)=-\frac{1}{2}$ is a relative minimum of g. Next, $g''(1)=-\frac{1}{2}<0$ and so $g(1)=\frac{1}{2}$ is a relative maximum of g.

71. $f(x)=\dfrac{x^4}{x-1}$.

$$f'(x)=\frac{(x-1)(4x^3)-x^4(1)}{(x-1)^2}=\frac{4x^4-4x^3-x^4}{(x-1)^2}$$
$$=\frac{3x^4-4x^3}{(x-1)^2}=\frac{x^3(3x-4)}{(x-1)^2}$$

and so $x=0$ and $x=4/3$ are critical points of f. Next,

$$f''(x)=\frac{(x-1)^2(12x^3-12x^2)-(3x^4-4x^3)(2)(x-1)}{(x-1)^4}$$
$$=\frac{(x-1)(12x^4-12x^3-12x^3+12x^2-6x^4+8x^3)}{(x-1)^4}$$
$$=\frac{6x^4-16x^3+12x^2}{(x-1)^3}=\frac{2x^2(3x^2-8x+6)}{(x-1)^3}.$$

Since $f''(\frac{4}{3})>0$, we see that $f(\frac{4}{3})=\frac{256}{27}$ is a relative minimum. Since $f''(0)=0$, the Second Derivative Test fails. Using the sign diagram for f',

and the First Derivative Test, we see that $f(0)=0$ is a relative maximum.

73. $g(x)=\dfrac{2-x}{(x+2)^3}$.

$$g'(x)=\frac{(x+2)^3(-1)-(2-x)(3)(x+2)^2}{(x+2)^6}$$
$$=\frac{(x+2)^2[-(x+2)-3(2-x)]}{(x+2)^6}=\frac{2(x-4)}{(x+2)^4}.$$

We see that $x=4$ is a critical point of g.

$$g''(x)=\frac{(x+2)^4(2)-2(x-4)(4)(x+2)^3}{(x+2)^8}$$
$$=\frac{2(x+2)^3[(x+2)-4(x-4)]}{(x+2)^8}=-\frac{2(3x-18)}{(x+2)^5}.$$

Since $g''(4)=\dfrac{12}{(6)^5}>0$, we see that g has a relative minimum at $(4,-\frac{1}{108})$.

75. a. $N'(t)$ is positive because N is increasing on $(0,12)$.
b. $N''(t) < 0$ on $(0,6)$ and $N''(t) > 0$ on $(6,12)$.
c. The rate of growth of the number of help-wanted advertisements was increasing over the first six months of the year and decreasing over the last six months.

77. $S(x) = -0.002x^3 + 0.6x^2 + x + 500$; $S'(x) = -0.006x^2 + 1.2x + 1$;
$S''(x) = -0.012x + 1.2$.
$x = 100$ is a candidate for an inflection point of S. The sign diagram for S'' is

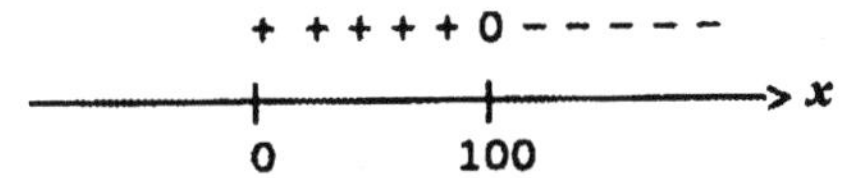

We see that $(100,4600)$ is an inflection point of S.

79. We wish to find the inflection point of the function $N(t) = -t^3 + 6t^2 + 15t$. Now, $N'(t) = -3t^2 + 12t + 15$ and $N''(t) = -6t + 12 = -6(t - 2)$ giving $t = 2$ as the only candidate for an inflection point of N. From the sign diagram

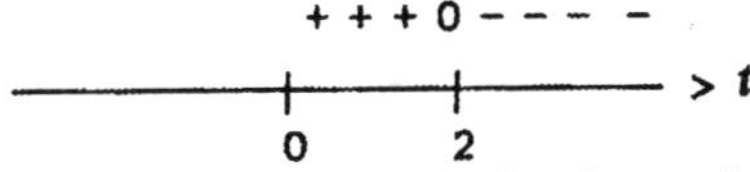

for N'', we conclude that $t = 2$ gives an inflection point of N. Therefore, the average worker is performing at peak efficiency at 10 A.M.

81. $s = f(t) = -t^3 + 54t^2 + 480t + 6$. The velocity of the rocket is
$$v = f'(t) = -3t^2 + 108t + 480 \text{ and}$$
its acceleration is
$$a = f''(t) = -6t + 108 = -6(t - 18).$$
From the sign diagram

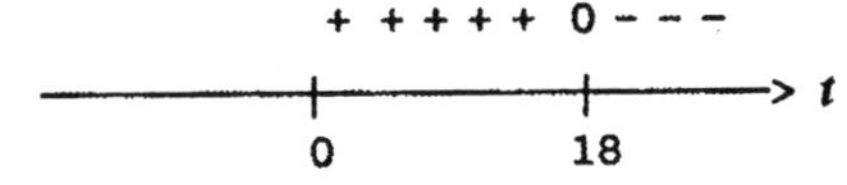

we see that $(18, 20{,}310)$ is an inflection point of f. Our computations reveal that the maximum velocity of the rocket is attained when $t = 18$. The maximum velocity is
$$f'(18) = -3(18)^2 + 108(18) + 480$$
$$= 1452, \text{ or } 1452 \text{ ft/sec.}$$

83. $f(x) = ax^2 + bx + c$. $f'(x) = 2ax + b$ and $f''(x) = 2a$. So $f''(x) > 0$ if $a > 0$, and the parabola opens upward. If $a < 0$, then $f''(x) < 0$ and the parabola opens downward.

85. $f(x) = ax^3 + bx^2 + cx$. $f'(x) = 3ax^2 + 2bx + c$ and $f''(x) = 6ax + 2b = 6a(x + \frac{b}{3a})$.
The sign diagram of f'' is

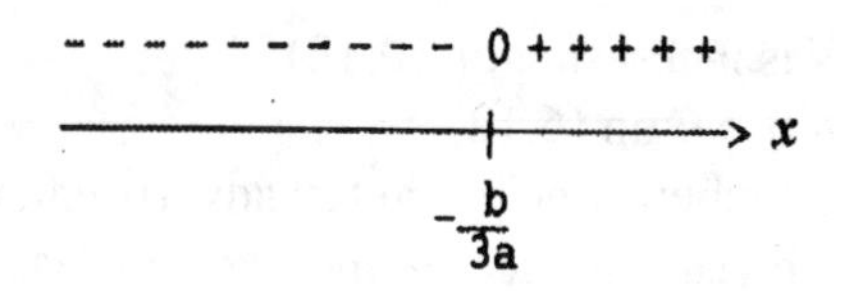

or

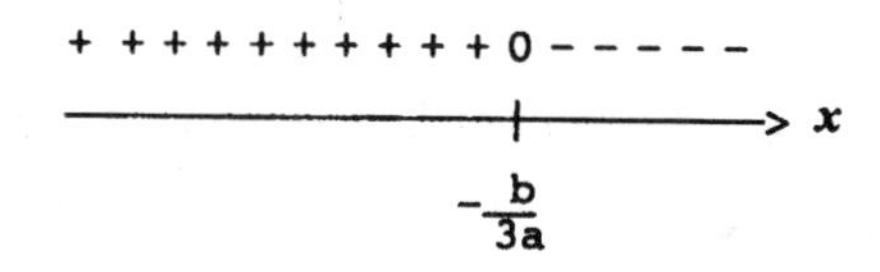

depending on whether $a > 0$ or $a < 0$. So $(-\frac{b}{3a}, f(-\frac{b}{3a}))$ is an inflection point of f.

USING TECHNOLOGY EXERCISES 12.2, page 817

1. a. f is concave upward on $(-\infty,0) \cup (1.1667, \infty)$ and concave downward on $(0, 1.1667)$.
   b. (1.1667, 1.1153); (0,2)

3. a. f is concave downward on $(-\infty,0)$ and concave upward on $(0, \infty)$.
   b. (0,2)

EXERCISES 12.3, page 829

1. $y = 0$ is a horizontal asymptote.

3. $y = 0$ is a horizontal asymptote and $x = 0$ is a vertical asymptote.

5. $y = 0$ is a horizontal asymptote and $x = -1$ and $x = 1$ are vertical asymptotes.

7. $y = 3$ is a horizontal asymptote and $x = 0$ is a vertical asymptote.

9. $y = 1$ and $y = -1$ are horizontal asymptotes.

11. $\lim_{x\to\infty} \frac{1}{x} = 0$ and so $y = 0$ is a horizontal asymptote. Next, since the numerator of the rational expression is not equal to zero and the denominator is zero at $x = 0$, we see that $x = 0$ is a vertical asymptote.

13. $f(x)=-\frac{2}{x^2}$. $\lim_{x\to\infty} -\frac{2}{x^2}=0$, so $y = 0$ is a horizontal asymptote. Next, the denominator of $f(x)$ is equal to zero at $x = 0$. Since the numerator of $f(x)$ is not equal to zero at $x = 0$, we see that $x = 0$ is a vertical asymptote.

15. $\lim_{x\to\infty}\frac{x-1}{x+1}=\lim_{x\to\infty}\frac{1-\frac{1}{x}}{1+\frac{1}{x}}=1$, and so $y = 1$ is a horizontal asymptote. Next, the denominator is equal to zero at $x = -1$ and the numerator is not equal to zero at this point, so $x = -1$ is a vertical asymptote.

17. $h(x) = x^3 - 3x^2 + x + 1$. $h(x)$ is a polynomial function and, therefore, it does not have any horizontal or vertical asymptotes.

19. $\lim_{t\to\infty}\frac{t^2}{t^2-9}=\lim_{t\to\infty}\frac{1}{1-\frac{9}{t^2}}=1$, and so $y = 1$ is a horizontal asymptote. Next, observe that the denominator of the rational expression $t^2 - 9 = (t + 3)(t - 3) = 0$ if $t = -3$ and $t = 3$. But the numerator is not equal to zero at these points. Therefore, $t = -3$ and $t = 3$ are vertical asymptotes.

21. $\lim_{x\to\infty}\frac{3x}{x^2-x-6}=\frac{\frac{3}{x}}{1-\frac{1}{x}-\frac{6}{x^2}}=0$ and so $y = 0$ is a horizontal asymptote. Next, observe that the denominator $x^2 - x - 6 = (x - 3)(x + 2) = 0$ if $x = -2$ or $x = 3$. But the numerator $3x$ is not equal to zero at these points. Therefore, $x = -2$ and $x = 3$ are vertical asymptotes.

23. $\lim_{t\to\infty}\left[2+\frac{5}{(t-2)^2}\right]=2$, and so $y = 2$ is a horizontal asymptote. Next observe that

$$\lim_{t\to 2} g(t)=\lim_{t\to 2}\left[2+\frac{5}{(t-2)^2}\right]=\infty$$

and so $t = 2$ is a vertical asymptote.

25. $\lim_{x\to\infty}\frac{x^2-2}{x^2-4}=\lim_{x\to\infty}\frac{1-\frac{2}{x^2}}{1-\frac{4}{x^2}}=1$ and so $y = 1$ is a horizontal asymptote. Next, observe that the denominator $x^2 - 4 = (x + 2)(x - 2) = 0$ if $x = -2$ or 2. Since the numerator $x^2 - 2$ is not equal to zero at these points, the lines $x = -2$ and $x = 2$ are vertical asymptotes.

27. $g(x)=\frac{x^3-x}{x(x+1)}$; Rewrite $g(x)=\frac{x^2-1}{x+1}$ $(x\neq 0)$ and note that

$\lim_{x\to-\infty} g(x) = \lim_{x\to-\infty} \frac{x-\frac{1}{x}}{1+\frac{1}{x}} = -\infty$ and $\lim_{x\to\infty} g(x) = \infty$. Therefore, there are no horizontal asymptotes. Next, note that the denominator of $g(x)$ is equal to zero at $x = 0$ and $x = -1$. However, since the numerator of $g(x)$ is also equal to zero when $x = 0$, we see that $x = 0$ is not a vertical asymptote. Also, the numerator of $g(x)$ is equal to zero when $x = -1$, so $x = -1$ is not a vertical asymptote.

29. f is the derivative function of the function g. Observe that at a relative maximum (relative minimum) of g, $f(x) = 0$.

31.

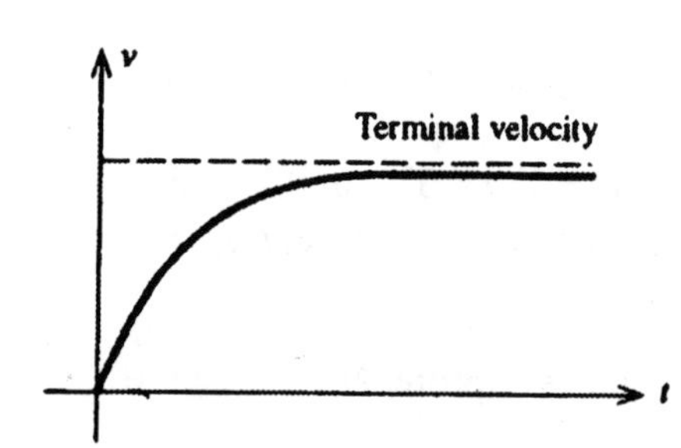

33.

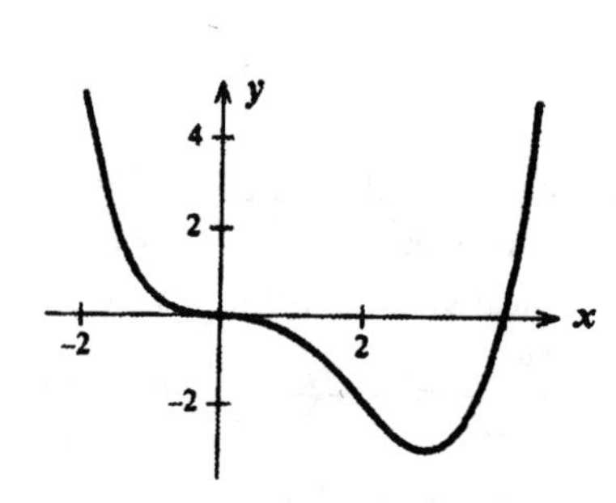

35.

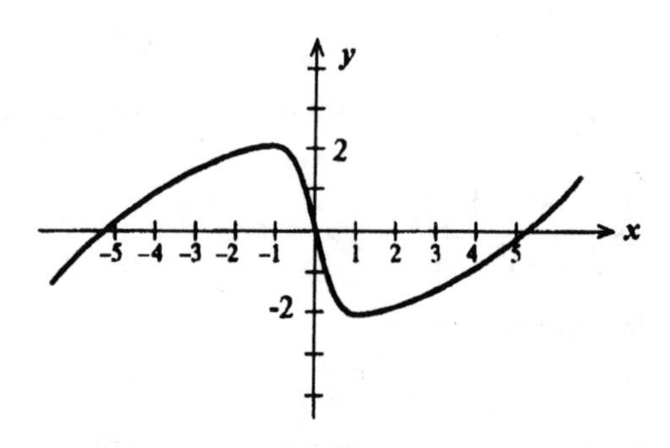

37. $g(x) = 4 - 3x - 2x^3$.
We first gather the following information on the graph of f.
1. The domain of f is $(-\infty, \infty)$.
2. Setting $x = 0$ gives $y = 4$ as the y-intercept. Setting $y = g(x) = 0$ gives a cubic equation which is not easily solved and we will not attempt to find the x-intercepts.
3. $\lim_{x\to-\infty} g(x) = \infty$ and $\lim_{x\to\infty} g(x) = -\infty$.

4. There are no asymptotes of g.
5. $g'(x) = -3 - 6x^2 = -3(2x^2 + 1) < 0$ for all values of x and so g is decreasing on $(-\infty, \infty)$.
6. The results of 5 show that g has no critical points and hence has no relative extrema.
7. $g''(x) = -12x$. Since $g''(x) > 0$ for $x < 0$ and $g''(x) < 0$ for $x > 0$, we see that g is concave upward on $(-\infty,0)$ and concave downward on $(0, \infty)$.
8. From the results of (7), we see that (0,4) is an inflection point of g.

The graph of g follows.

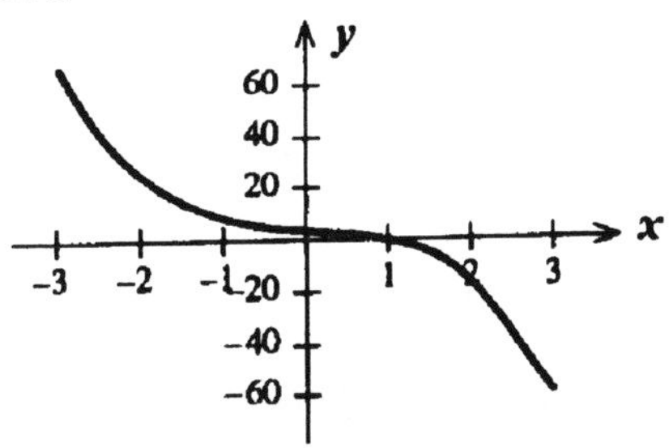

39. $h(x) = x^3 - 3x + 1$

We first gather the following information on the graph of h.
1. The domain of h is $(-\infty, \infty)$.
2. Setting $x = 0$ gives 1 as the y-intercept. We will not find the x-intercept.
3. $\lim_{x\to-\infty} (x^3 - 3x + 1) = -\infty$ and $\lim_{x\to\infty} (x^3 - 3x + 1) = \infty$
4. There are no asymptotes since $h(x)$ is a polynomial.
5. $h'(x) = 3x^2 - 3 = 3(x + 1)(x - 1)$, and we see that $x = -1$ and $x = 1$ are critical points. From the sign diagram

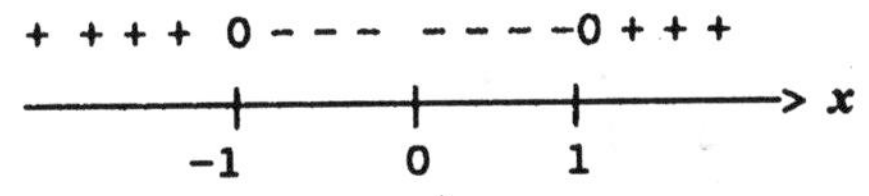

we see that h is increasing on $(-\infty,-1) \cup (1, \infty)$ and decreasing on $(-1,1)$.
6. The results of (5) shows that $(-1,3)$ is a relative maximum and $(1,-1)$ is a relative minimum.
7. $h''(x) = 6x$ and $h''(x) < 0$ if $x < 0$ and $h''(x) > 0$ if $x > 0$. So the graph of h is concave downward on $(-\infty,0)$ and concave upward on $(0, \infty)$.
8. The results of (7) show that (0,1) is an inflection point of h.

The graph of h follows.

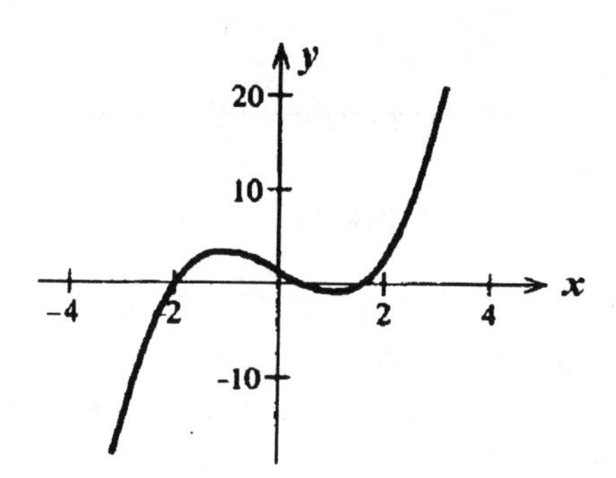

41. $f(x) = -2x^3 + 3x^2 + 12x + 2$

We first gather the following information on the graph of f.

1. The domain of f is $(-\infty, \infty)$.
2. Setting $x = 0$ gives 2 as the y-intercept.
3. $\lim_{x\to-\infty}(-2x^3+3x^2+12x+2)=\infty$ and $\lim_{x\to\infty}(-2x^3+3x^2+12x+2)=-\infty$
4. There are no asymptotes because $f(x)$ is a polynomial function.
5. $f'(x) = -6x^2 + 6x + 12 = -6(x^2 - x - 2) = -6(x-2)(x+1) = 0$ if $x = -1$ or $x = 2$, the critical points of f. From the sign diagram

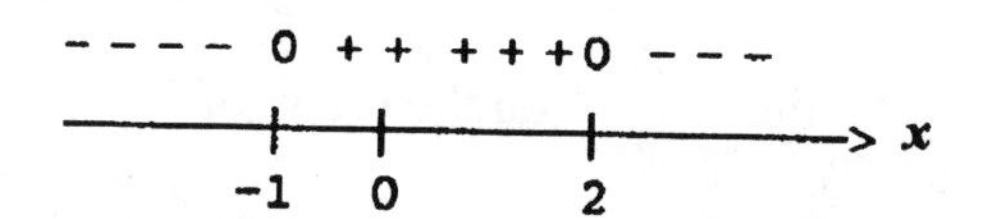

we see that f is decreasing on $(-\infty,-1) \cup (2, \infty)$ and increasing on $(-1,2)$.

6. The results of (5) show that $(-1,-5)$ is a relative minimum and $(2,22)$ is a relative maximum.
7. $f''(x) = -12x + 6 = 0$ if $x = 1/2$. The sign diagram of f''

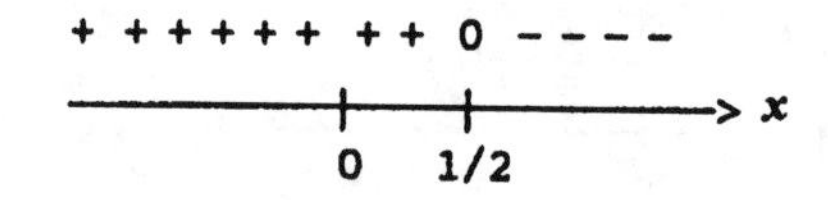

shows that the graph of f is concave upward on $(-\infty,1/2)$ and concave downward on $(1/2, \infty)$.

8. The results of (7) show that $(\frac{1}{2}, \frac{17}{2})$ is an inflection point.

The graph of f follows.

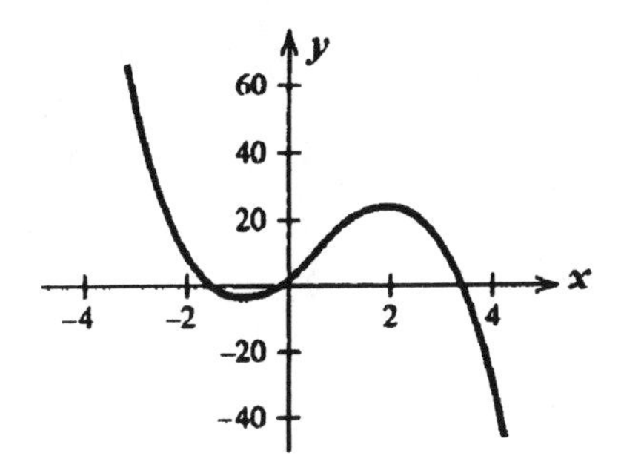

43. $h(x) = h(x) = \frac{3}{2}x^4 - 2x^3 - 6x^2 + 8$

We first gather the following information on the graph of h.

1. The domain of h is $(-\infty, \infty)$.
2. Setting $x = 0$ gives 8 as the y-intercept.
3. $\lim_{x\to-\infty} h(x) = \lim_{x\to\infty} h(x) = \infty$
4. There are no asymptotes.
5. $h'(x) = 6x^3 - 6x^2 - 12x = 6x(x^2 - x - 2) = 6x(x-2)(x+1) = 0$ if $x = -1$, 0, or 2, and these are the critical points of h. The sign diagram of h' is

- - - -0 + + 0 - - - 0 + + +

−1 0 2 → x

and this tells us that h is increasing on $(-1, 0) \cup (2, \infty)$ and decreasing on $(-\infty,-1) \cup (0,2)$.

6. The results of (5) show that $(-1, \frac{11}{2})$ and $(2,-8)$ are relative minima of h and $(0,8)$ is a relative maximum of h.
7. $h''(x) = 18x^2 - 12x - 12 = 6(3x^2 - 2x - 2)$. The zeros of h'' are

$$x = \frac{2 \pm \sqrt{4+24}}{6} \approx -0.5 \text{ or } 1.2.$$

The sign diagram of h'' is

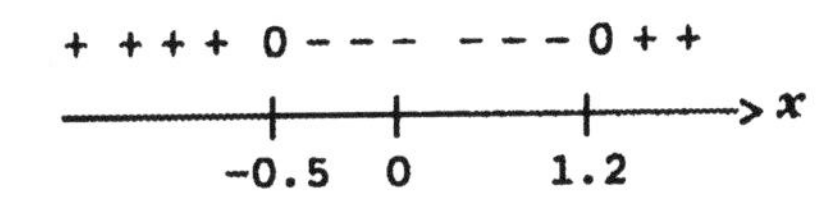

and tells us that the graph of h is concave upward on $(-\infty,-0.5) \cup (1.2, \infty)$ and is concave downward on $(0.5,1.2)$.

8. The results of (7) also show that $(-0.5,6.8)$ and $(1.2,-1)$ are inflection points.

The graph of h follows.

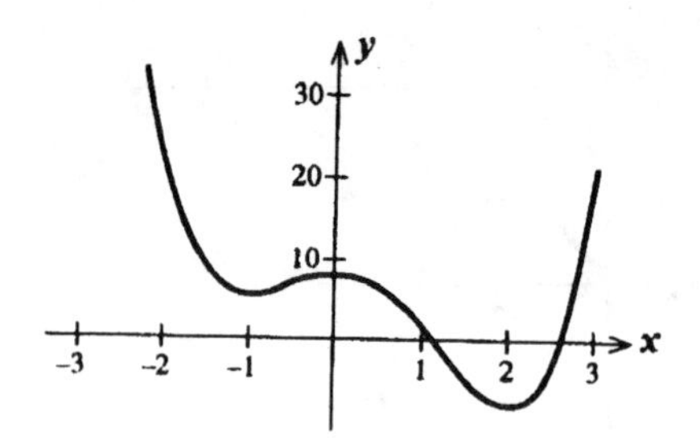

45. $f(t)=\sqrt{t^2-4}$.

We first gather the following information on f.

1. The domain of f is found by solving $t^2 - 4 \geq 0$ giving it as $(-\infty,-2] \cup [2,\infty)$.
2. Since $t \neq 0$, there is no y-intercept. Next, setting $y=f(t)=0$ gives the t-intercepts as -2 and 2.
3. $\lim\limits_{t\to-\infty} f(t)=\lim\limits_{t\to\infty} f(t)=\infty$
4. There are no asymptotes.
5. $f'(t)=\frac{1}{2}(t^2-4)^{-1/2}(2t)=t(t^2-4)^{-1/2}=\frac{t}{\sqrt{t^2-4}}$.

Setting $f'(t)=0$ gives $t=0$. But $t=0$ is not in the domain of f and so there are no critical points. The sign diagram for f' is

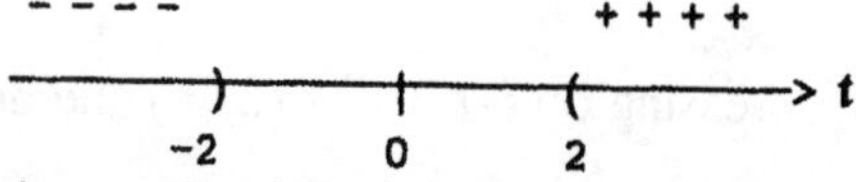

We see that f is increasing on $(2,\infty)$ and decreasing on $(-\infty,-2)$.

6. From the results of (5) we see that there are no relative extrema.
7. $f''(t)=(t^2-4)^{-1/2}+t(-\tfrac{1}{2})(t^2-4)^{-3/2}(2t)=(t^2-4)^{-3/2}(t^2-4-t^2)$

$$=-\frac{4}{(t^2-4)^{3/2}}.$$

8. Since $f''(t)<0$ for all t in the domain of f, we see that f is concave downward everywhere. From the results of (7), we see that there are no inflection points.

The graph of f follows.

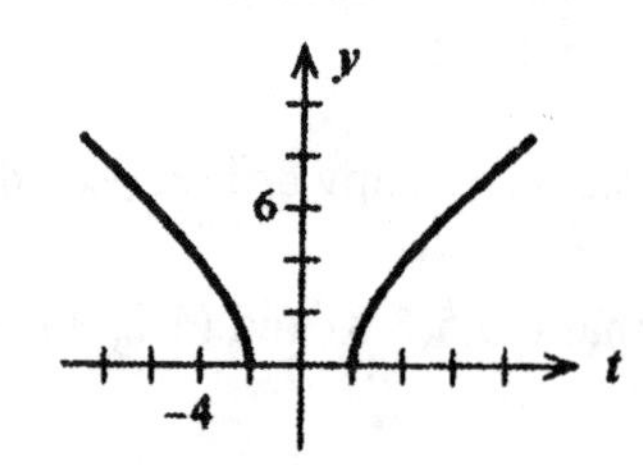

47. $g(x) = \frac{1}{2}x - \sqrt{x}$.

We first gather the following information on g.

1. The domain of g is $[0,\infty)$.
2. The y-intercept is 0. To find the x-intercept, set $y = 0$, giving

$$\frac{1}{2}x - \sqrt{x} = 0$$

$$x = 2\sqrt{x}$$

$$x^2 = 4x$$

$$x(x-4) = 0, \text{ and } x = 0 \text{ or } x = 4$$

3. $\lim_{x\to\infty} (\frac{1}{2}x - \sqrt{x}) = \lim_{x\to\infty} \frac{1}{2}x(1 - \frac{2}{\sqrt{x}}) = \infty$.
4. There are no asymptotes.
5. $g'(x) = \frac{1}{2} - \frac{1}{2}x^{-1/2} = \frac{1}{2}x^{-1/2}(x^{1/2} - 1) = \frac{\sqrt{x} - 1}{2\sqrt{x}}$

which is zero when $x = 1$. From the sign diagram for g'

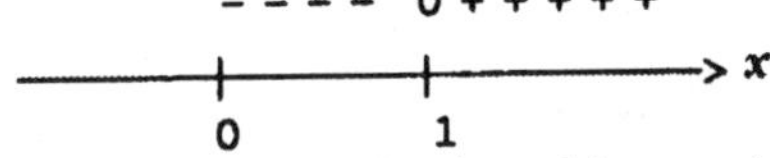

we see that g is decreasing on $(0,1)$ and increasing on $(1,\infty)$.

6. From the sign diagram of g', we see that $g(1) = -1/2$ is a relative minimum.
7. $g''(x) = (-\frac{1}{2})(-\frac{1}{2})x^{-3/2} = \frac{1}{4x^{3/2}} > 0$ for $x > 0$, and so g is concave upward on $(0,\infty)$.
8. There are no inflection points.

The graph of g follows.

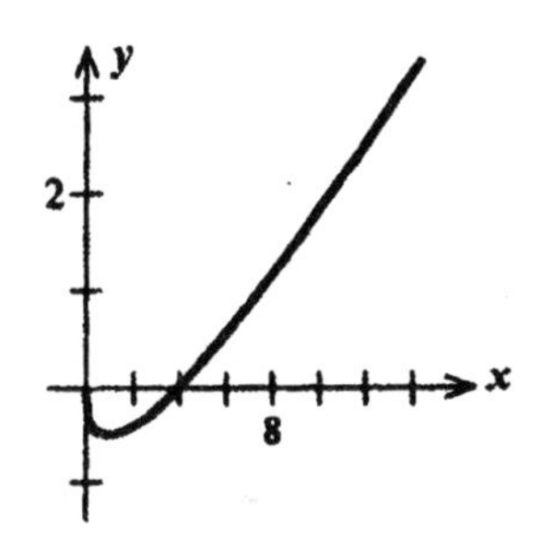

49. $g(x)=\dfrac{2}{x-1}$.

We first gather the following information on g.

1. The domain of g is $(-\infty,1) \cup (1,\infty)$.

2. Setting $x = 0$ gives -2 as the y-intercept. There are no x-intercepts since $\dfrac{2}{x-1} \neq 0$ for all values of x.

3. $\lim\limits_{x\to-\infty} \dfrac{2}{x-1} = 0$ and $\lim\limits_{x\to\infty} \dfrac{2}{x-1} = 0$

4. The results of (3) show that $y = 0$ is a horizontal asymptote. Furthermore, the denominator of $g(x)$ is equal to zero at $x = 1$ but the numerator is not equal to zero there. Therefore, $x = 1$ is a vertical asymptote.

5. $g'(x) = -2(x-1)^{-2} = -\dfrac{2}{(x-1)^2} < 0$ for all $x \neq 1$ and so g is decreasing on $(-\infty,1)$ and $(1,\infty)$.

6. Since g has no critical points, there are no relative extrema.

7. $g''(x) = \dfrac{4}{(x-1)^3}$ and so $g''(x) < 0$ *if* $x < 1$ and $g''(x) > 0$ if $x > 1$.

Therefore, the graph of g is concave downward on $(-\infty,1)$ and concave upward on $(1,\infty)$.

8. Since $g''(x) \neq 0$, there are no inflection points.

The graph of g follows.

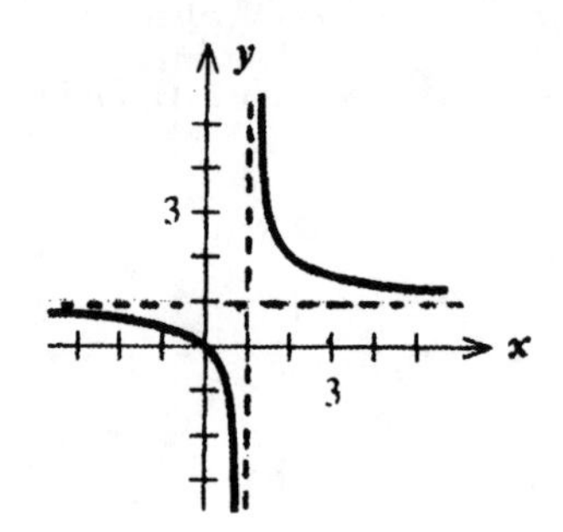

51. $h(x)=\dfrac{x+2}{x-2}$.

We first gather the following information on the graph of h.

1. The domain of h is $(-\infty,2) \cup (2,\infty)$.

2. Setting $x = 0$ gives $y = -1$ as the y-intercept. Next, setting $y = 0$ gives $x = -2$ as the x-intercept.

3. $\lim\limits_{x\to\infty} h(x) = \lim\limits_{x\to-\infty} h(x) = 1$.

4. Setting x - 2 = 0 gives $x = 2$. Furthermore,

$$\lim_{x\to 2^+} \frac{x+2}{x-2} = \infty \quad \text{and} \quad \lim_{x\to 2^+} \frac{x+2}{x-2} = -\infty$$

So $x = 2$ is a vertical asymptote of h. Also, from the resultsof (3), we see that $y = 1$ is a horizontal asymptote of h.

5. $h'(x) = \dfrac{(x-2)(1)-(x+2)(1)}{(x-2)^2} = -\dfrac{4}{(x-2)^2}$.

We see that there are no critical points of h. (Note $x = 2$ does not belong to the domain of h.) The sign diagram of h' follows.

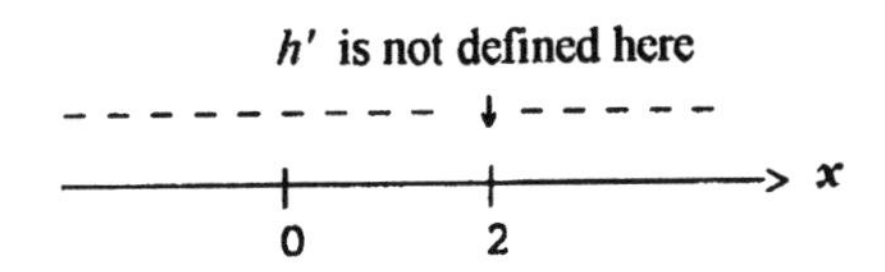

We see that h is decreasing on (-∞,2) ∪ (2,∞).

6. From the results of (5), we see that there is no relative extremum.

7. $h''(x) = \dfrac{8}{(x-2)^3}$. Note that $x = 2$ is not a candidate for an inflection point

because $h(2)$ is not defined. Since $h''(x) < 0$ for $x < 2$ and $h''(x) > 0$ for $x > 2$, we see that h is concave downward on (-∞,2) and concave upward on (2,∞).

8. From the results of (7), we see that there are no inflection points.

The graph of h follows.

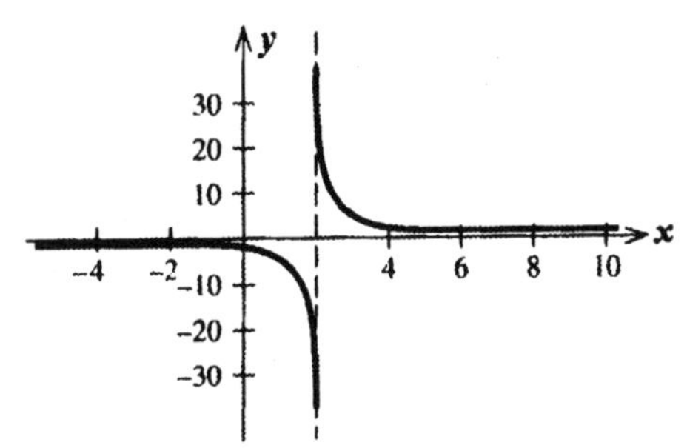

53. $f(t) = \dfrac{t^2}{1+t^2}$.

We first gather the following information on the graph of f.

1. The domain of f is (-∞, ∞).
2. Setting $t = 0$ gives the y-intercept as 0. Similarly, setting $y = 0$ gives the t-intercept as 0.

3. $\lim_{t\to-\infty}\frac{t^2}{1+t^2}=\lim_{t\to\infty}\frac{t^2}{1+t^2}=1.$

4. The results of (3) show that $y = 1$ is a horizontal asymptote. There are no vertical asymptotes since the denominator is not equal to zero.

5. $f'(t)=\frac{(1+t^2)(2t)-t^2(2t)}{(1+t^2)^2}=\frac{2t}{(1+t^2)^2}=0$, the only critical point of f. Since $f'(t)<0$ if $t<0$ and $f'(t)>0$ if $t>0$, we see that f is decreasing on$(-\infty,0)$ and increasing on $(0,\infty)$.

6. The results of (5) show that (0,0) is a relative minimum.

7. $f''(t)=\frac{(1+t^2)^2(2)-2t(2)(1+t^2)(2t)}{(1+t^2)^4}=\frac{2(1+t^2)[(1+t^2)-4t^2]}{(1+t^2)^4}$

$$=\frac{2(1-3t^2)}{(1+t^2)^3}=0 \text{ if } t=\pm\frac{\sqrt{3}}{3}.$$

The sign diagram of f'' is

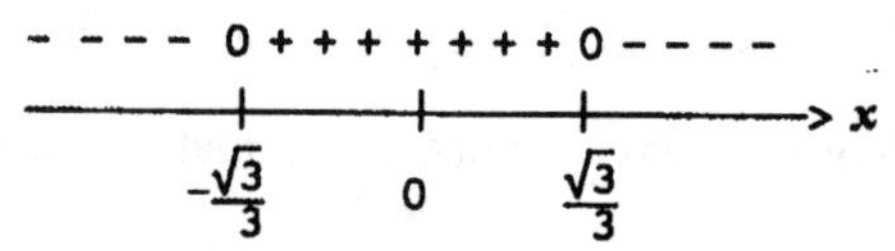

and shows that f is concave downward on $(-\infty,-\frac{\sqrt{3}}{3})\cup(\frac{\sqrt{3}}{3},\infty)$ and concave upward on $(-\frac{\sqrt{3}}{3},\frac{\sqrt{3}}{3})$.

8. The results of (7) show that $(-\frac{\sqrt{3}}{3},\frac{1}{4})$ and $(\frac{\sqrt{3}}{3},\frac{1}{4})$ are inflection points.

The graph of f appears below.

55. $g(t)=-\frac{t^2-2}{t-1}$.

First we obtain the following information on g.

1. The domain of g is $(-\infty, 1)\cup(1, \infty)$.
2. Setting $t = 0$ gives -2 as the y-intercept.
3. $\lim_{t\to-\infty}-\frac{t^2-2}{t-1}=\infty$ and $\lim_{t\to\infty}-\frac{t^2-2}{t-1}=-\infty$.
4. There are no horizontal asymptotes. The denominator is equal to zero at $t = 1$ at which point the numerator is not equal to zero. Therefore $t = 1$ is a vertical asymptote.
5. $g'(t)=-\frac{(t-1)(2t)-(t^2-2)(1)}{(t-1)^2}=-\frac{t^2-2t+2}{(t-1)^2}\neq 0$

for all values of t. The sign diagram of g'

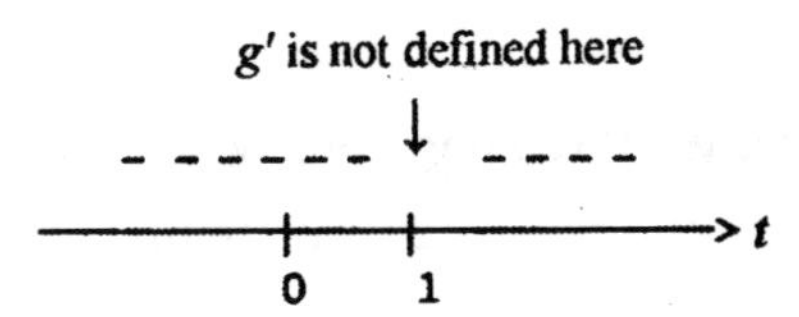

shows that g is decreasing on $(-\infty,1) \cup (1, \infty)$.

6. Since there are no critical points, g has no relative extrema.

7. $$g''(t) = -\frac{(t-1)^2(2t-2)-(t^2-2t+2)(2)(t-1)}{(t-1)^4}$$
$$= \frac{-2(t-1)(t^2-2t+1-t^2+2t-2)}{(t-1)^4} = \frac{2}{(t-1)^3}.$$

The sign diagram of g''

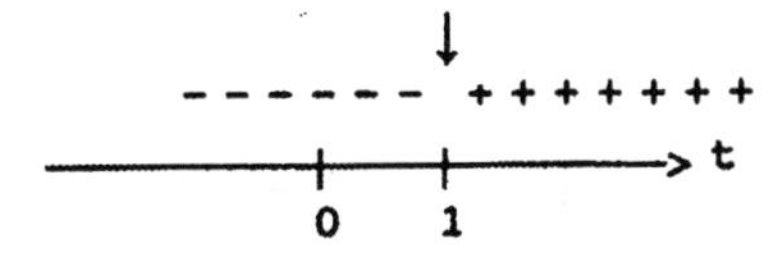

shows that the graph of g is concave upward on $(1,\infty)$ and concave downward on $(-\infty,1)$.

8. There are no inflection points since $g''(x) \neq 0$ for all x.

The graph of g follows.

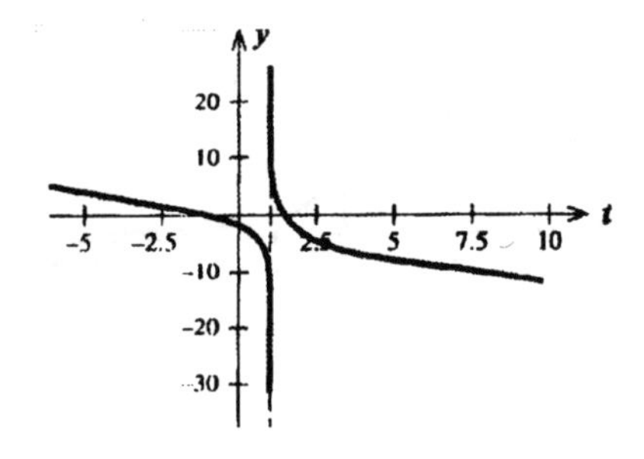

57. $g(t) = \dfrac{t+1}{t^2-2t-1}$.

We first gather some information on the graph of g.

1. Since $t^2 - 2t - 1 = 0$ if $t = \dfrac{2 \pm \sqrt{4+4}}{2} = 1 \pm \sqrt{2}$, we see that the domain of g is $(-\infty, 1-\sqrt{2}) \cup (1-\sqrt{2}, 1+\sqrt{2}) \cup (1+\sqrt{2}, \infty)$.

2. Setting $t = 0$ gives -1 as the y-intercept. Setting $y = 0$ gives -1 as the t-intercept.

3. $\lim_{t\to-\infty} g(t) = \lim_{t\to\infty} g(t) = 0.$

4. The results of (3) show that $y = 0$ is a horizontal asymptote. Since the denominator (but not the numerator) is zero at $t = 1 \pm \sqrt{2}$, we see that $t = 1 - \sqrt{2}$ and $t = 1 + \sqrt{2}$ are vertical asymptotes.

5. $g'(t) = \dfrac{(t^2 - 2t - 1)(1) - (t+1)(2t-2)}{(t^2 - 2t - 1)^2} = -\dfrac{(t^2 + 2t - 1)}{(t^2 - 2t - 1)^2} = 0$

if $\quad t = \dfrac{-2 \pm \sqrt{4+4}}{2} = -1 \pm \sqrt{2}$. The sign diagram of g' is

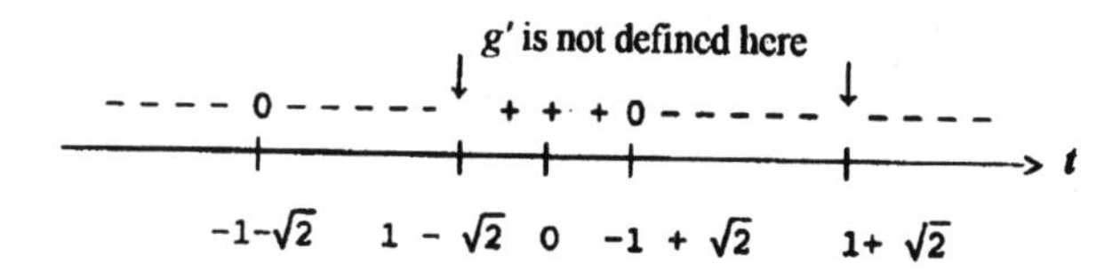

We see that g is decreasing on
$(-\infty, -1 - \sqrt{2}) \cup (-1 - \sqrt{2}, 1 - \sqrt{2}) \cup (-1 + \sqrt{2}, 1 + \sqrt{2}) \cup (1 + \sqrt{2}, \infty)$
and increasing on $(1 - \sqrt{2}, -1 + \sqrt{2})$.

6. From the results of (5), we see that $x = 1 - \sqrt{2}$ gives rise to a relative minimum and $x = -1 + \sqrt{2}$ gives rise to a relative maximum.
The graph of g follows.

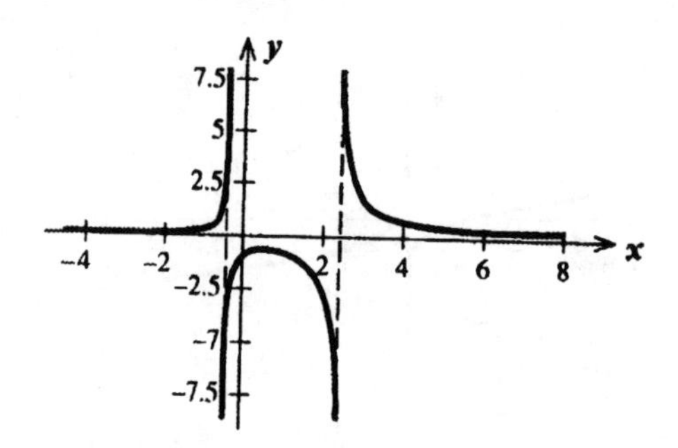

59. $h(x) = (x - 1)^{2/3} + 1.$

We begin by obtaining the following information on h.

1. The domain of h is $(-\infty, \infty)$.
2. Setting $x = 0$ gives 2 as the y-intercept; setting $h(x) = 0$ gives 2 as the x-intercept.
3. $\lim_{x\to\infty} [(x-1)^{2/3} + 1] = \infty.$ Similarly, $\lim_{x\to-\infty} [(x-1)^{2/3} + 1] = \infty$.
4. There are no asymptotes.

5. $h'(x) = \frac{2}{3}(x-1)^{-1/3}$ and is positive if $x > 1$ and negative if $x < 1$. So h is increasing on $(1,\infty)$, and decreasing on $(-\infty,1)$.
6. From (5), we see that h has a relative minimum at (1,1).
7. $h''(x) = \frac{2}{3}(-\frac{1}{3})(x-1)^{-4/3} = -\frac{2}{9}(x-1)^{-4/3} = -\frac{2}{(x-1)^{4/3}}$. Since $h''(x) < 0$ on $(-\infty,1) \cup (1,\infty)$, we see that h is concave downward on $(-\infty,1) \cup (1,\infty)$. Note that $h''(x)$ is not defined at $x = 1$.
8. From the results of (7), we see h has no inflection points.
The graph of h follows.

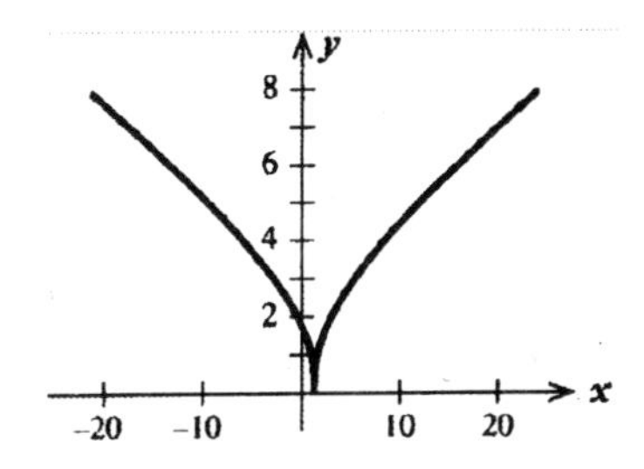

61. a. $\lim_{x\to\infty} \overline{C}(x) = \lim_{x\to\infty} (2.2 + \frac{2500}{x}) = 2.2$, and so $y = 2.2$ is a horizontal asymptote.
b. The limiting value is 2.2, or $2.20 per disc.

63. a. $\lim_{x\to\infty} \frac{ax}{x+b} = \lim_{x\to\infty} \frac{a}{1+\frac{b}{x}} = a.$
b. The initial speed of the reaction approaches a moles/liter/sec as the amount of substrate becomes arbitrarily large.

65. $G(t) = -0.2t^3 + 2.4t^2 + 60$.

We first gather the following information on the graph of G.
1. The domain of G is $(0,\infty)$.
2. Setting t = 0 gives 60 as the y-intercept.
Note that Step 3 is not necessary in this case because of the restricted domain.
4. There are no asymptotes since G is a polynomial function.
5. $G'(t) = -0.6t^2 + 4.8t = -0.6t(t - 8) = 0$, if $t = 0$ or $t = 8$. But these points do not lie in the interval (0,8), so they are not critical points. The sign diagram of G'

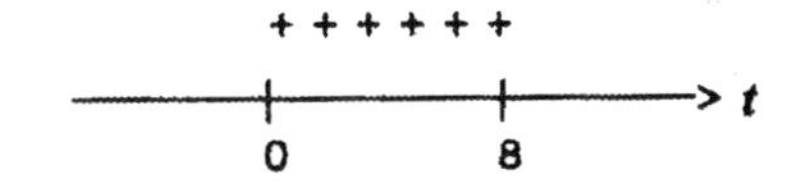

shows that G is increasing on (0,8).

6. The results of (5) tell us that there are no relative extrema.

7. $G''(t) = -1.2t + 4.8 = -1.2(t - 4)$. The sign diagram of G'' is

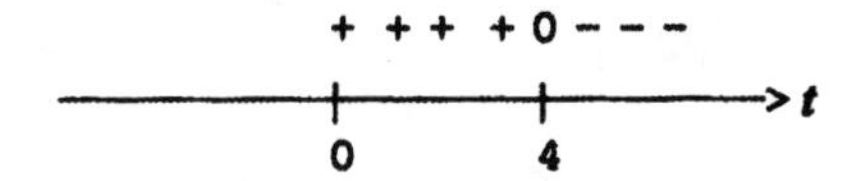

and shows that G is concave upward on (0,4) and concave downward on (4,8).

8. The results of (7) shows that (4,85.6) is an inflection point.

The graph of G is shown below.

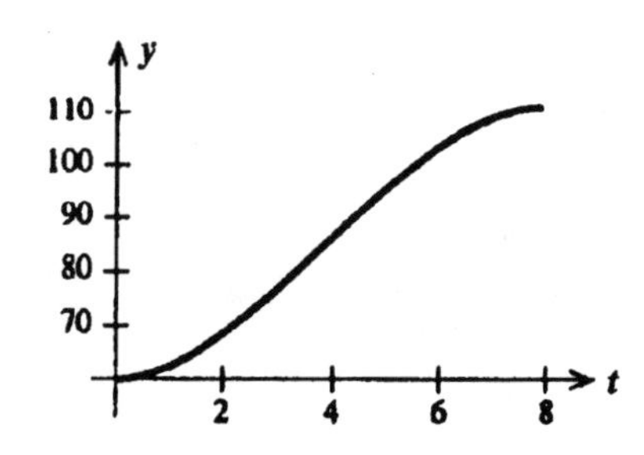

67. $f(t) = 100\left(\frac{t^2 - 4t + 4}{t^2 + 4}\right)$.

We first gather the following information on the function f.

1. The domain of f is restricted to $[0,\infty)$.
2. Setting $t = 0$ gives $y = 100$. Next, setting $y = 0$ gives $t = 2$.
3. $\lim_{t\to\infty} 100\left[\frac{t^2 - 4t + 4}{t^2 + 4}\right] = 100\lim_{t\to\infty}\left[\frac{1 - \frac{4}{t} + \frac{4}{t^2}}{\frac{4}{t^2}}\right] = 100$
4. From the results of (3), we see that $y = 100$ is a horizontal asymptote. There are no vertical asymptotes.
5. $f'(t) = 100\left[\frac{(t^2 + 4)(2t - 4) - (t^2 - 4t + 4)(2t)}{(t^2 + 4)^2}\right]$

$$= \frac{400(t^2 - 4)}{(t^2 + 4)^2} = \frac{400(t - 2)(t + 2)}{(t^2 + 4)^2}.$$

Setting $f'(t) = 0$ gives $t = 2$ as a critical point of f. ($t = -2$ is not in the domain of f.) Since $f'(t) < 0$ when $t < 2$ and $f'(t) > 0$ when $t > 2$, we see that f is decreasing on (0,2) and increasing on $(2,\infty)$.

6. The results of (5) imply that $f(2) = 0$ is a relative minimum.

7. $$f''(t) = 400\left[\frac{(t^2+4)^2(2t)-(t^2-4)2(t^2+4)(2t)}{(t^2+4)^4}\right]$$

$$= 400\left[\frac{(2t)(t^2+4)(t^2+4-2t^2+8)}{(t^2+4)^4}\right] = -\frac{800t(t^2-12)}{(t^2+4)^3}.$$

Setting $f''(t) = 0$ gives $t = 2\sqrt{3}$ as a candidate for a point of inflection. The sign diagram for f'' is

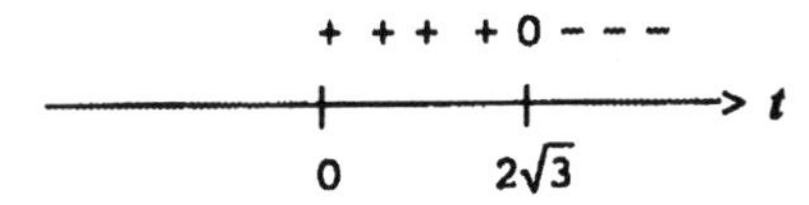

We see that f is concave upward on $(0, 2\sqrt{3})$ and concave downward on $(2\sqrt{3}, \infty)$.

8. From the results of (7), we see that $(2\sqrt{3}, 50(2 - \sqrt{3}))$ is an inflection point of f.

The graph of f follows.

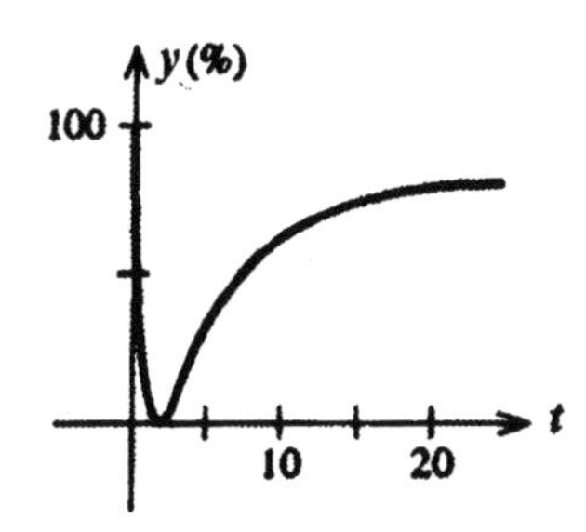

69. $$C(x) = \frac{0.5x}{100-x}.$$

We first gather the following information on the graph of C.

1. The domain of C is $[0,100)$.
2. Setting $x = 0$ gives the y-intercept as 0.

Because of the restricted domain, we omit steps 3 and 4.

5. $C'(x) = 0.5\left[\frac{(100-x)(1)-x(-1)}{(100-x)^2}\right] = \frac{50}{(100-x)^2} > 0$ for all $x \neq 100$. Therefore C is increasing on $(0,100)$.

6. There are no relative extrema.

7. $C''(x) = -\frac{100}{(100-x)^3}$. So $C''(x) > 0$ if $x < 100$ and the graph of C is concave

upward on (0,100).
8. There are no inflection points.

The graph of C follows.

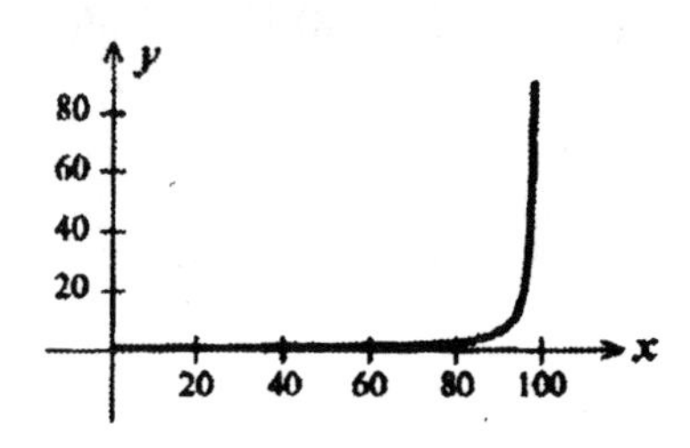

USING TECHNOLOGY EXERCISES 12.3, page 835

1.

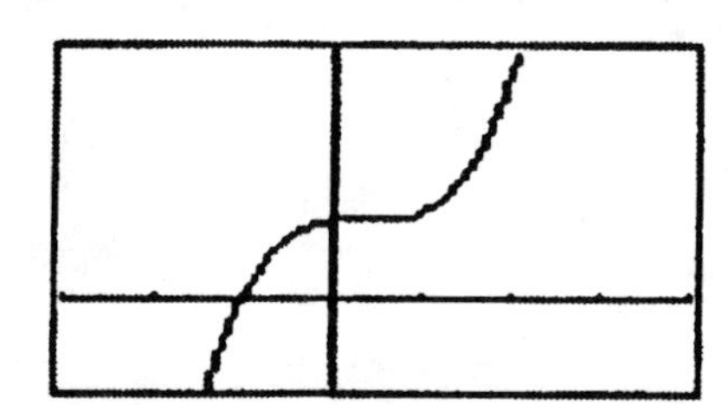

3.

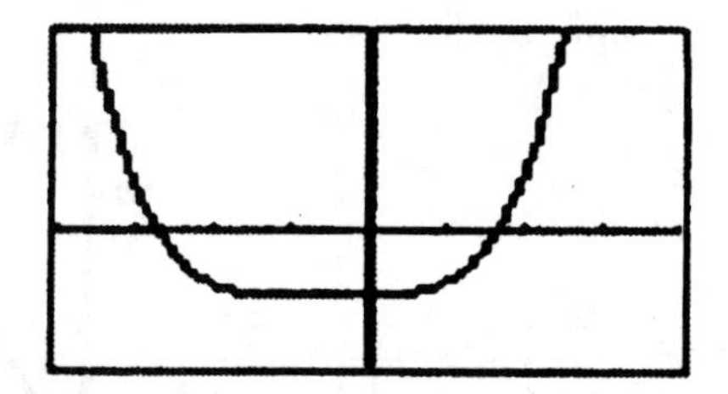

5. -0.9733; 2.3165, 4.6569

7. -1.1310; 2.9267

9. 1.5142

EXERCISES 12.4, page 848

1. f has no absolute extrema.

3. f has an absolute minimum at (0,0).

5. f has an absolute minimum at (0,-2) and an absolute maximum at (1,3).

7. f has an absolute minimum at $(\frac{3}{2}, -\frac{27}{16})$ and an absolute maximum at (-1,3).

9. The graph of $f(x) = 2x^2 + 3x - 4$ is a parabola that opens upward. Therefore, the vertex of the parabola is the absolute minimum of f. To find the vertex, we solve the equation

$f'(x) = 4x + 3 = 0$

giving $x = -3/4$. We conclude that the absolute minimum value is $f(-\frac{3}{4}) = -\frac{41}{8}$.

11. Since $\lim_{x \to -\infty} x^{1/3} = -\infty$ and $\lim_{x \to \infty} x^{1/3} = \infty$, we see that h is unbounded. Therefore it has no absolute extrema.

13. $f(x) = \dfrac{1}{1+x^2}$.

Using the techniques of graphing, we sketch the graph of f (see Fig. 12.41, page 807, in the text). The absolute maximum of f is $f(0) = 1$. Alternatively, observe that $1 + x^2 \geq 1$ for all real values of x. Therefore, $f(x) \leq 1$ for all x, and we see that the absolute maximum is attained when $x = 1$.

15. $f(x) = x^2 - 2x - 3$ and $f'(x) = 2x - 2 = 0$, so $x = 1$ is a critical point. From the table,

x	-2	1	3
$f(x)$	5	-4	0

we conclude that the absolute maximum value is $f(-2) = 5$ and the absolute minimum value is $f(1) = -4$.

17. $f(x) = -x^2 + 4x + 6$; The function f is continuous and defined on the closed interval $[0,5]$. $f'(x) = -2x + 4$ and $x = 2$ is a critical point. From the table

x	0	2	5
$f(x)$	6	10	1

we conclude that $f(2) = 10$ is the absolute maximum value and $f(5) = 1$ is the absolute minimum value.

19. The function $f(x) = x^3 + 3x^2 - 1$ is continuous and defined on the closed interval $[-3,2]$ and differentiable in $(-3,2)$. The critical points of f are found by solving

$f'(x) = 3x^2 + 6x = 3x(x + 2)$

giving $x = -2$ and $x = 0$. Next, we compute the values of f given in the following table.

x	-3 -2 0 2
$f(x)$	-1 3 -1 19

From the table, we see that the absolute maximum value of f is $f(2) = 19$ and the absolute minimum value is $f(-3) = -1$ and $f(0) = -1$.

21. The function $g(x) = 3x^4 + 4x^3$ is continuous and differentiable on the closed interval [-2,1] and differentiable in (-2,1). The critical points of g are found by solving

$$g'(x) = 12x^3 + 12x^2 = 12x^2(x + 1)$$

giving $x = 0$ and $x = -1$. We next compute the values of g shown in the following table.

x	-2 -1 0 1
$g(x)$	16 -1 0 7

From the table we see that $g(-2) = 16$ is the absolute maximum value of g and $g(-1) = -1$ is the absolute minimum value of g.

23. $f(x) = \dfrac{x+1}{x-1}$ on [2,4]. Next, we compute,

$$f'(x) = \frac{(x-1)(1)-(x+1)(1)}{(x-1)^2} = -\frac{2}{(x-1)^2}.$$

Since there are no critical points, ($x = 1$ is not in the domain of f), we need only test the endpoints. From the table

x	2 4
$g(x)$	3 5/3

we conclude that $f(4) = 5/3$ is the absolute minimum value and $f(2) = 3$ is the absolute maximum value.

25. $f(x) = 4x + \dfrac{1}{x}$ is continuous on [1,3] and differentiable in (1,3). To find the critical points of f, we solve $f'(x) = 4 - \dfrac{1}{x^2} = 0$, obtaining $x = \pm\frac{1}{2}$. Since these critical

points lie outside the interval [1,3], they are not candidates for the absolute extrema of f. Evaluating f at the endpoints of the interval [1,3], we find that the absolute maximum value of f is $f(3) = \frac{37}{3}$, and the absolute minimum of f is $f(1) = 5$.

27. $f(x) = \frac{1}{2}x^2 - 2\sqrt{x} = \frac{1}{2}x^2 - 2x^{1/2}$. To find the critical points of f, we solve

$$f'(x) = x - x^{-1/2} = 0, \quad \text{or} \quad x^{3/2} - 1 = 0,$$

obtaining $x = 1$. From the table

x	0	1	3
$f(x)$	0	$-\frac{3}{2}$	$\frac{9}{2} - 2\sqrt{3} \approx 1.04$

we conclude that $f(3) \approx 1.04$ is the absolute maximum value and $f(1) = -3/2$ is the absolute minimum value.

29. The graph of $f(x) = 1/x$ over the interval $(0,\infty)$ is shown below.

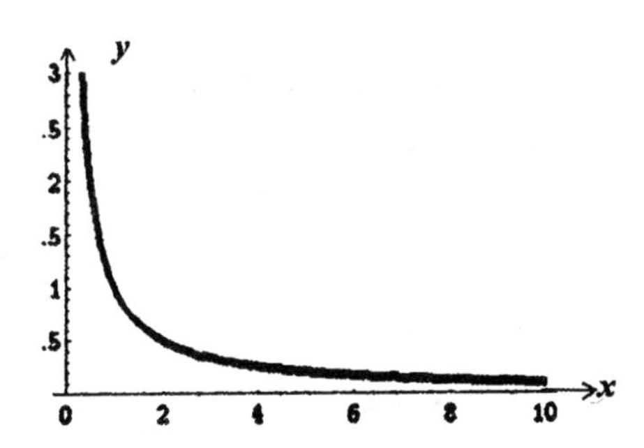

From the graph of f, we conclude that f has no absolute extrema.

31. $f(x) = 3x^{2/3} - 2x$. The function f is continuous on [0,3] and differentiable on (0,3). To find the critical points of f, we solve

$$f'(x) = 2x^{-1/3} - 2 = 0$$

obtaining $x = 1$ as the critical point. From the table,

x	0	1	3
$f(x)$	0	1	$3^{5/3} - 6 \approx 0.24$

we conclude that the absolute maximum value is $f(1) = 1$ and the absolute minimum value is $f(0) = 0$.

33. $f(x) = x^{2/3}(x^2 - 4)$. $f'(x) = x^{2/3}(2x) + \frac{2}{3}x^{-1/3}(x^2 - 4) = \frac{2}{3}x^{-1/3}[3x^2 + (x^2 - 4)]$

$$= \frac{8(x^2 - 1)}{3x^{1/3}} = 0.$$

Observe that f' is not defined at $x = 0$. Furthermore, $f'(x) = 0$ at $x \pm 1$. So the critical points of f are -1, 0, 1. From the following table,

x	-1 0 1 2
$f(x)$	-3 0 -3 0

we see that f has an absolute minimum at (-1,-3) and (1,-3) and absolute maxima at (0,0) and (2,0).

35. $f(x) = \dfrac{x}{x^2 + 2}$. To find the critical points of f, we solve

$$f'(x) = \frac{(x^2 + 2) - x(2x)}{(x^2 + 2)^2} = \frac{2 - x^2}{(x^2 + 2)^2} = 0$$

obtaining $x = \pm\sqrt{2}$. Since $x = -\sqrt{2}$ lies outside [-1,2], $x = \sqrt{2}$ is the only critical point in the given interval. From the table

x	-1	$\sqrt{2}$	2
$f(x)$	$-\frac{1}{3}$	$\sqrt{2}/4 \approx 0.35$	$\frac{1}{3}$

we conclude that $f(\sqrt{2})) = \sqrt{2}/4 \approx 0.35$ is the absolute maximum value and $f(-1) = -1/3$ is the absolute minimum value.

37. The function $f(x) = f(x) = \dfrac{x}{\sqrt{x^2 + 1}} = \dfrac{x}{(x^2 + 1)^{1/2}}$ is continuous and defined on the closed interval [-1,1] and differentiable on (-1,1). To find the critical points of f, we first compute

$$f'(x) = \frac{(x^2 + 1)^{1/2}(1) - x(\frac{1}{2})(x^2 + 1)^{-1/2}(2x)}{[(x^2 + 1)^{1/2}]^2}$$

$$= \frac{(x^2+1)^{-1/2}[x^2+1-x^2]}{x^2+1} = \frac{1}{(x^2+1)^{3/2}}$$

which is never equal to zero. Next, we compute the values of f shown in the following table.

x	-1	1
$f(x)$	$-\sqrt{2}/2$	$\sqrt{2}/2$

We conclude that $f(-1) = -\sqrt{2}/2$ is the absolute minimum value and $f(1) = \sqrt{2}/2$ is the absolute maximum value.

39. $h(t) = 16t^2 + 64t + 80$. To find the maximum value of h, we solve
$h'(t) = -32t + 64 = -32(t - 2) = 0$
giving $t = 2$ as the critical point of h. Furthermore, this value of t gives rise to the absolute maximum value of h since the graph of h is parabola that opens downward. The maximum height is given by
$$h(2) = -16(4) + 64(2) + 80 = 144, \text{ or } 144 \text{ feet.}$$

41. We compute $P'(x) = -0.08x + 240$. Setting $P'(x) = 0$ gives $x = 3000$. The graph of P is a parabola that opens downward and so $x = 3000$ gives rise to the absolute maximum of P. Thus, to maximize profits, the company should produce 3000 cameras per month.

43. $N'(t) = 0.81 - 1.14(\frac{1}{2}t^{-1/2}) = 0.81 - \frac{0.57}{t^{1/2}}$. Setting $N'(t) = 0$ gives $t^{1/2} = \frac{0.57}{0.81}$, or $t =$ 0.4952 as a critical point of N. Evaluating $N(t)$ at the endpoints $t = 0$ and $t = 6$ as well as at the critical point, we have

t	0	0.4952	6
$N(t)$	1.53	1.13	3.60

From the table, we see that the absolute maximum of N occurs at $t = 6$ and the absolute minimum occurs at $t \approx 0.5$. Our results tell us that the number of nonfarm full-time self- employed women over the time interval from 1963 to 1993 was the highest in 1993 and stood at approximately 3.6 million.

45. The revenue is $R(x) = px = -0.00042x^2 + 6x$. Therefore, the profit is

$$P(x) = R(x) - C(x) = -0.00042x^2 + 6x - (600 + 2x - 0.00002x^2)$$
$$= -0.0004x^2 + 4x - 600.$$
$$P'(x) = -0.0008x + 4 = 0$$

if $x = 5000$, a critical point of P. From the following table

x	0	5000	12000
$P(x)$	-600	9400	-10200

we see that Phonola should produce 5000 records/month.

47. $R(x) = px = -0.05x^2 + 600x$

$P(x) = R(x) - C(x) = -0.05x^2 + 600x - (0.000002x^3 - 0.03x^2 + 400x + 80000)$

$= -0.000002x^3 - 0.02x^2 + 200x - 80000.$

We want to maximize P on [0, 12,000]. $P'(x) = -0.000006x^2 - 0.04x + 200.$

Setting $P'(x) = 0$ gives $3x^2 + 20{,}000x - 100{,}000{,}000 = 0$

$$\text{or} \quad x = \frac{-20{,}000 \pm \sqrt{20{,}000^2 + 1{,}200{,}000{,}000}}{6} = -10{,}000, \text{ or } 3{,}333.3$$

So $x = 3{,}333.3$ is a critical point in the interval [0, 12,000].

x	0	3,333	12,000
$f(x)$	-80,000	290,370	-4,016,000

From the table, we see that a level of production of 3,333 units will yield a maximum profit.

49. $\overline{C}(x) = \dfrac{C(x)}{x} = 0.0025x + 80 + \dfrac{10{,}000}{x}.$

b. $\overline{C}'(x) = 0.0025 - \dfrac{10{,}000}{x^2} = 0$ if $0.0025x^2 = 10{,}000$, or $x = 2000$.

Since $\overline{C}''(x) = \dfrac{20{,}000}{x^3}$, we see that $\overline{C}''(x) > 0$ for $x > 0$ and so $\overline{C}$ is concave upward on $(0,\infty)$. Therefore, $x = 2000$ yields a minimum.

c. We solve $\overline{C}(x) = C'(x)$. $0.0025x + 80 + \dfrac{10{,}000}{x} = 0.005x + 80$, $0.0025x^2 = 10{,}000$, or $x = 2000$.

d. It appears that we can solve the problem in two ways.
REMARK This can be proved.

51. The demand equation is $p=\sqrt{800-x}=(800-x)^{1/2}$.
The revenue function is $R(x) = xp = x(800 - x)^{1/2}$.
To find the maximum of R, we compute

$$\begin{aligned} R'(x) &= \tfrac{1}{2}(800-x)^{-1/2}(-1)(x)+(800-x)^{1/2} \\ &= \tfrac{1}{2}(800-x)^{-1/2}[-x+2(800-x)] \\ &= \tfrac{1}{2}(800-x)^{-1/2}(1600-3x). \end{aligned}$$

Next, $R'(x) = 0$ implies $x = 800$ or $x = 1600/3$ are critical points of R. Next, we compute the values of R given in the following table.

x	0	800	1600/3
$R(x)$	0	0	8709

We conclude that $R(\frac{1600}{3}) = 8709$ is the absolute maximum value. Therefore, the revenue is maximized by producing $1600/3 \approx 533$ dresses.

53. $f(t)=100\left[\dfrac{t^2-4t+4}{t^2+4}\right]$.

a. $f'(t)=100\left[\dfrac{(t^2+4)(2t-4)-(t^2-4t+4)(2t)}{(t^2+4)^2}\right]=\dfrac{400(t^2-4)}{(t^2+4)^2}$

$$=\frac{400(t-2)(t+2)}{(t^2+4)^2}.$$

From the sign diagram for f'

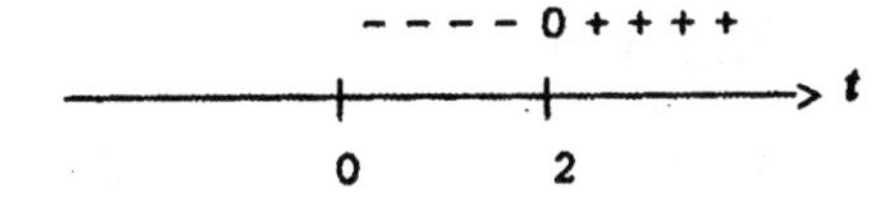

we see that $t = 2$ gives a relative minimum, and we conclude that the oxygen content is the lowest 2 days after the organic waste has been dumped into the pond.

b.

$$f''(t) = 400\left[\frac{(t^2+4)^2(2t)-(t^2-4)2(t^2+4)(2t)}{(t+4)^4}\right] = 400\left[\frac{(2t)(t^2+4)(t^2+4-2t^2+8)}{(t^2+4)^4}\right]$$

$$= -\frac{800t(t^2-12)}{(t^2+4)^3}$$

and $f''(t) = 0$ when $t = 0$ and $t = \pm 2\sqrt{3}$. We reject $t = 0$ and $t = -2\sqrt{3}$. From the sign diagram for f'',

```
         0 + + +0 - - - -
   ---------+-------+-----------> t
            0      2√3
```

we see that $f'(2\sqrt{3})$ gives an inflection point of f and we conclude that this is an absolute maximum. Therefore, the rate of oxygen regeneration is greatest 3.5 days after the organic waste has been dumped into the pond.

55. We compute $\overline{R}'(x) = \dfrac{xR'(x)-R(x)}{x^2}$. Setting $\overline{R}'(x) = 0$ gives $xR'(x)-R(x) = 0$ or

$$R'(x) = \frac{R(x)}{x} = \overline{R}(x),$$ so a critical point of $\overline{R}$ occurs when $\overline{R}(x) = R'(x)$.

Next, we compute

$$\overline{R}''(x) = \frac{x^2[R'(x)+xR''(x)-R'(x)-R'(x)]-[xR'(x)-R(x)](2x)}{x^4}$$

$$= \frac{R''(x)}{x} < 0.$$

So, by the Second Derivative Test, the critical point does give a maximum revenue.

57. The growth rate is $G'(t) = -0.6t^2 + 4.8t$. To find the maximum growth rate, we compute

$$G''(t) = -1.2t + 4.8.$$

Setting $G''(t) = 0$ gives $t = 4$ as a critical point.

t	0	4	8
$G'(t)$	0	9.6	0

From the table, we see that G is maximal at $t = 4$; that is, the growth rate is greatest in 1992.

59. $f(t) = -0.0129t^4 + 0.3087t^3 + 2.1760t^2 + 62.8466t + 506.2955$

To find the maximum of $f(t)$, we first compute

$$f'(t) = -0.0516t^3 + 0.9261t^2 + 4.352t + 62.8466.$$

Then

$f'(23.6811) = -0.0516(23.6811)^3 + 0.9261(23.6811)^2 + 4.352(23.6811) + 62.8466$

Next, we compute $f'(23) \approx 25.03$ and $f'(24) = -15.18$. From the sign diagram

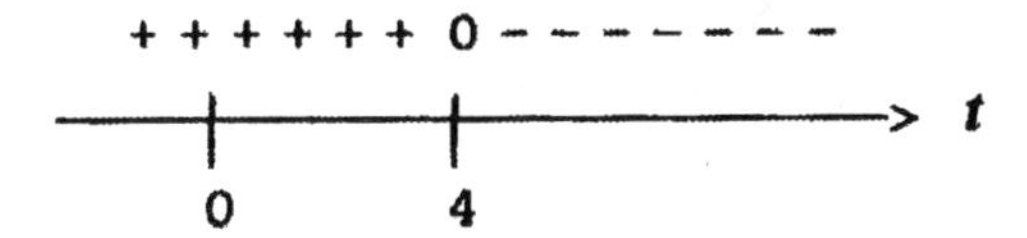

we conclude that $A(t)$ is maximized when $t = 4$. Therefore, the social security surplus will be maximized 4 decades from 1990, that is, in 2030.

61. Since $f(x) = c$ for all x, the function f satisfies $f(x) \leq c$ for all x and so f has an absolute maximum at all points of x. Similarly, f has an absolute minimum at all points of x.

63. a. g is not continuous at $x = 0$ because $\lim_{x \to 0} g(x)$ does not exist.

b. $\lim_{x \to 0} g(x) = \lim_{x \to 0^-} \frac{1}{x} = -\infty$ and $\lim_{x \to 0^+} g(x) = \lim_{x \to 0^+} \frac{1}{x} = \infty$

c.

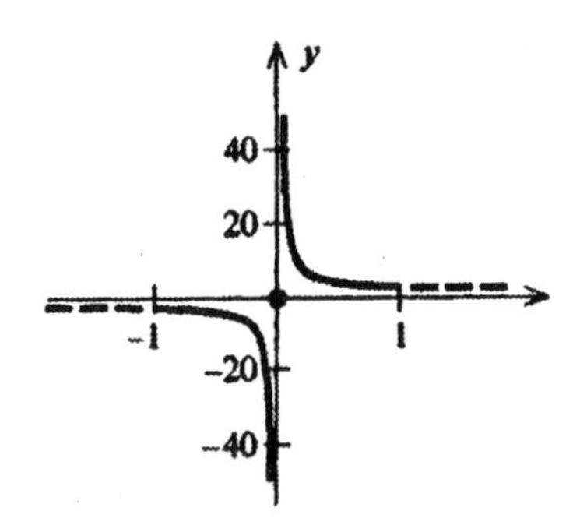

USING TECHNOLOGY EXERCISES 12.4, page 853

1. Absolute maximum value: 145.8985; absolute minimum value: -4.3834

3. Absolute maximum value: 16; absolute minimum value: -0.1257

5. Absolute maximum value: 2.8889; absolute minimum value: 0

7. a. b. 200.1410 banks/yr

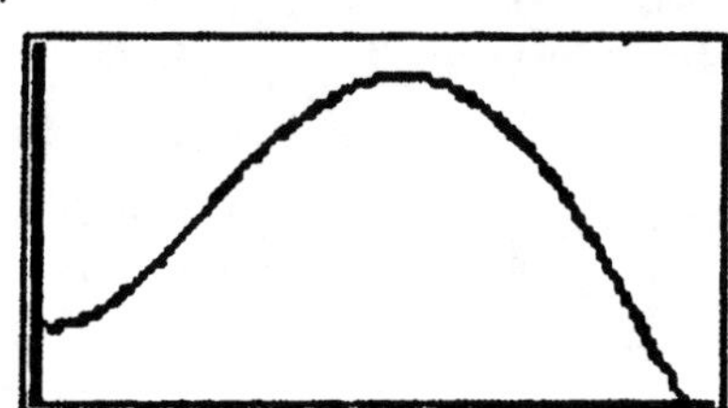

9. a.

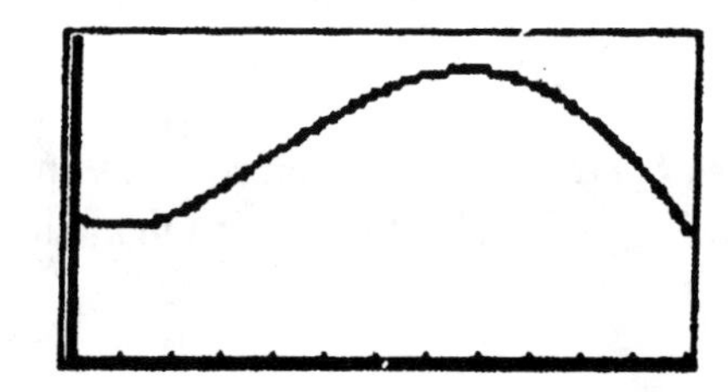

b. Absolute maximum value: 108.8756;
absolute minimum value: 49.7773

EXERCISES 12.5, page 763

1. Refer to the following figure.

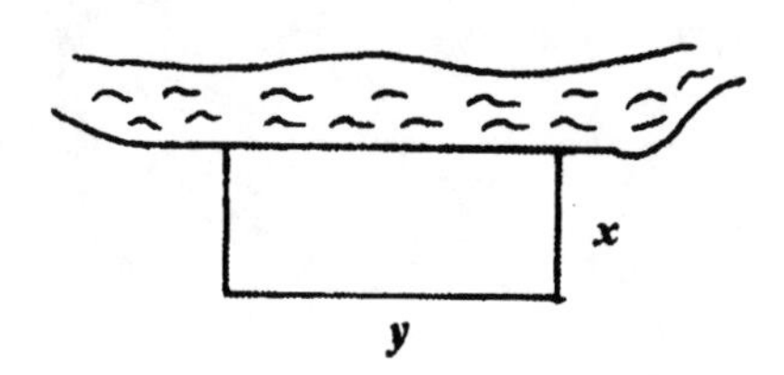

We have $2x + y = 3000$ and we want to maximize the function

$$A = f(x) = xy = x(3000 - 2x) = 3000x - 2x^2$$

on the interval [0,1500]. The critical point of A is obtained by solving $f'(x) = 3000 - 4x = 0$, giving $x = 750$. From the table of values

x	0	750	1500
$f(x)$	0	1,125,000	0

we conclude that $x = 750$ yields the absolute maximum value of A. Thus, the required dimensions are 750×1500 yards. The maximum area is 1,125,000 square yards.

3. Let x denote the length of the side made of wood and y the length of the side made of steel. The cost of construction will be $C = 6(2x) + 3y$.

   But $xy = 800$. So $y = 800/x$ and therefore $C = f(x) = 12x + 3\left(\frac{800}{x}\right) = 12x + \frac{2400}{x}$.

   To minimize C, we compute $f'(x) = 12 - \frac{2400}{x^2} = \frac{12x^2 - 2400}{x^2} = \frac{12(x^2 - 200)}{x^2}$.

   Setting $f'(x) = 0$ gives $x = \pm\sqrt{200}$ as critical points of f. The sign diagram of f'

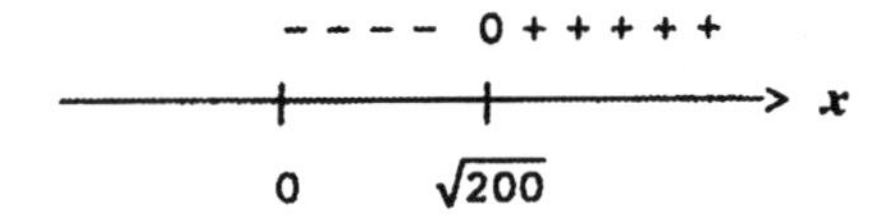

   shows that $x = \pm\sqrt{200}$ gives a relative minimum of f.

$$f''(x) = \frac{4800}{x^3} > 0$$

   if $x > 0$ and so f is concave upward for $x > 0$. Therefore $x = \sqrt{200} = 10\sqrt{2}$ actually yields the absolute minimum. So the dimensions of the enclosure should be

$$10\sqrt{2}\text{ ft} \times \frac{800}{10\sqrt{2}}\text{ ft, or } 14.1\text{ ft } x \text{ } 56.6\text{ ft.}$$

5. Let the dimensions of each square that is cut out be $x'' \times x''$. Refer to the following diagram.

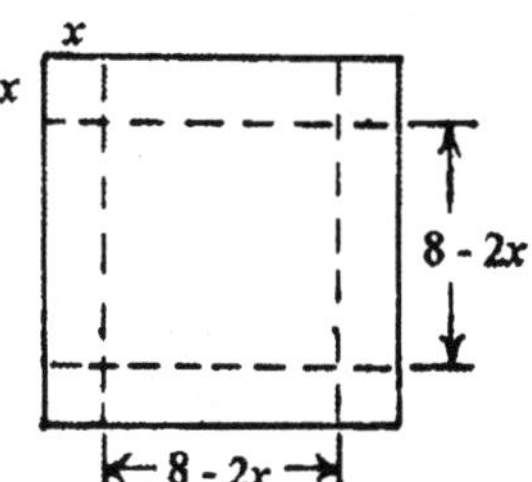

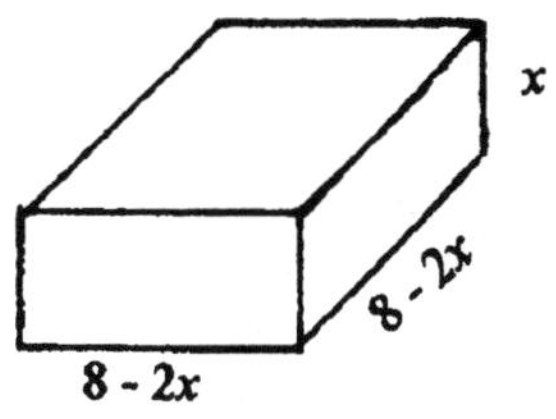

   Then the dimensions of the box will be $(8 - 2x)''$ by $(8 - 2x)''$ by x''. Its volume will be

$V = f(x) = x(8 - 2x)^2$. We want to maximize f on $[0,4]$.

$$f'(x) = (8 - 2x)^2 + x(2)(8 - 2x)(-2) \qquad \text{[Using the Product Rule.]}$$
$$= (8 - 2x)[(8 - 2x) - 4x] = (8 - 2x)(8 - 6x) = 0$$

if $x = 4$ or $4/3$. The latter is a critical point in $(0,4)$.

x	0	4/3	4
$f(x)$	0	1024/27	0

We see that $x = 4/3$ yields an absolute maximum for f. So the dimensions of the box should be $\frac{16}{3}" \times \frac{16}{3}" \times \frac{4}{3}"$.

7. Let x denote the length of the sides of the box and y denote its height. Referring to

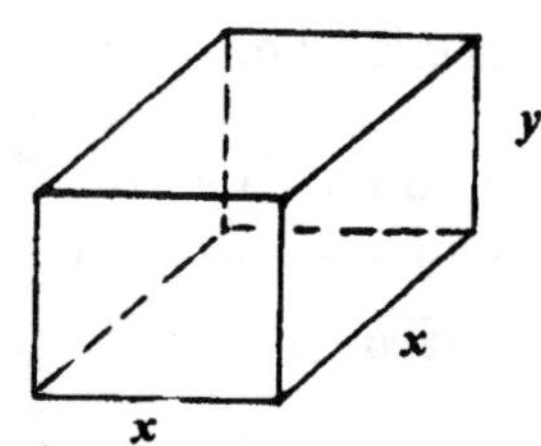

the following figure, we see that the volume of the box is given by $x^2y = 128$. The amount of material used is given by

$$S = f(x) = 2x^2 + 4xy$$
$$= 2x^2 + 4x\left(\frac{128}{x^2}\right)$$
$$= 2x^2 + \frac{512}{x} \text{ square inches.}$$

We want to minimize f subject to the condition that $x > 0$. Now

$$f'(x) = 4x - \frac{512}{x^2} = \frac{4x^3 - 512}{x^2} = \frac{4(x^3 - 128)}{x^2}.$$

Setting $f'(x) = 0$ yields $x = 5.04$, a critical point of f. Next,

$$f''(x) = 4 + \frac{1024}{x^3} > 0$$

for all $x > 0$. Thus, the graph of f is concave upward and so $x = 5.04$ yields an absolute minimum of f. Thus, the required dimensions are $5.04" \times 5.04" \times 5.04"$.

9. The length plus the girth of the box is $4x + h = 108$ and $h = 108 - 4x$. Then

$$V = x^2h = x^2(108 - 4x) = 108x^2 - 4x^3$$

and $V' = 216x - 12x^2$. We want to maximize V on the interval [0,27]. Setting $V'(x) = 0$ and solving for x, we obtain $x = 18$ and $x = 0$. Evaluating $V(x)$ at $x = 0$, $x = 18$, and $x = 27$, we obtain

$$V(0) = 0,\ V(18) = 11{,}664, \text{ and } V(27) = 0$$

Thus, the dimensions of the box are 18" × 18" × 36" and its maximum volume is approximately 11664 cu in.

11. We take $2\pi r + \ell = 108$. We want to maximize

$$V = \pi r^2 \ell = \pi r^2(-2\pi r + 108) = -2\pi^2 r^3 + 108\pi r^2$$

subject to the condition that $0 \le r \le \frac{54}{\pi}$. Now

$$V'(r) = -6\pi^2 r^2 + 216\pi r = -6\pi r(\pi r - 36).$$

Since $V' = 0$, we find $r = 0$ or $r = 36/\pi$, the critical points of V. From the table

r	0	$36/\pi$	$54/\pi$
V	0	$46{,}656/\pi$	0

we conclude that the maximum volume occurs when $r = 36/\pi \approx 11.5$ inches and $\ell = 108 - 2\pi\left(\frac{36}{\pi}\right) = 36$ inches and its volume is $46{,}656/\pi$ cu in .

13. Let y denote the height and x the width of the cabinet. Then $y = (3/2)x$. Since the volume is to be 2.4 cu ft, we have $xyd = 2.4$, where d *is* the depth of the cabinet. We have $x\left(\frac{3}{2}x\right)d = 2.4$ or $d = \frac{2.4(2)}{3x^2} = \frac{1.6}{x^2}$.

The cost for constructing the cabinet is

$$C = 40(2xd + 2yd) + 20(2xy) = 80\left[\frac{1.6}{x} + \left(\frac{3}{2}x\right)\left(\frac{1.6}{x^2}\right)\right] + 40x\left(\frac{3}{2}x\right)$$

$$= \frac{320}{x} + 60x^2.$$

$$C'(x) = -\frac{320}{x^2} + 120x = \frac{120x^3 - 320}{x^2} = 0 \text{ if } x = \sqrt[3]{\frac{8}{3}} = \frac{2}{\sqrt[3]{3}} = \frac{2}{3}\sqrt[3]{9}$$

and $x = \frac{2}{3}\sqrt[3]{9}$ point of C. The sign diagram shows that $x = \frac{2}{3}\sqrt[3]{9}$ gives a relative

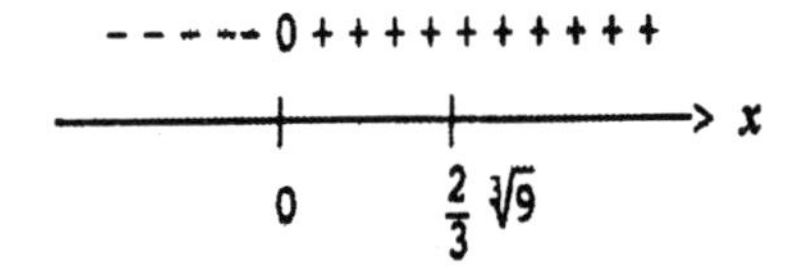

minimum. Next,

$$C''(x) = \frac{640}{x^3} + 120 > 0$$

for all $x > 0$ tells us that the graph of C is concave upward. So $x = \frac{2}{\sqrt[3]{9}}$ yields an absolute minimum. The required dimensions are $\frac{2}{3}\sqrt[3]{9}\,' \times \sqrt[3]{9}\,' \times \frac{2}{5}\sqrt[3]{9}\,'$.

15. We want to maximize the function

$$R(x) = (200 + x)(300 - x) = -x^2 + 100x + 60000.$$
$$R'(x) = -2x + 100 = 0$$

gives $x = 50$ and this is a critical point of R. Since $R''(x) = -2 < 0$, we see that $x = 50$ gives an absolute maximum of R. Therefore, the number of passengers should be 250. The fare will then be \$250/passenger and the revenue will be \$62,500.

17. We want to maximize $S = kh^2w$. But $h^2 + w^2 = 24^2$ or $h^2 = 576 - w^2$. So $S = f(w) = kw(576 - w^2) = k(576w - w^3)$. Now, setting

$$f'(w) = k(576 - 3w^2) = 0$$

gives $w = \pm\sqrt{192} \approx \pm 13.86$. Only the positive root is a critical point of interest. Next, we find $f''(w) = -6kw$, and in particular,

$$f''(\sqrt{192}) = -6\sqrt{192}\,k < 0,$$

so that $w = \pm\sqrt{192} \approx \pm 13.86$ gives a relative maximum of f. Since $f''(w) < 0$ for $w > 0$, we see that the graph of f is concave downward on $(0,\infty)$ and so, $w = \sqrt{192}$ gives an absolute maximum of f. We find $h^2 = 576 - 192 = 384$ or $h \approx 19.60$. So the width and height of the log should be approximately 13.86 inches and 19.60 inches, respectively.

19. We want to minimize $C(x) = (10{,}000 - x) + 3\sqrt{3000^2 + x^2}$ subject to $0 \le x \le 10{,}000$.

Now $$C'(x) = -1 + 3(\tfrac{1}{2})(9{,}000{,}000 + x^2)^{-1/2}(2x)$$
$$= -1 + \frac{3x}{\sqrt{9{,}000{,}000 + x^2}} = 0.$$

$$C'(x) = 0 \Rightarrow 3x = \sqrt{9{,}000{,}000 + x^2}$$
$$9x^2 = 9{,}000{,}000 + x^2$$

or $$x = \frac{3000}{\sqrt{8}} \approx 750\sqrt{2}.$$

x	0	$750\sqrt{2}$	10000
$f(x)$	19000	18485	31321

From the table, we see that $x = 750\sqrt{2}$ gives the absolute minimum.

21. The fuel cost is $x/400$ dollars per mile, and the labor cost is $8/x$ dollars per mile. Therefore, the total cost is $C(x) = \frac{8}{x} + \frac{x}{400}$; $C'(x) = -\frac{8}{x^2} + \frac{1}{400}$.

Setting $C'(x) = 0$ gives $-\frac{8}{x^2} = -\frac{1}{400}$; $x^2 = 3200$, and $x = 56.57$.

Next, $C''(x) = \frac{16}{x^3} > 0$ for all $x > 0$ so C is concave upward. Therefore, $x = 56.57$ gives the absolute minimum. So the most economical speed is 56.57 mph.

23. Let x denote the number of bottles in each order. We want to minimize

$$C(x) = 200\left(\frac{2{,}000{,}000}{x}\right) + \frac{x}{2}(0.40) = \frac{400{,}00{,}000}{x} + 0.2x.$$

We compute $C'(x) = -\frac{400{,}000{,}000}{x^2} + 0.2$. Setting $C'(x) = 0$ gives

$$x^2 = \frac{400{,}000{,}000}{0.2} = 2{,}000{,}000{,}000$$

or $x = 44{,}721$, a critical point of C.

$$C'(x) = \frac{800{,}000{,}000}{x^3} > 0 \text{ for all } x > 0,$$

and we see that the graph of C is concave upward and so $x = 44{,}721$ gives an absolute minimum of C. Therefore, there should be 45 orders per year, each order of 44,721 bottles.

CHAPTER 12 REVIEW, page 868

1. a. $f(x) = \frac{1}{3}x^3 - x^2 + x - 6$. $f'(x) = x^2 - 2x + 1 = (x - 1)^2$. $f'(x) = 0$ gives $x = 1$, the critical point of f. Now, $f'(x) > 0$ for all $x \neq 1$. Thus, f is increasing on $(-\infty, 1) \cup (1, \infty)$.

b. Since $f'(x)$ does not change sign as we move across the critical point $x = 1$, the First Derivative Test implies that $x = 1$ does not give rise to a relative extremum of f.

c. $f''(x) = 2(x - 1)$. Setting $f''(x) = 0$ gives $x = 1$ as a candidate for an inflection point of f. Since $f''(x) < 0$ for $x < 1$, and $f''(x) > 0$ for $x > 1$, we see that f is concave downward on $(-\infty,1)$ and concave upward on $(1,\infty)$.

d. The results of (c) imply that $(1, -\frac{17}{3})$ is an inflection point.

3. a. $f(x) = x^4 - 2x^2$. $f'(x) = 4x^3 - 4x = 4x(x^2 - 1) = 4x(x + 1)(x - 1)$. The sign diagram of f'

- - -0 + + 0 - - 0 + + +

−1 0 1 x

shows that f is decreasing on $(-\infty,-1) \cup (0,1)$ and increasing on $(-1,0) \cup (1,\infty)$.

b. The results of (a) and the First Derivative Test show that $(-1,-1)$ and $(1,-1)$ are relative minima and $(0,0)$ is a relative maximum.

c. $f''(x) = 12x^2 - 4 = 4(3x^2 - 1) = 0$ if $x = \pm\sqrt{3}/3$. The sign diagram

+ + + + 0 - - - - - - - 0 + + + +

$-\frac{\sqrt{3}}{3}$ 0 $\frac{\sqrt{3}}{3}$ x

shows that f is concave upward on $(-\infty, -\sqrt{3}/3) \cup (\sqrt{3}/3, \infty)$ and concave downward on $(-\sqrt{3}/3, \sqrt{3}/3)$.

d. The results of (c) show that $(-\sqrt{3}/3,-5/9)$ and $(\sqrt{3}/3,-5/9)$ are inflection points.

5. a. $f(x) = \dfrac{x^2}{x-1}$. $f'(x) = \dfrac{(x-1)(2x)-x^2(1)}{(x-1)^2} = \dfrac{x^2-2x}{(x-1)^2} = \dfrac{x(x-2)}{(x-1)^2}$.

The sign diagram of f'

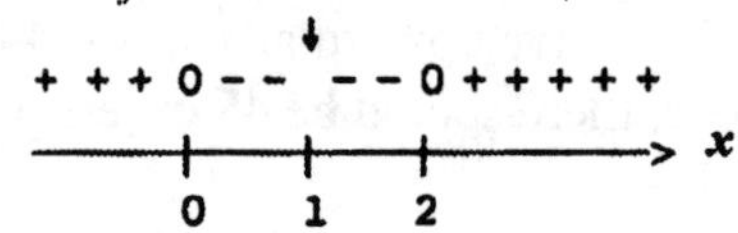

shows that f is increasing on $(-\infty,0) \cup (2,\infty)$ and decreasing on $(0,1) \cup (1,2)$.

b. The results of (a) show that $(0,0)$ is a relative maximum and $(2,4)$ is a relative minimum.

c. $$f''(x) = \frac{(x-1)^2(2x-2)-x(x-2)2(x-1)}{(x-1)^4} = \frac{2(x-1)[(x-1)^2 - x(x-2)]}{(x-1)^4}$$

$$= \frac{2}{(x-1)^3}.$$

Since $f''(x) < 0$ if $x < 1$ and $f''(x) > 0$ if $x > 1$, we see that f is concave downward on $(-\infty,1)$ and concave upward on $(1,\infty)$.
d. Since $x = 1$ is not in the domain of f, there are no inflection points.

7. $f(x) = (1 - x)^{1/3}$. $f'(x) = -\frac{1}{3}(1-x)^{-2/3} = -\frac{1}{3(1-x)^{2/3}}$.

The sign diagram for f' is

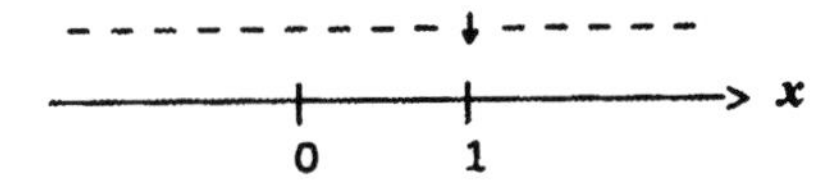

a. f is decreasing on $(-\infty,1) \cup (1,\infty)$.
b. There are no relative extrema.

Next, we compute $f''(x) = -\frac{2}{9}(1-x)^{-5/3} = -\frac{2}{9(1-x)^{5/3}}$.

The sign diagram for f'' is

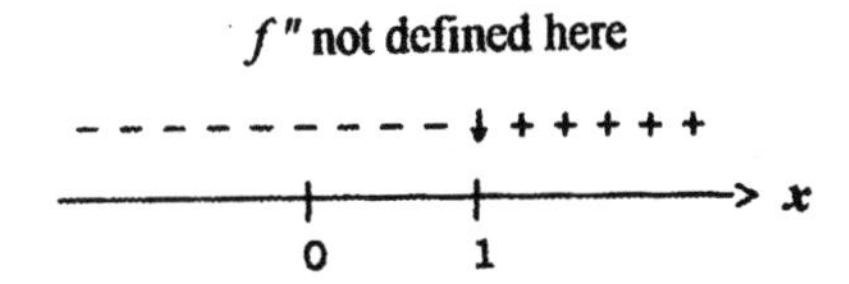

We find
c. f is concave downward on $(-\infty,1)$ and concave upward on $(1,\infty)$.
d. $x = 1$ is a candidate for an inflection point of f. Referring to the sign diagram for f'' we see that $(1,0)$ is an inflection point.

9. a. $f(x) = \frac{2x}{x+1}$. $f'(x) = \frac{(x+1)(2)-2x(1)}{(x+1)^2} = \frac{2}{(x+1)^2} > 0$ if $x \neq -1$.
Therefore f is increasing on $(-\infty,-1) \cup (-1,\infty)$.
b. Since there are no critical points, f has no relative extrema.
c. $f''(x) = -4(x+1)^{-3} = -\frac{4}{(x+1)^3}$.
Since $f''(x) > 0$ if $x < -1$ and $f''(x) < 0$ if $x > -1$, we see that f is concave upward on $(-\infty,-1)$ and concave downward on $(-1,\infty)$.
d. There are no inflection points since $f''(x) \neq 0$ for all x in the domain of f.

11. $f(x) = x^2 - 5x + 5$

1. The domain of f is $(-\infty, \infty)$.
2. Setting $x = 0$ gives 5 as the y-intercept.
3. $\lim_{x \to -\infty} (x^2 - 5x + 5) = \lim_{x \to \infty} (x^2 - 5x + 5) = \infty$.
4. There are no asymptotes because f is a quadratic function.
5. $f'(x) = 2x - 5 = 0$ if $x = 5/2$. The sign diagram

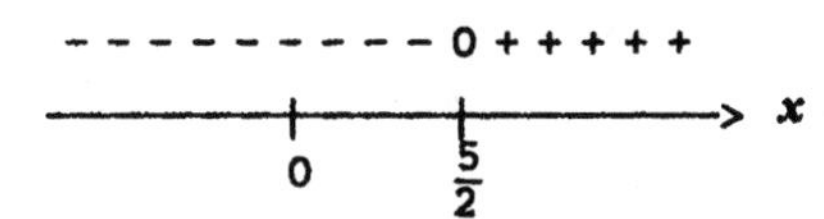

shows that f is increasing on $(\frac{5}{2}, \infty)$ and decreasing on $(-\infty, \frac{5}{2})$.

6. The First Derivative Test implies that $(\frac{5}{2}, -\frac{5}{4})$ is a relative minimum.
7. $f''(x) = 2 > 0$ and so f is concave upward on $(-\infty, \infty)$.
8. There are no inflection points.

The graph of f follows.

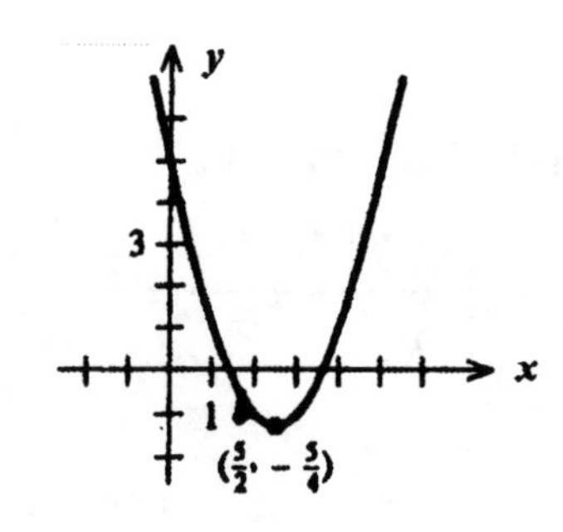

13. $g(x) = 2x^3 - 6x^2 + 6x + 1$.

1. The domain of g is $(-\infty, \infty)$.
2. Setting $x = 0$ gives 1 as the y-intercept.
3. $\lim_{x \to -\infty} g(x) = -\infty$, $\lim_{x \to \infty} g(x) = \infty$.
4. There are no vertical or horizontal asymptotes.
5. $g'(x) = 6x^2 - 12x + 6 = 6(x^2 - 2x + 1) = 6(x - 1)^2$. Since $g'(x) > 0$ for all $x \neq 1$, we see that g is increasing on $(-\infty, 1) \cup (1, \infty)$.
6. $g'(x)$ does not change sign as we move across the critical point $x = 1$, so there is no extremum.
7. $g''(x) = 12x - 12 = 12(x - 1)$. Since $g'(x) < 0$ if $x < 1$ and $g''(x) > 0$ if $x > 1$, we see that g is concave upward on $(1, \infty)$ and concave downward on $(-\infty, 1)$.
8. The point $x = 1$ gives rise to the inflection point $(1,3)$.
9. The graph of g follows.

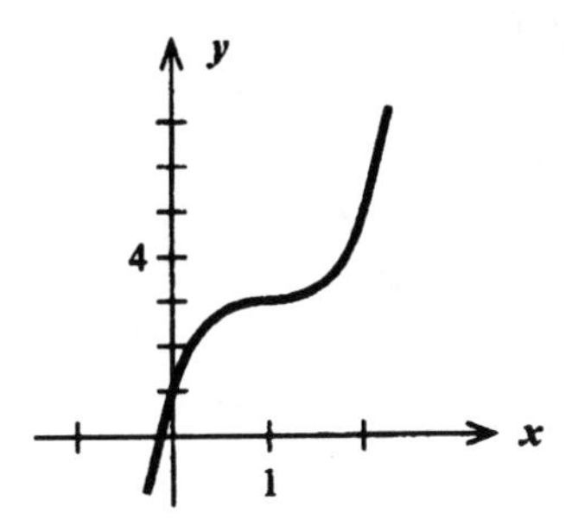

15. $h(x) = x\sqrt{x-2}$.

1. The domain of h is $[2,\infty)$.
2. There are no y-intercepts. Next, setting $y = 0$ gives 2 as the x-intercept.
3. $\lim_{x\to\infty} x\sqrt{x-2} = \infty$.
4. There are no asymptotes.
5. $h'(x) = (x-2)^{1/2} + x(\frac{1}{2})(x-2)^{-1/2} = \frac{1}{2}(x-2)^{-1/2}[2(x-2)+x]$

$$= \frac{3x-4}{2\sqrt{x-2}} > 0 \text{ on } [2,\infty)$$

and so h is increasing on $[2,\infty)$.

6. Since h has no critical points in $(2,\infty)$, there are no relative extrema.

7. $$h''(x) = \frac{1}{2}\left[\frac{(x-2)^{1/2}(3) - (3x-4)\frac{1}{2}(x-2)^{-1/2}}{x-2}\right]$$

$$= \frac{(x-2)^{-1/2}[6(x-2)-(3x-4)]}{4(x-2)} = \frac{3x-8}{4(x-2)^{3/2}}.$$

The sign diagram for h''

- - .- 0 + + + + +

2 $\frac{8}{3}$ x

shows that h is concave downward on $(2,\frac{8}{3})$ and concave upward on $(\frac{8}{3},\infty)$.

8. The results of (7) tell us that $(\frac{8}{3},\frac{8\sqrt{6}}{9})$ is an inflection point.

The graph of h follows.

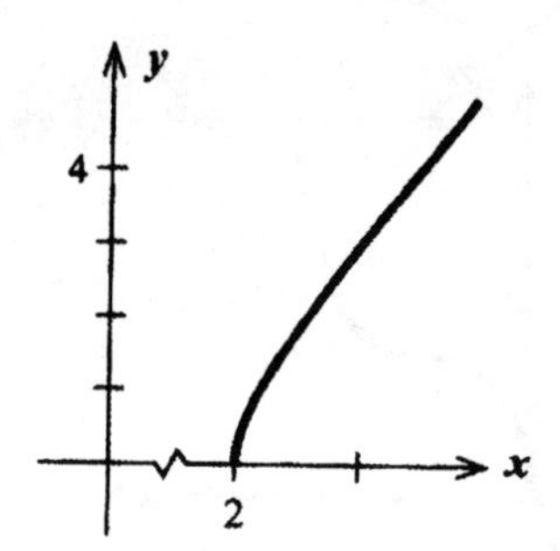

17. $f(x) = \dfrac{x-2}{x+2}$.

1. The domain of f is $(-\infty,-2) \cup (-2,\infty)$.
2. Setting $x = 0$ gives -1 as the y-intercept. Setting $y = 0$ gives 2 as the x-intercept.
3. $\lim\limits_{x\to-\infty} \dfrac{x-2}{x+2} = \lim\limits_{x\to\infty} \dfrac{x-2}{x+2} = 1$.
4. The results of (3) tell us that $y = 1$ is a horizontal asymptote. Next, observe that the denominator of $f(x)$ is equal to zero at $x = -2$, but its numerator is not equal to zero there. Therefore, $x = -2$ is a vertical asymptote.
5. $$f'(x) = \frac{(x+2)(1)-(x-2)(1)}{(x+2)^2} = \frac{4}{(x+2)^2}.$$

The sign diagram of f'

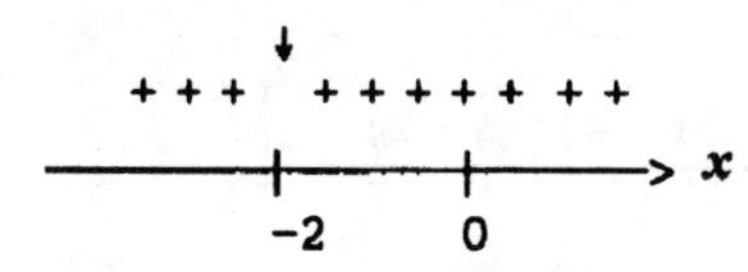

tells us that f is increasing on $(-\infty,-2) \cup (-2,\infty)$.

6. The results of (5) tells us that there are no relative extrema.
7. $f''(x) = -\dfrac{8}{(x+2)^3}$. The sign diagram of f'' follows

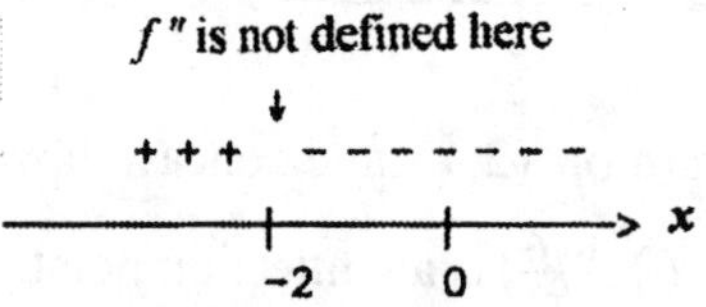

and it shows that f is concave upward on $(-\infty,-2)$ and concave downward on $(-2,\infty)$.

8. There are no inflection points.
The graph of f follows.

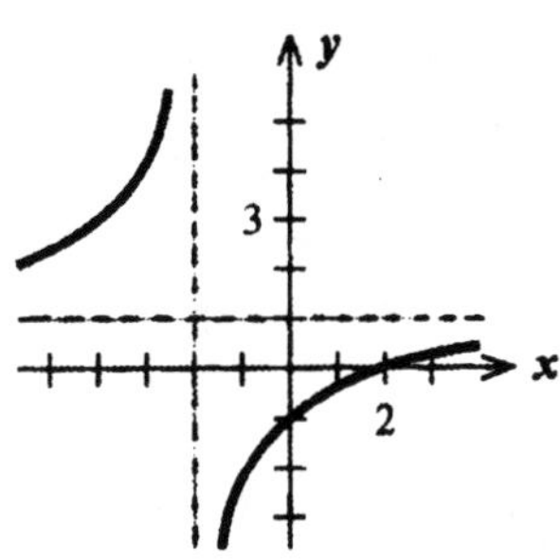

19. $\lim_{x\to-\infty} \frac{1}{2x+3} = \lim_{x\to\infty} \frac{1}{2x+3} = 0$ and so $y = 0$ is a horizontal asymptote. Since the denominator is equal to zero at $x = -3/2$, but the numerator is not equal to zero there, we see that $x = -3/2$ is a vertical asymptote.

21. $\lim_{x\to-\infty} \frac{5x}{x^2-2x-8} = \lim_{x\to\infty} \frac{5x}{x^2-2x-8} = 0$ and so $y = 0$ is a horizontal asymptote. Next,

note that the denominator is zero if $x^2 - 2x - 8 = (x - 4)(x + 2) = 0$, or $x = -2$ or $x = 4$. Since the numerator is not equal to zero at these points, we see that $x = -2$ and $x = 4$ are vertical asymptotes.

23. $f(x) = 2x^2 + 3x - 2$; $f'(x) = 4x + 3$. Setting $f'(x) = 0$ gives $x = -3/4$ as a critical point of f. Next, $f''(x) = 4 > 0$ for all x, so f is concave upward on $(-\infty, \infty)$. Therefore, $f(-\frac{3}{4}) = -\frac{25}{8}$ is an absolute minimum of f. There is no absolute maximum.

25. $g(t) = \sqrt{25-t^2} = (25-t^2)^{1/2}$. Differentiating $g(t)$, we have

$$g'(t) = \tfrac{1}{2}(25-t^2)^{-1/2}(-2t) = -\frac{t}{\sqrt{25-t^2}}.$$

Setting $g'(t) = 0$ gives $t = 0$ as a critical point of g. The domain of g is given by solving the inequality $25 - t^2 \geq 0$ or $(5 - t)(5 + t) \geq 0$ which implies that $t \in [-5,5]$. From the table

t	-5 0 5
$g(t)$	0 5 0

we conclude that $g(0) = 5$ is the absolute maximum of g and $g(-5) = 0$ and $g(5) = 0$ is the absolute minimum value of g.

27. $h(t) = t^3 - 6t^2$. $h'(t) = 3t^2 - 12t = 3t(t-4) = 0$ if $t = 0$ or $t = 4$, critical points of h. But only $t = 4$ lies in (2,5).

t	2	4	5
$h(t)$	-16	-32	-25

From the table, we see that there is an absolute minimum at (4,-32) and an absolute maximum at (2,-16).

29. $f(x) = x - \frac{1}{x}$ on [1,3]. $f'(x) = 1 + \frac{1}{x^2}$. Since $f'(x)$ is never zero, f has no critical point.

x	1	3
$f(x)$	0	$\frac{8}{3}$

We see that $f(1) = 0$ is the absolute minimum value and $f(3) = 8/3$ is the absolute maximum value.

31. $f(x) = s\sqrt{1-s^2}$ on [-1,1]. The function f is continuous on [-1,1] and differentiable on (-1,1). Next,

$$f'(s) = (1-s^2)^{1/2} + s(\tfrac{1}{2})(1-s^2)^{-1/2}(-2s) = \frac{1-2s^2}{\sqrt{1-s^2}}.$$

Setting $f'(s) = 0$, we have $s = \pm\sqrt{2}/2$, giving the critical points of f. From the table

x	-1	$-\sqrt{2}/2$	$\sqrt{2}/2$	1
$f(x)$	0	-1/2	1/2	0

we see that $f(-\sqrt{2}/2) = -1/2$ is the absolute minimum value and $f(\sqrt{2}/2) = 1/2$ is the absolute maximum value of f.

33. We want to maximize $P(x) = -x^2 + 8x + 20$. Now, $P'(x) = -2x + 8 = 0$ if $x = 4$, a critical point of P. Since $P''(x) = -2 < 0$, the graph of P is concave downward. Therefore, the critical point $x = 4$ yields an absolute maximum. So, to maximize profit, the company should spend \$4000 on advertising per month.

35. The revenue is $R(x) = px = x(-0.0005x^2 + 60) = -0.0005x^3 + 60x$. Therefore, the total profit is $P(x) = R(x) - C(x) = -0.0005x^3 + 0.001x^2 + 42x - 4000$.

$$P'(x) = -0.0015x^2 + 0.002x + 42.$$

Setting $P'(x) = 0$, we have $3x^2 - 4x - 84{,}000 = 0$. Solving for x, we find

$$x = \frac{4 \pm \sqrt{16 - 4(3)(84{,}000)}}{2(3)} = \frac{4 \pm 1004}{6} = 168, \text{ or } -167.$$

We reject the negative root. Next,

$$P''(x) = -0.003x + 0.002 \text{ and } P''(168) = -0.003(168) + 0.002 = -0.502 < 0.$$

By the Second Derivative Test, $x = 168$ gives a relative maximum. Therefore, the required level of production is 168 video discs.

37. $N(t) = -2t^3 + 12t^2 + 2t$. We wish to find the inflection point of the function N. Now, $N'(t) = -6t^2 + 24t + 2$ and $N''(t) = -12t + 24 = -12(t - 2)$.
Setting $N''(t) = 0$ gives $t = 2$. Furthermore, $N''(t) > 0$ when $t < 2$ and $N''(t) < 0$ when $t > 2$. Therefore, $t = 2$ is an inflection point of N. Thus, the average worker is performing at peak efficiency at 10 A.M.

39. Let x denote the number of cases in each order. Then the average number of cases of beer in storage during the year is $x/2$. The storage cost is $2(x/2)$, or x dollars. Next, we see that the number of orders required is $800{,}000/x$, and so the ordering cost is

$$\frac{500(800{,}000)}{x} = \frac{400{,}000{,}000}{x}$$

dollars. Thus, the total cost incurred by the company per year is given by

$$C(x) = x + \frac{400{,}000{,}000}{x}.$$

We want to minimize C in the interval $(0, \infty)$. Now

$$C'(x) = 1 - \frac{400{,}000{,}000}{x^2}.$$

Setting $C'(x) = 0$ gives $x^2 = 400{,}000{,}000$, or $x = 20{,}000$ (we reject $x = -20{,}000$).

Next, $C''(x) = \dfrac{800{,}000{,}000}{x^3} > 0$ for all x, so C is concave upward. Thus, $x = 20{,}000$ gives rise to the absolute minimum of C. Thus, the company should order 20,000 cases of beer per order.

CHAPTER 13

EXERCISES 13.1, page 880

1. a. $4^{-3} \times 4^5 = 4^{-3+5} = 4^2 = 16$ b. $3^{-3} \times 3^6 = 3^{6-3} = 3^3 = 27.$

3. a. $9(9)^{-1/2} = \dfrac{9}{9^{1/2}} = \dfrac{9}{3} = 3.$ b. $5(5)^{-1/2} = 5^{1/2} = \sqrt{5}.$

5. a. $\dfrac{(-3)^4(-3)^5}{(-3)^8} = (-3)^{4+5-8} = (-3)^1 = -3.$
   b. $\dfrac{(2^{-4})(2^6)}{2^{-1}} = 2^{-4+6+1} = 2^3 = 8.$

7. a. $(64x^9)^{1/3} = 64^{1/3}(x^{9/3}) = 4x^3.$
   b. $(25x^3y^4)^{1/2} = 25^{1/2}(x^{3/2})(y^{4/2}) = 5x^{3/2}y^2 = 5xy^2\sqrt{x}.$

9. a. $\dfrac{6a^{-5}}{3a^{-3}} = 2a^{-5+3} = 2a^{-2} = \dfrac{2}{a^2}.$ b. $\dfrac{4b^{-4}}{12b^{-6}} = \dfrac{1}{3}b^{-4+6} = \dfrac{1}{3}b^2.$

11. a. $(2x^3y^2)^3 = 2^3 \times x^{3(3)} \times y^{2(3)} = 8x^9y^6.$
    b. $(4x^2y^2z^3)^2 = 4^2 \times x^{2(2)} \times y^{2(2)} \times z^{3(2)} = 16x^4y^4z^6.$

13. $6^{2x} = 6^4$ if and only if $2x = 4$ or $x = 2$.

15. $3^{3x-4} = 3^5$ if and only if $3x - 4 = 5$, $3x = 9$, or $x = 3$.

17. $(2.1)^{x+2} = (2.1)^5$ if and only if $x + 2 = 5$, or $x = 3$.

19. $8^x = (\frac{1}{32})^{x-2}$, $(2^3)^x = (32)^{2-x} = (2^5)^{2-x}$, so $2^{3x} = 2^{5(2-x)}$, $3x = 10 - 5x$, $8x = 10$, or $x = 5/4$.

21. $y = 2^x$, $y = 3^x$, and $y = 4^x$

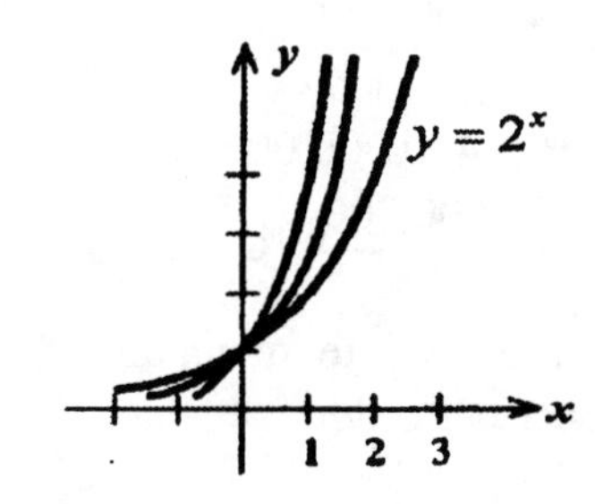

23. $y = 2^{-x}, y = 3^{-x}$, and $y = 4^{-x}$

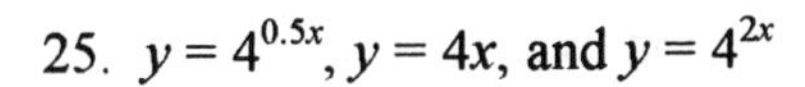
25. $y = 4^{0.5x}, y = 4x$, and $y = 4^{2x}$

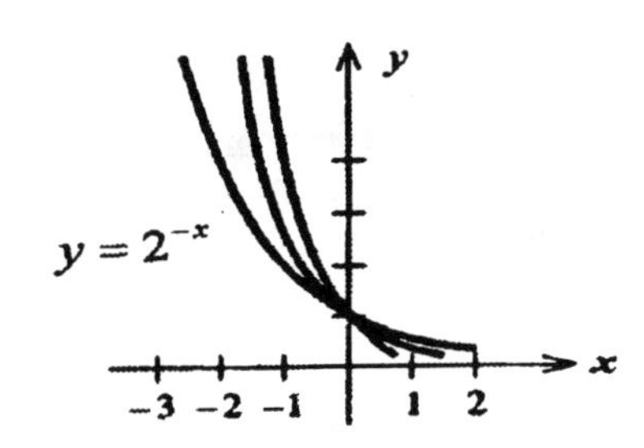

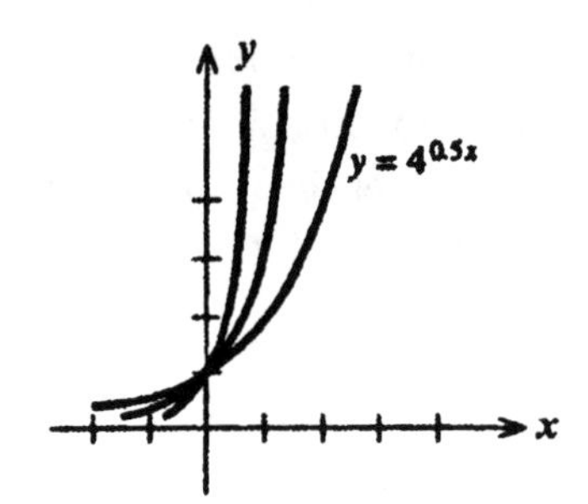

27. $y = e^{0.5x}, y = e^{x}, y = e^{1.5x}$

29. $y = 0.5e^{-x}, y = e^{-x}$, and $y = 2e^{-x}$

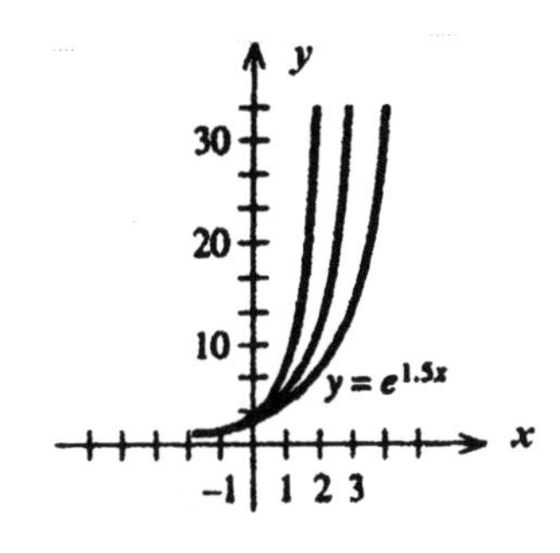

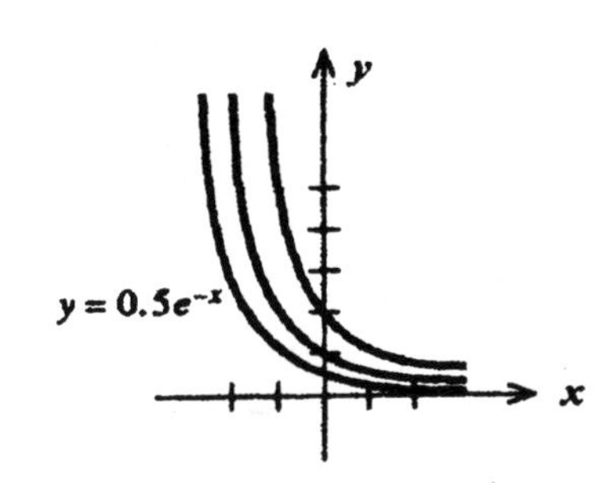

31. Suppose \$1 is invested in each investment.

Investment A: Accumulated amount is $\left(1+\dfrac{0.1}{2}\right)^{8} \approx 1.47746$.

Investment B: Accumulated amount is $e^{0.0975(4)} \approx 1.47698$.
So Investment A has a higher rate of return.

33. a. If they invest the money at 10.5 percent compunded quarterly, they should set aside $P = 70{,}000\left(1+\dfrac{0.105}{4}\right)^{-28} = 33{,}885.14$, or \$33,815.14.

b. If they invest the money at 10.5 percent compounded continuously, they should set aside $P = 70{,}000e^{-0.735} = 33{,}565.38$, or \$33,565.38.

35. a. If inflation over the next 15 years is 6 percent, then Ms. Lindtrom's first year's pension will be worth $P = 40{,}000e^{-0.9} = 16{,}262.79$, or \$16,262.79.
b. If inflation over the next 15 years is 8 percent, then Ms. Lindtrom's first year's pension will be worth $P = 40{,}000e^{-1.2} = 12{,}047.77$, or \$12,047.77.
c. If inflation over the next 15 years is 12 percent, then Ms. Lindtrom's first year's pension will be worth $P = 40{,}000e^{-1.8} = 6611.96$, or \$6,611.96.

37. $r_{eff} = \lim_{m\to\infty}\left(1+\dfrac{r}{m}\right)^m - 1 = e^r - 1.$

39. The effective rate of interest at Bank A is given by $R = \left(1+\dfrac{0.07}{4}\right)^4 - 1 = 0.07186,$

or 7.186 percent. The effective rate of interest at Bank B is given by
$R = e^e - 1 = e^{0.07125} - 1 = 0.07385$, or 7.385 percent.
We conclude that Bank B has the higher effective rate of interest.

USING TECHNOLOGY EXERCISES 13.1, page 883

1.

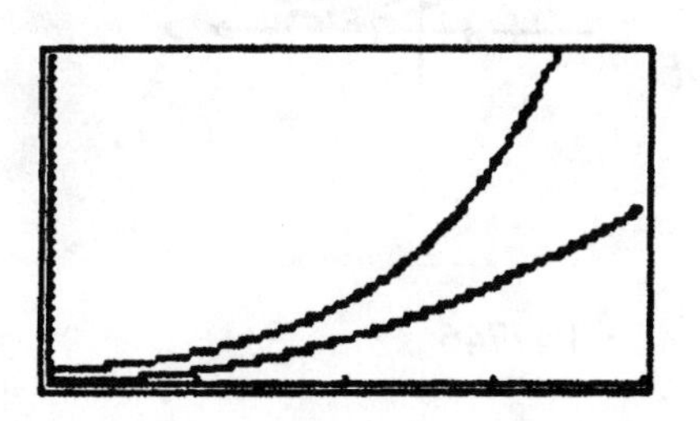

3.

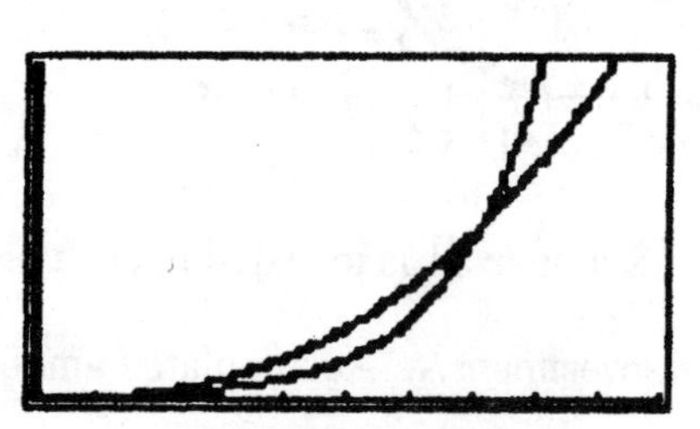

5.

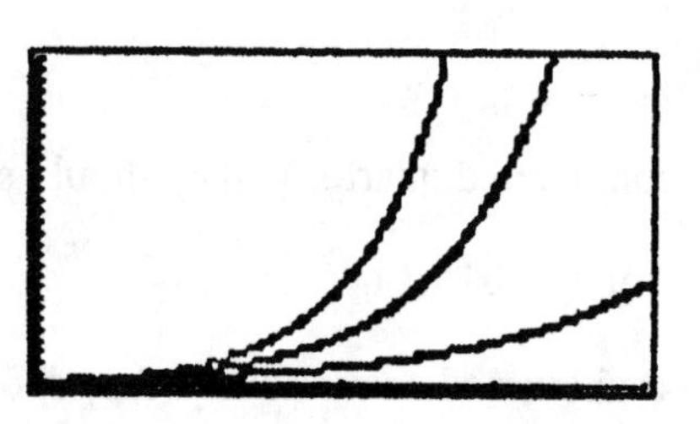

7.

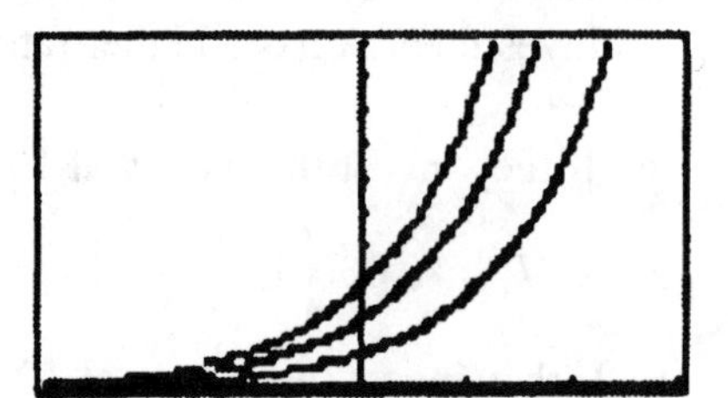

9.

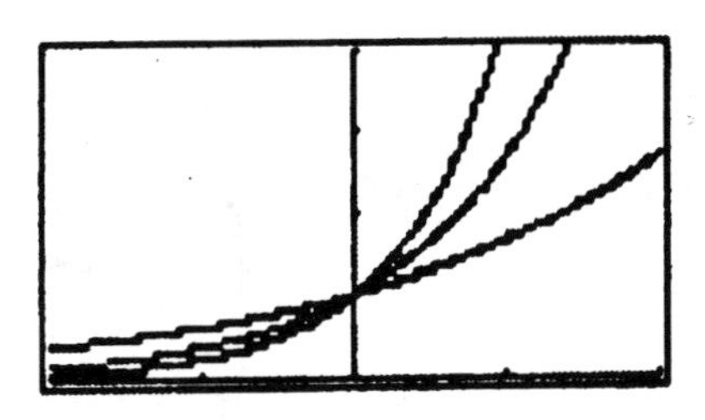

EXERCISES 13.2 , page 892

1. $\log_2 64 = 6$

3. $\log_3 \dfrac{1}{9} = -2$

5. $\log_{1/3} \dfrac{1}{3} = 1$

7. $\log_{32} 8 = \dfrac{3}{5}$

9. $\log_{10} 0.001 = -3$

11. $\log 12 = \log 4 \times 3 = \log 4 + \log 3 = 0.6021 + 0.4771 = 1.0792.$

13. $\log 16 = \log 4^2 = 2 \log 4 = 2(0.6021) = 1.2042.$

15. $\log 48 = \log 3 \times 4^2 = \log 3 + 2 \log 4 = 0.4771 + 2(0.6021) = 1.6813.$

17. $\log x(x + 1)^4 = \log x + \log (x + 1)^4 = \log x + 4 \log (x + 1).$

19. $\log \dfrac{\sqrt{x+1}}{x^2+1} = \log (x + 1)^{1/2} - \log(x^2 + 1) = \frac{1}{2} \log (x + 1) - \log (x^2 + 1)$

21. $\ln xe^{-x^2} = \ln x - x^2.$

23. $\ln \left(\dfrac{x^{1/2}}{x^2\sqrt{1+x^2}} \right) = \ln x^{1/2} - \ln x^2 - \ln (1 + x^2)^{1/2}$

$= \frac{1}{2} \ln x - 2 \ln x - \frac{1}{2} \ln (1 + x^2) = -\frac{3}{2} \ln x - \frac{1}{2} \ln (1 + x^2).$

25. $\ln x^x = x \ln x.$

27. $y = \log_3 x$

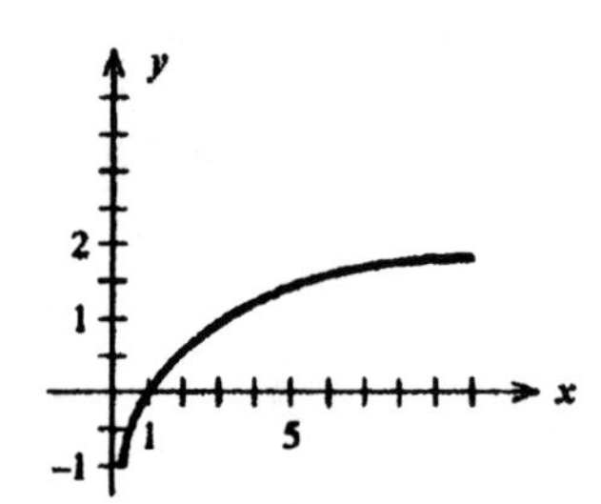

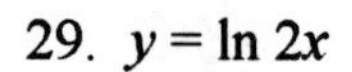

29. $y = \ln 2x$

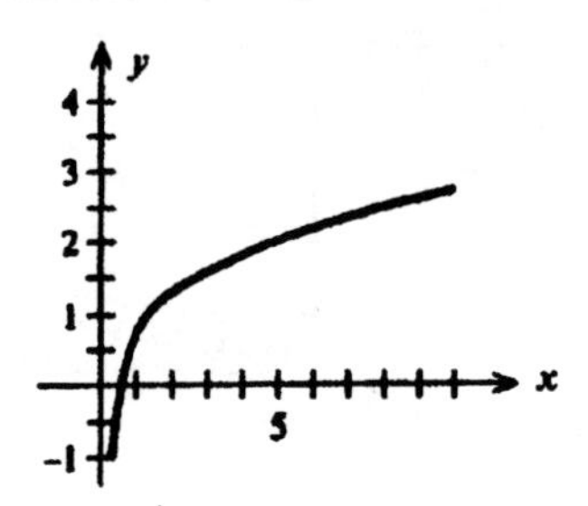

31. $y = 2^x$ and $y = \log_2 x$

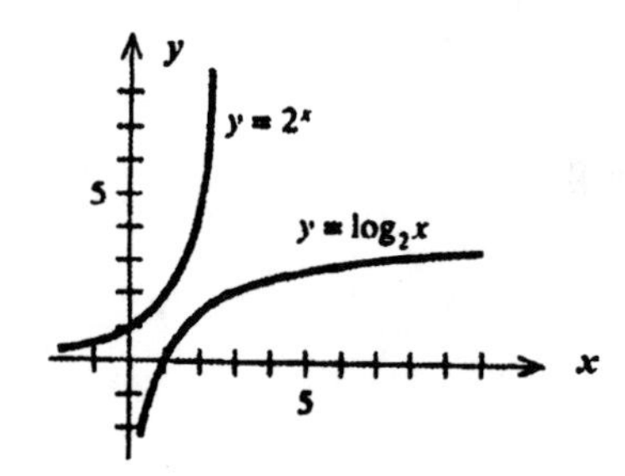

33. $e^{0.4t} = 8$, $0.4t \ln e = \ln 8$, and $0.4t = \ln 8 (\ln e = 1.)$ So, $t = \dfrac{\ln 8}{0.4} = 5.1986$.

35. $5e^{-2t} = 6$, $e^{-2t} = \dfrac{6}{5} = 1.2$. Taking the logarithm, we have

$$-2t \ln e = \ln 1.2, \text{ or } t = t = -\frac{\ln 1.2}{2} \approx -0.0912.$$

37. $2e^{-0.2t} - 4 = 6$, $2e^{-0.2t} = 10$. Taking the logarithm, we have

$$\ln e^{-0.2t} = \ln 5, \; -0.2t \ln e = \ln 5, \; -0.2t = \ln 5,$$

and $\quad t = -\dfrac{\ln 5}{0.2} \approx -8.0472.$

39. $\dfrac{50}{1+4e^{0.2t}} = 20$, $1+4e^{0.2t} = \dfrac{50}{20} = 2.5$, $4e^{0.2t} = 1.5$,

$e^{0.2t} = \dfrac{1.5}{4} = 0.375$, $\ln e^{0.2t} = \ln 0.375$, $0.2t = \ln 0.375$. So $t = \dfrac{\ln 0.375}{0.2} \approx -4.9041$.

41. Taking the logarithm on both sides, we obtain

$\ln A = \ln Be^{-t/2}$, $\ln A = \ln B + \ln e^{-t/2}$, $\ln A - \ln B = -t/2 \ln e$,

$\ln \frac{A}{B} = -\frac{t}{2}$ or $t = -2 \ln \frac{A}{B} = 2 \ln \frac{B}{A}$

43. $p(x) = 19.4 \ln x + 18$. For a child weighing 92 lb, we find
$p(92) = 19.4 \ln 92 + 18 = 105.72$ millimeters of mercury.

45. a. $30 = 10 \log \frac{I}{I_0}$; $3 = \log \frac{I}{I_0}$; $\frac{I}{I_0} = 10^3 = 1000.$

So $I = 1000\, I_0$.

b. When $D = 80$, $I = 10^8 I_0$ and when $D = 30$, $I = 10^3 I_0$. Therefore, an 80–decibel sound is $10^8/10^3$ or $10^5 = 100{,}000$ times louder than a 30–decibel sound.

c. It is $10^{15}/10^8 = 10^7$, or 10,000,000, times louder.

47. With $T_0 = 70$, $T_1 = 98.6$, and $T = 80$, we have

$$80 = 70 + (98.6 - 70)(0.97)^t$$
$$28.6(0.97)^t = 10,$$
$$(0.97)^t = 0.34965.$$

Taking logarithms, we have

$$\ln (0.97)^t = \ln 0.34965, \text{ or } t = \frac{\ln 0.34965}{\ln 0.97} \approx 34.50.$$

So he was killed 34½ hours earlier at 1:30 P.M.

49. a. Let $p = \log_b m$ and $q = \log_b n$ so that $m = b^p$ and $n = b^q$. Then $mn = b^p b^q = b^{p+q}$ and by definition, $p + q = \log_b mn$; that is, $\log_b mn = \log_b m + \log_b n$.

b. $\frac{m}{n} = \frac{b^p}{b^q} = b^{p-q}$. So, by definition, $p - q = \log_b \frac{m}{n}$; that is $\log_b \frac{m}{n} = \log_b m - \log_b n$.

51. a. By definition $\log_b 1 = 0$ means $1 = b^0 = 1$.
b. By definition $\log_b b = 1$ means $b = b^1 = b$.

EXERCISES 13.3 , page 901

1. $f(x) = e^{3x}$; $f'(x) = 3e^{3x}$

3. $g(t) = e^{-t}$; $g'(t) = -e^{-t}$

5. $f(x) = e^x + x;\ f'(x) = e^x + 1$

7. $f(x) = x^3e^x,\ f'(x) = x^3e^x + e^x(3x^2) = x^2e^x(x + 3).$

9. $f(x) = \dfrac{2e^x}{x},\ f'(x) = \dfrac{x(2e^x) - 2e^x(1)}{x^2} = \dfrac{2e^x(x-1)}{x^2}.$

11. $f(x) = 3(e^x + e^{-x});\ f'(x) = 3(e^x - e^{-x}).$

13. $f(w) = \dfrac{e^w + 1}{e^w} = 1 + \dfrac{1}{e^w} = 1 + e^{-w}.\ f'(w) = -e^{-w} = -\dfrac{1}{e^w}.$

15. $f(x) = 2e^{3x-1},\ f'(x) = 2e^{3x-1}(3) = 6e^{3x-1}.$

17. $h(x) = e^{-x^2};\ h'(x) = e^{-x^2}(-2x) = -2xe^{-x^2}.$

19. $f(x) = 3e^{-1/x};\ f'(x) = 3e^{-1/x} \cdot \dfrac{d}{dx}\left(-\dfrac{1}{x}\right) = 3e^{-1/x}\left(\dfrac{1}{x^2}\right) = \dfrac{3e^{-1/x}}{x^2}.$

21. $f(x) = (e^x + 1)^{25},\ f'(x) = 25(e^x + 1)^{24}e^x = 25e^x(e^x + 1)^{24}.$

23. $f(x) = e^{\sqrt{x}};\ f'(x) = e^{\sqrt{x}}\dfrac{d}{dx}x^{1/2} = e^{\sqrt{x}}\dfrac{1}{2}x^{-1/2} = \dfrac{e^{\sqrt{x}}}{2\sqrt{x}}.$

25. $f(x) = (x - 1)e^{3x+2};\ f'(x) = (x - 1)(3)e^{3x+2} + e^{3x+2} = e^{3x+2}(3x - 3 + 1) = e^{3x+2}(3x - 2).$

27. $f(x) = \dfrac{e^x - 1}{e^x + 1};\ f'(x) = \dfrac{(e^x+1)(e^x) - (e^x - 1)(e^x)}{(e^x+1)^2} = \dfrac{e^x(e^x + 1 - e^x + 1)}{(e^x+1)^2} = \dfrac{2e^x}{(e^x+1)^2}.$

29. $f(x) = e^{-4x} + 2e^{3x};\ f(x) = -4e^{-4x} + 6e^{3x}$ and $f\ ''(x) = 16e^{-4x} + 18e^{3x} = 2(8e^{-4x} + 9e^{3x}).$

31. $f(x) = 2xe^{3x};\ f'(x) = 2e^{3x} + 2xe^{3x}(3) = 2(3x + 1)e^{3x}.$
$f''(x) = 6e^{3x} + 2(3x + 1)e^{3x}(3) = 6(3x + 2)e^{3x}.$

33. $y = f(x) = e^{2x-3}.\ f'(x) = 2e^{2x-3}.$ To find the slope of the tangent line to the graph of f at $x = 3/2$, we compute $f'(\tfrac{3}{2}) = 2e^{3-3} = 2.$
Next, using the point–slope form of the equation of a line, we find that

$$y - 1 = 2(x - \tfrac{3}{2})$$
$$= 2x - 3, \qquad \text{or} \quad y = 2x - 2.$$

35. $f(x) = e^{-x^2/2}$, $f'(x) = e^{-x^2/2}(-x) = -xe^{-x^2/2}$. Setting $f'(x) = 0$, gives $x = 0$ as the only critical point of f. From the sign diagram,

+ + + + + + + 0 - - - - -

——————+——————> x

0

we conclude that f is increasing on $(-\infty,0)$ and decreasing on $(0,\infty)$.

37. $f(x) = \frac{1}{2}e^{x} - \frac{1}{2}e^{-x}$, $f'(x) = \frac{1}{2}(e^{x} + e^{-x})$, $f''(x) = \frac{1}{2}(e^{x} - e^{-x})$. Setting $f''(x) = 0$, gives $e^{x} = e^{-x}$ or $e^{2x} = 1$, and $x = 0$. From the sign diagram for f'',

- - - - - - -0 + + + ++ + + +

——————+——————> x

0

we conclude that f is concave upward on $(0,\infty)$ and concave downward on $(-\infty,0)$.

39. $f(x) = xe^{-2x}$. $f'(x) = e^{-2x} + xe^{-2x}(-2) = (1 - 2x)e^{-2x}$.

$f''(x) = -2e^{-2x} + (1 - 2x)e^{-2x}(-2) = 4(x - 1)e^{-2x}$.
Observe that $f''(x) = 0$ if $x = 1$. The sign diagram of f''

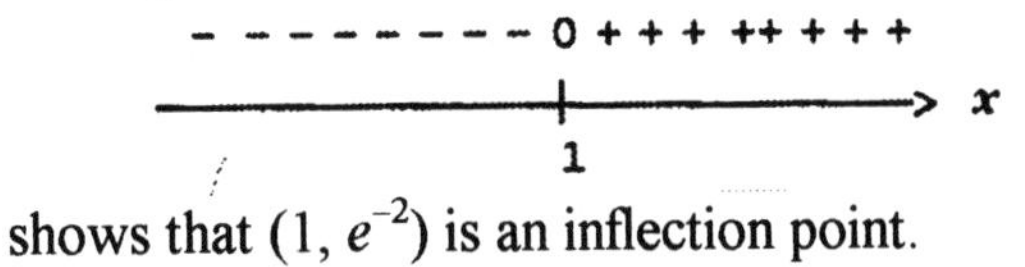

shows that $(1, e^{-2})$ is an inflection point.

41. $f(x) = e^{-x^2}$. $f'(x) = -2xe^{-x^2} = 0$ if $x = 0$, the only critical point of f.

x	-1 0 1
$f(x)$	e^{-1} 1 e^{-1}

From the table, we see that f has an absolute minimum value of e^{-1} attained at $x = -1$ and $x = 1$. It has an absolute maximum at $(0,1)$.

43. $g(x) = (2x - 1)e^{-x}$; $g'(x) = 2e^{-x} + (2x - 1)e^{-x}(-1) = (3 - 2x)e^{-x} = 0$, if $x = 3/2$. The graph of g shows that $(\frac{3}{2}, 2e^{-3/2})$ is an absolute maximum, and $(0,-1)$ is an absolute minimum.

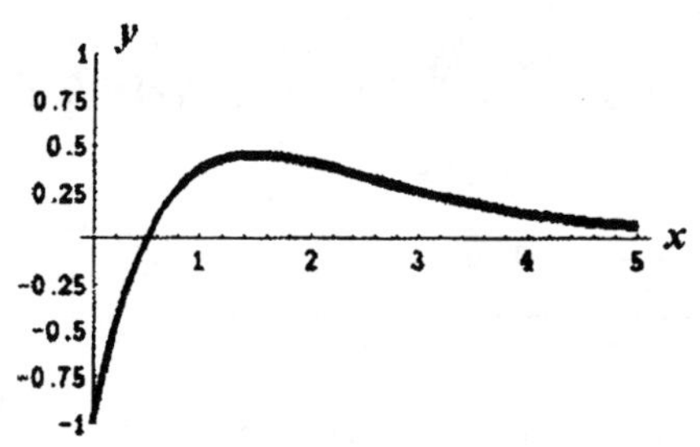

45. $f(t) = e^t - t$;

We first gather the following information on f.

1. The domain of f is $(-\infty,\infty)$.
2. Setting $t = 0$ gives 1 as the y–intercept.
3. $\lim_{t\to-\infty} (e^t - t) = \infty$ and $\lim_{t\to\infty} (e^t - t) = \infty$.
4. There are no asymptotes.
5. $f'(t) = e^t - 1$ if $t = 0$, a critical point of f. From the sign diagram for f'

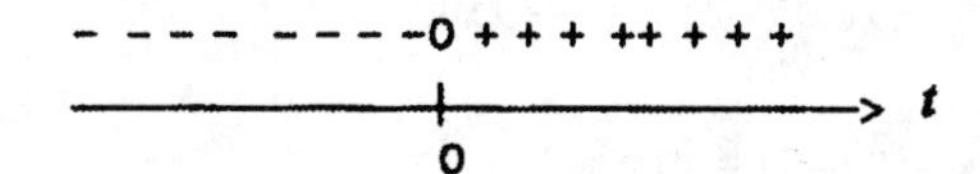

we see that f is decreasing on $(-\infty,0)$ and increasing on $(0,\infty)$.

6. From the results of (5), we see that $(0,1)$ is a relative minimum of f.
7. $f''(t) = e^t > 0$ for all t in $(-\infty,\infty)$. So the graph of f is concave upward on $(-\infty,\infty)$.
8. There are no inflection points.

The graph of f follows.

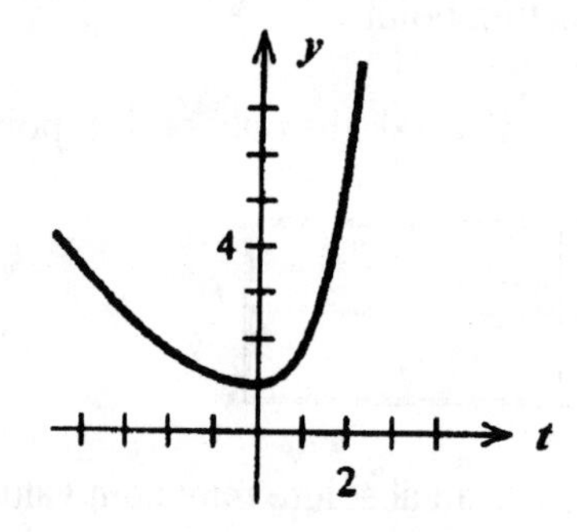

47. $f(x) = 2 - e^{-x}$.

We first gather the following information on f.

1. The domain of f is $(-\infty,\infty)$.
2. Setting $x = 0$ gives 1 as the y-intercept.
3. $\lim_{x\to-\infty} (2 - e^{-x}) = -\infty$ and $\lim_{x\to\infty} (2 - e^{-x}) = 2$,
4. From the results of (3), we see that $y = 2$ is a horizontal asymptote of f.
5. $f'(x) = e^{-x}$. Observe that $f'(x) > 0$ for all x in $(-\infty,\infty)$ and so f is increasing on $(-\infty,\infty)$.
6. Since there are no critical points, f has no relative extrema.
7. $f''(x) = -e^{-x} < 0$ for all x in $(-\infty,\infty)$ and so the graph of f is concave downward on $(-\infty,\infty)$.
8. There are no inflection points

The graph of f follows.

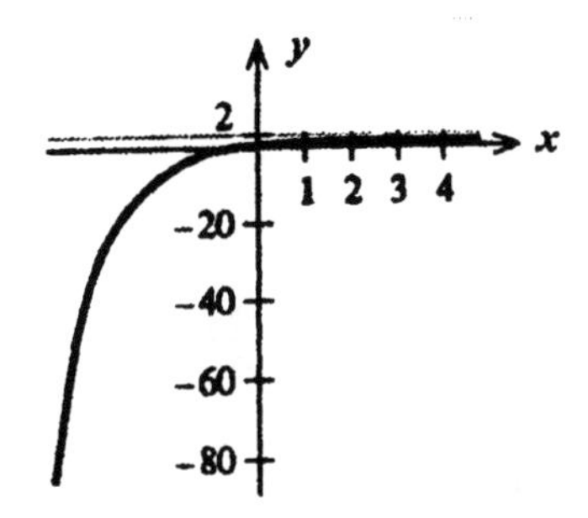

49. $S(t) = 20{,}000(1 + e^{-0.5t})$; $S'(t) = 20{,}000(-0.5e^{-0.5t}) = -10{,}000e^{-0.5t}$;

$S'(1) = -10{,}000e^{-0.5} = -6065$, or $-\$6065$/day.

$S'(2) = -10{,}000e^{-1} = -3679$, or $-\$3679$/day.

$S'(3) = -10{,}000(e^{-1.5}) = -2231$, or $-\$2231$/day.

$S'(4) = -10{,}000e^{-2} = -1353$, or $-\$1353$/day.

51. $N(t) = 5.3e^{0.095t^2 - 0.85t}$.

a. $N'(t) = 5.3e^{0.095t^2-0.85t}(0.19t - 0.85)$. Since $N'(t)$ is negative for $(0 \le t \le 4)$, we see that $N(t)$ is decreasing over that interval.

b. To find the rate at which the number of polio cases was decreasing at the beginning of 1959, we compute

$$N'(0) = 5.3e^{0.095(0^2)-0.85(0)}(0.85) = 5.3(-0.85) = -4.505$$

(t is measured in thousands), or 4,505 cases per year. To find the rate at which the number of polio cases was decreasing at the beginning of 1962, we compute

$$N'(3) = 5.3e^{0.095(9)-0.85(3)}(0.57 - 0.85)$$
$$= (-0.28)(0.9731) \approx -0.273, \text{ or 273 cases per year.}$$

53. $p = 240\left(1 - \dfrac{3}{3 + e^{-0.0005x}}\right) = 240[1 - 3(3 + e^{-0.0005x})^{-1}].$

$$p' = 720(3 + e^{-0.0005x})^{-2}(-0.0005e^{-0.0005x})$$
$$p'(1000) = 720(3 + e^{-0.0005(1000)})^{-2}(-0.0005e^{-0.005(1000)})$$
$$= -\frac{0.36(0.606531)}{(3 + 0.606531)^2} \approx -0.0168, \quad \text{or } -1.68 \text{ cents per case.}$$

$$p(1000) = 240(1 - \frac{3}{3.606531}) \approx 40.36, \quad \text{or \$40.36/case.}$$

55. We want to find the maximum of dT/dt:

$$T'(t) = -1000\frac{d}{dt}(t + 10)e^{-0.1t} = -1000[e^{-0.1t} + (t + 10)e^{-0.1t}(-0.1)]$$
$$= 100te^{-0.1t}.$$
$$T''(t) = 100\frac{d}{dt}(te^{-0.1t}) = 100[e^{-0.1t} + te^{-0.1t}(-0.1)]$$
$$= 100e^{-0.1t}(1 - 0.1t).$$

Observe that $T''(t) = 0$ if $t = 10$, a critical point of T From the sign diagram of

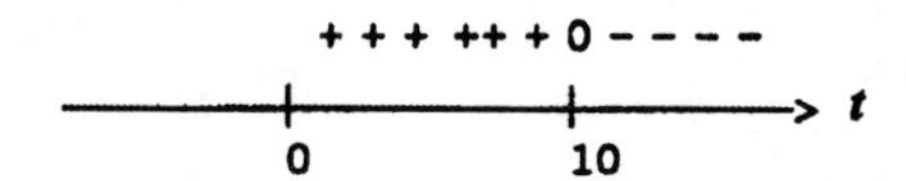

T'' see that $t = 10$ gives a relative maximum of T'. This is, in fact, an absolute maximum. So the maximum production will be reached in the 10th year of operation.

57. $f(t) = 1.5 + 1.8te^{-1.2t}$

$$f'(t) = 1.8\frac{d}{dt}(te^{-1.2t}) = 1.8[e^{-1.2t} + te^{-1.2t}(-1.2)]$$
$$= 1.8e^{-1.2t}(1 - 1.2t).$$

$f'(0) = 1.8$, $f'(1) = -0.11$, $f'(2) = -0.23$, and $f'(3) = -0.13$,
and this tells us that the rate of change of the amount of oil used in 1.8 barrels per \$1000 of output per decade in 1965; it is decreasing at the rate of 0.11 barrels per \$1000 of output per decade in 1966, and so on.

USING TECHNOLOGY EXERCISES 13.3, page 904

1. 5.4366 3. 12.3929 5. 0.1861

7. a. The initial population of crocodiles is $P(0) = \frac{300}{6} = 50$.

   b. $\lim_{t \to 0} P(t) = \lim_{t \to 0} \dfrac{300e^{-0.024t}}{5e^{-0.024t} + 1} = \dfrac{0}{0+1} = 0.$

   c.

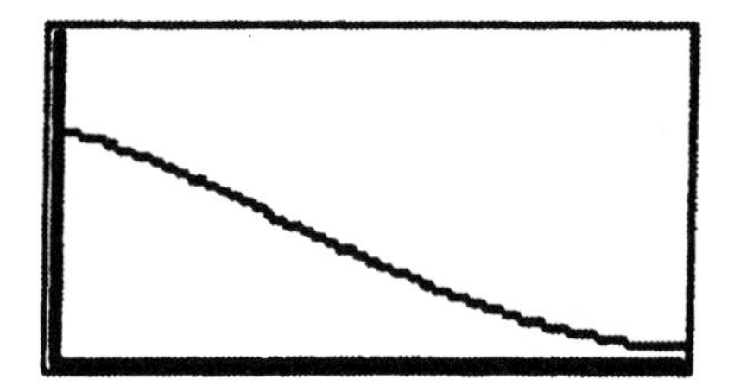

9. a.

b. 4.2720 billion/half century

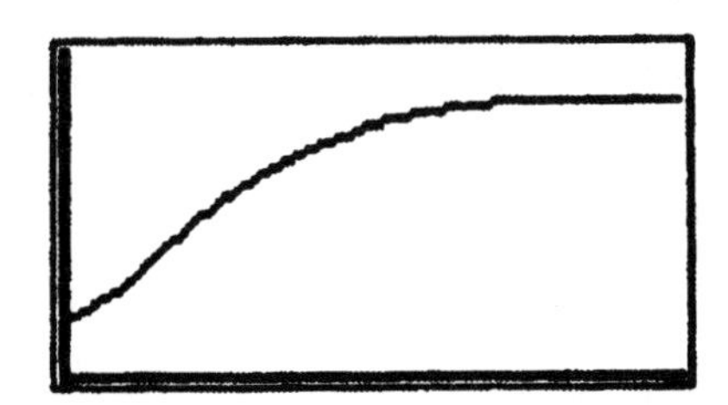

11. a. At the beginning of June, there are $F(1) = 196.20$ or approximately 196 aphids in a typical bean stem. At the beginning of July the number is $F(2) = 180.02$, or approximately 180 aphids per bean stem.
    b. At the beginning of June, the population of aphids is changing at the rate of $F'(1) = 226.02$; that is, it is increasing at the rate of 226 aphids on a typical bean stem per month. At the beginning of July, the population is changing at the rate of $F'(2) = -238.3$; that is, it is decreasing at the rate of 238 aphids per month.

EXERCISES 13.4 , page 912

1. $f(x) = 5 \ln x; f'(x) = 5\left(\dfrac{1}{x}\right) = \dfrac{5}{x}.$

3. $f(x) = \ln(x+1); f'(x) = \dfrac{1}{x+1}$.

5. $f(x) = \ln x^8; f'(x) = \dfrac{8x^7}{x^8} = \dfrac{8}{x}$.

7. $f(x) = \ln x^{1/2};\quad f'(x) = \dfrac{\frac{1}{2}x^{-1/2}}{x^{1/2}} = \dfrac{1}{2x}$.

9. $f(x) = \ln\left(\dfrac{1}{x^2}\right) = \ln x^{-2} = -2\ln x;\quad f'(x) = -\dfrac{2}{x}$.

11. $f(x) = \ln(4x^2 - 6x + 3);\quad f'(x) = \dfrac{8x-6}{4x^2-6x+3} = \dfrac{2(4x-3)}{4x^2-6x+3}$.

13. $f(x) = \ln\left(\dfrac{2x}{x+1}\right) = \ln 2x - \ln(x+1)$.

$$f'(x) = \frac{2}{2x} - \frac{1}{x+1} = \frac{2(x+1)-2x}{2x(x+1)} = \frac{2x+2-2x}{2x(x+1)}$$
$$= \frac{2}{2x(x+1)} = \frac{1}{x(x+1)}.$$

15. $f(x) = x^2\ln x;\ f'(x) = x^2\left(\frac{1}{x}\right) + (\ln x)(2x) = x + 2x\ln x = x(1 + 2\ln x)$

17. $f(x) = \dfrac{2\ln x}{x}.\quad f'(x) = \dfrac{x\left(\frac{2}{x}\right) - 2\ln x}{x^2} = \dfrac{2(1-\ln x))}{x^2}$.

19. $f(u) = \ln(u-2)^3;\ f'(u) = \dfrac{3(u-2)^2}{(u-2)^3} = \dfrac{3}{u-2}$.

21. $f(x) = (\ln x)^{1/2}$ and $f'(x) = \dfrac{1}{2}(\ln x)^{-1/2}\left(\dfrac{1}{x}\right) = \dfrac{1}{2x\sqrt{\ln x}}$.

23. $f(x) = (\ln x)^3;\quad f'(x) = 3(\ln x)^2\left(\dfrac{1}{x}\right) = \dfrac{3(\ln x)^2}{x}$.

25. $f(x) = \ln(x^3 + 1);\quad f'(x) = \dfrac{3x^2}{x^3+1}$.

27. $f(x) = e^x \ln x$. $f'(x) = e^x \ln x + e^x\left(\dfrac{1}{x}\right) = \dfrac{e^x(x \ln x + 1)}{x}$.

29. $f(t) = e^{2t} \ln(t+1)$

$$f'(t) = e^{2t}\left(\frac{1}{t+1}\right) + \ln(t+1)\cdot(2e^{2t})$$
$$= \frac{[2(t+1)\ln(t+1)+1]e^{2t}}{t+1}.$$

31. $f(x)\ \dfrac{\ln x}{x}$. $f'(x) = \dfrac{x(\frac{1}{x}) - \ln x}{x^2} = \dfrac{1-\ln x}{x^2}$.

33. $f(x) = \ln 2 + \ln x$; So $f'(x) = \dfrac{1}{x}$ and $f''(x) = -\dfrac{1}{x^2}$.

35. $f(x) = \ln(x^2+2)$; $f'(x) = \dfrac{2x}{(x^2+2)}$ and

$$f''(x) = \frac{(x^2+2)(2) - 2x(2x)}{(x^2+2)^2} = \frac{2(2-x^2)}{(x^2+2)^2}.$$

37. $y = (x+1)^2(x+2)^3$

$$\ln y = \ln(x+1)^2(x+2)^3 = \ln(x+1)^2 + \ln(x+2)^3$$
$$= 2\ln(x+1) + 3\ln(x+2).$$
$$\frac{y'}{y} = \frac{2}{x+1} + \frac{3}{x+2} = \frac{2(x+2)+3(x+1)}{(x+1)(x+2)} = \frac{5x+7}{(x+1)(x+2)}.$$
$$y' = \frac{(5x+7)(x+1)^2(x+2)^3}{(x+1)(x+2)} = (5x+7)(x+1)(x+2)^2.$$

39. $y = (x-1)^2(x+1)^3(x+3)^4$

$$\ln y = 2\ln(x-1) + 3\ln(x+1) + 4\ln(x+3)$$
$$\frac{y'}{y} = \frac{2}{x-1} + \frac{3}{x+1} + \frac{4}{x+3}$$
$$= \frac{2(x+1)(x+3) + 3(x-1)(x+3) + 4(x-1)(x+1)}{(x-1)(x+1)(x+3)}$$
$$= \frac{2x^2+8x+6+3x^2+6x-9+4x^2-4}{(x-1)(x+1)(x+3)} = \frac{9x^2+14x-7}{(x-1)(x+1)(x+3)}.$$

Therefore,

$$y' = \frac{9x^2 + 14x - 7}{(x-1)(x+1)(x+3)} \cdot y$$

$$= \frac{(9x^2 + 14x - 7)(x-1)^2(x+1)^3(x+3)^4}{(x-1)(x+1)(x+3)}$$

$$= (9x^2 + 14x - 7)(x-1)(x+1)^2(x+3)^3.$$

41. $y = \dfrac{(2x^2-1)^5}{\sqrt{x+1}}$.

$$\ln y = \ln\frac{(2x^2-1)^5}{(x+1)^{1/2}} = 5\ln(2x^2-1) - \frac{1}{2}\ln(x+1)$$

So $\dfrac{y'}{y} = \dfrac{20x}{2x^2-1} - \dfrac{1}{2(x+1)} = \dfrac{40x(x+1)-(2x^2-1)}{2(2x^2-1)(x+1)}$

$$= \frac{38x^2+40x+1}{2(2x^2-1)(x+1)}.$$

$$y' = \frac{38x^2+40x+1}{2(2x^2-1)(x+1)} \cdot \frac{(2x^2-1)^5}{\sqrt{x+1}} = \frac{(38x^2+40x+1)(2x^2-1)^4}{2(x+1)^{3/2}}.$$

43. $y = 3^x$

$$\ln y = x\ln 3$$

$$\frac{1}{y}\cdot\frac{dy}{dx} = \ln 3$$

$$\frac{dy}{dx} = y\ln 3 = 3^x \ln 3.$$

45. $y = (x^2+1)^x$; $\ln y = \ln(x^2+1)^x = x\ln(x^2+1)$. So

$$\frac{y'}{y} = \ln(x^2+1) + x\left(\frac{2x}{x^2+1}\right) = \frac{(x^2+1)\ln(x^2+1)+2x^2}{x^2+1}.$$

$$y' = \frac{[(x^2+1)\ln(x^2+1)+2x^2](x^2+1)^x}{x^2+1}$$

47. $y = x\ln x$. The slope of the tangent line at any point is

$$y' = \ln x + x\left(\tfrac{1}{x}\right) = \ln x + 1.$$

In particular, the slope of the tangent line at (1,0) where $x = 1$ is $m = \ln 1 + 1 = 1$. So, an equation of the tangent line is

$$y - 0 = 1(x - 1) \text{ or } y = x - 1.$$

49. $f(x) = \ln x^2 = 2 \ln x$ and so $f'(x) = 2/x$. Since $f'(x) < 0$ if $x < 0$, and $f'(x) > 0$ if $x > 0$, we see that f is decreasing on $(-\infty, 0)$ and increasing on $(0, \infty)$.

51. $f(x) = x^2 + \ln x^2$; $f'(x) = 2x + \dfrac{2x}{x^2} = 2x + \dfrac{2}{x}$.

$f''(x) = 2 - \dfrac{2}{x^2}$.

To find the intervals of concavity for f, we first set $f''(x) = 0$ giving

$$2 - \frac{2}{x^2} = 0, \quad 2 = \frac{2}{x^2}, \quad 2x^2 = 2$$

or $\qquad x^2 = 1$ and $x = \pm 1$.

Next, we construct the sign diagram for f''

f'' is not defined here

+ + + + 0 - - - - - - 0+ + +

-1 0 1 x

and conclude that f is concave upward on $(-\infty, -1) \cup (1, \infty)$ and concave downward on $(-1, 0) \cup (0, 1)$.

53. $f(x) = \ln(x^2 + 1)$.

$$f'(x) = \frac{2x}{x^2 + 1}; \; f''(x) = \frac{(x^2 + 1)(2) - (2x)(2x)}{(x^2 + 1)^2} = -\frac{2(x^2 - 1)}{(x^2 + 1)^2}.$$

Setting $f''(x) = 0$ gives $x = \pm 1$ as candidates for inflection points of f.
From the sign diagram for f''

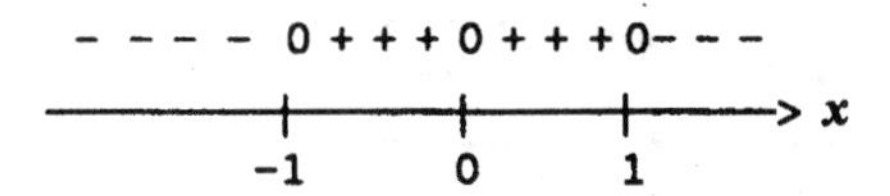

we see that $(-1, \ln 2)$ and $(1, \ln 2)$ are inflection points of f.

55. $f(x) = x - \ln x$; $f'(x) = 1 - \dfrac{1}{x} = \dfrac{x - 1}{x} = 0$ if $x = 1$, a critical point of f.

x	1/2	1	3
$f(x)$	$1/2 + \ln 2$	1	$3 - \ln 3$

From the table, we see that f has an absolute minimum at (1,1) and an absolute maximum at $(3, 3 - \ln 3)$.

57. $f(x) = \ln(x - 1)$.
1. The domain of f is obtained by requiring that $x - 1 > 0$. We find the domain to be $(1,\infty)$.
2. Since $x \neq 0$, there are no y-intercepts. Next, setting $y = 0$ gives $x - 1 = 1$ or $x = 2$ as the x-intercept.
3. $\lim_{x\to 1^+} \ln(x - 1) = -\infty$.
4. There are no horizontal asymptotes. Observe that $\lim_{x\to 1^+} \ln(x - 1) = -\infty$ so $x = 1$ is a vertical asymptote.
5. $f'(x) = \dfrac{1}{x-1}$. The sign diagram for f' is

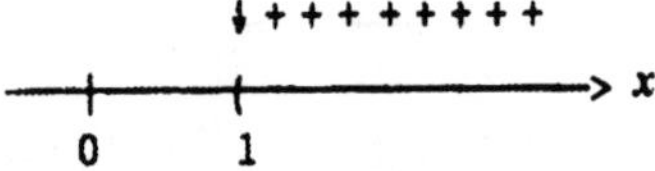

We conclude that f is increasing on $(1,\infty)$.
6. The results of (5) show that f is increasing on $(1,\infty)$.
7. $f''(x) = -\dfrac{1}{(x-1)^2}$. Since $f''(x) < 0$ for $x > 1$, we see that f is concave downward on $(1,\infty)$.
8. From the results of (7), we see that f has no inflection points.

The graph of f follows.

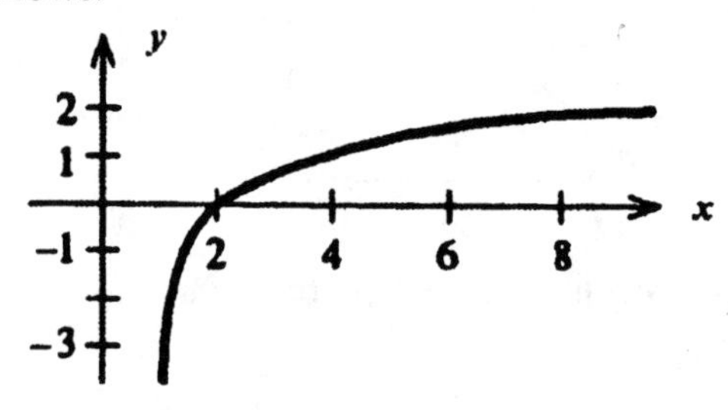

59. If $x \le 0$, then $|x| = -x$. Therefore, $\ln |x| = \ln (-x)$. Writing $f(x) = \ln |x|$ we have

$$|x| = -x = e^{f(x)}.$$

Differentiating both sides with respect to x and using the Chain Rule, we have

$$-1 = e^{f(x)} \cdot f'(x) \quad \text{or} \quad f'(x) = -\frac{1}{e^{f(x)}} = -\frac{1}{-x} = \frac{1}{x}.$$

EXERCISES 13.5 , page 922

1. a. The growth constant is $k = 0.05$. b. Initially, the quantity present is 400 units.

   c.

t	0	10	20	100	1000
Q	400	660	1087	59365	2.07×10^{24}

3. a. $Q(t) = Q_0e^{kt}$. Here $Q_0 = 100$ and so $Q(t) = 100e^{kt}$. Since the number of cells doubles in 20 minutes, we have

$$Q(20) = 100e^{20k} = 200, \ e^{20k} = 2, \ 20k = \ln 2, \text{ or } k = \tfrac{1}{20} \ln 2 \approx 0.03466.$$

$$Q(t) = 100e^{0.03466t}$$

b. We solve the equation $100e^{0.03466t} = 1{,}000{,}000$. We obtain

$$e^{0.03466t} = 10000 \text{ or } 0.03466t = \ln 10000,$$

$$t = \frac{\ln 10{,}000}{0.03466} \approx 266, \qquad \text{or 266 minutes.}$$

c. $Q(t) = 1000e^{0.03466t}$.

5. a. We solve the equation

$$5.3e^{0.0198t} = 3(5.3) \text{ or } e^{0.0198t} = 3,$$

or $$0.0198t = \ln 3 \text{ and } t = \frac{\ln 3}{0.0198} \approx 55.5.$$

So the world population will triple in approximately 55.5 years.

b. If the growth rate is 1.8 percent, then proceeding as before, we find

$$k = \ln 1.018 \approx 0.0178.$$

So $N(t) = 5.3e^{0.0178t}$. If $t = 55.5$, the population would be

$$N(55.5) = 5.3e^{0.0178(55.5)} \approx 14.25,$$

or approximately 14.25 billion.

7. $P(h) = p^0e^{-kh}$, $P(0) = 15$, therefore, $p_0 = 15$.

$$P(4000) = 15e^{-4000k} = 12.5$$

$$e^{-4000k} = \frac{12.5}{15},$$

$$-4000k = \ln\left(\frac{12.5}{15}\right) \quad \text{and } k = 0.00004558.$$

Therefore, $P(12{,}000) = 15e^{-0.00004558(12{,}000)} = 8.68$, or 8.7 lb/sq in.

9. Suppose the amount of phosphorus 32 at time t is given by

   $Q(t) = Q_0e^{-kt}$

   where Q_0 is the amount present initially and k is the decay constant. Since this element has a half–life of 14.2 days, we have

   $$\tfrac{1}{2}Q_0 = Q_0e^{-14.2k}, \quad e^{-14.2k} = \tfrac{1}{2}, \; -14.2k = \ln\tfrac{1}{2}, \; k = \frac{\frac{1}{2}}{14.2} \approx 0.0488.$$

   Therefore, the amount of phosphorus 32 present at any time t is given by

   $Q(t) = 100e^{-0.0488t}$

   The amount left after 7.1 days is given by

   $Q(7.1) = 100e^{-0.0488(7.1)} = 100e^{-0.3465}$

   $= 70.617$, or 70.617 grams.

11. We solve the equation

    $0.2Q_0 = Q_0e^{-0.00012t}$

    obtaining $\quad t = \dfrac{\ln 0.2}{-0.00012} \approx 13{,}412$, or approximately 13,412 years.

13. The graph of $Q(t)$ follows.

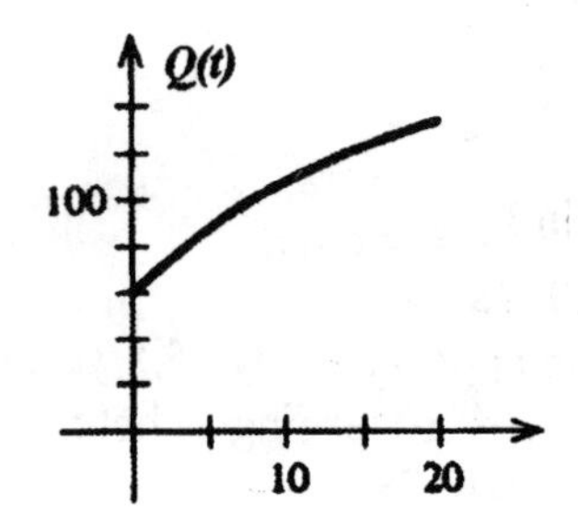

a. $Q(0) = 120(1 - e^0) + 60 = 60$, or 60 w.p.m.
b. $Q(10) = 120(1 - e^{-0.5}) + 60 = 107.22$, or approximately 107 w.p.m.
c. $Q(20) = 120(1 - e^{-1}) + 60 = 135.65$, or approximately 136 w.p.m.

15. The graph of $D(t)$ follows.

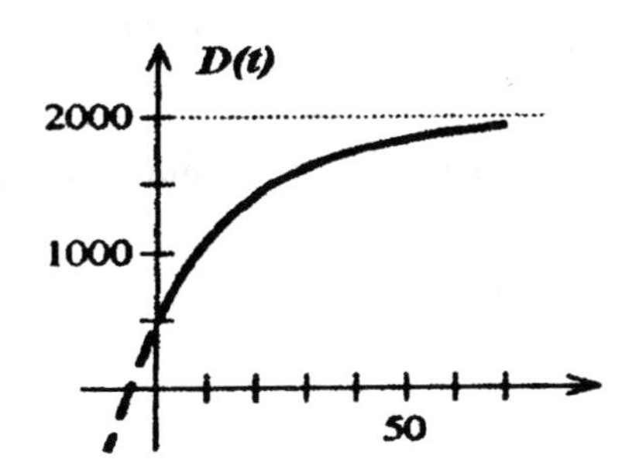

a. After one month, the demand is $D(1) = 2000 - 1500e^{-0.05} \approx 573$.
After twelve months, the demand is $D(12) = 2000 - 1500e^{-0.6} \approx 1177$.
After twenty-four months the demand is

$$D(24) = 2000 - 1500e^{-1.2} \approx 1548.$$

After sixty months, the demand is

$$D(60) = 2000 - 1500e^{-3} \approx 1925.$$

b. $$\lim_{t\to\infty} D(t) = \lim_{t\to\infty} 2000 - 1500e^{-0.05t} = 2000$$

and we conclude that the demand is expected to stabilize at 2000 computers per month.

c. $D'(t) = -1500e^{-0.05t}(-0.05) = 75e^{-0.05t}$.
Therefore, the rate of growth after ten months is given by

$$D'(10) = 75e^{-0.5} \approx 45.49,$$

or approximately 46 computers per month.

17. a. $Q(1) = \dfrac{1000}{1+199e^{-0.8}} \approx 11.06$, or 11 children.

b. $Q(10) = \dfrac{1000}{1+199e^{-8}} \approx 937.4$, or 937 children.

c. $\lim\limits_{t\to\infty} \dfrac{1000}{1+199e^{-0.8t}} = 1000$, or 1000 children.

19. $P(t) = \dfrac{68}{1+21.67e^{-0.62t}}$.
The percentage of households that owned VCRs at the beginning of 1985 is given by

$$P(0) = \frac{68}{1+21.67e^{-0.62(0)}} = \frac{68}{22.67} \approx 3,$$

or approximately 3 percent. The percentage of households that owned VCRs at the beginning of 1995 is given by

$$P(10)=\frac{68}{1+21.67e^{-0.62(10)}}\approx 65.14, \quad \text{or approximately 65.14 percent.}$$

21. The first of the given conditions implies that $f(0)=300$, that is,

$$300=\frac{3000}{1+Be^{0}}=\frac{3000}{1+B}.$$

So $1+B=10$, or $B=9$. Therefore,

$$f(t)=\frac{3000}{1+9e^{-kt}}.$$

Next, the condition $f(2)=600$ gives the equation

$$600=\frac{3000}{1+9e^{-2k}},\ 1+9e^{-2k}=5,\ e^{-2k}=\frac{4}{9},\ \text{or } k=-\frac{1}{2}\ln\left(\frac{4}{9}\right).$$

Therefore, $f(t)=\dfrac{3000}{1+9e^{(1/2)t\cdot\ln(4/9)}}=\dfrac{3000}{1+9(\frac{4}{9})^{t/2}}$.

The number of students who had heard about the policy four hours later is given by

$$f(4)=\frac{3000}{1+9(\frac{4}{9})^{2}}=1080, \quad \text{or 1080 students.}$$

23. a. $Q'(t)=Ce^{-Ae^{-kt}}\dfrac{d}{dt}(-Ae^{-kt})=-ACe^{-Ae^{-kt}}\cdot e^{-kt}(-k)=ACke^{(-Ae^{-kt}-kt)}$.

b. $Q''(t)=ACke^{(-Ae^{-kt}-kt)}\cdot[-k-Ae^{-kt}(-k)]=0$, if $Ae^{-kt}=1$.

$$e^{-kt}=\frac{1}{A},\quad -kt=\ln\frac{1}{A},\quad \text{or } t=-\frac{1}{k}\ln\frac{1}{A}=\frac{1}{k}\ln A.$$

The sign diagram

```
0 + + +  0 - - -   - - -
+--------+--------------> t
0       1/k ln A
```

shows that $t=\dfrac{1}{k}$ ln A is an inflection point and so the growth is most rapid at this time.

c. $\lim_{t\to\infty} Q(t)=C$.

CHAPTER 13 REVIEW EXERCISES, page 926

1. a-b

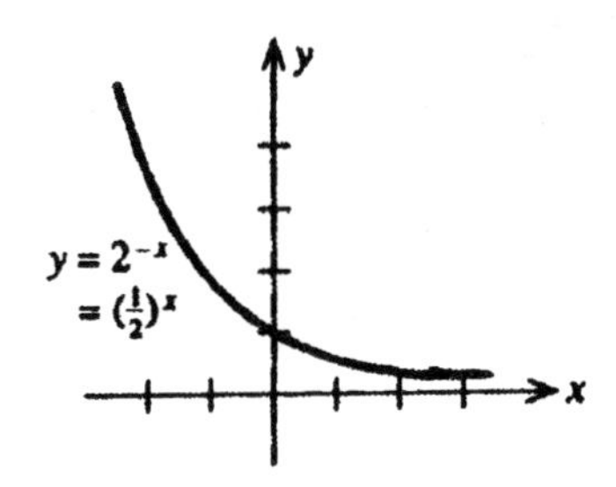

Since $y = \left(\frac{1}{2}\right)^x = \frac{1}{2^x} = 2^{-x}$, it has the same graph as that of $y = 2^{-x}$.

3. $16^{-3/4} = 0.125$ is equivalent to $-\frac{3}{4} = \log_{16} 0.125$.

5.
$$\ln(x-1) + \ln 4 = \ln(2x+4) - \ln 2$$
$$\ln(x-1) - \ln(2x+4) = -\ln 2 - \ln 4 = -(\ln 2 + \ln 4)$$
$$\ln\left(\frac{x-1}{2x+4}\right) = -\ln 8 = \ln \tfrac{1}{8}.$$
$$\left(\frac{x-1}{2x+4}\right) = \frac{1}{8}$$
$$8x - 8 = 2x + 4$$
$$6x = 12$$
$$x = 2.$$

CHECK: l.h.s. $\ln(2-1) + \ln 4 = \ln 4$
r.h.s $\ln(4+4) - \ln 2 = \ln 8 - \ln 2 = \ln \frac{8}{2} = \ln 4$.

7. $\ln 3.6 = \ln \frac{36}{10} - \ln 36 = \ln 10 = \ln 6^2 - \ln 2 \cdot 5 = 2 \ln 6 - \ln 2 - \ln 5$
$= 2(\ln 2 + \ln 3) - \ln 2 - \ln 5 = 2(x + y) - x - z = x + 2y - z.$

9. We first sketch the graph of $y = 2^x - 3$. Then we take the reflection of this graph with respect to the line $y = x$.

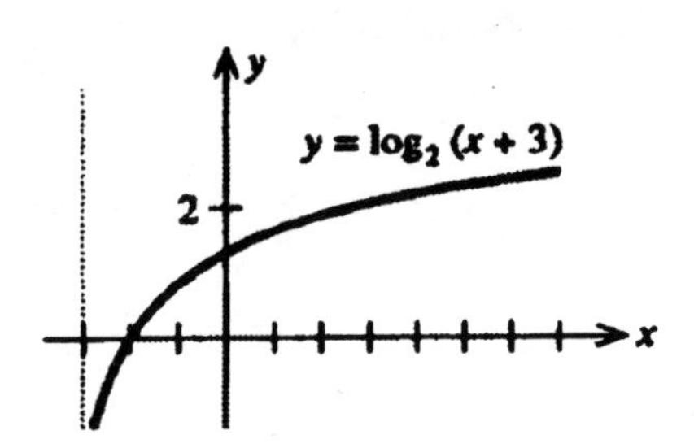

11. $f(x) = xe^{2x}; f'(x) = e^{2x} + xe^{2x}(2) = (1 + 2x)e^{2x}.$

13. $g(t) = \sqrt{t}e^{-2t};\ g'(t) = \frac{1}{2}t^{-1/2}e^{-2t} + \sqrt{t}e^{-2t}(-2) = \dfrac{1-4t}{2\sqrt{t}e^{2t}}.$

15. $y = \dfrac{e^{2x}}{1+e^{-2x}};\ y' = \dfrac{(1+e^{-2x})e^{2x}(2) - e^{2x} \cdot e^{-2x}(-2)}{(1+e^{-2x})^2} = \dfrac{2(e^{2x}+2)}{(1+e^{-2x})^2}.$

17. $f(x) = xe^{-x^2}; f'(x) = e^{-x^2} + xe^{-x^2}(-2x) = (1 - 2x^2)e^{-x^2}.$

19. $f(x) = x^2e^x + e^x;$
$f'(x) = 2xe^x + x^2e^x + e^x = (x^2 + 2x + 1)e^x = (x + 1)^2e^x.$

21. $f(x) = \ln(e^{x^2} + 1);\ f'(x) = \dfrac{e^{x^2}(2x)}{e^{x^2}+1} = \dfrac{2xe^{x^2}}{e^{x^2}+1}.$

23. $f(x) = \dfrac{\ln x}{x+1}.\ \ f'(x) = \dfrac{(x+1)\left(\dfrac{1}{x}\right) - \ln x}{(x+1)^2} = \dfrac{1+\dfrac{1}{x} - \ln x}{(x+1)^2} = \dfrac{x - x\ln x + 1}{x(x+1)^2}.$

25. $y = \ln(e^{4x}+3);\ y' = \dfrac{e^{4x}(4)}{e^{4x}+3} = \dfrac{4e^{4x}}{e^{4x}+3}.$

27. $f(x) = \dfrac{\ln x}{1+e^x};$

$$f'(x) = \frac{(1+e^x)\frac{d}{dx}\ln x - \ln x \frac{d}{dx}(1+e^x)}{(1+e^x)^2} = \frac{(1+e^x)\left(\frac{1}{x}\right) - (\ln x)e^x}{(1+e^x)^2}$$

$$= \frac{1+e^x - xe^x \ln x}{x(1+e^x)^2} = \frac{1+e^x(1-x\ln x)}{x(1+e^x)^2}.$$

29. $y = \ln(3x+1)$; $y' = \dfrac{3}{3x+1}$;

$$y'' = 3\frac{d}{dx}(3x+1)^{-1} = -3(3x+1)^{-2}(3) = -\frac{9}{(3x+1)^2}.$$

31. $h'(x) = g'(f(x))f'(x)$. But $g'(x) = 1 - \dfrac{1}{x^2}$ and $f'(x) = e^x$.

So $f(0) = e^0 = 1$ and $f'(0) = e^0 = 1$. Therefore,

$h'(0) = g'(f(0))f'(0) = g'(1)f'(0) = 0 \cdot 1 = 0.$

33. $y = (2x^3+1)(x^2+2)^3$. $\ln y = \ln(2x^3+1) + 3\ln(x^2+2)$.

$$\frac{y'}{y} = \frac{6x^2}{2x^3+1} + \frac{3(2x)}{x^2+2} = \frac{6x^2(x^2+2)+6x(2x^3+1)}{(2x^3+1)(x^2+2)}$$

$$= \frac{6x^4+12x^2+12x^4+6x}{(2x^3+1)(x^2+2)} = \frac{18x^4+12x^2+6x}{(2x^3+1)(x^2+2)}.$$

Therefore, $y' = 6x(3x^3+2x+1)(x^2+2)^2$.

35. $y' = -2e^{-2x}$ and this gives the slope of the tangent line to the graph of $y = e^{-2x}$ at any point (x, y). In particular, the slope of the tangent line at $(1, e^{-2})$ is $y'(1) = -2e^{-2}$. The required equation is $y - e^{-2} = -2e^{-2}(x-1)$ or $y = \dfrac{1}{e^2}(-2x+3)$.

37. $f(x) = xe^{-2x}$.

We first gather the following information on f.

1. The domain of f is $(-\infty,\infty)$.
2. Setting $x = 0$ gives 0 as the y-intercept.
3. $\lim_{x\to-\infty} xe^{-2x} = -\infty$ and $\lim_{x\to\infty} xe^{-2x} = 0$.
4. The results of (3) show that $y = 0$ is a horizontal asymptote.
5. $f'(x) = e^{-2x} + xe^{-2x}(-2) = (1-2x)e^{-2x}$. Observe that $f'(x) = 0$ if $x = 1/2$, a critical point of f. The sign diagram of f'

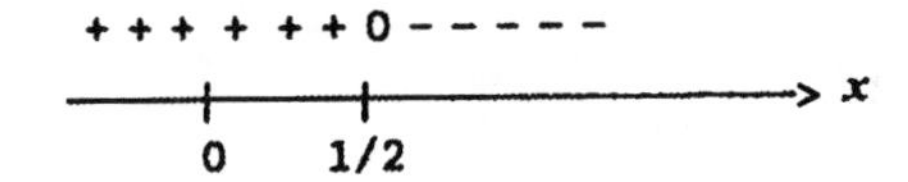

shows that f is increasing on $(-\infty, \frac{1}{2})$ and decreasing on $(\frac{1}{2}, \infty)$.

6. The results of (5) show that $(\frac{1}{2}, \frac{1}{2}e^{-1})$ is a relative maximum.

7. $f''(x) = -2e^{-2x} + (1-2x)e^{-2x}(-2) = 4(x-1)e^{-2x}$ and is equal to zero if $x = 1$. The sign diagram of f ''

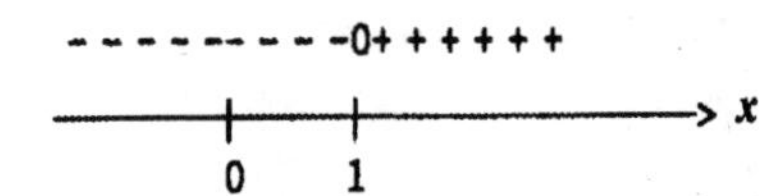

shows that the graph of f is concave downward on $(-\infty,1)$ and concave upward on $(1,\infty)$.
The graph of f follows.

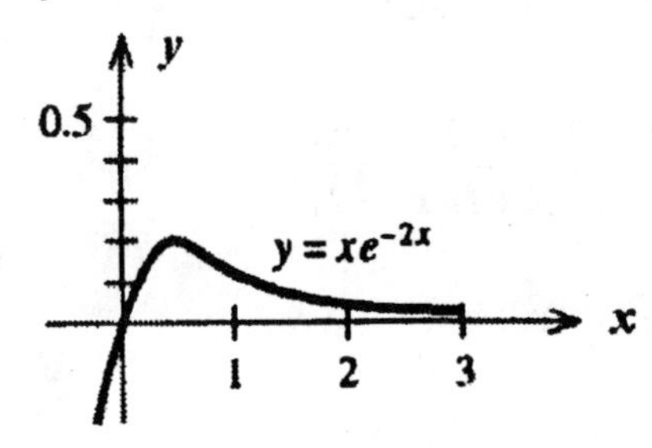

39. $f(t) = te^{-t}$. $f'(t) = e^{-t} + t(-e^{-t}) = e^{-t}(1-t)$. Setting $f'(t) = 0$ gives $t = 1$ as the only critical point of f. From the sign diagram of f'

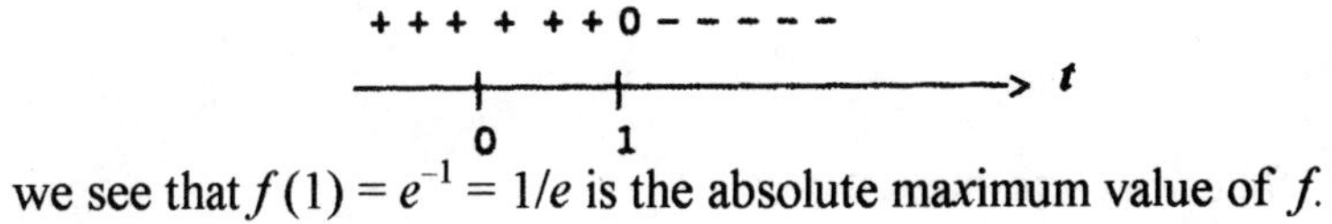

we see that $f(1) = e^{-1} = 1/e$ is the absolute maximum value of f.

41. We want to find r where r satisfies the equation $8.2 = 4.5\,e^{r(5)}$. We have

$$e^{5r} = \frac{8.2}{4.5} \quad \text{or} \quad r = \frac{1}{5}\ln\left(\frac{8.2}{4.5}\right) \approx 0.12$$

and so the annual rate of return is 12 percent per year.

43. a. $Q(t) = 2000e^{kt}$. Now $Q(120) = 18{,}000$ gives $2000e^{120k} = 18{,}000$, $e^{120k} = 9$,

or $120k = \ln 9$. So $k = \frac{1}{120}\ln 9 \approx 0.01831$ and $Q(t) = 2000e^{0.01831t}$.

b. $Q(4) = 2000e^{0.01831(240)} \approx 161{,}992$, or approximately 162,000.

45.

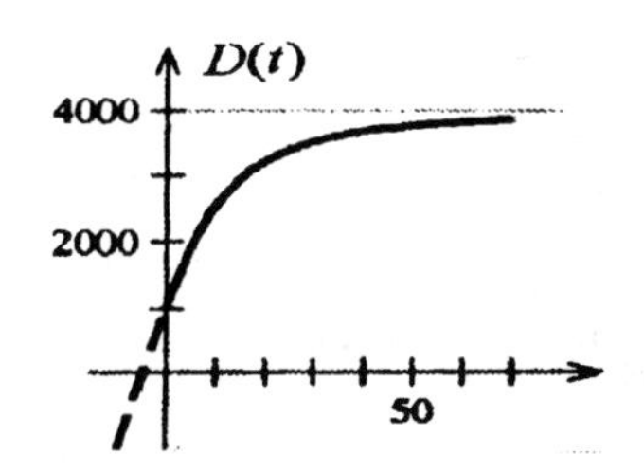

a. $D(1) = 4000 - 3000\,e^{-0.06} = 1175$, $D(12) = 4000 - 3000\,e^{-0.72} = 2540$, and $D(24) = 4000 - 3000\,e^{-1.44} = 3289$.

b. $\lim_{t\to\infty} D(t) = \lim_{t\to\infty} (4000 - 3000e^{-0.06t}) = 4000$.

CHAPTER 14

EXERCISES 14.1, page 940

1. $F(x) = \frac{1}{3}x^3 + 2x^2 - x + 2$; $F'(x) = x^2 + 4x - 1 = f(x)$.

3. $F(x) = (2x^2 - 1)^{1/2}$; $F'(x) = \frac{1}{2}(2x^2 - 1)^{-1/2}(4x) = 2x(2x^2 - 1)^{-1/2} = f(x)$.

5. a. $G'(x) = \dfrac{d}{dx}(2x) = 2 = f(x)$ b. $F(x) = G(x) + C = 2x + C$

   c.

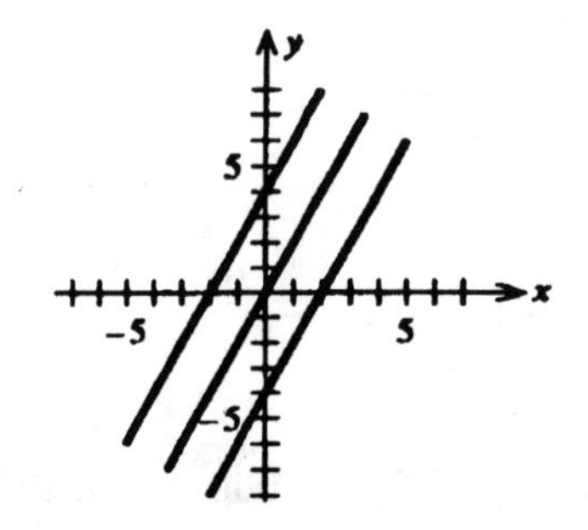

7. a. $G'(x) = \dfrac{d}{dx}(\frac{1}{3}x^3) = x^2 = f(x)$ b. $F(x) = G(x) + C = \frac{1}{3}x^3 + \text{C}$

   c.

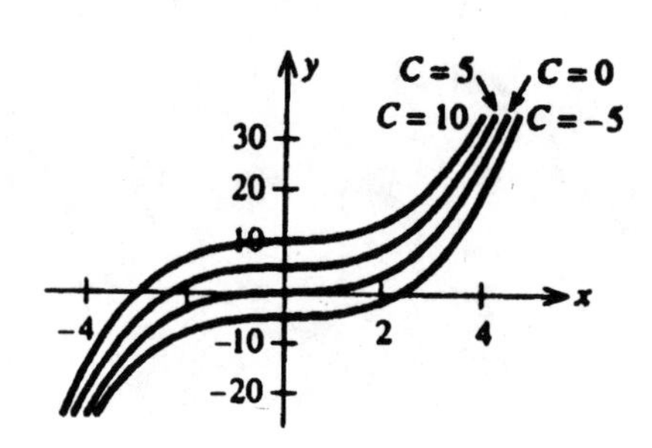

9. $\int 6\,dx = 6x + C.$

11. $\int x^3 dx = \frac{1}{4}x^4 + C$

13. $\int x^{-4} = -\frac{1}{3}x^{-3} + C$

15. $\int x^{2/3} dx = \frac{3}{5}x^{5/3} + C$

17. $\int x^{-5/4}dx = -4x^{-1/4} + C$

19. $\int \frac{2}{x^2}\,dx = 2\int x^{-2}dx = 2(-1x^{-1}) + C = -\frac{2}{x} + C$

21. $\int \pi\sqrt{t}\,dt = \pi\int t^{1/2}dt = \pi(\tfrac{2}{3}t^{3/2}) + C = \frac{2\pi}{3}t^{3/2} + C$

23. $\int (3-2x)dx = \int 3\,dx - 2\int x\,dx = 3x - x^2 + C$

25. $\int (x^2 + x + x^{-3})dx = \int x^2\,dx + \int x\,dx + \int x^{-3}\,dx = \tfrac{1}{3}x^3 + \tfrac{1}{2}x^2 - \tfrac{1}{2}x^{-2} + C$

27. $\int 4e^x\,dx = 4e^x + C$

29. $\int (1 + x + e^x)dx = x + \tfrac{1}{2}x^2 + e^x + C$

31. $\int (4x^3 - \frac{2}{x^2} - 1)\,dx = \int (4x^3 - 2x^{-2} - 1)\,dx = x^4 + 2x^{-1} - x + C = x^4 + \frac{2}{x} - x + C$

33. $\int (x^{5/2} + 2x^{3/2} - x)\,dx = \tfrac{2}{7}x^{7/2} + \tfrac{4}{5}x^{5/2} - \tfrac{1}{2}x^2 + C$

35. $\int (x^{1/2} + 3x^{-1/2})\,dx = \tfrac{2}{3}x^{3/2} + 6x^{1/2} + C$

37. $\int \left(\frac{u^3 + 2u^2 - u}{3u}\right)du = \frac{1}{3}\int (u^2 + 2u - 1)\,du = \frac{1}{9}u^3 + \frac{1}{3}u^2 - \frac{1}{3}u + C$

39. $\int (2t+1)(t-2)\,dt = \int (2t^2 - 3t - 2)\,dt = \tfrac{2}{3}t^3 - \tfrac{3}{2}t^2 - 2t + C$

41. $\int \frac{1}{x^2}(x^4 - 2x^2 + 1)\,dx = \int (x^2 - 2 + x^{-2})\,dx = \frac{1}{3}x^3 - 2x - x^{-1} + C = \frac{1}{3}x^3 - 2x - \frac{1}{x} + C$

43. $\displaystyle\int \frac{ds}{(s+1)^{-2}} = \int (s+1)^2\,ds = \int (s^2+2s+1)\,ds = \tfrac{1}{3}s^3 + s^2 + s + C$

45. $\displaystyle\int (e^t + t^e)\,dt = e^t + \frac{1}{e+1}t^{e+1} + C$

47. $\displaystyle\int \left(\frac{x^3+x^2-x+1}{x^2}\right)dx = \int \left(x+1-\frac{1}{x}+\frac{1}{x^2}\right)dx = \frac{1}{2}x^2 + x - \ln|x| - x^{-1} + C$

49. $\displaystyle\int \left(\frac{(x^{1/2}-1)^2}{x^2}\right)dx = \int \left(\frac{x-2x^{1/2}+1}{x^2}\right)dx = \int (x^{-1}-2x^{-3/2}+x^{-2})\,dx$

$\displaystyle= \ln|x| + 4x^{-1/2} - x^{-1} + C = \ln|x| + \frac{4}{\sqrt{x}} - \frac{1}{x} + C$

51. $\displaystyle\int f'(x)\,dx = \int (2x+1)\,dx = x^2 + x + C.$ The condition $f(1) = 3$ gives

$f(1) = 1 + 1 + C = 3$, or $C = 1$. Therefore, $f(x) = x^2 + x + 1$.

53. $f'(x) = 3x^2 + 4x - 1$; $f(x) = x^3 + 2x^2 - x + C$. Using the given initial condition, we have $f(2) = 8 + 2(4) - 2 + C = 9$, so $16 - 2 + C = 9$, or $C = -5$. Therefore, $f(x) = x^3 + 2x^2 - x - 5$.

55. $\displaystyle f(x) = \int f'(x)\,dx = \int \left(1+\frac{1}{x^2}\right)dx = \int (1+x^{-2})\,dx = x - \frac{1}{x} + C.$

Using the given initial condition, we have $f(1) = 1 - 1 + C = 2$, or $C = 2$. Therefore, $\displaystyle f(x) = x - \frac{1}{x} + 2.$

57. $\displaystyle f(x) = \int \frac{x+1}{x}\,dx = \int \left(1+\frac{1}{x}\right)dx = x + \ln|x| + C.$ Using the initial condition, we have $f(1) = 1 + \ln 1 + C = 1 + C = 1$, or $C = 0$. So $f(x) = x + \ln|x|$.

59. $f(x) = \int f'(x)\,dx = \int \tfrac{1}{2}x^{-1/2}\,dx = \tfrac{1}{2}(2x^{1/2}) + C = x^{1/2} + C$; $f(2) = \sqrt{2} + C = \sqrt{2}$ implies $C = 0$. So $f(x) = \sqrt{x}$.

61. $f'(x) = e^x + x;\ f(x) = e^x + \frac{1}{2}x^2 + C; f(0) = e^0 + \frac{1}{2}(0) + C = 1 + C$

So $3 = 1 + C$ or $2 = C$. Therefore, $f(x) = e^x + \frac{1}{2}x^2 + 2$.

63. The position of the car is

$S(t) = \int f(t)\,dt = \int 2\sqrt{t}\,dt = \int 2t^{1/2}\,dt = 2(\frac{2}{3}t^{3/2}) + C = \frac{4}{3}t^{3/2} + C.$

$S(0) = 0$ implies $S(0) = C = 0$. So $S(t) = \frac{4}{3}t^{3/2}$.

65. $C(x) = \int C'(x)\,dx = \int (0.000009x^2 - 0.009x + 8)\,dx$

$= 0.000003x^3 - 0.0045x^2 + 8x + k.$

$C(0) = k = 120$ and so $C(x) = 0.000003x^3 - 0.0045x^2 + 8x + 120.$

$C(500) = 0.000003(500)^3 - 0.0045(500)^2 + 8(500) + 120$, or \$3370.

67. $P'(x) = -0.004x + 20$, $P(x) = -0.002x^2 + 20x + C$. Since $C = -16{,}000$, we find that $P(x) = -0.002x^2 + 20x - 16{,}000$. The company realizes a maximum profit when $P'(x) = 0$, that is, when $x = 5000$ units. Next,

$P(5000) = -0.002(5000)^2 + 20(5000) - 16{,}000 = 34{,}000.$

Thus, a maximum profit of \$34,000 is realized at a production level of 5000 units.

69. $N(t) = \int N'(t)\,dt = \int (-3t^2 + 12t + 45)\,dt = -t^3 + 6t^2 + 45t + C$. But $N(0) = C = 0$ and so $N(t) = -t^3 + 6t^2 + 45t$. The number is $N(4) = -4^3 + 6(4)^2 + 45(4) = 212$.

71. The number of new subscribers at any time is

$N(t) = \int (100 + 210t^{3/4})\,dt = 100t + 120t^{7/4} + C.$

The given condition implies that $N(0) = 5000$. Using this condition, we find $C = 5000$. Therefore, $N(t) = 100t + 120t^{7/4} + 5000$. The number of subscribers 16 months from now is

$N(16) = 100(16) + 120(16)^{7/4} + 5000$, or 21,960.

73. The rate of change of the population at any time t is $P'(t) = 4500t^{1/2} + 1000$. Therefore, $P(t) = 3000t^{3/2} + 1000t + C$. But $P(0) = 30{,}000$ and this implies that

$P(t) = 3000t^{3/2} + 1000t + 30{,}000.$

Finally, the projected population 9 years after the construction has begun is

$P(9) = 3000(9)^{3/2} + 1000(9) + 30{,}000 = 120{,}000.$

75. $S'(W) = 0.131773W^{-0.575}$; $S = \int 0.131773W^{-0.575}\,dW = 0.310054W^{0.425} + C$

$S(70) = 0.310054(70)^{0.425} + C = 1.8867 + C = 1.886277.$
Therefore $C = -0.000007 \approx 0$. $S(75) = 0.310054(75)^{0.425} \approx 1.9424$.

77. $v(r) = \int v'(r)\,dr = \int -kr\,dr = -\frac{1}{2}kr^2 + C.$

But $v(R) = 0$ and so $v(R) = -\frac{1}{2}kR^2 + C = 0$, or $C = \frac{1}{2}kR^2$. Therefore,
$v(R) = -\frac{1}{2}kr^2 + \frac{1}{2}kR^2 = \frac{1}{2}k(R^2 - r^2)$.

79. Denote the constant deceleration by k (ft/sec^2). Then $f''(t) = -k$, so $f'(t) = v(t) = -kt + C_1$. Next, the given condition implies that $v(0) = 88$. This gives $C_1 = 88$, or $f'(t) = -kt + 88$.

$$s = f(t) = \int f'(t)\,dt = \int (-kt + 88)\,dt = -\frac{1}{2}kt^2 + 88t + C_2.$$

Also, $f(0) = 0$ gives $s = f(t) = -\frac{1}{2}kt^2 + 88t$. Since the car is brought to rest in 9 seconds, we have $v(9) = -9k + 88 = 0$, or $k = \frac{88}{9}$, or $9\frac{7}{9}$. So the deceleration is $9\frac{7}{9}$ ft/sec^2. The distance covered is

$$s = f(9) = -\frac{1}{2}\left(\frac{88}{9}\right)(81) + 88(9) = 396.$$

So the stopping distance is 396 ft.

81. Suppose the acceleration is k ft/sec^2. The distance covered is $s = f(t)$ and satisfies $f''(t) = k$. So $f'(t) = v(t) = \int k\,dt = kt + C_1$. Next, $v(0) = 0$ gives $v(t) = kt$, and $s = f(t) = \int kt\,dt = \frac{1}{2}kt^2 + C_2$. Also, $f(0) = 0$ gives $s = \frac{1}{2}kt^2$. If it travelled 800 ft, we have $800 = \frac{1}{2}kt^2$, or $t = \frac{40}{\sqrt{k}}$. Its speed at this time is

$$v(t) = kt = k\left(\frac{40}{\sqrt{k}}\right) = 40\sqrt{k}.$$

We want the speed to be at least 240 ft/sec. So we require $40\sqrt{k} > 240$, or $k > 36$. In other words, the minimum acceleration must be 36 ft/sec^2.

EXERCISES 14.2, page 953

1. Put $u = 4x + 3$ so that $du = 4\,dx$, or $dx = \frac{1}{4}du$. Then

$$\int 4(4x+3)^4\,dx = \int u^4\,du = \tfrac{1}{5}u^5 + C = \tfrac{1}{5}(4x+3)^5 + C.$$

3. Let $u = x^3 - 2x$ so that $du = (3x^2 - 2)\,dx$. Then

$$\int (x^3-2x)^2(3x^2-2)\,dx = \int u^2\,du = \tfrac{1}{3}u^3 + C = \tfrac{1}{3}(x^3-2x)^3 + C.$$

5. Let $u = 2x^2 + 3$ so that $du = 4x\,dx$. Then

$$\int \frac{4x}{(2x^2+3)^3}\,dx = \int \frac{1}{u^3}\,du = \int u^{-3}\,du = -\tfrac{1}{2}u^{-2} + C = -\frac{1}{2(2x^2+3)^2} + C.$$

7. Put $u = t^3 + 2$ so that $du = 3t^2\,dt$ or $t^2\,dt = \tfrac{1}{3}du$. Then

$$\int 3t^2\sqrt{t^3+2}\,dt = \int u^{1/2}\,du = \tfrac{2}{3}u^{3/2} + C = \tfrac{2}{3}(t^3+2)^{3/2} + C$$

9. Let $u = x^2 - 1$ so that $du = 2x\,dx$ and $x\,dx = \tfrac{1}{2}du$. Then,

$$\int (x^2-1)^9\,x\,dx = \int \tfrac{1}{2}u^9\,du = \tfrac{1}{20}u^{10} + C = \tfrac{1}{20}(x^2-1)^{10} + C.$$

11. Let $u = 1 - x^5$ so that $du = -5x^4\,dx$ or $x^4\,dx = -\tfrac{1}{5}du$. Then

$$\int \frac{x^4}{1-x^5}\,dx = -\frac{1}{5}\int \frac{du}{u} = -\frac{1}{5}\ln|u| + C = -\frac{1}{5}\ln\left|1-x^5\right| + C.$$

13. Let $u = x - 2$ so that $du = dx$. Then

$$\int \frac{2}{x-2}\,dx = 2\int \frac{du}{u} = 2\ln|u| + C = \ln u^2 + C = \ln(x-2)^2 + C$$

15. Let $u = 0.3x^2 - 0.4x + 2$. Then $du = (0.6x - 0.4)\,dx = 2(0.3x - 0.2)\,dx$.

$$\int \frac{0.3x-0.2}{0.3x^2-0.4x+2}\,dx = \int \frac{1}{2u}\,du = \frac{1}{2}\ln|u| + C = \frac{1}{2}\ln(0.3x^2-0.4x+2) + C.$$

17. Let $u = 3x^2 - 1$ so that $du = 6x\,dx$, or $x\,dx = \tfrac{1}{6}du$. Then

$$\int \frac{x}{3x^2-1}\,dx = \frac{1}{6}\int \frac{du}{u} = \frac{1}{6}\ln|u| + C = \frac{1}{6}\ln\left|3x^2-1\right| + C.$$

19. Let $u = -2x$ so that $du = -2\,dx$ or $dx = -\frac{1}{2}du$. Then

$$\int e^{-2x}\,dx = -\frac{1}{2}\int e^{u}\,du = -\frac{1}{2}e^{u} + C = -\frac{1}{2}e^{-2x} + C.$$

21. Let $u = 2 - x$ so that $du = -\,dx$ or $dx = -\,du$. Then

$$\int e^{2-x}dx = -\int e^{u}\,du = -e^{u} + C = -e^{2-x} + C.$$

23. Let $u = -x^2$, then $du = -2x\,dx$ or $x\,dx = -\frac{1}{2}du$.

$$\int xe^{-x^2}\,dx = \int -\frac{1}{2}e^{u}du = -\frac{1}{2}e^{u} + C = -\frac{1}{2}e^{-x^2} + C.$$

25. $\int (e^{x} - e^{-x})dx = \int e^{x}\,dx - \int e^{-x}\,dx = e^{x} - \int e^{-x}\,dx.$

To evaluate the second integral on the right, let $u = -x$ so that $du = -dx$ or $dx = -du$. Therefore,

$$\int (e^{x} - e^{-x})dx = e^{x} + \int e^{u}\,du = e^{x} + e^{u} + C = e^{x} + e^{-x} + C.$$

27. Let $u = 1 + e^{x}$ so that $du = e^{x}\,dx$. Then

$$\int \frac{e^{x}}{1+e^{x}}dx = \int \frac{du}{u} = \ln|u| + C = \ln(1+e^{x}) + C.$$

29. Let $u = \sqrt{x} = x^{1/2}$. Then $du = \frac{1}{2}x^{-1/2}\,dx$ or $2\,du = x^{-1/2}\,dx$.

$$\int \frac{e^{\sqrt{x}}}{\sqrt{x}}dx = \int 2e^{u}\,du = 2e^{u} + C = 2e^{\sqrt{x}} + C.$$

31. Let $u = e^{3x} + x^3$ so that $du = (3e^{3x} + 3x^2)\,dx = 3(e^{3x} + x^2)\,dx$ or $(e^{3x} + x^2)dx = \frac{1}{3}du$.

Then

$$\int \frac{e^{3x} + x^2}{(e^{3x} + x^3)^3}dx = \frac{1}{3}\int \frac{du}{u^3} = \frac{1}{3}\int u^{-3}\,du = -\frac{1}{6}u^{-2} + C = -\frac{1}{6(3^{3x} + x^3)^2} + C.$$

33. Let $u = e^{2x} + 1$, so that $du = 2e^{2x}\,dx$, or $\frac{1}{2}du = e^{2x}\,dx$.

$$\int e^{2x}(e^{2x} + 1)^3\,dx = \int \frac{1}{2}u^3\,du = \frac{1}{8}u^4 + C = \frac{1}{8}(e^{2x} + 1)^4 + C.$$

35. Let $u = \ln 5x$ so that $du = \dfrac{1}{x}\,dx$. Then

$$\int \frac{\ln 5x}{x}\,dx = \int u\,du = \tfrac{1}{2}u^2 + C = \tfrac{1}{2}(\ln 5x)^2 + C.$$

37. Let $u = \ln x$ so that $du = \frac{1}{x}dx$. Then

$$\int \frac{1}{x\ln x}\,dx = \int \frac{du}{u} = \ln|u| + C = \ln\left|\ln x\right| + C.$$

39. Let $u = \ln x$ so that $du = \frac{1}{x}dx$. Then

$$\int \frac{\sqrt{\ln x}}{x}\,dx = \int \sqrt{u}\,du = \tfrac{2}{3}u^{3/2} + C = \tfrac{2}{3}(\ln x)^{3/2} + C$$

41. $$\int\left(xe^{x} - \frac{x}{x^2+2}\right)dx = \int xe^{x^2} - \int \frac{x}{x^2+2}\,dx.$$

To evaluate the first integral, let $u = x^2$ so that $du = 2x\,dx$, or $x\,dx = \frac{1}{2}du$. Then

$$\int xe^{x^2}\,dx = \tfrac{1}{2}\int e^u\,du + C_1 = \tfrac{1}{2}e^u + C_1 = \tfrac{1}{2}e^{x^2} + C_1.$$

To evaluate the second integral, let $u = x^2 + 2$ so that $du = 2x\,dx$, or $x\,dx = \frac{1}{2}\,du$. Then

$$\int \frac{x}{x^2+2}\,dx = \frac{1}{2}\int \frac{du}{u} = \frac{1}{2}\ln|u| + C_2 = \frac{1}{2}\ln(x^2+2) + C_2.$$

Therefore, $\displaystyle\int\left(xe^{x^2} - \frac{x}{x^2+2}\right)dx = \frac{1}{2}e^{x^2} - \frac{1}{2}\ln(x^2+2) + C.$

43. Let $u = \sqrt{x} - 1$ so that $du = \frac{1}{2}x^{-1/2}\,dx = \dfrac{1}{2\sqrt{x}}\,dx$ or $dx = 2\sqrt{x}\,du$.

Also, we have $\sqrt{x} = u + 1$, so that $x = (u+1)^2 = u^2 + 2u + 1$ and $dx = 2(u+1)\,du$. So

$$\int \frac{x+1}{\sqrt{x}-1}\,dx = \int \frac{u^2+2u+2}{u}\cdot 2(u+1)\,du = 2\int \frac{(u^3+3u^2+4u+2)}{u}\,du$$

$$= 2\int\left(u^2 + 3u + 4 + \frac{2}{u}\right)du = 2\left(\frac{1}{3}u^3 + \frac{3}{2}u^2 + 4u + 2\ln|u|\right) + C$$

$$= 2\left[\frac{1}{3}(\sqrt{x}-1)^3 + \frac{3}{2}(\sqrt{x}-1)^2 + 4(\sqrt{x}-1) + 2\ \ln\left|\sqrt{x}-1\right|\right] + C.$$

45. Let $u = x - 1$ so that $du = dx$. Also, $x = u + 1$ and so

$$\int x(x-1)^5\,dx = \int (u+1)u^5\,du = \int (u^6 + u^5)\,du$$

$$= \frac{1}{7}u^7 + \frac{1}{6}u^6 + C = \frac{1}{7}(x-1)^7 + \frac{1}{6}(x-1)^6 + C$$

$$= \frac{(6x+1)(x-1)^6}{42} + C.$$

47. Let $u = 1 + \sqrt{x}$ so that $du = \frac{1}{2}x^{-1/2}\,dx$ and $dx = 2\sqrt{x} = 2(u-1)\,du$

$$\int \frac{1-\sqrt{x}}{1+\sqrt{x}}\,dx = \int \left(\frac{1-(u-1)}{u}\right)\cdot 2(u-1)\,du = 2\int \frac{(2-u)(u-1)}{u}\,du$$

$$= 2\int \frac{-u^2+3u-2}{u}\,du = 2\int \left(-u+3-\frac{2}{u}\right)du = -u^2 + 6u - 4\ln|u| + C$$

$$= -(1+\sqrt{x})^2 + 6(1+\sqrt{x}) - 4\ln(1+\sqrt{x}) + C$$

$$= -1 - 2\sqrt{x} - x + 6 + 6\sqrt{x} - 4\ln(1+\sqrt{x}) + C$$

$$= -x + 4\sqrt{x} + 5 - 4\ln(1+\sqrt{x}) + C.$$

49. $I = \int v^2(1-v)^6\,dv$. Let $u = 1 - v$, then $du = -dv$. Also, $1 - u = v$, and $(1-u)^2 = v^2$.

Therefore,

$$I = \int -(1-2u+u^2)u^6\,du = \int -(u^6 - 2u^7 + u^8)\,du = -\left(\frac{u^7}{7} - \frac{2u^8}{8} + \frac{u^9}{9}\right) + C$$

$$= -u^7\left(\frac{1}{7} - \frac{1}{4}u + \frac{1}{9}u^2\right) + C = -\frac{1}{252}(1-v)^7[36 - 63(1-v) + 28(1-2v+v^2)]$$

$$= -\frac{1}{252}(1-v)^7[36 - 63 + 63v + 28 - 56v + 28v^2]$$

$$= -\frac{1}{252}(1-v)^7(28v^2 + 7v + 1) + C.$$

51. $f(x) = \int f'(x)\,dx = 5\int (2x-1)^4\,dx$. Let $u = 2x - 1$ so that $du = 2x-1$ so that $du = 2\ dx$, or $dx = \frac{1}{2}du$. Then

$$f(x)=\frac{5}{2}\int u^4\,du=\frac{1}{2}u^5+C=\frac{1}{2}(2x-1)^5+C.$$

Next, $f(1)=3$ implies $\frac{1}{2}+C=3$ or $C=\frac{5}{2}$. Therefore, $f(x)=\frac{1}{2}(2x-1)^5+\frac{5}{2}$.

53. $f(x)=\int -2xe^{-x^2+1}\,dx$. Let $u=-x^2+1$ so that $du=-2x\,dx$. Then

$f(x)=\int e^u\,du=e^u+C=e^{-x^2+1}+C$. The condition $f(1)=0$ implies

$f(1)=1+C=0$, or $C=-1$. Therefore, $f(x)=e^{-x^2+1}-1$.

55. $N'(t)=2000(1+0.2t)^{-3/2}$. Let $u=1+0.2t$. Then $du=0.2\,dt$ and $5\,du=dt$.

Therefore, $N(t)=(5)(2000)\int u^{-3/2}\,du=-20{,}000u^{-1/2}+C=-20{,}000(1+0.2t)^{-1/2}+C$.

Next, $N(0)=-20{,}000(1)^{-1/2}+C=1000$. Therefore, $C=21{,}000$ and

$N(t)=-\dfrac{20{,}000}{\sqrt{1+0.2t}}+21{,}000$. In particular, $N(5)=-\dfrac{20{,}000}{\sqrt{2}}+21{,}000\approx 6{,}858$.

57. $p(x)=\displaystyle\int -\frac{250x}{(16+x^2)^{3/2}}\,dx=-250\int\frac{x}{(16+x^2)^{3/2}}\,dx$.

Let $u=16+x^2$ so that $du=2x\,dx$ and $x\,dx=\frac{1}{2}\,du$.

Then $p(x)=-\frac{250}{2}\int u^{-3/2}\,du=(-125)(-2)u^{-1/2}+C=\dfrac{250}{\sqrt{16+x^2}}+C$.

$p(3)=\dfrac{250}{\sqrt{16+9}}+C=50$ implies $C=0$ and $p(x)=\dfrac{250}{\sqrt{16+x^2}}$.

59. Let $u=2t+4$, so that $du=2\,dt$. Then

$$r(t)=\int\frac{30}{\sqrt{2t+4}}\,dt=30\int\frac{1}{2}u^{-1/2}\,du=30u^{1/2}+C=30\sqrt{2t+4}+C.$$

$r(0)=60+C=0$, and $C=-60$. Therefore, $r(t)=30\left(\sqrt{2t+4}-2\right)$. Then

$r(16)=30\left(\sqrt{36}-2\right)=120$ ft. Therefore, the polluted area is

$$\pi r^2=\pi(120)^2=14{,}400\pi,\qquad\text{or } 14{,}400\pi \text{ sq ft.}$$

61. Let $u=2.449e^{-0.3277t}$ so that $du=-0.73565373e^{-0.3277t}\,dt$ and $e^{-0.3277t}\,dt=-1.359335\,du$.

Then $h(t) = \int \frac{52.8706e^{-0.3277t}}{(1+2.449e^{-0.3277t})^2} dt = (52.8706)(-1.359335)\int \frac{du}{u^2}$

$$= 71.86887u^{-1} + C = \frac{71.86887}{1+2.449e^{-0.3277t}} + C.$$

$$h(0) = \frac{71.86887}{1+2.449} + C = 19.4, \text{ and } C = -1.43760.$$

Therefore, $h(t) = \dfrac{71.86887}{1+2.449e^{-0.3277t}} - 1.43760,$

and $h(8) = \dfrac{71.86887}{1+2.449e^{-0.3277(8)}} - 1.43760 \approx 59.6$, or 59.6 inches.

63. The number of speakers sold at the end of t years is

$$f(t) = \int f'(t)dt = \int 2000(3-2e^{-t})dt = 2000(3t+2e^{-t}) + C$$

But 2000 pairs of speakers were sold in the first year, and this implies that $f(1) = 2000$. So $2000 = 2000(3 + 2e^{-1}) + C$ and $C = -5472$.Therefore,

$$f(t) = 2000(3t + 2e^{-t}) - 5472.$$

The number of pairs sold in the first five years is

$$f(5) = 2000(15 + 2e^{-5}) - 5472 = 24{,}555, \text{ or } 24{,}555 \text{ pairs.}$$

65. $x(t) = \int x'(t)dt = \int \frac{1}{V}(ac - bx_0)e^{-bt/V}\,dt = \frac{1}{V}(ac - bx_0)\int e^{-bt/V}dt$.

Let $u = -bt/V$, so that $du = -b/V\,dt$ and $dt = -V/b\,du$. Then

$$x(t) = \frac{1}{V}(ac - bx_0)\int -\frac{V}{b}e^u\,du = \left(-\frac{ac}{b} + x_0\right)e^u = \left(-\frac{ac}{b} + x_0\right)e^{-bt/V} + C.$$

Since $x(0) = \left(-\dfrac{ac}{b} + x_0\right) + C = x_0$, $C = \dfrac{ac}{b}$, and

$$x(t) = \frac{ac}{b} + \left(x_0 - \frac{ac}{b}\right)e^{-bt/V}.$$

EXERCISES 14.3, page 965

1. $\frac{1}{3}(1.9 + 1.5 + 1.8 + 2.4 + 2.7 + 2.5) = \frac{12.8}{3} \approx 4.27.$

3. a. $A = \frac{1}{2}(2)(6) = 6$ sq units.

b. $\Delta x = \frac{2}{4} = \frac{1}{2}$; $x_1 = 0$, $x_2 = \frac{1}{2}$, $x_3 = 1$, $x_4 = \frac{3}{2}$.

$A \approx \frac{1}{2}[3(0) + 3(\frac{1}{2}) + 3(1) + 3(\frac{3}{2})] = \frac{9}{2}$
$= 4.5$ sq units.

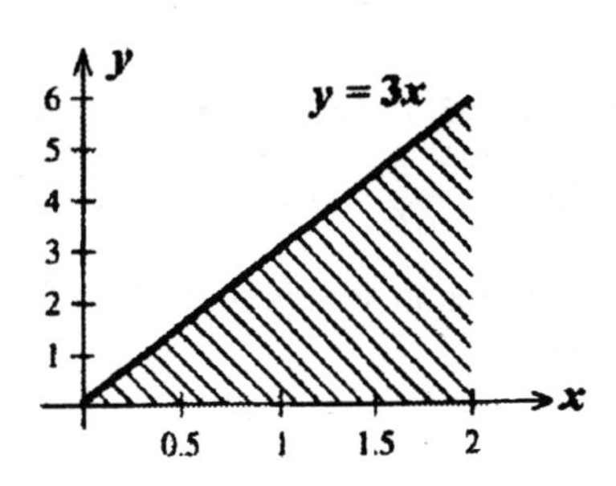

c. $\Delta x = \frac{2}{8} = \frac{1}{4}$. $x_1 = 0, \ldots, x_8 = \frac{7}{4}$.

$$A \approx \frac{1}{4}\left[3(0) + 3(\tfrac{1}{4}) + 3(\tfrac{1}{2}) + 3(\tfrac{3}{4}) + 3(1) + 3(\tfrac{5}{4}) + 3(\tfrac{3}{2}) + 3(\tfrac{7}{4})\right]$$
$$= \frac{21}{4} = 5.25 \text{ sq units.}$$

d. Yes.

5. a. $A = 4$

b. $\Delta x = \frac{2}{5} = 0.4$; $x_1 = 0$, $x_2 = 0.4$, $x_3 = 0.8$, $x_4 = 1.2$
$x_5 = 1.6$,
$A \approx 0.4\{[4 - 2(0)] + [4 - 2(0.4)] + [4 - 2(0.8)]$
$+ [4 - 2(1.2)] + [4 - 2(1.6)]\}$
$= 4.8$

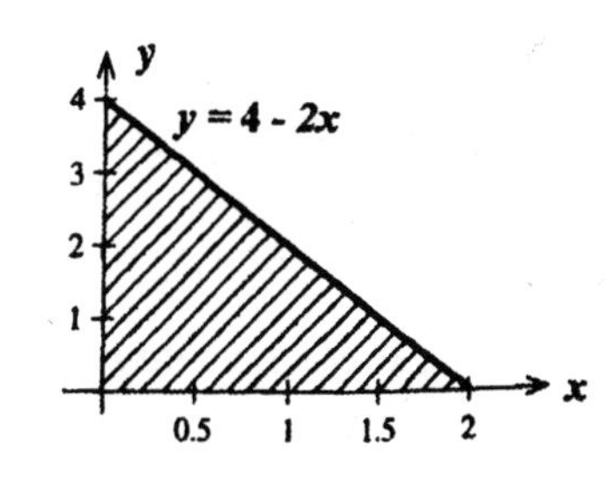

c. $\Delta x = \frac{2}{10} = 0.2$, $x_1 = 0$, $x_2 = 0.2$, $x_3 = 0.4$, ..., $x_{10} = 1.8$.
$A \approx 0.2\{[4 - 2(0)] + [4 - 2(0.2)] + [4 - 2(0.4)]$
$+ \cdots + [4 - 2(1.8)]\} = 4.4$

d. Yes.

7. a. $\Delta x = \dfrac{4-2}{2} = 1$; $x_1 = 2.5$, $x_2 = 3.5$; The Riemann sum is $[(2.5)^2 + (3.5)^2] = 18.5$.

b. $\Delta x = \dfrac{4-2}{5} = 0.4$; $x_1 = 2.2$, $x_2 = 2.6$, $x_3 = 3.0$, $x_4 = 3.4$, $x_5 = 3.8$
The Riemann sum is $0.4[2.2^2 + 2.6^2 + 3.0^2 + 3.4^2 + 3.8^2] = 18.64$.

c. $\Delta x = \dfrac{4-2}{10} = 0.2$; $x_1 = 2.1$, $x_2 = 2.3$, $x_2 = 2.5$, ..., $x_{10} = 3.9$
The Riemann sum is $0.2[2.1^2 + 2.3^2 + 2.5^2 + \cdots + 3.9^2] = 18.66$.
The area seems to be $18\frac{2}{3}$ sq units.

9. a. $\Delta x = \dfrac{4-2}{2} = 1$; $x_1 = 3$, $x_2 = 4$. The Riemann sum is $(1)[3^2 + 4^2] = 25$.

b. $\Delta x = \dfrac{4-2}{5} = 0.4$; $x_1 = 2.4$, $x_2 = 2.8$, $x_3 = 3.2$, $x_4 = 3.6$, $x_5 = 4$.

The Riemann sum is $0.4[2.4^2 + 2.8^2 + \cdots + 4^2] = 21.12$.

c. $\Delta x = \dfrac{4-2}{10} = 0.2$; $x_1 = 2.2$, $x_2 = 2.4$, $x_3 = 2.6$, ..., $x_{10} = 4$.

The Riemann sum is $0.2[2.2^2 + 2.4^2 + 2.6^2 + \cdots + 4^2] = 19.88$.

d. 19.9

11. a. $\Delta x = \dfrac{1}{2}$, $x_1 = 0$, $x_2 = \dfrac{1}{2}$. The Riemann sum is

$f(x_1)\Delta x + f(x_2)\Delta x = \left[(0)^3 + (\tfrac{1}{2})^3\right]\tfrac{1}{2} = \tfrac{1}{16} = 0.0625.$

b. $\Delta x = \dfrac{1}{5}$, $x_1 = 0$, $x_2 = \dfrac{1}{5}$, $x_3 = \dfrac{2}{5}$, $x_4 = \dfrac{3}{5}$, $x_5 = \dfrac{4}{5}$. The Riemann sum

is $f(x_1)\Delta x + f(x_2)\Delta x + \cdots f(x_5)\Delta x = \left[(\tfrac{1}{5})^3 + (\tfrac{2}{5})^3 + \cdots + (\tfrac{4}{5})^3\right]\tfrac{1}{5} = \tfrac{100}{625} = 0.16.$

c. $\Delta x = \dfrac{1}{10}$; $x_1 = 0$, $x_2 = \dfrac{1}{10}$, $x_3 = \dfrac{2}{10}$, $\cdots$, $x_{10} = \dfrac{9}{10}$.

The Riemann sum is

$$f(x_1)\Delta x + f(x_2)\Delta x + \cdots + f(x_{10})\Delta x = \left[(\tfrac{1}{10})^3 + (\tfrac{2}{10})^3 + \cdots + (\tfrac{9}{10})^3\right]\tfrac{1}{10}$$

$$= \tfrac{2025}{10{,}000} = 0.2025 \approx 0.2 \text{ sq units.}$$

The Riemann sum seems to approach 0.2.

13. $\Delta x = \dfrac{2-0}{5} = \dfrac{2}{5}$; $x_1 = \dfrac{1}{5}$, $x_2 = \dfrac{3}{5}$, $x_3 = \dfrac{5}{5}$, $x_4 = \dfrac{7}{5}$, $x_5 = \dfrac{9}{5}$.

$$A \approx \left\{\left[(\tfrac{1}{5})^2 + 1\right] + \left[(\tfrac{3}{5})^2 + 1\right] + \left[(\tfrac{5}{5})^2 + 1\right] + \left[(\tfrac{7}{5})^2 + 1\right] + \left[(\tfrac{9}{5})^2 + 1\right](\tfrac{2}{5})\right\}$$

$$= \tfrac{580}{125} = 4.64.$$

15. $\Delta x = \dfrac{3-1}{4} = \dfrac{1}{2}$; $x_1 = \dfrac{3}{2}$, $x_2 = \dfrac{4}{2}$, $x_3 = \dfrac{5}{2}$, $x_4 = 3$.

$$A \approx \left[\frac{1}{\frac{3}{2}} + \frac{1}{\frac{4}{2}} + \frac{1}{\frac{5}{2}} + \frac{1}{3}\right]\frac{1}{2} \approx 0.95 \text{ sq units.}$$

17. $A = 20[f(10) + f(30) + f(50) + f(70) + f(90)]$
$= 20(80 + 100 + 110 + 100 + 80) = 9400$ sq ft.

EXERCISES 14.4, page 976

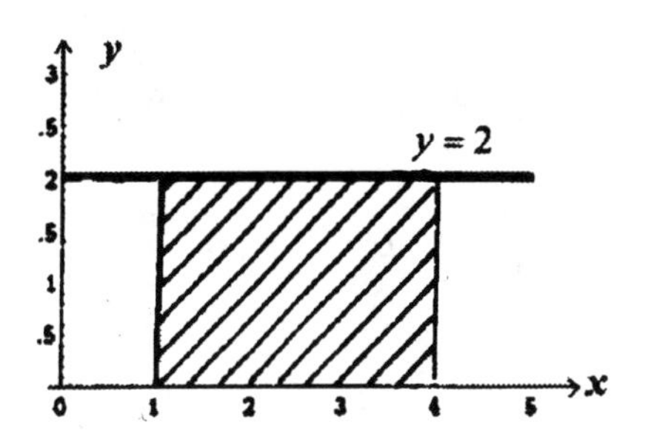

1. $A = \int_2^4 2\,dx = 2x\Big|_1^4 = 2(4-1) = 6$, or 6 square units. The region is a rectangle whose area is $3 \cdot 2$, or 6, square units.

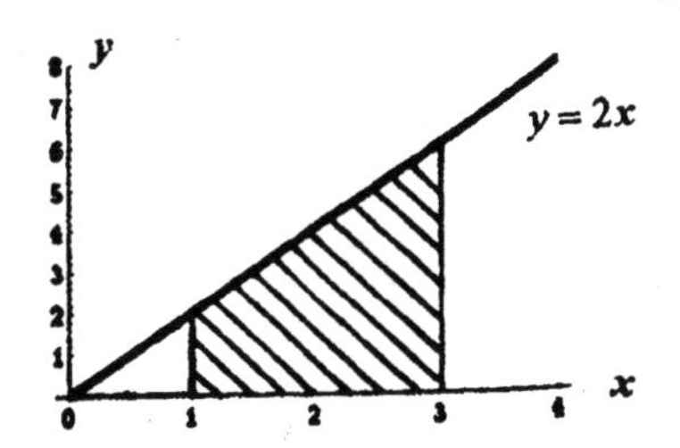

3. $A = \int_1^3 2x\,dx = x^2\Big|_1^3 = 9 - 1 = 8$, or 8 sq units. The region is a parallelogram of area $(1/2)(3-1)(2+6) = 8$ sq units.

5. $A = \int_{-1}^2 (2x+3)\,dx = x^2 + 3x\Big|_{-1}^2 = (4+6) - (1-3) = 12$, or 12 sq. units.

7. $A = \int_{-1}^4 (-x^2+4)\,dx = -\dfrac{1}{3}x^3 + 4x\Big|_{-1}^2 = \left(-\dfrac{8}{3}+8\right) - \left(\dfrac{1}{3}-4\right) = 9$, or 9 sq units.

9. $A = \int_1^2 \dfrac{1}{x}\,dx = \ln|x|\Big|_1^2 = \ln 2 - \ln 1 = \ln 2$, or ln 2 sq units.

11. $A = \int_1^9 \sqrt{x}\,dx = \dfrac{2}{3}x^{3/2}\Big|_1^9 = \dfrac{2}{3}(27-1) = \dfrac{52}{3}$, or $17\frac{1}{3}$ sq units.

13. $A = \int_{-8}^{-1} (1 - x^{1/3})\,dx = x - \frac{3}{4}x^{4/3}\Big|_{-8}^{-1} = (-1 - \frac{3}{4}) - (-8 - 12) = 18\frac{1}{4}$, or $18\frac{1}{4}$ sq units.

15. $A = \int_0^2 e^x\,dx = e^x\Big|_0^2 = (e^2 - 1)$, or approximately 6.39 sq units.

17. $\int_2^4 3\,dx = 3x\Big|_2^4 = 3(4-2) = 6.$

19. $\int_1^3 (2x+3)\,dx = x^2 + 3x\Big|_1^3 = (9+9) - (1+3) = 14.$

21. $\int_{-1}^3 2x^2\,dx = \frac{2}{3}x^3\Big|_{-1}^3 = \frac{2}{3}(27) - \frac{2}{3}(-1) = \frac{56}{3}.$

23. $\int_{-2}^{2}(x^2-1)\,dx = \frac{1}{3}x^3 - x\Big|_{-2}^{2} = \left(\frac{8}{3}-2\right)-\left(-\frac{8}{3}+2\right) = \frac{4}{3}.$

25. $\int_{1}^{8} 4x^{1/3}\,dx = (4)(\frac{3}{4})x^{4/3}\Big|_{1}^{8} = 3(16-1) = 45.$

27. $\int_{0}^{1}(x^3-2x^2+1)\,dx = \frac{1}{4}x^4 - \frac{2}{3}x^3 + x\Big|_{0}^{1} = \frac{1}{4}-\frac{2}{3}+1 = \frac{7}{12}$

29. $\int_{2}^{4}\frac{1}{x}\,dx = \ln|x|\Big|_{2}^{4} = \ln 4 - \ln 2 = \ln(\frac{4}{2}) = \ln 2.$

31. $\int_{0}^{4} x(x^2-1)\,dx = \int_{0}^{4}(x^3-x)\,dx = \frac{1}{4}x^4 - \frac{1}{2}x^2\Big|_{0}^{4} = 64-8 = 56.$

33. $\int_{0}^{3}(t^2-t)^2\,dt = \int_{1}^{3} t^4 - 2t^3 + t^2)\,dt = \frac{1}{5}t^5 - \frac{1}{2}t^4 + \frac{1}{3}t^3\Big|_{1}^{3}$

$$= \left(\frac{243}{5}-\frac{81}{2}+\frac{27}{3}\right)-\left(\frac{1}{5}-\frac{1}{2}+\frac{1}{3}\right) = \frac{512}{30} = \frac{256}{15}.$$

35. $\int_{-3}^{1} x^{-2}\,dx = -\frac{1}{x}\Big|_{-3}^{-1} = 1-\frac{1}{3} = \frac{2}{3}.$

37. $\int_{1}^{4}\left(\sqrt{x}-\frac{1}{\sqrt{x}}\right)dx = \int_{1}^{4}(x^{1/2}-x^{-1/2})\,dx = \frac{2}{3}x^{3/2} - 2x^{1/2}\Big|_{1}^{4}$

$$= \left(\frac{16}{3}-4\right)-\left(\frac{2}{3}-2\right) = \frac{8}{3}.$$

39. $\int_{1}^{4}\frac{3x^{32}-2x^{2.}+4}{x^2}\,dx = \int_{1}^{4}(3x-2+4x^{-2})\,dx = \frac{3}{2}x^2 - 2x - \frac{4}{x}\Big|_{1}^{4}$

$$= (24-8-1)-(\tfrac{3}{2}-2-40 = \tfrac{39}{2}.$$

41. *a.* $C(300)-C(0) = \int_{0}^{300}(0.0003x^2 - 0.12x + 20)\,dx = 0.0001x^3 - 0.06x^2 + 20x\Big|_{0}^{300}$

$$= 0.0001(300)^3 - 0.06(300)^2 + 20(300) = 3300.$$

Therefore C(300) = 3300 + C(0) = 3300 + 800 = 4100, or $4100.

b. $\int_{200}^{300} C'(x)\,dx = (0.0001x^3 - 0.06x^2 + 20x)\Big|_{200}^{300}$

$$= [0.0001(300)^3 - 0.06(300)^2 + 20(300)]$$
$$-[0.0001(200)^3 - 0.06(200)^2 + 20(200)]$$
$$= 900 \text{ or } \$900.$$

43. a. The profit is $\int_0^{200} (-0.0003x^2 + 0.02x + 20)\,dx + P(0)$

$$= -0.0001x^3 + 0.01x^2 + 20x\Big|_0^{200} + P(0)$$
$$= 3600 + P(0) = 3600 - 800, \text{ or } \$2800.$$

b. $\int_{200}^{220} P'(x)\,dx = P(220) - P(200) = -0.0001x^3 + 0.01x^2 + 20x\Big|_{200}^{220}$

$$= 219.20, \text{ or } \$219.20.$$

45. The distance is

$$\int_0^{20} v(t)\,dt = \int_0^{20} (-t^2 + 20t + 440)\,dt = -\tfrac{1}{3}t^3 + 10t^2 + 440t\Big|_0^{20}$$
$$\approx 10{,}133\tfrac{1}{3} \text{ ft.}$$

47. The amount of smoke left after 5 minutes is

$$100 - \int_0^5 R(t)\,dt = 100 - \int_0^5 (0.00032t^4 - 0.01872t^3 + 0.3948t^2 - 3.83t + 17.63)\,dt$$
$$= 100 - (0.000064t^5 - 0.00468t^4 + 0.1316t^3 - 1.915t^2$$
$$+ 17.63t)\Big|_0^5 = 46, \text{ or 46 percent.}$$

The amount of smoke left after 10 minutes is

$$100 - \int_0^{10} R(t)\,dt$$
$$= 100 - (0.000064t^5 - 0.00468t^4 + 0.1316t^3 - 1.915t^2 + 17.63\,t\Big|_0^{10}$$
$$= 24 \text{ percent.}$$

USING TECHNOLOGY EXERCISES 14.4, PAGE 979

1. 6.1787
3. 0.7873
5. −0.5888
7. 2.7044
9. 3.9973
11. 37.7 million

EXERCISES 14.5, page 988

1. Let $u = x^2 - 1$ so that $du = 2x\,dx$ or $x\,dx = \frac{1}{2}\,du$. Also, if $x = 0$, then $u = -1$ and if $x = 2$, then $u = 3$. So

$$\int_0^2 x(x^2-1)\,dx = \frac{1}{2}\int_{-1}^3 u^3\,du = \frac{1}{8}u^4\Big|_{-1}^3 = \frac{1}{8}(81) - \frac{1}{8}(1) = 10.$$

3. Let $u = 5x^2 + 4$ so that $du = 10x\,dx$ or $x\,dx = \frac{1}{10}\,du$. Also, if $x = 0$, then $u = 4$, and if $x = 1$, then $u = 9$. So

$$\int_0^1 x\sqrt{5x^2+4}\,dx = \frac{1}{10}\int_4^9 u^{1/2}\,du = \frac{1}{15}u^{3/2}\Big|_4^9 = \frac{1}{15}(27) - \frac{1}{15}(8) = \frac{19}{15}.$$

5. Let $u = x^3 + 1$ so that $du = 3x^2\,dx$ or $x^2\,dx = \frac{1}{3}\,du$. Also, if $x = 0$, then $u = 1$, and if $x = 2$, then $u = 9$. So,

$$\int_0^2 x^2(x^3+1)^{3/2}\,dx = \frac{1}{3}\int_1^9 u^{3/2}\,du = \frac{2}{15}u^{5/2}\Big|_1^9 = \frac{2}{15}(243) - \frac{2}{15}(1) = \frac{484}{15}.$$

7. Let $u = 2x + 1$ so that $du = 2\,dx$ or $dx = \frac{1}{2}\,du$. Also, if $x = 0$, then $u = 1$ and if $x = 1$ then $u = 3$. So

$$\int_0^1 \frac{1}{\sqrt{2x+1}}\,dx = \frac{1}{2}\int_1^3 \frac{1}{\sqrt{u}}\,du = \frac{1}{2}\int_1^3 u^{-1/2}\,du = u^{1/2}\Big|_1^3 = \sqrt{3} - 1.$$

9. $\int_1^2 (2x-1)^4\,dx$. Put $u = 2x - 1$ so that $du = 2\,dx$ or $dx = \frac{1}{2}\,du$.

Then $\int_1^2 (2x-1)^4\,dx = \frac{1}{2}\int_1^3 u^4\,du = \frac{1}{10}u^5\Big|_1^3 = \frac{1}{10}(243-1) = \frac{121}{5} = 24\frac{1}{5}$.

11. Let $u = x^3 + 1$ so that $du = 3x^2\,dx$ or $x^2\,dx = \frac{1}{3}\,du$. Also, if $x = -1$, then $u = 0$ and if $x = 1$, then $u = 2$. So

$$\int_{-1}^1 x^2(x^3+1)^4\,dx = \frac{1}{3}\int_0^2 u^4\,du = \frac{1}{15}u^5\Big|_0^2 = \frac{32}{15}.$$

13. Let $u = x - 1$ so that $du = dx$. Then if $x = 1$, $u = 0$, and if $x = 5$, then $u = 4$.

$$\int_1^5 x\sqrt{x-1}\,dx = \int_0^4 (u+1)u^{1/2}\,du = \int_0^4 (u^{3/2} + u^{1/2})\,du$$

$$= \frac{2}{5}u^{5/2} + \frac{2}{3}u^{3/2}\Big|_0^4 = \frac{2}{5}(32) + \frac{2}{3}(8) = 18\frac{2}{15}.$$

15. Let $u = x^2$ so that $du = 2x\,dx$ or $x\,dx = \frac{1}{2}du$. If $x = 0$, $u = 0$ and if $x = 2$, $u = 4$. So

$$\int_0^2 xe^{x^2}dx = \frac{1}{2}\int_0^4 e^u\,du = \frac{1}{2}e^u\Big|_0^4 = \frac{1}{2}(e^4 - 1).$$

17. $$\int (e^{2x} + x^2 + 1)dx = \frac{1}{2}e^{2x} + \frac{1}{3}x^3 + x\Big|_0^1 = (\frac{1}{2}e^2 + \frac{1}{3} + 1) - \frac{1}{2}$$
$$= \frac{1}{2}e^2 + \frac{5}{6}.$$

19. Put $u = x^2 + 1$ so that $du = 2x\,dx$ or $x\,dx = \frac{1}{2}du$. Then

$$\int_{-1}^1 xe^{x^2+1}dx = \frac{1}{2}\int_2^2 e^u\,du = \frac{1}{2}e^u\Big|_2^2 = 0$$

(Since the upper and lower limits are equal.)

21. Let $u = x - 2$ so that $du = dx$. If $x = 3$, $u = 1$ and if $x = 6$, $u = 4$. So

$$\int_3^6 \frac{2}{x-2}dx = 2\int_1^4 \frac{du}{u} = 2\ln|u|\Big|_1^4 = 2\ln 4.$$

23. Let $u = x^3 + 3x^2 - 1$ so that $du = (3x^2 + 6x)dx = 3(x^2 + 2x)dx$. If $x = 1$, $u = 3$, and if $x = 2$, $u = 19$. So

$$\int_1^2 \frac{x^2 + 2x}{x^3 + 3x^2 - 1}dx = \frac{1}{3}\int_3^{19} \frac{du}{u} = \frac{1}{3}\ln u\Big|_3^{19} = \frac{1}{3}(\ln 19 - \ln 3).$$

25. $$\int_1^2 \left(4e^{2u} - \frac{1}{u}\right)du = 2e^{2u} - \ln u\Big|_1^2 = (2e^4 - \ln 2) - (2e^2 - 0) = 2e^4 - 2e^2 - \ln 2.$$

27. $$\int (2e^{-4x} - x^{-2})dx = -\frac{1}{2}e^{-4x} + \frac{1}{x}\Big|_1^2 = (-\frac{1}{2}e^{-8} + \frac{1}{2}) - (-\frac{1}{2}e^{-4} + 1)$$
$$= -\frac{1}{2}e^{-8} + \frac{1}{2}e^{-4} - \frac{1}{2} = \frac{1}{2}(e^{-4} - e^{-8} - 1).$$

29. $$\text{AV} = \frac{1}{2}\int_0^2 (2x+3)\,dx = \frac{1}{2}(x^2 + 3x)\Big|_0^2 = \frac{1}{2}(10) = 5.$$

31. $$\text{AV} = \frac{1}{2}\int_1^3 (2x^2 - 3)\,dx = \frac{1}{2}(\frac{2}{3}x^3 - 3x)\Big|_1^3 = \frac{1}{2}(9 + \frac{7}{3}) = \frac{17}{3}.$$

33. $AV = \frac{1}{3}\int_{-1}^{2}(x^2+2x-3)\,dx = \frac{1}{3}(\frac{1}{3}x^3+x^2-3x)\Big|_{-1}^{2}$
$= \frac{1}{3}[(\frac{8}{3}+4-6)-(-\frac{1}{3}+1+3)] = \frac{1}{3}(\frac{8}{3}-2+\frac{1}{3}-4) = -1.$

35. $AV = \frac{1}{2}\int_{0}^{4}(2x+1)^{1/2}\,dx = (\frac{1}{4})(\frac{1}{2})(\frac{2}{3})(2x+1)^{3/2}\Big|_{0}^{4} = \frac{1}{12}(27-1) = \frac{13}{6}$

37. $AV = \dfrac{1}{2}\int_{0}^{2} xe^{x^2}\,dx = \dfrac{1}{4}e^{x^2}\Big|_{0}^{2} = \dfrac{1}{4}(e^4-1).$

39. The amount produced will be

$$\int_0^{20} 3.5e^{0.05t}dt = \frac{3.5}{0.05}e^u\Big|_0^1 \qquad \text{(Use the substitution } u = 0.05t.\text{)}$$
$$= 70(e-1) \approx 120.3,$$

or 120.3 billion metric tons.

41. The amount is $\int_1^2 t(\frac{1}{2}t^2+1)^{1/2}\,dt$. Let $u = \frac{1}{2}t^2+1$, so that $du = t\,dt$. Therefore,

$$\int_1^2 t(\tfrac{1}{2}t^2+1)^{1/2}\,dt = \int_{3/2}^{3} u^{1/2}\,du = \tfrac{2}{3}u^{3/2}\Big|_{3/2}^{3} = \tfrac{2}{3}[(3)^{3/2}-(\tfrac{3}{2})^{3/2}]$$
$$\approx 2.24 \text{ million dollars.}$$

43. The tractor will depreciate

$$\int_0^5 13388.61e^{-0.22314t}\,dt = \frac{13388.61}{-0.22314}e^{-0.22314t}\Big|_0^5$$
$$= -60{,}000.94e^{-0.22314t}\Big|_0^5 = -60{,}000.94(-0.672314)$$
$$= 40{,}339.47, \quad \text{or } \$40{,}339.$$

45. The average temperature is $\frac{1}{12}\int_0^{12}(-0.05t^3+0.4t^2+3.8t+5.6)\,dt$

$= \frac{1}{12}(-\frac{0.05}{4}t^4+\frac{0.4}{3}t^3+1.9t^2+5.6t)\Big|_0^{12} = 26°\,F.$

47. The average number is

$$\frac{1}{5}\int_0^5\left(-\frac{40{,}000}{\sqrt{1+0.2t}}+50{,}000\right)dt = -8000\int_0^5(1+0.2t)^{-1/2}\,dt + 10{,}000\int_0^5 dt$$

Integrating the first integral by substitution with $u = 1 + 0.2t$, so that $du = 0.2\,dt$ or

$dt = 5\,du$, we find that the average value is

$$-8000\int_1^2 5u^{-1/2}\,du + \int_0^5 10{,}000\,dt = -40{,}000(2u^{1/2})\Big|_1^2 + 10{,}000t\Big|_0^5$$

$$= -40{,}000(2\sqrt{2} - 2) + 50{,}000 = 16{,}863$$

or 16,863 subscribers.

49. The average concentration of the drug is

$$\frac{1}{4}\int_0^4 \frac{0.2t}{t^2+1}\,dt = \frac{0.2}{4}\int_0^4 \frac{t}{t^2+1}\,dt = \frac{0.2}{(4)(2)}\ln(t^2+1)\Big|_0^4$$

$$= 0.025\ln 17 \approx 0.071,$$

or 0.071 milligrams per cubic centimeter.

51. The average velocity of the blood is

$$\frac{1}{R}\int_0^R k(R^2 - r^2)\,dr = \frac{k}{R}\int_0^R (R^2 - r^2)\,dr = \frac{k}{r}(R^2 r - \tfrac{1}{3}r^3)\Big|_0^R$$

$$= \frac{k}{R}(R^3 - \tfrac{1}{3}R^3) = \frac{k}{R}\cdot\frac{2}{3}R^3 = \frac{2k}{3}R^2 \text{ cm / sec.}$$

53. $$\int_a^b f(x)\,dx = F(x)\Big|_a^b = F(b) - F(a) = -[F(a) - F(b)]$$

$$= -F(x)\Big|_b^a = -\int_b^a f(x)\,dx$$

55. $$\int_a^b cf(x)\,dx = xF(x)\Big|_a^b = c[F(b) - F(a)] = c\int_a^b f(x)\,dx.$$

57. $$\int_0^1 (1 + x - e^x)\,dx = x + \tfrac{1}{2}x^2 - e^x\Big|_0^1 = (1 + \tfrac{1}{2} - e) + 1 = \tfrac{5}{2} - e.$$

$$\int_0^1 dx + \int_0^1 x\,dx - \int_0^1 e^x\,dx = x\Big|_0^1 + \tfrac{1}{2}x^2\Big|_0^1 - e^x\Big|_0^1$$

$$= (1 - 0) + (\tfrac{1}{2} - 0) - (e - 1) = \tfrac{5}{2} - e.$$

59. $$\int_0^3 (1 + x^3)\,dx = x + \tfrac{1}{4}x^4\Big|_0^3 = 3 + \tfrac{81}{4} = \tfrac{93}{4}.$$

$$\int_0^1 (1 + x^3)\,dx + \int_1^2 (1 + x^3)\,dx + \int_2^3 (1 + x^3)\,dx$$

$$= (x + \tfrac{1}{4}x^4)\Big|_0^1 + (x + \tfrac{1}{4}x^4)\Big|_1^2 + (x + \tfrac{1}{4}x^4)\Big|_2^3$$

$$= (1 + \tfrac{1}{4}) + (2 + 4) - (1 + \tfrac{1}{4}) + (3 + \tfrac{81}{4}) - (2 + 4) = \tfrac{93}{4}.$$

EXERCISES 14.6, page 999

1. $-\int_0^6 (x^3 - 6x^2)\,dx = -\frac{1}{4}x^4 + 2x^3\Big|_0^6 = -\frac{1}{4}(6^4) + 2(6^3) = 108$ sq units.

3. $A = -\int_{-1}^0 x\sqrt{1-x^2}\,dx + \int_0^1 x\sqrt{1-x^2}\,dx = 2\int_0^1 x(1-x^2)^{1/2}\,dx$ (by symmetry). Let $u = 1 - x^2$ so that $du = -2x\,dx$ or $x\,dx = -\frac{1}{2}\,du$. Also, if $x = 0$, then $u = 1$ and if $x = 1$, $u = 0$. So $A = (2)(-\frac{1}{2})\int_0^1 u^{1/2}\,du = -\frac{2}{3}u^{3/2}\Big|_1^0 = \frac{2}{3}$, or $\frac{2}{3}$ sq units.

5. $A = -\int_0^4 (x - 2\sqrt{x})\,dx = \int_0^4 (-x + 2x^{1/2})\,dx = -\frac{1}{2}x^2 + \frac{4}{3}x^{3/2}\Big|_0^4$
 $= 8 + \frac{32}{3} = \frac{8}{3}$ sq units.

7. The required area is given by

$$\int_{-1}^0 (x^2 - x^{1/3})\,dx + \int_0^1 (x^{1/3} - x^2)\,dx = \frac{1}{3}x^3 - \frac{3}{4}x^{4/3}\Big|_{-1}^0 + \frac{3}{4}x^{4/3} - \frac{1}{3}x^3\Big|_0^1$$
$$= -(-\frac{1}{3} - \frac{3}{4}) + (\frac{3}{4} - \frac{1}{3}) = 1\frac{1}{2} \quad \text{sq units.}$$

9. The required area is given by
 $-\int_{-1}^2 -x^2\,dx = \frac{1}{3}x^3\Big|_{-1}^2 = \frac{8}{3} + \frac{1}{3} = 3$ sq units.

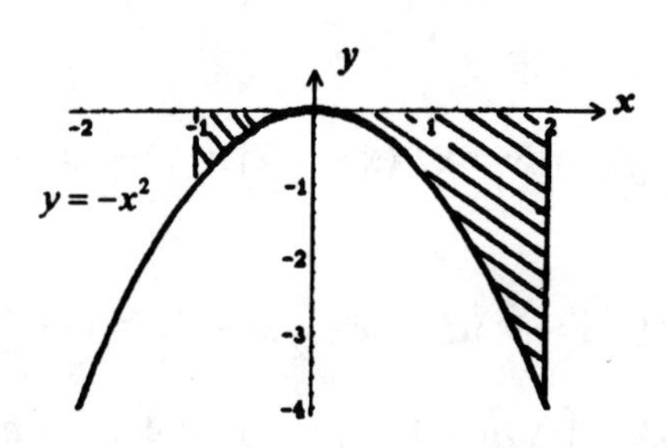

11. $y = x^2 - 5x + 4 = (x-4)(x-1) = 0$
 if $x = 1$ or 4. These give the x-intercepts.
 $A = -\int_1^3 (x^2 - 5x + 4)\,dx = -\frac{1}{3}x^3 + \frac{5}{2}x^2 - 4x\Big|_1^3$
 $= (-9 + \frac{45}{2} - 12) - (-\frac{1}{3} + \frac{5}{2} - 4) = \frac{10}{3} = 3\frac{1}{3}$.

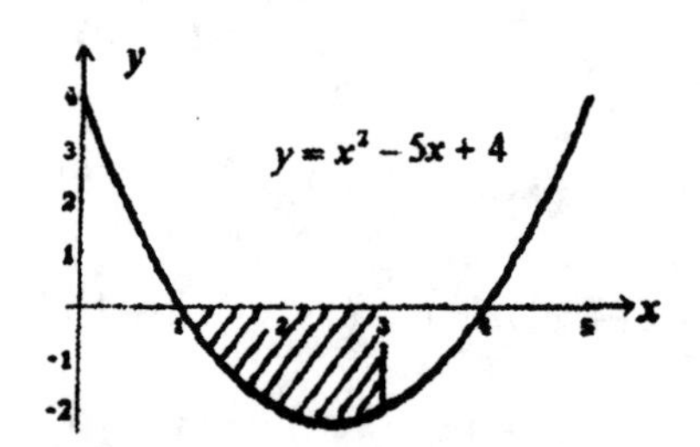

13. The required area is given by

$$-\int_0^9 -(1+\sqrt{x})\,dx = x + \tfrac{2}{3}x^{3/2}\Big|_0^9 = 9 + 18 = 27.$$

15. $$-\int_{-2}^{4} -e^{(1/2)x}\,dx = 2e^{(1/2)x}\Big|_{-2}^{4}$$

$$= 2(e^2 - e^{-1})\text{ sq units.}$$

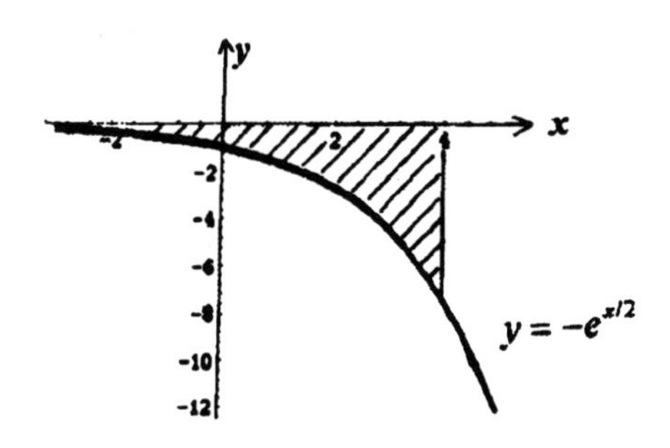

17. 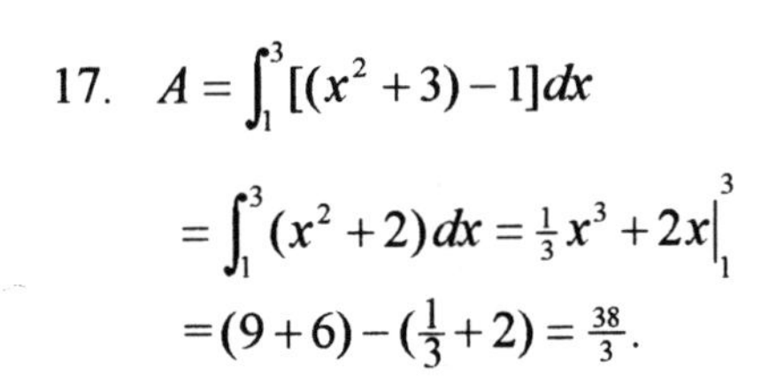

$$A = \int_1^3 [(x^2+3) - 1]\,dx$$

$$= \int_1^3 (x^2+2)\,dx = \tfrac{1}{3}x^3 + 2x\Big|_1^3$$

$$= (9+6) - (\tfrac{1}{3} + 2) = \tfrac{38}{3}.$$

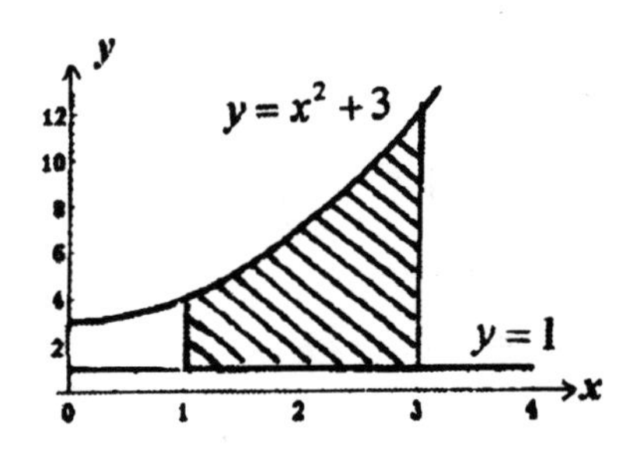

19. 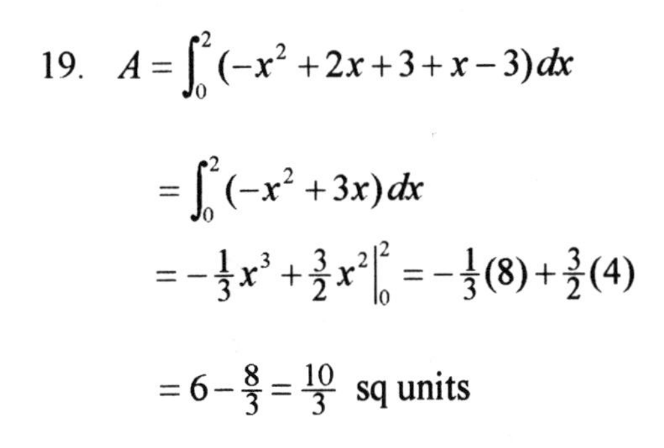

$$A = \int_0^2 (-x^2 + 2x + 3 + x - 3)\,dx$$

$$= \int_0^2 (-x^2 + 3x)\,dx$$

$$= -\tfrac{1}{3}x^3 + \tfrac{3}{2}x^2\Big|_0^2 = -\tfrac{1}{3}(8) + \tfrac{3}{2}(4)$$

$$= 6 - \tfrac{8}{3} = \tfrac{10}{3}\text{ sq units}$$

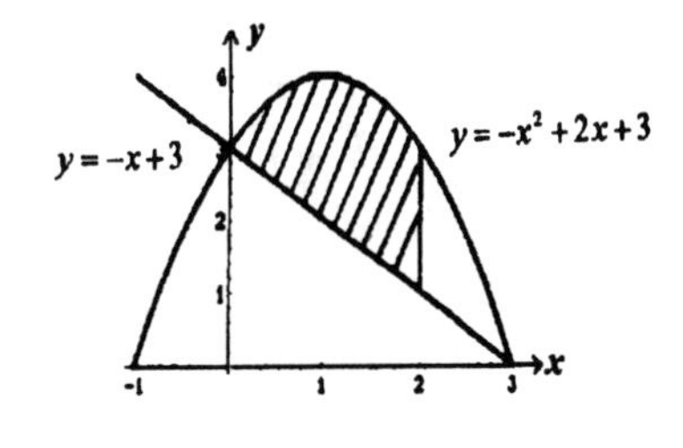

21. $A = \int_{-1}^{2} [(x^2+1) - \frac{1}{3}x^3]\,dx$

$= \int_{-1}^{2} (-\frac{1}{3}x^3 + x^2 + 1)\,dx$

$= -\frac{1}{12}x^4 + \frac{1}{3}x^3 + x\Big|_{-1}^{2}$

$= (-\frac{4}{3} + \frac{8}{3} + 2) - (-\frac{1}{12} - \frac{1}{3} - 1) = 4\frac{3}{4}$ sq units.

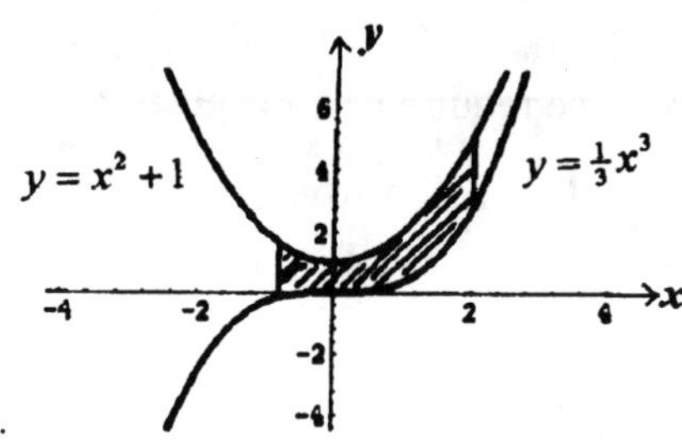

23. $A = \int_{1}^{4} \left[(2x-1) - \frac{1}{x}\right] dx = \int_{1}^{4} \left(2x - 1 - \frac{1}{x}\right) dx$

$= (x^2 - x - \ln x)\Big|_{1}^{4}$

$= (16 - 4 - \ln 4) - (1 - 1 - \ln 1)$

$= 12 - \ln 4 \approx 10.6$ sq units.

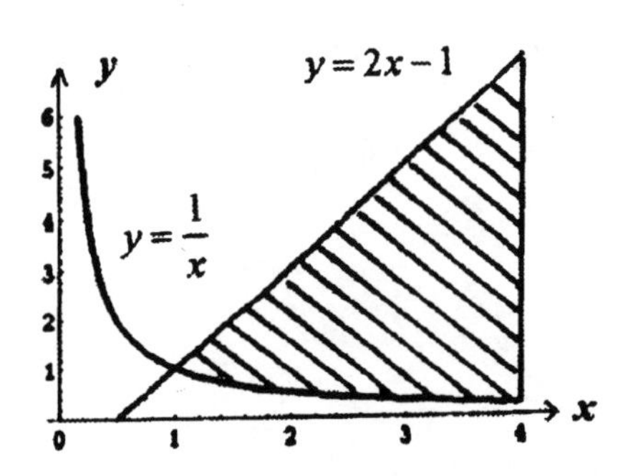

25. $A = \int_{1}^{2} \left(e^x - \frac{1}{x}\right) dx = e^x - \ln x\Big|_{1}^{2}$

$= (e^2 - \ln 2) - e = (e^2 - e - \ln 2)$ sq units.

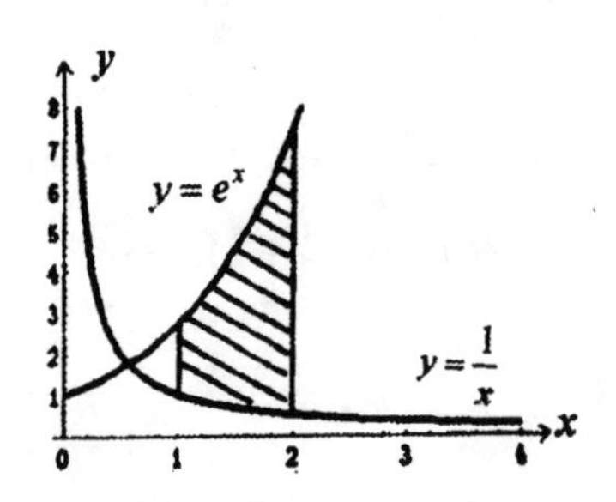

27.

$A = -\int_{-1}^{0} x\,dx + \int_{0}^{2} x\,dx$

$= -\frac{1}{2}x^2\Big|_{-1}^{0} + \frac{1}{2}x^2\Big|_{0}^{2}$

$= \frac{1}{2} + 2 = 2\frac{1}{2}$ sq units.

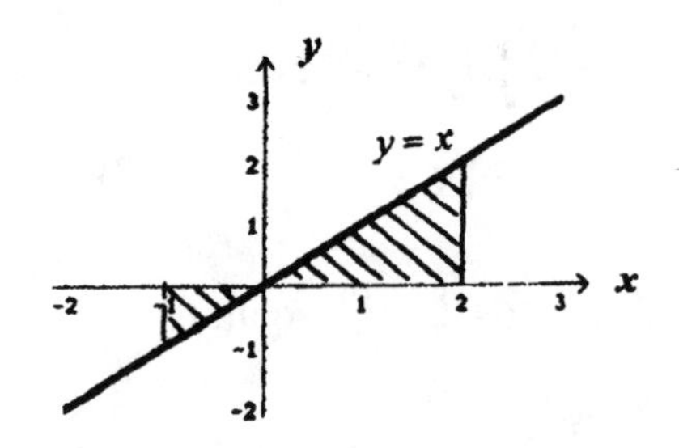

29. The x–intercepts are found by solving $x^2 - 4x + 3 = (x-3)(x-1) = 0$ giving $x = 1$ or 3. The region is shown in the figure.

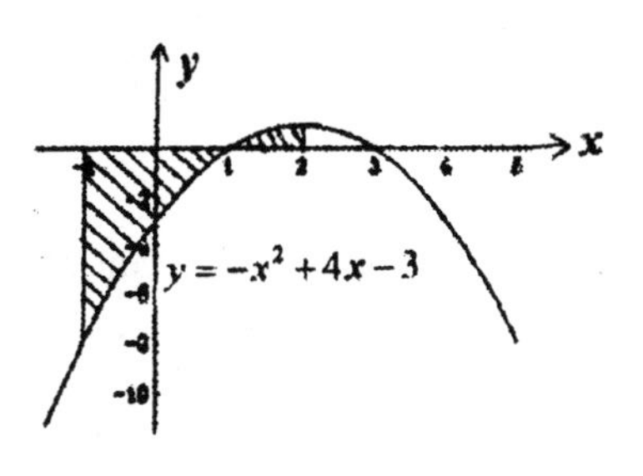

$$A = -\int_{-1}^{1}[(-x^2+4x-3)\,dx + \int_{1}^{2}(-x^2+4x-3)\,dx$$
$$= \tfrac{1}{3}x^3 - 2x^2 + 3x\Big|_{-1}^{1} + (-\tfrac{1}{3}x^3 + 2x^2 - 3x)\Big|_{1}^{2}$$
$$= (\tfrac{1}{3} - 2 + 3) - (-\tfrac{1}{3} - 2 - 3)$$
$$+(-\tfrac{8}{3} + 8 - 6) - (-\tfrac{1}{3} + 2 - 3) = \tfrac{22}{3} \text{ sq units.}$$

31. The region is shown in the figure at the right.

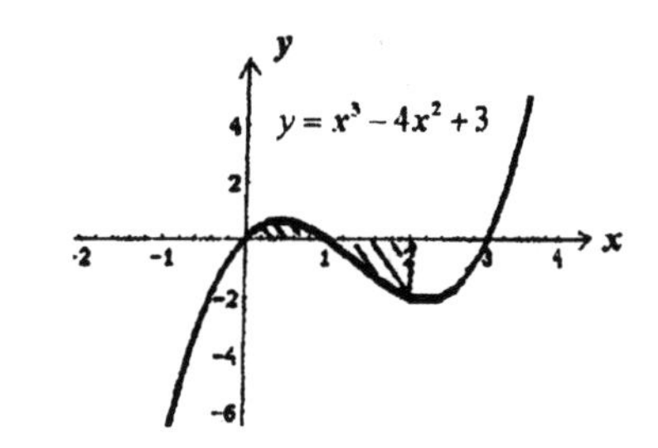

$$A = \int_{0}^{1}(x^3 - 4x^2 + 3x)\,dx - \int_{1}^{2}(x^3 - 4x^2 + 3x)\,dx$$
$$= (\tfrac{1}{4}x^4 - \tfrac{4}{3}x^3 + \tfrac{3}{2}x^2)\Big|_{0}^{1}$$
$$-(\tfrac{1}{4}x^4 - \tfrac{4}{3}x^3 + \tfrac{3}{2}x^2)\Big|_{1}^{2} = \tfrac{3}{2} \text{ sq units.}$$

33. The region is shown in the figure at the right.

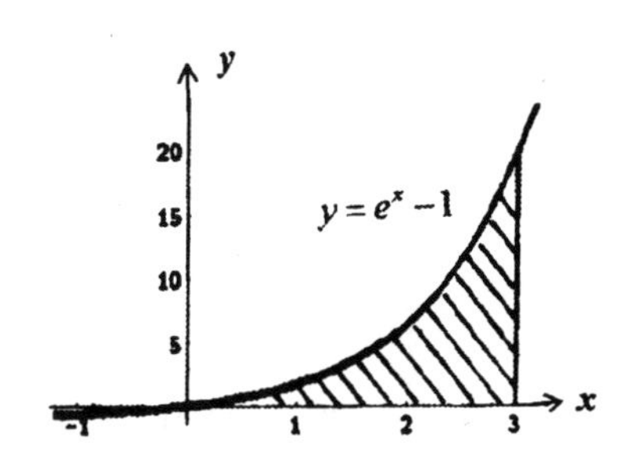

$$A = -\int_{-1}^{0}(e^x - 1)\,dx + \int_{0}^{3}(e^x - 1)\,dx$$
$$= (-e^x + x)\Big|_{-1}^{0} + (e^x - x)\Big|_{0}^{3}$$
$$= -1 - (-e^{-1} - 1) + (e^3 - 3) - 1$$
$$= e^3 - 4 + \tfrac{1}{e} \approx 16.5 \quad \text{sq units.}$$

35. To find the points of intersection of the two curves, we solve the equation

$x^2 - 4 = x + 2$

$x^2 - x - 6 = (x-3)(x+2) = 0$, obtaining

$x = -2$ or $x = 3$. The region is shown in the figure at the right.

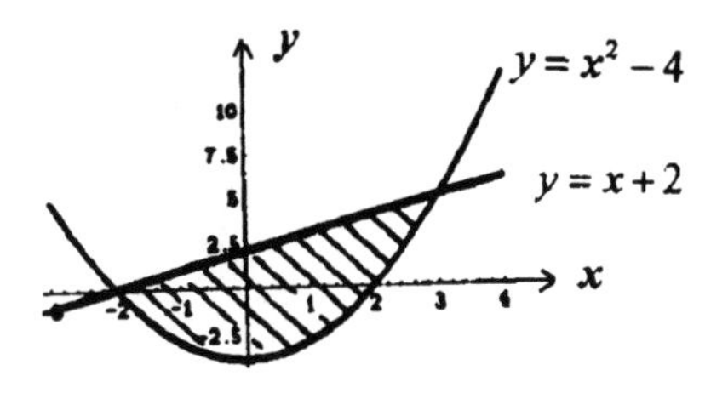

$$A = \int_{-2}^{3} [(x+2) - (x^2 - 4)]dx = \int_{-2}^{3} (-x^2 + x + 6)\,dx = (-\tfrac{1}{3}x^3 + \tfrac{1}{2}x^2 + 6x)\Big|_{-2}^{3}$$
$$= (-9 + \tfrac{9}{2} + 18) - (\tfrac{8}{3} + 2 - 12) = \tfrac{125}{6} \text{ sq units.}$$

37. To find the points of intersection of the two curves, we solve the equation $x^3 = x^2$ or $x^3 - x^2 = x^2(x - 1) = 0$ giving $x = 0$ or 1. The region is shown in the figure.

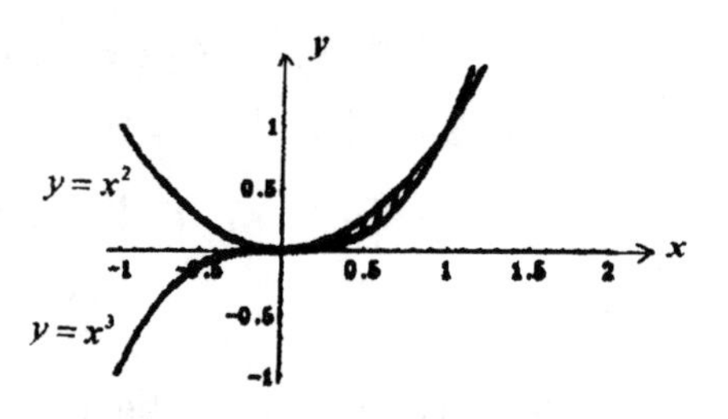

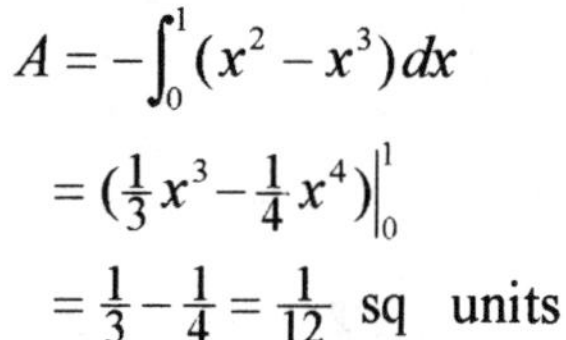

$$A = -\int_0^1 (x^2 - x^3)\,dx$$
$$= (\tfrac{1}{3}x^3 - \tfrac{1}{4}x^4)\Big|_0^1$$
$$= \tfrac{1}{3} - \tfrac{1}{4} = \tfrac{1}{12} \text{ sq units}$$

39. The graphs intersect at the points where $\sqrt{x} = x^2$ or $x = x^4$ and $x(x^3 - 1) = 0$; that is, when $x = 0$ and $x = 1$. The required area is

$$A = \int_0^1 (x^{1/2} - x^2)\,dx = \tfrac{2}{3}x^{3/2} - \tfrac{1}{3}x^3\Big|_0^1 = \tfrac{2}{3} - \tfrac{1}{3} = \tfrac{1}{3}.$$

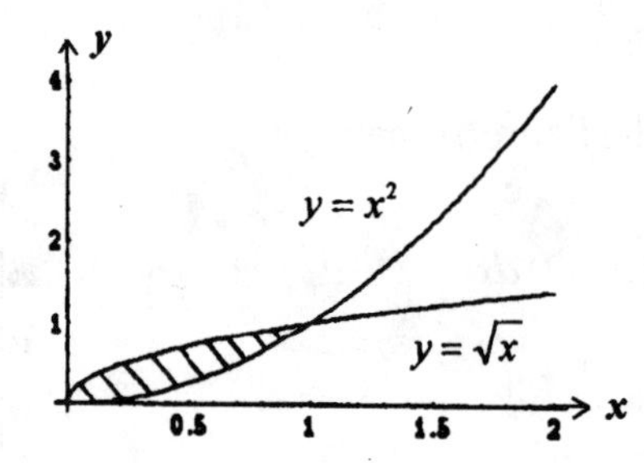

41. S gives the additional revenue that the company would realize if it used a different advertising agency.

$$S = \int_0^b [g(x) - f(x)]dx.$$

43. a. S gives the difference in the amount of smoke removed by the two brands over the same time interval $[a, b]$.

b. $$S = \int_a^b [f(t) - g(t)]dt$$

45. The additional amount of coal that will be produced is

$$\int_0^{20}(3.5e^{0.05t} - 3.5e^{0.01t})\,dt = 3.5\int_0^{20}(e^{0.05t} - e^{0.01t})\,dt$$

$$= 3.5(20e^{0.05t} - 100e^{0.01t})\Big|_0^{20} = 3.5[20e - 100e^{0.2}) - (20 - 100)]$$

$$= 42.8 \text{ billion metric tons.}$$

47. If the campaign is mounted, there will be

$$\int_0^5 (60e^{0.02t} + t^2 - 60)\,dt = 3000e^{0.02t} + \tfrac{t^3}{3} - 60t\Big|_0^5$$

$$= 3315.5 + \frac{125}{3} - 300 - 3000 \approx 57.179,$$

or 57,179 fewer people. (Remember t is measured in thousands.)

USING TECHNOLOGY EXERCISES 14.6, page 1005

1. a.

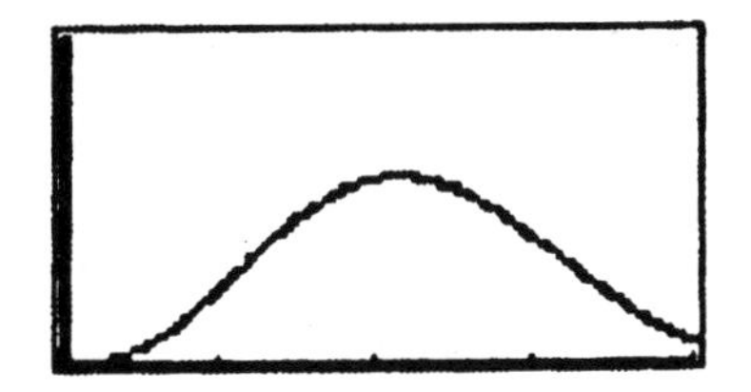

b. 1074.2857 sq units

3. a.

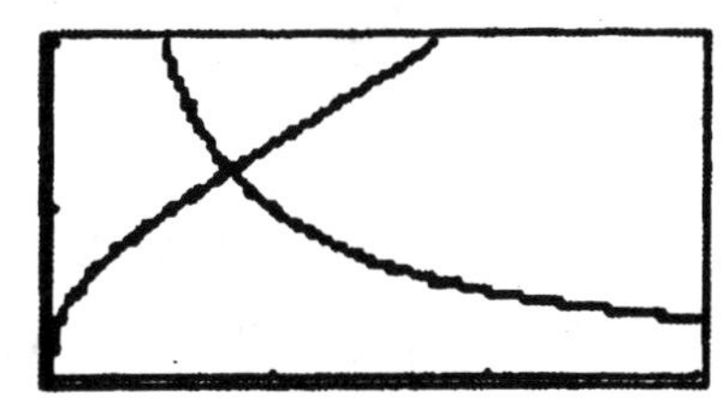

b. 0.9961 sq units

5. a.

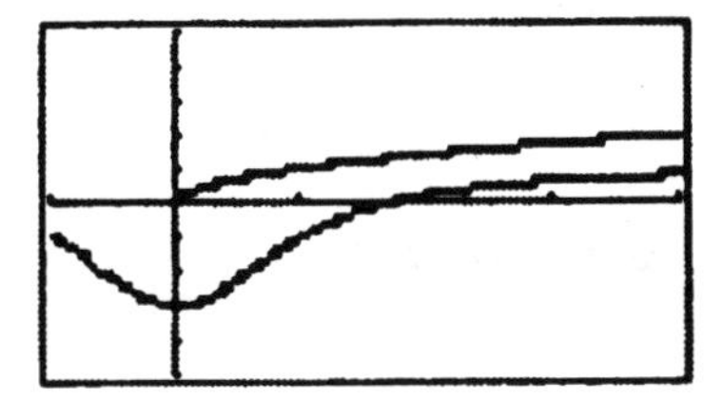

b. 5.4603 sq units

7. a.

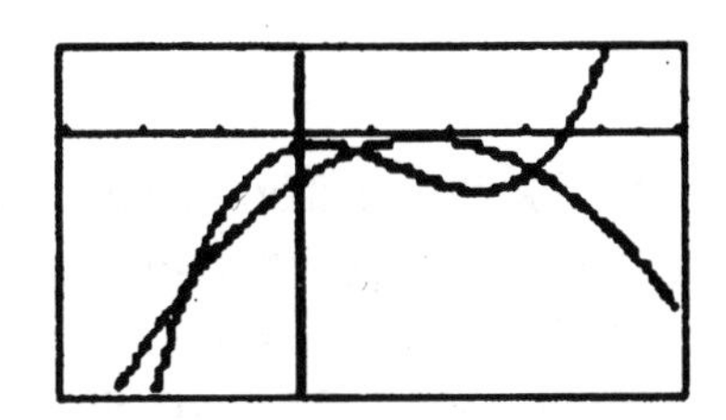

b. 25.8549 sq units

9. a.

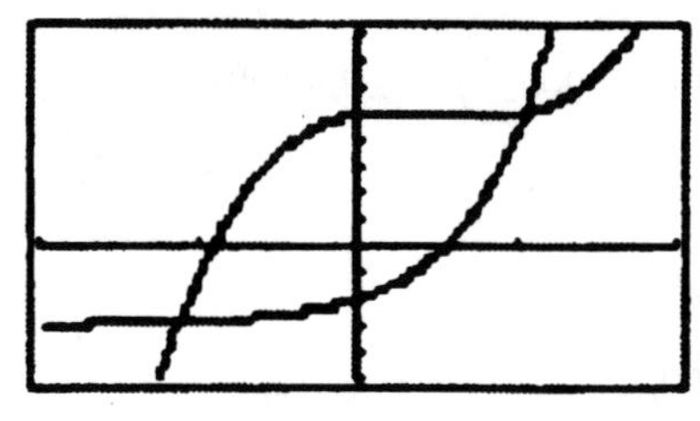

b. 10.5144 sq units

11. a.

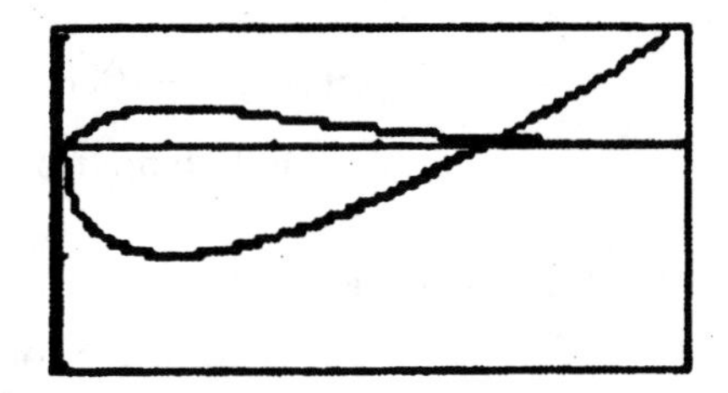

b. 3.5799 sq units

EXERCISES 14.7, page 1017

1. When $p = 4$, $-0.01x^2 - 0.1x + 6 = 4$ or $x^2 + 10x - 200 = 0$, $(x - 10)(x + 20) = 0$ and $x = 10$ or -20. We reject the root $x = -20$. The consumers' surplus is

$$CS = \int_0^{10} (-0.01x^2 - 0.1x + 6)\,dx - (4)(10)$$
$$= -\frac{0.01}{3}x^3 - 0.05x^2 + 6x\Big|_0^{10} - 40 \approx 11.667\text{, or \$11,667.}$$

3. Setting $p = 10$, we have $\sqrt{225 - 5x} = 10$, $225 - 5x = 100$, or $x = 25$.
Then $CS = \int_0^{25} \sqrt{225 - 5x}\,dx - (10)(25) = \int_0^{25} (225 - 5x)^{1/2}\,dx - 250.$
To evaluate the integral, let $u = 225 - 5x$ so that $du = -5\,dx$ or $dx = -\frac{1}{5}du$. If $x = 0$, $u = 225$ and if $x = 25$, $u = 100$. So

$$CS = -\tfrac{1}{5}\int_{225}^{100} u^{1/2}\,du - 250 = -\tfrac{2}{15}u^{3/2}\Big|_{225}^{100} - 250$$
$$= -\tfrac{2}{15}(1000 - 3375) - 250 = 66.667\text{, or \$6,667.}$$

5. To find the equilibrium point, we solve
$0.01x^2 + 0.1x + 3 = -0.01x^2 - 0.2x + 8$, or $0.02x^2 + 0.3x - 5 = 0$,
$2x^2 + 30x - 500 = (2x - 20)(x + 25) = 0$
obtaining $x = -25$ or 10. So the equilibrium point is $(10,5)$. Then

$$PS = (5)(10) - \int_0^{10} (0.01x^2 + 0.1x + 3)\,dx$$
$$= 50 - \left(\frac{0.01}{3}x^3 + 0.05x^2 + 3x\right)\Big|_0^{10} = 50 - \frac{10}{3} - 5 - 30 = \frac{35}{3},$$

or approximately \$11,667.

7. To find the market equilibrium, we solve

$$-0.2x^2 + 80 = 0.1x^2 + x + 40, \quad 0.3x^2 + x - 40 = 0,$$
$$3x^2 + 10x - 400 = 0, (3x + 40)(x - 10) = 0$$

giving $x = -\frac{40}{3}$ or $x = 10$. We reject the negative root. The corresponding equilibrium price is \$60. The consumers' surplus is

$$CS = \int_0^{10} (-0.2x^2 + 80)dx - (60)(10) = -\frac{0.2}{3}x^3 + 80x\Big|_0^{10} - 600 = 133\tfrac{1}{3},$$

or \$13,333. The producers' surplus is

$$PS = 600 - \int_0^{10} (0.1x^2 + x + 40)dx = 600 - [\tfrac{0.1}{3}x^3 + \tfrac{1}{2}x^2 + 40x]\Big|_0^{10}$$
$$= 116\tfrac{2}{3}, \text{ or } \$11,667.$$

9. Here $P = 200{,}000$, $r = 0.08$, and $T = 5$. So

$$PV = \int_0^5 200{,}000e^{-0.08t}dt = -\frac{200{,}000}{0.08}e^{-0.08t}\Big|_0^5 = -2{,}500{,}000(e^{-0.4} - 1)$$
$$\approx 824{,}199.85, \text{ or } \$824{,}200.$$

11. Here $P = 250$, $m = 12$, $T = 20$, and $r = 0.08$. So

$$A = \frac{mP}{r}(e^{rT} - 1) = \frac{12(250)}{0.08}(e^{1.6} - 1) \approx 148{,}238.70$$

or approximately \$148,239.

13. Here $P = 150$, $m = 12$, $T = 15$, and $r = 0.08$. So

$$A = \frac{12(150)}{0.08}(e^{1.2} - 1) \approx 52{,}202.60, \text{ or approximately } \$52{,}203.$$

15. Here $P = 2000$, $m = 1$, $T = 15.75$, and $r = 0.1$. So

$$A = \frac{1(2000)}{0.1}(e^{1.575} - 1) \approx 76{,}615, \text{ or approximately } \$76{,}615.$$

17. Here $P = 1200$, $m = 12$, $T = 15$, and $r = 0.1$. So

$$PV = \frac{12(1200)}{0.1}(1 - e^{-1.5}) \approx 111{,}869, \text{ or approximately } \$111{,}869.$$

19. We want the present value of an annuity with $P = 300$, $m = 12$, $T = 10$, and $r = 0.12$. So

$$PV = \frac{12(300)}{0.12}(1 - e^{-1.2}) \approx 20{,}964, \text{ or approximately } \$20{,}964.$$

21. a.

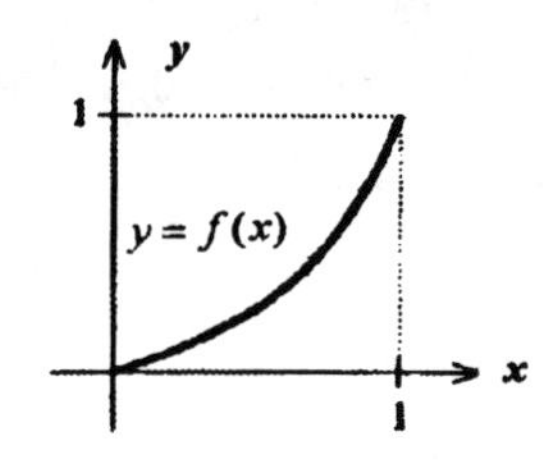

b. $f(0.4) = \frac{15}{16}(0.16) + \frac{1}{16}(0.4) \approx 0.175$; $f(0.9) = \frac{15}{16}(0.81) + \frac{1}{16}(0.9) \approx 0.816$.
So, the lowest 40 percent of the people receive 17.5 percent of the total income and the lowest 90 percent of the people receive 81.6 percent of the income.

23. a.

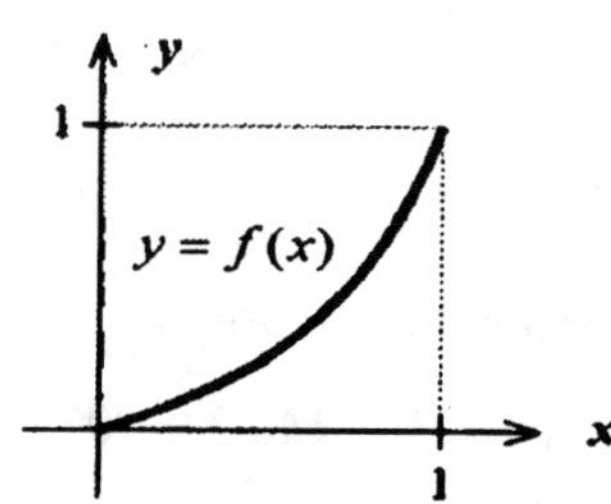

b. $f(0.3) = \frac{14}{15}(0.09) + \frac{1}{15}(0.3) = 0.104$
$f(0.7) = \frac{14}{15}(0.49) + \frac{1}{15}(0.7) \approx 0.504.$

USING TECHNOLOGY EXERCISES 14.7, page 1020

1. Consumer's surplus: \$18,000,000; producer's surplus: \$11,700,000.

3. Consumer's surplus: \$33,120; producer's surplus: \$2,880.

CHAPTER 14 REVIEW EXERCISES, page 1023

1. $\int (x^3 + 2x^2 - x)\,dx = \frac{1}{4}x^4 + \frac{2}{3}x^3 - \frac{1}{2}x^2 + C.$

3. $\int \left(x^4 - 2x^3 + \frac{1}{x^2}\right) dx = \frac{x^5}{5} - \frac{1}{2}x^4 - \frac{1}{x} + C$

5. $\int x(2x^2 + x^{1/2})dx = \int (2x^3 + x^{3/2})dx = \frac{1}{2}x^4 + \frac{2}{5}x^{5/2} + C.$

7. $\int (x^2 - x + \frac{2}{x} + 5)dx = \int x^2\,dx - \int x\,dx + 2\int \frac{dx}{x} + 5\int dx$

$$= \frac{1}{3}x^3 - \frac{1}{2}x^2 + 2\ln|x| + 5x + C.$$

9. Let $u = 3x^2 - 2x + 1$ so that $du = (6x - 2)\,dx = 2(3x - 1)\,dx$ or $(3x - 1)\,dx = \frac{1}{2}du$.

So $\int (3x-1)(3x^2 - 2x + 1)^{1/3}\,dx = \frac{1}{2}\int u^{1/3}\,du = \frac{3}{8}u^{4/3} + C = \frac{3}{8}(3x^2 - 2x + 1)^{4/3} + C.$

11. Let $u = x^2 - 2x + 5$ so that $du = 2(x - 1)\,dx$ or $(x - 1)\,dx = \frac{1}{2}du$.

$$\int \frac{x-1}{x^2 - 2x + 5}dx = \frac{1}{2}\int \frac{du}{u} = \frac{1}{2}\ln|u| + C = \frac{1}{2}\ln(x^2 - 2x + 5) + C.$$

13. Put $u = x^2 + x + 1$ so that $du = (2x + 1)\,dx = 2(x + \frac{1}{2})\,dx$ and $(x + \frac{1}{2})dx = \frac{1}{2}du$.

$$\int (x + \tfrac{1}{2})e^{x^2+x+1}dx = \tfrac{1}{2}\int e^u\,du = \tfrac{1}{2}e^u + C = \tfrac{1}{2}e^{x^2+x+1} + C.$$

15. Let $u = \ln x$ so that $du = \frac{1}{x}dx$. Then

$$\int \frac{(\ln x)^5}{x}dx = \int u^5\,du = \frac{1}{6}u^6 + C = \frac{1}{6}(\ln x)^6 + C.$$

17. Let $u = x^2 + 1$ so that $du = 2x\,dx$ or $x\,dx = \frac{1}{2}du$. Then

$$\int x^3(x^2 + 1)^{10}\,dx = \tfrac{1}{2}\int (u-1)u^{10}\,du \qquad (x^2 = u - 1)$$

$$= \tfrac{1}{2}\int (u^{11} - u^{10})\,du = \tfrac{1}{2}(\tfrac{1}{12}u^{12} - \tfrac{1}{11}u^{11}) + C$$

$$= \tfrac{1}{264}u^{11}(11u - 12) + C = \tfrac{1}{264}(x^2 + 1)^{11}(11x^2 - 1) + C.$$

19. Put $u = x - 2$ so that $du = dx$. Then $x = u + 2$ and

$$\int \frac{x}{\sqrt{x-2}}dx = \int \frac{u+2}{\sqrt{u}}du = \int (u^{1/2} + 2u^{-1/2})\,du = \int u^{1/2}\,du + 2\int u^{-1/2}\,du$$

$$= \tfrac{2}{3}u^{3/2} + 4u^{1/2} + C = \tfrac{2}{3}u^{1/2}(u+6) + C = \tfrac{2}{3}\sqrt{x-2}(x - 2 + 6) + C$$

$$= \tfrac{2}{3}(x+4)\sqrt{x-2}+C.$$

21. $\int_0^1 (2x^3 - 3x^2 + 1)\,dx = \tfrac{1}{2}x^4 - x^3 + x\Big|_0^1 = \tfrac{1}{2} - 1 + 1 = \tfrac{1}{2}.$

23. $\int_1^4 (x^{1/2} + x^{-3/2})\,dx = \tfrac{2}{3}x^{3/2} - 2x^{-1/2}\Big|_1^4 = \tfrac{2}{3}x^{3/2} - \dfrac{2}{\sqrt{x}}\Big|_1^4 = (\tfrac{16}{3} - 1) - (\tfrac{2}{3} - 2) = \tfrac{17}{3}.$

25. Put $u = x^3 - 3x^2 + 1$ so that $du = (3x^2 - 6x)\,dx = 3(x^2 - 2x)\,dx$ or $(x^2 - 2x)\,dx = \tfrac{1}{3}du$. Then if $x = -1$, $u = -3$, and if $x = 0$, $u = 1$,

$$\int_{-1}^{0} 12(x^2 - 2x)(x^3 - 3x^2 + 1)^3\,dx = (12)(\tfrac{1}{3})\int_{-3}^{1} u^3\,du = 4(\tfrac{1}{4})u^4\Big|_{-3}^{1}$$
$$= 1 - 81 = -80.$$

27. Let $u = x^2 + 1$ so that $du = 2x\,dx$ or $x\,dx = \tfrac{1}{2}du$. Then, if $x = 0$, $u = 1$, and if $x = 2$, $u = 5$, so
$$\int_0^2 \frac{x}{x^2+1}\,dx = \frac{1}{2}\int_1^5 \frac{du}{u} = \frac{1}{2}\ln u\Big|_1^5 = \frac{1}{2}\ln 5.$$

29. Let $u = 1 + 2x^2$ so that $du = 4x\,dx$ or $x\,dx = \tfrac{1}{4}\,du$. If $x = 0$, then $u = 1$ and if $x = 2$, then $u = 9$.
$$\int_0^2 \frac{4x}{\sqrt{1+2x^2}}\,dx = \int_1^9 \frac{du}{u^{1/2}} = 2u^{1/2}\Big|_1^9 = 2(3-1) = 4.$$

31. Let $u = 1 + e^{-x}$ so that $du = -e^{-x}\,dx$ and $e^{-x}\,dx = -\,du$. Then
$$\int_{-1}^{0} \frac{e^{-x}}{(1+e^{-x})^2}\,dx = -\int_{1+e}^{2} \frac{du}{u^2} = \frac{1}{u}\Big|_{1+e}^{2} = \frac{1}{2} - \frac{1}{1+e} = \frac{e-1}{2(1+e)}.$$

33. $f(x) = \int f'(x)\,dx = \int (3x^2 - 4x + 1)\,dx = 3\int x^2\,dx - 4\int x\,dx + \int dx$
$$= x^3 - 2x^2 + x + C.$$
The given condition implies that $f(1) = 1$ or $1 - 2 + 1 + C = 1$ and $C = 1$. Therefore, the required function is
$$f(x) = x^3 - 2x^2 + x + 1.$$

35. $f(x)=\int f'(x)\,dx=\int (1-e^{-x})\,dx=x+e^{-x}+C$, $f(0)=2$ implies $0+1+C=2$ or $C=1$. So $f(x)=x+e^{-x}+1$.

37. $\Delta x=\frac{2-1}{5}=\frac{1}{5}$; $x_1=\frac{6}{5}$, $x_2=\frac{7}{5}$, $x_3=\frac{8}{5}$, $x_4=\frac{9}{5}$, $x_5=\frac{10}{5}$. The Riemann sum is

$$f(x_1)\Delta x+\cdots+f(x_5)\Delta x=\left\{\left[-2(\tfrac{6}{5})^2+1\right]+\left[-2(\tfrac{7}{5})^2+1\right]+\cdots+\left[-2(\tfrac{10}{5})^2+1\right]\right\}(\tfrac{1}{5})$$
$$=-4.28.$$

39. a. $R(x)=\int R'(x)\,dx=\int (-0.03x+60)\,dx=-0.015x^2+60x+C$.
$R(0)=0$ implies that $C=0$. So, $R(x)=-0.015x^2+60x$.
b. From $R(x)=px$, we have $-0.015x^2+60x=px$ or $p=-0.015x+60$.

41. The total number of systems that Vista may expect to sell t months from the time they are put on the market is given by $f(t)=3000t-50{,}000(1-e^{-0.04t})$.
The number is

$$\int_0^{12}(3000-2000e^{-0.04t})\,dt=\left(3000t-\frac{2000}{-0.04}e^{-0.04t}\right)\Bigg|_0^{12}=3000(12)+50{,}000e^{-0.48}-50{,}000$$
$$=16{,}939.$$

43. $C(x)=\int C'(x)\,dx=\int (0.00003x^2-0.03x+10)\,dx$
$=0.00001x^3-0.015x^2+10x+k.$
But $C(0)=600$ and this implies that $k=600$. Therefore,
$C(x)=0.00001x^3-0.015x^2+10x+600.$
The total cost incurred in producing the first 500 corn poppers is
$C(500)=0.00001(500)^3-0.015(500)^2+10(500)+600$
$=3{,}100$, or \$3,100.

45. $A=\int_{-1}^{2}(3x^2+2x+1)\,dx=x^3+x^2+x\big|_{-1}^{2}=[2^3+2^2+2]-[(-1)^3+1-1]$
$=14-(-1)=15.$

47. $$A=\int_1^3\frac{1}{x^2}\,dx=\int_1^3 x^{-2}\,dx=-\frac{1}{x}\bigg|_1^3=-\frac{1}{3}+1=\frac{2}{3}.$$

49.

$$A = \int_a^b [f(x) - g(x)]dx$$
$$= \int_0^2 (e^x - x)dx$$
$$= \left(e^x - \frac{1}{2}x^2\right)\Big|_0^2$$
$$= (e^2 - 2) - (1 - 0) = e^2 - 3.$$

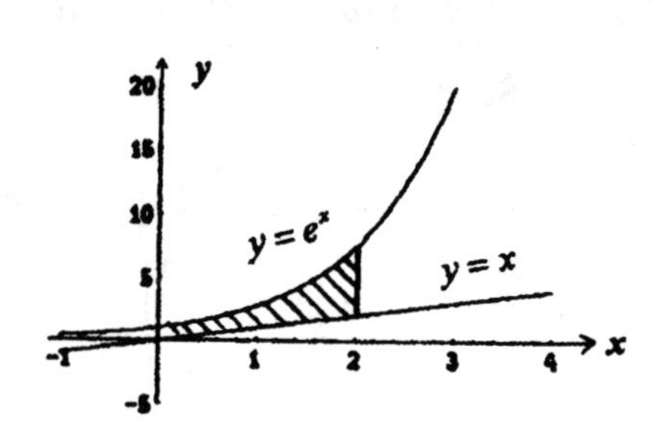

51. $A = \int_0^1 (x^3 - 3x^2 + 2x)dx - \int_1^2 (x^3 - 3x^2 + 2x)dx$

$$= \tfrac{x^4}{4} - x^3 + x^2\Big|_0^1 - \left(\tfrac{x^4}{4} - x^3 + x^2\right)\Big|_1^2$$
$$= \tfrac{1}{4} - 1 + 1 - [(4 - 8 + 4) - (\tfrac{1}{4} - 1 + 1)]$$
$$= \tfrac{1}{4} + \tfrac{1}{4} = \tfrac{1}{2}.$$

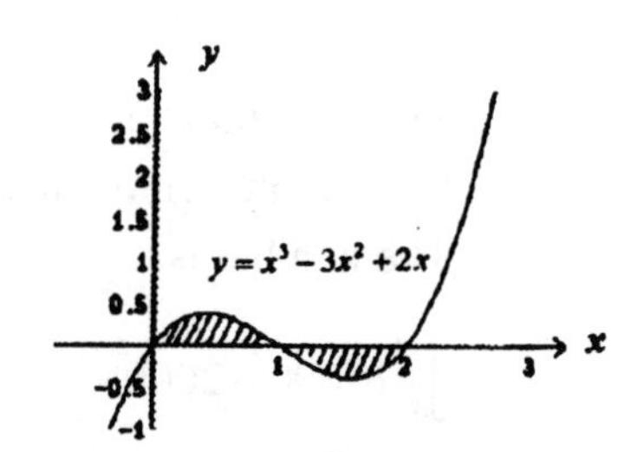

53.

$$A = \frac{1}{3}\int_0^3 \frac{x}{\sqrt{x^2 + 16}}dx = \frac{1}{3}\cdot\frac{1}{2}\cdot 2(x^2 + 16)^{1/2}\Big|_0^3$$
$$= \frac{1}{3}(x^2 + 16)^{1/2}\Big|_0^3 = \frac{1}{3}(5 - 4) = \frac{1}{3}.$$

55. To find the equilibrium point, we solve $0.1x^2 + 2x + 20 = -0.1x^2 - x + 40$
$0.2x^2 + 3x - 20 = 0$, $x^2 + 15x - 100 = 0$, $(x + 20)(x - 5) = 0$, or $x = 5$.
Therefore, $p = -0.1(25) - 5 + 40 = 32.5$.

$$CS = \int_0^5 (-0.1x^2 - x + 40)dx - (5)(32.5) = -\frac{0.1}{3}x^3 - \frac{1}{2}x^2 + 40x\Big|_0^5 - 162.5$$
$$= 20.833, \text{ or } \$2083.$$
$$PS = (5)(32.5) - \int_0^5 (0.1x^2 + 2x + 20)dx = 162.5 - \tfrac{0.1}{3}x^3 + x^2 + 20x)\Big|_0^5$$
$$= 33.333, \text{ or } \$3,333.$$

57. Use Equation (18) with $P = 925$, $m = 12$, $T = 30$, and $r = 0.12$, obtaining

$$PV = \frac{mP}{r}(1 - e^{-rT}) = \frac{(12)(925)}{(0.12)}(1 - e^{-0.12(30)}) = 89,972.56,$$

and we conclude that the present value of the purchase price of the house is $89,972.56 + $9000, or $98,972.56.

59. a.

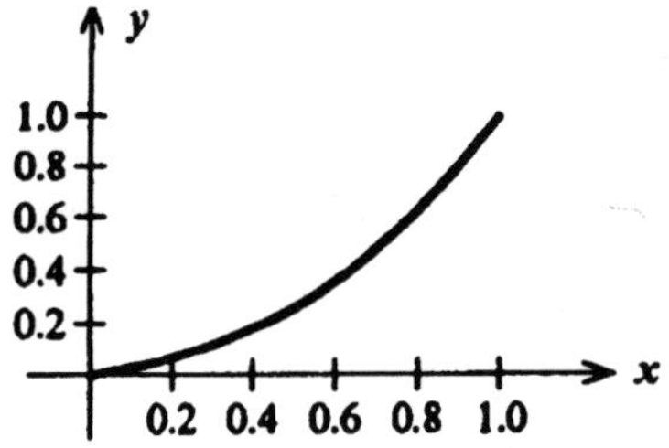

1.0
0.8
0.6
0.4
0.2
0.2 0.4 0.6 0.8 1.0

b. $f(0.3) = \frac{17}{18}(0.3)^2 + \frac{1}{18}(0.3) \approx 0.1$

so that 30 percent of the people receive 10 percent of the total income.

$f(0.6) = \frac{17}{18}(0.6)^2 + \frac{1}{18}(0.6) \approx 0.37$

so that 60 percent of the people receive 37 percent of the total revenue.

c. The coefficient of inequality for this curve is

$$L = 2\int_0^1 [x - \tfrac{17}{18}x^2 - \tfrac{1}{18}x]dx = \tfrac{17}{9}\int_0^1 (x - x^2)dx = \tfrac{17}{9}\left(\tfrac{1}{2}x^2 - \tfrac{1}{3}x^3\right)\Big|_0^1 = \tfrac{17}{54} \approx 0.315.$$

CHAPTER 15

EXERCISES 15.1, page 1033

1. $I = \int xe^{2x}\,dx$. Let $u = x$ and $dv = e^{2x}\,dx$. Then $du = dx$ and $v = \frac{1}{2}e^{2x}$. Therefore,

$$I = uv - \int v\,du = \frac{1}{2}xe^{2x} - \int \frac{1}{2}e^{2x}\,dx = \frac{1}{2}xe^{2x} - \frac{1}{4}e^{2x} = \frac{1}{4}e^{2x}(2x-1) + C.$$

3. Let $u = x$ and $dv = e^{x/4}\,dx$. Then $du = dx$ and $v = 4e^{x/4}$.

$$\int xe^{x/4}\,dx = uv - \int v\,du = 4xe^{x/4} - 4\int e^{x/4}\,dx = 4xe^{x/4} - 16e^{x/4} + C$$
$$= 4(x-4)e^{x/4} + C.$$

5. $\int (e^x - x)^2\,dx = \int (e^{2x} - 2xe^x + x^2)\,dx = \int e^{2x}\,dx - 2\int xe^x\,dx + \int x^2\,dx.$.

Using the result $\int xe^x\,dx = (x-1)e^x + k$, from Example 1, we see that

$\int (e^x - x)^2\,dx = \frac{1}{2}e^{2x} - 2(x-1)e^x + \frac{1}{3}x^3 + C.$

7. $I = \int (x+1)e^x\,dx$. Let $u = x + 1$, $dv = e^x\,dx$. Then $du = dx$ and $v = e^x$. Therefore,

$I = (x+1)e^x - \int e^x\,dx = (x+1)e^x - e^x + C = xe^x + C.$

9. Let $u = x$ and $dv = (x + 1)^{-3/2}\,dx$. Then $du = dx$ and $v = -2(x + 1)^{-1/2}$.

$$\int x(x+1)^{-3/2}\,dx = uv - \int v\,du = -2x(x+1)^{-1/2} + 2\int (x+1)^{-1/2}\,dx$$
$$= -2x(x+1)^{-1/2} + 4(x+1)^{1/2} + C$$
$$= 2(x+1)^{-1/2}[-x + 2(x+1)] + C = \frac{2(x+2)}{\sqrt{x+1}} + C.$$

11. $I = \int x(x-5)^{1/2}\,dx$. Let $u = x$ and $dv = (x - 5)^{1/2}\,dx$. Then $du = dx$ and

$v = \frac{2}{3}(x - 5)^{3/2}$. Therefore,

$$I = \frac{2}{3}x(x-5)^{3/2} - \int \frac{2}{3}(x-5)^{3/2}\,dx = \frac{2}{3}x(x-5)^{3/2} - \frac{2}{3}\cdot\frac{2}{5}(x-5)^{5/2} + C$$

$$= \tfrac{2}{3}(x-5)^{3/2}[x - \tfrac{2}{5}(x-5)] + C = \tfrac{2}{15}(x-5)^{3/2}(5x - 2x + 10) + C$$
$$= \tfrac{2}{15}(x-5)^{3/2}(3x+10) + C.$$

13. $I = \int x \ln 2x\, dx$. Let $u = \ln 2x$ and $dv = x\, dx$. Then $du = \frac{1}{x}\, dx$ and $v = \frac{1}{2}x^2$.

Therefore,

$$I = \tfrac{1}{2}x^2 \ln 2x - \int \tfrac{1}{2}x\, dx = \tfrac{1}{2}x^2 \ln 2x - \tfrac{1}{4}x^2 + C = \tfrac{1}{4}x^2(2\ln 2x - 1) + C.$$

15. Let $u = \ln x$ and $dv = x^3\, dx$, then $du = \frac{1}{x}dx$, and $v = \frac{1}{4}x^4$.

$$\int x^3 \ln x\, dx = \tfrac{1}{4}x^4 \ln x - \tfrac{1}{4}\int x^3\, dx = \tfrac{1}{4}x^4 \ln x - \tfrac{1}{16}x^4 + C$$
$$= \tfrac{1}{16}x^4(4 \ln x - 1) + C.$$

17. Let $u = \ln x^{1/2}$ and $dv = x^{1/2}\, dx$. Then $du = \frac{1}{2x}dx$ and $v = \frac{2}{3}x^{3/2}$,

and $\int \sqrt{x} \ln \sqrt{x}\, dx = uv - \int v\, du = \frac{2}{3}x^{3/2} \ln x^{1/2} - \frac{1}{3}\int x^{1/2}\, dx$

$$= \tfrac{2}{3}x^{3/2} \ln x^{1/2} - \tfrac{2}{9}x^{3/2} + C = \tfrac{2}{9}x\sqrt{x}(3 \ln \sqrt{x} - 1) + C.$$

19. Let $u = \ln x$ and $dv = x^{-2}\, dx$. Then $du = \frac{1}{x}dx$ and $v = -x^{-1}$,

$$\int \frac{\ln x}{x^2}dx = uv - \int v\, du = -\frac{\ln x}{x} + \int x^{-2}\, dx = -\frac{\ln x}{x} - \frac{1}{x} + C$$
$$= -\frac{1}{x}(\ln x + 1) + C.$$

21. Let $u = \ln x$ and $dv = dx$. Then $du = \frac{1}{x}dx$ and $v = x$ and

$$\int \ln x\, dx = uv - \int v\, du = x \ln x - \int dx = x \ln x - x + C = x(\ln x - 1) + C.$$

23. Let $u = x^2$ and $dv = e^{-x}\, dx$. Then $du = 2x\, dx$ and $v = -e^{-x}$, and

$$\int x^2 e^{-x}\, dx = uv - \int v\, du = -x^2 e^{-x} + 2\int xe^{-x}\, dx.$$

We can integrate by parts again, or, using the result of Problem 2, we find

$$\int x^2 e^{-x}\, dx = -x^2 e^{-x} + 2[-(x+1)e^{-x}] + C = -x^2 e^{-x} - 2(x+1)e^{-x} + C$$
$$= -(x^2 + 2x + 2)e^{-x} + C.$$

25. $I = \int x(\ln x)^2\,dx$. Let $u = (\ln x)^2$ and $dv = x\,dx$, so that

$du = 2(\ln x)\left(\frac{1}{x}\right) = \frac{2\ln x}{x}$ and $v = \frac{1}{2}x^2$. Then

$$I = \tfrac{1}{2}x^2(\ln x)^2 - \int x\ln x\,dx.$$

Next, we evaluate $\int x\ln x\,dx$, by letting $u = \ln x$ and $dv = x\,dx$, so that $du = \frac{1}{x}dx$ and $v = \frac{1}{2}x^2$. Then

$$\int x\ln x\,dx = \tfrac{1}{2}x^2(\ln x) - \tfrac{1}{2}\int x\,dx = \tfrac{1}{2}x^2\ln x - \tfrac{1}{4}x^2 + C.$$

Therefore, $\int x(\ln x)^2\,dx = \frac{1}{2}x^2(\ln x)^2 - \frac{1}{2}x^2\ln x + \frac{1}{4}x^2 + C$

$$= \tfrac{1}{4}x^2[2(\ln x)^2 - 2\ln x + 1] + C.$$

27. $\int_0^{\ln 2} xe^x\,dx = (x-1)e^x\Big|_0^{\ln 2}$ (Using the results of Example 1.)

$= (\ln 2 - 1)e^{\ln 2} - (-e^0) = 2(\ln 2 - 1) + 1$ (Recall $e^{\ln 2} = 2$.)

$= 2\ln 2 - 1.$

29. We first integrate $I = \int \ln x\,dx$. Integrating by parts with $u = \ln x$ and $dv = dx$ so that $du = \frac{1}{x}\,dx$ and $v = x$, we find

$$I = x\ln x - \int dx = x\ln x - x + C = x(\ln x - 1) + C.$$

Therefore, $\int_1^4 \ln x\,dx = x(\ln x - 1)\Big|_1^4 = 4(\ln 4 - 1) - 1(\ln 1 - 1) = 4\ln 4 - 3.$

31. Let $u = x$ and $dv = e^{2x}\,dx$. Then $u = dx$ and $v = \frac{1}{2}e^{2x}$ and

$$\int_0^2 xe^{2x}\,dx = \tfrac{1}{2}xe^{2x}\Big|_0^2 - \tfrac{1}{2}\int_0^2 e^{2x}\,dx = e^4 - \tfrac{1}{4}e^{2x}\Big|_0^2$$
$$= e^4 - \tfrac{1}{4}e^4 + \tfrac{1}{4} = \tfrac{1}{4}(3e^4 + 1).$$

33. Let $u = x$ and $dv = e^{-2x}\,dx$, so that $du = dx$ and $v = -\frac{1}{2}e^{-2x}$.

$$f(x) = \int xe^{-2x}\,dx = -\tfrac{1}{2}xe^{-2x} - \tfrac{1}{4}e^{-2x} + C$$

$$f(0) = -\tfrac{1}{4} + C = 3 \text{ and } C = \tfrac{13}{4}.$$

Therefore, $y = -\frac{1}{2}xe^{-2x} - \frac{1}{4}e^{-2x} + \frac{13}{4}$.

35. The required area is given by $\int_1^5 \ln x\, dx$.

We first find $\int \ln x\, dx$. Using the technique of integration by parts with $u = \ln x$ and $dv = dx$ so that $du = \frac{1}{x}dx$ and $v = x$, we have

$$\int \ln x\, dx = x\ln x - \int dx = x\ln x - x = x(\ln x - 1) + C.$$

Therefore,

$$\int_1^5 \ln x\, dx = x(\ln x - 1)\Big|_1^5 = 5(\ln 5 - 1) - 1(\ln 1 - 1) = 5\ \ln 5 - 4$$

and the required area is (5 ln 5 - 4) sq units.

37. The distance covered is given by $\int_0^{10} 100te^{-0.2t}\, dt = 100\int_0^{10} te^{-0.2t}\, dt$.

We integrate by parts, letting $u = t$ and $dv = e^{-0.2t}\, dt$ so that $du = dt$ and $v = -\dfrac{1}{0.2}e^{-0.2t} = -5e^{-0.2t}$. Therefore,

$$100\int_0^{10} te^{-0.2t}\, dt = 100\left[-5te^{-0.2t}\Big|_0^{10}\right] + 5\int_0^{10} e^{-0.2t}\, dt$$

$$= 100[-5te^{-0.2t} - 25e^{-0.2t}]\Big|_0^{10} = -500e^{-0.2t}(t+5)\Big|_0^{10}$$

$$= -500e^{-2}(15) + 500(5) = 1485, \text{ or } 1485 \text{ feet.}$$

39. The average concentration is $C = \dfrac{1}{12}\int_0^{12} 3te^{-t/3}\, dt = \dfrac{1}{4}\int_0^{12} te^{-t/3}\, dt$.

Let $u = t$ and $dv = e^{-t/3}\, dt$. So $du = dt$ and $v = -3e^{-t/3}$. Then

$$C = \frac{1}{4}\left[-3te^{-t/3}\Big|_0^{12} + 3\int_0^{12} e^{-t/3}\, dt\right] = \frac{1}{4}\left\{-36e^{-4} - \left[9e^{-t/3}\Big|_0^{12}\right]\right\}$$

$$= \frac{1}{4}(-36e^{-4} - 9e^{-4} + 9) \approx 2.04 \text{ mg/ml.}$$

41. $N = 2\int te^{-0.1t}\, dt$. Let $u = t$ and $dv = e^{-0.1t}$, so that $du = dt$ and $v = -10e^{-0.1t}$. Then

$v = -10e^{-0.1t}$. Then

$$N(t) = 2[-10te^{-0.1t} + 10\int e - 0.1t\, dt] = 2(-10te^{-0.1t} - 100e^{-0.1t}) + C$$

$$= -20e^{-0.1t}(t+10) + 200. \qquad [N(0) = 0]$$

43. The present value of the franchise is

$$PV = \int_0^T P(t)e^{-rt}\,dt = \int_0^{15} (50{,}000 + 3000t)e^{-0.1t}dt$$

$$= 50{,}000\int_0^{15} e^{-0.1t}dt + 3000\int_0^{15} te^{-0.1t}dt.$$

The first integral is

$$50{,}000\int_0^{15} e^{-0.1t}\,dt = \frac{50{,}000}{-0.1}e^{-0.1t}\Big|_0^{15} = -500{,}000(e^{-1.5} - 1) \approx 388{,}435.$$

The second integral is evaluated by integration by parts . Let $u = t$ and $dv = e^{-0.1t}\,dt$ so that $du = dt$ and $v = -10e^{-0.1t}$.

$$3000\int_0^{15} te^{-0.1t}dt = 3000[-10te^{-0.1t}\Big|_0^{15} + 10\int_0^{15} e^{-0.1t}\,dt]$$

$$= 3000\left[-150e^{-1.5} - 100e^{-0.1t}\Big|_0^{15}\right]$$

$$= 3000(-150e^{-1.5} - 100e^{-1.5} + 100) \approx 132{,}652.$$

Therefore, $PV \approx 388{,}435 + 132{,}652 = 521{,}087$ or \$521,087.

EXERCISES 15.2, page 1048

1. $\Delta x = \frac{2}{6} = \frac{1}{3}, x_0 = 0, x_1 = \frac{1}{3}, x_2 = \frac{2}{3}, x_3 = 1, x_4 = \frac{4}{3}, x_5 = \frac{5}{3}, x_6 = 2.$

Trapezoidal Rule:

$$\int_0^2 x^2\,dx \approx \tfrac{1}{6}\left[0 + 2(\tfrac{1}{3})^2 + 2(\tfrac{2}{3})^2 + 2(1)^2 + 2(\tfrac{4}{3})^2 + 2(\tfrac{5}{3})^2 + 2^2\right]$$

$$\approx \tfrac{1}{6}\,(0.22222 + 0.88889 + 2 + 3.55556 + 5.55556 + 4)$$

$$\approx 2.7037.$$

Simpson's Rule:

$$\int x^2\,dx = \tfrac{1}{9}[0 + 4(\tfrac{1}{3})^2 + 2(\tfrac{2}{3})^2 + 4(1)^2 + 2(\tfrac{4}{3})^2 + 4(\tfrac{5}{3})^2 + 2^2]$$

$$\approx \tfrac{1}{9}\,(0.44444 + 0.88889 + 4 + 3.55556 + 11.11111 + 4) \approx 2.6667.$$

Exact Value: $\int_0^2 x^2\,dx = \frac{1}{3}x^3\Big|_0^2 = \frac{8}{3} = 2\frac{2}{3}.$

3. $\Delta x = \frac{b-a}{n} = \frac{1-0}{4} = \frac{1}{4}; x_0 = 0, x_1 = \frac{1}{4}, x_2 = \frac{1}{2}, x_3 = \frac{3}{4}, x_4 = 1.$

Trapezoidal Rule:

$$\int_0^1 x^3\,dx \approx \tfrac{1}{8}\left[0 + 2(\tfrac{1}{4})^3 + 2(\tfrac{1}{2})^3 + 2(\tfrac{3}{4})^3 + 1^3\right] \approx \tfrac{1}{8}(0 + 0.3125 + 0.25 + 0.8)$$

$$\approx\ 0.265625.$$

Simpson's Rule:

$$\int_0^1 x^3\,dx \approx \tfrac{1}{12}\left[0+4(\tfrac{1}{4})^3+2(\tfrac{1}{2})^3+4(\tfrac{3}{4})^3+1\right] \approx \tfrac{1}{12}[0+0.625+0.25+1.6875+1] \approx 0.25.$$

Exact Value:

$$\int_0^1 x^3\,dx = \tfrac{1}{4}x^4\Big|_0^1 = \tfrac{1}{4}-0=\tfrac{1}{4}.$$

5. a. Here $a = 1$, $b = 2$, and $n = 4$; so $\Delta x = \frac{2-1}{4} = \frac{1}{4} = 0.25$, and $x_0 = 1$, $x_1 = 1.25$, $x_2 = 1.5$, $x_3 = 1.75$, $x_4 = 2$.

Trapezoidal Rule:

$$\int_1^2 \frac{1}{x}dx \approx \frac{0.25}{2}\left[1+2\left(\frac{1}{1.25}\right)+2\left(\frac{1}{1.5}\right)+2\left(\frac{1}{1.75}\right)+\frac{1}{2}\right] \approx 0.697.$$

Simpson's Rule:

$$\int_1^2 \frac{1}{x}dx \approx \frac{0.25}{3}\left[1+4\left(\frac{1}{1.25}\right)+2\left(\frac{1}{1.5}\right)+4\left(\frac{1}{1.75}\right)+\frac{1}{2}\right] \approx 0.6933.$$

$$\int_1^2 \frac{1}{x}dx = \ln x\Big|_1^2 = \ln 2 - \ln 1 \approx 0.6931.$$

7. $\Delta x = \frac{1}{4}$, $x_0 = 1$, $x_1 = \frac{5}{4}$, $x_2 = \frac{3}{2}$, $x_3 = \frac{7}{4}$, $x_4 = 2$.

Trapezoidal Rule:

$$\int_1^2 \frac{1}{x^2}\,dx \approx \tfrac{1}{8}\left[1+2(\tfrac{4}{5})^2+2(\tfrac{2}{3})^2+2(\tfrac{4}{7})^2+(\tfrac{1}{2})^2\right] \approx 0.5090.$$

Simpson's Rule:

$$\int_1^2 \frac{1}{x^2}\,dx \approx \tfrac{1}{12}\left[1+4(\tfrac{4}{5})^2+2(\tfrac{2}{3})^2+4(\tfrac{4}{7})^2+(\tfrac{1}{2})^2\right] \approx 0.5004.$$

Exact Value:

$$\int_1^2 \frac{1}{x^2}\,dx = -\frac{1}{x}\Big|_1^2 = -\frac{1}{2}+1 = \frac{1}{2}.$$

9. $\Delta x = \frac{b-a}{n} = \frac{4-0}{8} = \frac{1}{2}; x_0 = 0, x_1 = \frac{1}{2}, x_2 = \frac{2}{2}, x_3 = \frac{3}{2}, \ldots, x_8 = \frac{8}{2}.$

Trapezoidal Rule:

$$\int_0^4 \sqrt{x}\,dx \approx \frac{\frac{1}{2}}{2}\left[0+2\sqrt{0.5}+2\sqrt{1}+2\sqrt{1.5}+\cdots+2\sqrt{3.5}+\sqrt{4}\right] \approx 5.26504.$$

Simpson's Rule:

$$\int_0^4 \sqrt{x}\,dx \approx \frac{\frac{1}{2}}{3}\left[0+4\sqrt{0.5}+2\sqrt{1}+4\sqrt{1.5}+\cdots+4\sqrt{3.5}+\sqrt{4}\right] \approx 5.30463.$$

The actual value is

$$\int_0^4 \sqrt{x}\,dx \approx \frac{2}{3}x^{3/2}\Big|_0^4 = \frac{2}{3}(8) = \frac{16}{3} \approx 5.333333.$$

11. $\Delta x = \frac{1-0}{6} = \frac{1}{6}; x_0 = 0, x_1 = \frac{1}{6}, x_2 = \frac{2}{6}, \ldots, x_6 = \frac{6}{6}.$

Trapezoidal Rule:

$$\int_0^1 e^{-x}\,dx \approx \frac{\frac{1}{6}}{2}[1+2e^{-1/6}+2e^{-2/6}+\cdots+2e^{-5/6}+e^{-1}] \approx 0.633583.$$

Simpson's Rule:

$$\int_0^1 e^{-x}\,dx \approx \frac{\frac{1}{6}}{3}[1+4e^{-1/6}+2e^{-2/6}+\cdots+4e^{-5/6}+e^{-1}] \approx 0.632123.$$

The actual value is

$$\int_0^1 e^{-x}\,dx = -e^{-x}\Big|_0^1 = -e^{-1}+1 \approx 0.632121.$$

13. $\Delta x = \frac{1}{4}; x_0 = 0, x_1 = \frac{5}{4}, x_2 = \frac{3}{2}, x_3 = \frac{7}{4}, x_4 = 2.$

Trapezoidal Rule:

$$\int_1^2 \ln x\,dx \approx \frac{1}{8}[\ln 1+2\ln\tfrac{5}{4}+2\ln\tfrac{3}{2}+2\ln\tfrac{7}{4}+\ln 2] \approx 0.38370.$$

Simpson's Rule:

$$\int_1^2 \ln x\,dx \approx \frac{1}{12}[\ln 1+4\ln\tfrac{5}{4}+2\ln\tfrac{3}{2}+4\ln\tfrac{7}{4}+\ln 2] \approx 0.38626.$$

Exact Value:

$$\int_1^2 \ln x\,dx \approx x(\ln x-1)\Big|_1^2 = 2(\ln 2-1)+1 = 2\ln 2-1.$$

15. $\Delta x = \frac{1-0}{4} = \frac{1}{4}; x_0 = 0, x_1 = \frac{1}{4}, x_2 = \frac{2}{4}, x_3 = \frac{3}{4}, x_4 = \frac{4}{4}.$

Trapezoidal Rule:

$$\int_0^1 \sqrt{1+x^3}\,dx \approx \frac{\frac{1}{4}}{2}\left[\sqrt{1}+2\sqrt{1+(\tfrac{1}{4})^3}+\cdots+2\sqrt{1+(\tfrac{3}{4})^3}+\sqrt{2}\right] \approx 1.1170.$$

Simpson's Rule:

$$\int_0^1 \sqrt{1+x^3}\,dx \approx \frac{\frac{1}{4}}{3}\left[\sqrt{1}+4\sqrt{1+(\tfrac{1}{4})^3}+2\sqrt{1+(\tfrac{2}{4})^3}\cdots+4\sqrt{1+(\tfrac{3}{4})^3}+\sqrt{2}\right] \approx 1.1114.$$

17. $\Delta x = \frac{2-0}{4} = \frac{1}{2}; x_0 = 0, x_1 = \frac{1}{2}, x_2 = \frac{2}{2}, x_3 = \frac{3}{2}, x_4 = \frac{4}{2}$.

Trapezoidal Rule:

$$\int_0^2 \frac{1}{\sqrt{x^3+1}}\,dx = \frac{\frac{1}{2}}{2}\left[1 + \frac{2}{\sqrt{(\frac{1}{2})^3+1}} + \frac{2}{\sqrt{(1)^3+1}} + \frac{2}{\sqrt{(\frac{3}{2})^3+1}} + \frac{1}{\sqrt{(2)^3+1}}\right]$$

$$\approx 1.3973$$

Simpson's Rule:

$$\int_0^2 \frac{1}{\sqrt{x^3+1}}\,dx = \frac{\frac{1}{2}}{3}\left[1 + \frac{4}{\sqrt{(\frac{1}{2})^3+1}} + \frac{2}{\sqrt{(1)^3+1}} + \frac{4}{\sqrt{(\frac{3}{2})^3+1}} + \frac{1}{\sqrt{(2)^3+1}}\right]$$

$$\approx 1.4052$$

19. $\Delta x = \frac{2}{4} = \frac{1}{2}; x_0 = 0, x_1 = \frac{1}{2}, x_2 = 1, x_3 = \frac{3}{2},\ x_4 = 2.$

Trapezoidal Rule:

$$\int_0^2 e^{-x^2}\,dx = \frac{1}{4}[e^{-0} + 2e^{-(1/2)^2} + 2e^{-1} + 2e^{-(3/2)^2} + e^{-4}] \approx 0.8806.$$

Simpson's Rule:

$$\int_0^2 e^{-x^2}\,dx = \frac{1}{6}[e^{-0} + 4e^{-(1/2)^2} + 2e^{-1} + 4e^{-(3/2)^2} + e^{-4}] \approx 0.8818.$$

21. $\Delta x = \frac{2-1}{4} = \frac{1}{4}; x_0 = 1,\ x_1 = \frac{5}{4}, x_2 = \frac{6}{4}, x_3 = \frac{7}{4},\ x_4 = \frac{8}{4}$.

Trapezoidal Rule:

$$\int_1^2 x^{-1/2}e^x\,dx = \frac{\frac{1}{4}}{2}\left[e + \frac{2e^{5/4}}{\sqrt{\frac{5}{4}}} + \cdots + \frac{2e^{7/4}}{\sqrt{\frac{7}{4}}} + \frac{e^2}{\sqrt{2}}\right] \approx 3.7757.$$

Simpson's Rule:

$$\int_1^2 x^{-1/2}e^x\,dx = \frac{\frac{1}{4}}{3}\left[e + \frac{4e^{5/4}}{\sqrt{\frac{5}{4}}} + \cdots + \frac{4e^{7/4}}{\sqrt{\frac{7}{4}}} + \frac{e^2}{\sqrt{2}}\right] \approx 3.7625.$$

23. a. Here $a = -1$, $b = 2$, $n = 10$, and $f(x) = x^5$. $f'(x) = 5x^4$ and $f''(x) = 20x^3$. Because $f'''(x) = 60x^2 > 0$ on (-1,0) ∪ (0,2), we see that $f''(x)$ is increasing

on (-1,0) ∪ (0,2), we see that $f''(x)$ is increasing on (-1,0) ∪ (0,2). So, we take

$$M = f''(2) = 20(2^3) = 160.$$

Using (7), we see that the maximum error incurred is

$$\frac{M(b-a)^3}{12n^2} = \frac{160[2-(-1)]^3}{12(100)} = 3.6.$$

b. We compute $f''' = 60x^2$ and $f^{(iv)}(x) = 120x$. $f^{(iv)}(x)$ is clearly increasing on (-1,2), so we can take $M = f^{(iv)}(2) = 240$. Therefore, using (8), we see that an error bound is

$$\frac{M(b-a)^3}{180n^4} = \frac{240(3)^5}{180(10^4)} \approx 0.0324.$$

25. a. Here $a = 1$, $b = 3$, $n = 10$, and $f(x) = \frac{1}{x}$. We find

$$f'(x) = -\frac{1}{x^2},\ f'''(x) = \frac{2}{x^3}.$$

Since $f'''(x) = -\frac{6}{x^4} < 0$ on (1,3), we see that $f''(x)$ is decreasing there. We may take $M = f''(1) = 2$. Using (7), we find an error bound is

$$\frac{M(b-a)^3}{12n^2} = \frac{2(3-1)^3}{12(100)} \approx 0.013.$$

b. $f'''(x) = -\frac{6}{x^4}$ and $f^{(iv)}(x) = \frac{24}{x^5}$. $f^{(iv)}(x)$ is decreasing on (1,3), so we can take $M = f^{(iv)}(1) = 24$. Using (8), we find an error bound is

$$\frac{24(3-1)^5}{180(10^4)} \approx 0.00043.$$

27. a. Here $a = 0$, $b = 2$, $n = 8$, and $f(x) = (1+x)^{-1/2}$. We find

$$f'(x) = -\tfrac{1}{2}(1+x)^{-3/2},\ f''(x) = \tfrac{3}{4}(1+x)^{-5/2}.$$

Since f'' is positive and decreasing on (0,2), we see that

$$|f''(x)| \le \tfrac{3}{4}.$$

So the maximum error is $\frac{\frac{3}{4}(2-0)^3}{12(8)^2} = 0.0078125.$

b. $f''' = -\frac{15x}{8}(1+x)^{-7/2}$ and $f^{(4)}(x) = \frac{105}{16}(1+x)^{-9/2}$. Since $f^{(4)}$ is positive and decreasing on (0,2), we find $|f^{(4)}(x)| \le \frac{105}{16}$.

Therefore, the maximum error is $\frac{\frac{105}{16}(2-0)^5}{180(8)^4} = 0.000285.$

29. The distance covered is given by

$$d = \int_0^2 V(t)\,dt = \frac{\frac{1}{4}}{2}\left[V(0) + 2V(\tfrac{1}{4}) + \cdots + 2V(\tfrac{7}{4}) + V(2)\right]$$

$$= \tfrac{1}{8}[19.5 + 2(24.3) + 2(34.2) + 2(40.5) + 2(38.4) + 2(26.2)$$

$$+ 2(18) + 2(16) + 8] \approx 52.84, \text{ or } 52.84 \text{ miles.}$$

31. $$\frac{1}{13}\int_0^{13} f(t)\,dt = (\tfrac{1}{13})(\tfrac{1}{2})\{13.2 + 2[14.8 + 16.6 + 17.2 + 18.7 + 19.3 + 22.6 + 24.2 + 25$$

$$+ 24.6 + 25.6 + 26.4 + 26.6] + 26.6\}$$

$$\approx 21.65.$$

33. Solving the equation $25 = \dfrac{50}{0.01x^2 + 1}$, we see that $0.01x^2 + 1 = 2$, $0.01x^2 = 1$, and $x = 10$. Therefore, $CS = \displaystyle\int_0^{10} \frac{50}{0.01x^2 + 1}dx - (25)(10)$ and $\Delta x = \frac{10}{8} = 1.25$, $x_0 = 0$, $x_1 = 1.25$, $x_2 = 2.50$, ..., $x_8 = 10$.

a. $$CS = \frac{1.25}{2}\left\{50 + 2\left[\frac{50}{0.01(1.25)^2 + 1}\right] + \cdots + \left[\frac{50}{0.01(10)^2 + 1}\right]\right\} - 250$$

$$\approx 142{,}373.56, \text{ or } \$142{,}373.56.$$

b. $$CS = \frac{1.25}{3}\left\{50 + 4\left[\frac{50}{0.01(1.25)^2 + 1}\right] + \cdots + \left[\frac{50}{0.01(10)^2 + 1}\right]\right\} - 250$$

$$\approx 142{,}698.12, \text{ or } \$142{,}698.12.$$

35. $\Delta t = \frac{5}{10} = 0.5$, $t_0 = 0$, $t_1 = 0.5$, $t_2 = 1$, ..., $t_{10} = 5$.

$$PSI = \frac{1}{5}\int_0^5 \left[\frac{136}{1 + 0.25(t - 4.5)^2} + 28\right]dt = 27.2\int_0^5 \frac{1}{1 + 0.25(t - 4.5)^2} + 28\,dt$$

$$\approx \frac{(0.5)(27.2)}{2}\left[\frac{1}{1 + 0.25(-4.5)^2} + \frac{2}{1 + 0.25(5 - 4.5)^2} + \cdots + \frac{1}{1 + 0.25(5 - 4.5)^2}\right] + 28$$

$$\approx 103.9$$

37. $\Delta x = \frac{21-19}{10} = 0.2$; $x_0 = 19$, $x_1 = 19.2$ $x_2 = 19.4$, ..., $x_{10} = 21$

$$P = \frac{100}{2.6\sqrt{2\pi}}\int_{19}^{21} e^{-0.5[(x-20)/2.6]^2}dx$$

$$\approx \frac{100}{2.6\sqrt{2\pi}}\left(\frac{0.2}{3}\right)[e^{-0.5[(19-20)/2.6]^2} + 4e^{-0.5[19.2-20)/2.6]^2} + \cdots + e^{-0.5[(21-20)/2.6]^2}]$$

$\approx$ 29.94, or 30 percent.

39. $R = \dfrac{60D}{\int_0^T C(t)\,dt} = \dfrac{480}{\int_0^{24} C(t)\,dt}$. Now,

$$\int_0^{24} C(t)\,dt \approx \tfrac{24}{12}\cdot\tfrac{1}{3}[0+4(0)+2(2.8)+4(6.1)+2(9.7)+4(7.6)+2(4.8)$$
$$+4(3.7)+2(1.9)+4(0.8)+2(0.3)+4(0.1)+0]$$
$$\approx 74.8$$

and $R = \frac{480}{74.8} \approx$ 6.42, or 6.42 liters/min.

EXERCISES 15.3, page 1060

1. The required area is given by

$$\int_3^\infty \frac{2}{x^2}dx = \lim_{b\to\infty}\int_3^b \frac{2}{x^2}dx = \lim_{b\to\infty}\left(-\frac{2}{x}\right)\Bigg|_3^b = \lim_{b\to\infty}\left(-\frac{2}{b}+\frac{2}{3}\right) = \frac{2}{3} \text{ or } 2/3 \text{ sq units.}$$

3. $$A = \int \frac{1}{(x-2)^2}dx = \lim_{b\to\infty}\int_3^b (x-2)^{-2}\,dx = \lim_{b\to\infty} -\frac{1}{x-2}\Bigg|_3^b = \lim_{b\to\infty}\left(-\frac{1}{b-2}+1\right) = 1.$$

5. $$A = \int_1^\infty \frac{1}{x^{3/2}}dx = \lim_{b\to\infty}\int_1^b x^{-3/2}\,dx = \lim_{b\to\infty} -\frac{2}{\sqrt{x}}\Bigg|_1^b = \lim_{b\to\infty}\left(-\frac{2}{\sqrt{b}}+2\right) = 2.$$

7. $$A = \int_0^\infty \frac{1}{(x+1)^{5/2}}dx = \lim_{b\to\infty}\int_1^b (x+1)^{-5/2}\,dx = \lim_{b\to\infty} -\frac{2}{3}(x+1)^{-3/2}\Bigg|_0^b$$
$$= \lim_{b\to\infty}\left[-\frac{2}{3(b+1)^{3/2}}+\frac{2}{3}\right] = \frac{2}{3}, \text{ or } 2/3 \text{ sq units.}$$

9. $$A = \int_{-\infty}^2 e^{2x}\,dx = \lim_{a\to-\infty}\int_a^2 e^{2x}\,dx = \lim_{a\to-\infty}\tfrac{1}{2}e^{2x}\Big|_a^2 = \lim_{a\to-\infty}\left(\tfrac{1}{2}e^4 - \tfrac{1}{2}e^{2a}\right) = \tfrac{1}{2}e^4.$$

11. Using symmetry, the required area is given by

$$2\int_0^{\infty}\frac{x}{(1+x^2)^2}dx = 2\lim_{b\to\infty}\int_0^{\infty}\frac{x}{(1+x^2)^2}dx.$$

To evaluate the indefinite integral $\int\frac{x}{(1+x^2)^2}dx$, put $u = 1 + x^2$ so that

$du = 2x\,dx$ or $x\,dx = \frac{1}{2}\,du$.

Then $\displaystyle\int\frac{x}{(1+x^2)^2}dx = \frac{1}{2}\int\frac{du}{u^2} = -\frac{1}{2u} + C = -\frac{1}{2(1+x^2)} + C.$

Therefore, $\displaystyle 2\lim_{b\to\infty}\int_0^b\frac{x}{(1+x^2)}dx = \lim_{b\to\infty} -\frac{1}{(1+x^2)^2}\Big|_0^b = \lim_{b\to\infty}\left[-\frac{1}{(1+b^2)}+1\right] = 1,$

or 1 sq unit.

13. a. $I(b) = \int_0^b \sqrt{x}\,dx = \frac{2}{3}x^{3/2}\Big|_0^b = \frac{2}{3}b^{3/2}.$

b. $\lim_{b\to\infty} I(b) = \lim_{b\to\infty}\frac{2}{3}b^{3/2} = \infty.$

15. $$\int_1^{\infty}\frac{3}{x^4}dx = \lim_{b\to\infty}\int_1^b 3x^{-4}\,dx = \lim_{b\to\infty}\left(-\frac{1}{x^3}\right)\Big|_1^b = \lim_{b\to\infty}\left(-\frac{1}{b^3}+1\right) = 1.$$

17. $$A = \int_4^{\infty}\frac{2}{x^{3/2}}dx = \lim_{b\to\infty}\int_4^b 2x^{-3/2}\,dx = \lim_{b\to\infty} -4x^{-1/2}\Big|_4^b = \lim_{b\to\infty}\left(-\frac{4}{\sqrt{b}}+2\right) = 2.$$

19. $$\int_1^{\infty}\frac{4}{x}dx = \lim_{b\to\infty}\int_1^b\frac{4}{x}dx = \lim_{b\to\infty} 4\ln x\Big|_1^b = \lim_{b\to\infty}(4\ln b) = \infty.$$

21. $$\int_{-\infty}^0 (x-2)^{-3}\,dx = \lim_{a\to-\infty}\int_a^0 (x-2)^{-3}\,dx = \lim_{a\to-\infty} -\frac{1}{2(x-2)^2}\Big|_a^0 = -\frac{1}{8}.$$

23. $$\int_1^{\infty}\frac{1}{(2x-1)^{3/2}}dx = \lim_{b\to\infty}\int_1^b (2x-1)^{-3/2}\,dx = \lim_{b\to\infty} -\frac{1}{(2x-1)^{1/2}}\Big|_1^b$$

$$= \lim_{b\to\infty}\left(-\frac{1}{\sqrt{2b-1}}+1\right) = 1.$$

25. $\int_0^{\infty} e^{-x}\,dx = \lim_{b\to\infty}\int_0^b e^{-x}\,dx = \lim_{b\to\infty} -e^{-x}\Big|_0^b = \lim_{b\to\infty}(-e^{-b}+1) = 1.$

27. $\int_{-\infty}^{0} e^{2x}\,dx = \lim_{a\to-\infty} \tfrac{1}{2}e^{2x}\Big|_a^0 = \lim_{a\to-\infty}\left(\tfrac{1}{2}-\tfrac{1}{2}e^{2a}\right) = \tfrac{1}{2}.$

29. $\int_1^{\infty} \frac{e^{\sqrt{x}}}{\sqrt{x}}\,dx = \lim_{b\to\infty}\int_1^{\infty} \frac{e^{\sqrt{x}}}{\sqrt{x}}\,dx = \lim_{b\to\infty} -2e^{\sqrt{x}}\Big|_1^b$ (Integrate by substitution: $u = \sqrt{x}$.)

$= \lim_{b\to\infty}(2e^{\sqrt{b}} - 2e) = \infty$, , and so it diverges.

31. $\int_{-\infty}^{0} xe^x\,dx = \lim_{a\to-\infty}\int_a^0 xe^x\,dx = \lim_{a\to-\infty}(x-1)e^x\Big|_a^0 = \lim_{a\to-\infty}[-1+(a-1)e^a] = -1.$

Note: We have used integration by parts to evaluate the integral.

33. $\int x\,dx = \lim_{a\to-\infty} \tfrac{1}{2}x^2\Big|_a^0 + \lim_{b\to\infty}\tfrac{1}{2}x^2\Big|_0^{\infty}$ both of which diverge and so the integral diverges.

35. $\int x^2(1+x^3)^{-2}\,dx = \lim_{a\to-\infty} -\frac{1}{3(1+x^3)}\Big|_a^0 + \lim_{b\to\infty} -\frac{1}{3(1+x^3)}\Big|_0^b$

$= -\tfrac{1}{3}+\tfrac{1}{3} = 0.$

37. $\int_{-\infty}^{\infty} xe^{1-x^2}\,dx = \lim_{a\to-\infty}\int_a^0 xe^{1-x^2}\,dx + \lim_{b\to\infty}\int_0^b xe^{1-x^2}\,dx$

$= \lim_{a\to-\infty} -\tfrac{1}{2}e^{1-x^2}\Big|_a^0 + \lim_{b\to\infty} -\tfrac{1}{2}e^{1-x^2}\Big|_0^b$

$= \lim_{a\to-\infty}\left(-\tfrac{1}{2}e+\tfrac{1}{2}e^{1-a^2}\right) + \lim_{b\to\infty}\left(-\tfrac{1}{2}e^{1-b^2}+\tfrac{1}{2}e\right) = 0.$

39. $\int_{-\infty}^{\infty} \frac{e^{-x}}{1+e^{-x}}\,dx = \lim_{a\to-\infty} -\ln(1+e^{-x})\Big|_a^0 + \lim_{b\to\infty} -\ln(1+e^{-x})\Big|_0^b = \infty$, and it is divergent.

41. We want the present value PV of a perpetuity with $m = 1$, $P = 1500$, and $r = 0.08$.

We find $PV = \frac{(1)(1500)}{0.08} = 18{,}750$, or \$18,750.

43. $$PV = \int_0^\infty (10{,}000 + 4000t)e^{-rt}dt = 10{,}000\int_0^\infty e^{-rt}dt + 4000\int_0^\infty te^{-rt}dt$$

$$= \lim_{b\to\infty}\left(-\frac{10{,}000}{r}e^{-rt}\Big|_0^b\right) + 4000\left(\frac{1}{r^2}\right)\left(-rt-1)e^{-rt}\Big|_0^b\right)$$

(Integrating by parts.)

$$= \frac{10{,}000}{r} + \frac{4000}{r^2} = \frac{10{,}000r + 4000}{r^2} \text{ dollars.}$$

45. a. $CV \approx \int_0^\infty Re^{-it}\,dt = \lim_{b\to\infty}\int_0^b Re^{-it}\,dt = \lim_{b\to\infty} -\frac{R}{i}e^{-it}\Big|_0^b = \lim_{b\to\infty}\left(-\frac{R}{i}e^{-ib} + \frac{R}{i}\right) = \frac{R}{i}.$

b. $\text{CV} \approx \dfrac{10{,}000}{0.12} \approx 83{,}333$, or \$83,333.

EXERCISES 15.4, page 1071

1. $f(x) \geq 0$ on [2,6]. Next $\int_2^6 \frac{2}{32}x\,dx = \frac{1}{32}x^2\Big|_2^6 = \frac{1}{32}(36-4) = 1,$

and so f is a probability density function on [2,6].

3. $f(x) = \frac{3}{8}x^2$ is nonnegative on [0,2]. Next, we compute

$$\int_0^2 \tfrac{3}{8}x^2\,dx = \tfrac{1}{8}x^3\Big|_0^2 = 1$$

and so f is a probability density function.

5. $\int_0^1 20(x^3 - x^4)dx = 20(\frac{1}{4}x^4 - \frac{1}{5}x^5)\Big|_0^1 = 20(\frac{1}{4} - \frac{1}{5}) = 20(\frac{1}{20}) = 1.$
Furthermore, $f(x) = 20(x^3 - x^4) = 20x^3(1 - x) \geq 0$ on [0,1].
Therefore, f is a density function on [0,1] as asserted.

7. Clearly $f(x) \geq 0$ on [1,4]. Next,

$$\int_1^4 f(x)\,dx = \tfrac{3}{14}\int_1^4 x^{1/2}\,dx = (\tfrac{3}{14})(\tfrac{2}{3})x^{3/2}\Big|_1^4 = \tfrac{1}{7}(8-1) = 1,$$

and so f is a probability density function on [1,3].

9. a. $\int_0^4 k(4-x)\,dx = k\int_0^4 (4-x)\,dx = k(4x - \frac{1}{2}x^2)\Big|_0^4 = k(16-8) = 8k = 1$
implies that $k = 1/8$.

b. $P(1 \le x \le 3) = \frac{1}{8}\int_1^3 (4-x)\,dx = \frac{1}{8}(4x - \frac{1}{2}x^2)\Big|_1^3 = \frac{1}{8}\left[(12-\frac{9}{2})-(4-\frac{1}{2})\right]$

$= \frac{1}{2}.$

11. a. $P(x \le 100) = \dfrac{1}{100}\int_0^{100} e^{-x/100}\,dx = -e^{-x/100}\Big|_0^{100} = -e^{-1} + 1 \approx 0.63.$

b. $P(x \ge 120) = \dfrac{1}{100}\int_{120}^{\infty} e^{-x/100}\,dx = \lim_{b\to\infty}(-e^{-b/100} + e^{-1.2})$

$= e^{-1.2} \approx 0.30.$

c. $P(60 \le x \le 140) = \dfrac{1}{100}\int_{60}^{140} e^{-x/100}dx = -e^{-x/100}\Big|_{60}^{140} = (-e^{-1.4} + e^{-0.6})$

$\approx 0.30.$

13. $E(x) = \int_0^6 x\cdot\frac{1}{36}(6x - x^2)\,dx = \frac{1}{36}\int_0^6 (6x^2 - x^3)\,dx = \frac{1}{36}(2x^3 - \frac{1}{4}x^4)\Big|_0^6$

$= \frac{1}{36}(432 - 324) = 3.$

15. The probability function is $f(x) = \frac{1}{8}e^{-x/8}$. The required probability is

$P(x \ge 8) = \frac{1}{8}\int_8^{\infty} e^{-x/8}\,dx = \lim_{b\to\infty} -e^{-x/8}\Big|_8^b = \lim_{b\to\infty}(-e^{-b/8} + e^{-1}) = e^{-1} \approx 0.37.$

17. The probability function is $f(x) = 0.00001e^{-0.00001x}$. The required probability is

$P(x \le 20{,}000) = 0.00001\int_0^{20{,}000} e^{-0.00001x}\,dx = -e^{-0.00001x}\Big|_0^{20{,}000} = -e^{0.2} + 1 \approx 0.18.$

CHAPTER 15 REVIEW EXERCISES, page 1074

1. Let $u = 2x$ and $dv = e^{-x}\,dx$ so that $du = 2\,dx$ and $v = -e^{-x}$. Then

$$\int 2xe^{-x}\,dx = uv - \int v\,du = -2xe^{-x} + 2\int e^{-x}\,dx$$

$$= -2xe^{-x} - 2e^{-x} + C = -2(1+x)e^{-x} + C.$$

3. Let $u = \ln 5x$ and $dv = dx$, so that $du = \frac{1}{x}dx$ and $v = x$. Then

$\int \ln 5x\,dx = x\ln 5x\,dx - \int dx = x\ln 5x - x + C = x(\ln 5x - 1) + C.$

5. Let $u = x$ and $dv = e^{-2x}\,dx$ so that $du = dx$ and $v = -\frac{1}{2}e^{-2x}$. Then

$$\int_0^1 xe^{-2x}\,dx = -\tfrac{1}{2}xe^{-2x}\Big|_0^1 + \tfrac{1}{2}\int_0^1 e^{-2x}\,dx = -\tfrac{1}{2}e^{-2} - \tfrac{1}{4}e^{-2x}\Big|_0^1$$
$$= -\tfrac{1}{2}e^{-2} - \tfrac{1}{4}e^{-2} + \tfrac{1}{4} = \tfrac{1}{4}(1 - 3e^{-2}).$$

7. $f(x) = \int f'(x)\,dx = \int \frac{\ln x}{\sqrt{x}}\,dx$. To evaluate the integral, we integrate by parts with

$u = \ln x$, $dv = x^{-1/2}\,dx$, $du = \frac{1}{x}\,dx$ and $v = 2x^{1/2}\,dx$. Then

$$\int \frac{\ln x}{x^{1/2}}\,dx = 2x^{1/2}\ln x - \int 2x^{-1/2}\,dx = 2x^{1/2}\ln x - 4x^{1/2} + C$$
$$= 2x^{1/2}(\ln x - 2) + C = 2\sqrt{x}(\ln x - 2) + C.$$

But $f(1) = -2$ and this gives $2\sqrt{1}(\ln 1 - 2) + C = -2$, or $C = 2$. Therefore, $f(x) = 2\sqrt{x}(\ln x - 2) + 2$.

9. $$\int_0^\infty e^{-2x}\,dx = \lim_{b\to\infty}\int_0^b e^{-2x}\,dx = \lim_{b\to\infty}(-\tfrac{1}{2}e^{-2x})\Big|_0^b = \lim_{b\to\infty}(-\tfrac{1}{2}e^{-2b} + \tfrac{1}{2}) = \tfrac{1}{2}.$$

11. $$\int_3^\infty \frac{2}{x}\,dx = \lim_{b\to\infty}\int_3^b \frac{2}{x}\,dx = \lim_{b\to\infty} 2\ln x\Big|_3^b = \lim_{b\to\infty}(2\ln b - 2\ln 3) = \infty.$$

13. $$\int_2^\infty \frac{dx}{(1+2x)^2} = \lim_{b\to\infty}\int_2^b (1+2x)^{-2}\,dx = \lim_{b\to\infty}(\tfrac{1}{2})(-1)(1+2x)^{-1}\Big|_2^b$$
$$= \lim_{b\to\infty}\left(-\frac{1}{2(1+2b)} + \frac{1}{2(5)}\right) = \frac{1}{10}.$$

15. $\Delta x = \frac{b-a}{n} = \frac{3-1}{4} = \frac{1}{2}$; $x_0 = 1$, $x_1 = \frac{3}{2}$, $x_2 = 2$, $x_3 = \frac{5}{2}$, $x_4 = 3$.

Trapezoidal Rule:

$$\int_1^3 \frac{dx}{1+\sqrt{x}} \approx \frac{\frac{1}{2}}{2}\left[\frac{1}{2} + \frac{2}{1+\sqrt{1.5}} + \frac{2}{1+\sqrt{2}} + \frac{2}{1+\sqrt{2.5}} + \frac{1}{1+\sqrt{3}}\right] \approx 0.8421.$$

Simpson's Rule

$$\int_1^3 \frac{dx}{1+\sqrt{x}} \approx \frac{\frac{1}{2}}{3}\left[\frac{1}{2} + \frac{4}{1+\sqrt{1.5}} + \frac{2}{1+\sqrt{2}} + \frac{4}{1+\sqrt{2.5}} + \frac{1}{1+\sqrt{3}}\right]$$

$$\approx 0.8404.$$

17. $\Delta x = \frac{1-(-1)}{4} = \frac{1}{2};\ x_0 = -1,\ x_1 = -\frac{1}{2},\ x_2 = 0,\ x_3 = \frac{1}{2},\ x_4 = 1.$

Trapezoidal Rule:

$$\int_{-1}^{1} \sqrt{1+x^4}\, dx \approx \frac{0.5}{2}\left[\sqrt{2} + 2\sqrt{1+(-0.5)^4} + 2 + 2\sqrt{1+(0.5)^4} + \sqrt{2}\right]$$
$$\approx 2.2379.$$

Simpson's Rule:

$$\int_{-1}^{1} \sqrt{1+x^4}\, dx \approx \frac{0.5}{3}\left[\sqrt{2} + 4\sqrt{1+(-0.5)^4} + 2 + 4\sqrt{1+(0.5)^4} + \sqrt{2}\right]$$
$$\approx 2.1791.$$

19. $\frac{3}{128}\int_0^4 (16-x^2)\,dx = \frac{3}{128}(16x - \frac{1}{3}x^3)\Big|_0^4 = \frac{3}{128}(64 - \frac{64}{3}) \ = \frac{3}{128}\left(\frac{192-64}{3}\right) = 1.$
Also, $f(x) \geq 0$ on $[0,4]$.

21. a. $\int_0^2 kx\sqrt{4-x^2}\,dx = k\int_0^2 x(4-x^2)^{1/2}\,dx = k(-\frac{1}{2})(\frac{2}{3})(4-x^2)^{3/2}\Big|_0^2$
$= (-\frac{k}{3})(0-4^{3/2}) = \frac{k}{3}(8) = 1,$

or $k = 3/8$.

b. $\int_1^2 \frac{3}{8}x\sqrt{4-x^2}\,dx = \frac{3}{8}(-\frac{1}{3})(4-x^2)^{3/2}\Big|_1^2 = -\frac{1}{8}(0-3^{3/2}) = 0.6495.$

23. a. $\int_0^3 kx^2(3-x)\,dx = k\int_0^3 (3x^2 - x^3)\,dx = k(x^3 - \frac{1}{4}x^4)\Big|_0^3$
$= k(27 - \frac{81}{4}) = k(\frac{108-81}{4}) = k(\frac{27}{4}) = 1,$

or $k = 4/27$.

b. $\int_1^2 \frac{4}{27}x^2(3-x)\,dx = \frac{4}{27}\int_1^2 (3x^2 - x^3)\,dx = \frac{4}{27}(x^3 - \frac{1}{4}x^4)\Big|_1^2$
$= \frac{4}{27}(8-4) - (1 - \frac{1}{4}) = \frac{13}{27} \approx 0.4815.$

25. Let $u = t$ and $dv = e^{-0.05t}$ so that $du = 1$ and $v = -20e^{-0.05t}$, and integrate by parts obtaining $S(t) = -20te^{-0.05t} + \int 20e^{-0.05t}\,dt = -20te^{-0.05t} - 400e^{-0.05t} + C$
$= -20te^{-0.05t} - 400e^{-0.05t} + C = -20e^{-0.05t}(t+20) + C.$
The initial condition implies $S(0) = 0$ giving $-20(20) + C = 0$, or $C = 400$. Therefore, $S(t) = -20e^{-0.05t}(t + 20) + 400$. By the end of the first year, the number of units sold is given by $S(12) = -20e^{-0.6}(32) + 400 = 48.761$, or 48,761 cartridges.

27. **Trapezoidal Rule:**

$$A = \frac{100}{2}[0 + 480 + 520 + 600 + 680 + 680 + 800 + 680 + 600 + 440 + 0]$$
$$= 274{,}000, \text{ or } 274{,}000 \text{ sq ft.}$$

Simpson's Rule:

$$A = \frac{100}{3}[0 + 960 + 520 + 1200 + 680 + 1360 + 800 + 1360 + 600 + 880 + 0]$$
$$= 278{,}667, \text{ or } 278{,}667 \text{ sq ft.}$$

CHAPTER 16

EXERCISES 16.1, page 1085

1. $f(0,0) = 2(0) + 3(0) - 4 = -4.$ $f(1,0) = 2(1) + 3(0) - 4 = -2.$
   $f(0,1) = 2(0) + 3(1) - 4 = -1.$ $f(1,2) = 2(1) + 3(2) - 4 = 4.$
   $f(2,-1) = 2(2) + 3(-1) - 4 = -3.$

3. $f(1,2) = 1^2 + 2(1)(2) - 1 + 3 = 7;\ f(2,1) = 2^2 + 2(2)(1) - 2 + 3 = 9$
   $f(-1,2) = (-1)^2 + 2(-1)(2) - (-1) + 3 = 1 \quad f(2,-1) = 2^2 + 2(2)(-1) - 2 + 3 = 1.$

5. $g(s,t) = 3s\sqrt{t} + t\sqrt{s} + 2;\quad g(1,2) = 3(1)\sqrt{2} + 2\sqrt{1} + 2 = 4 + 3\sqrt{2}$
   $g(2,1) = 3(2)\sqrt{1} + \sqrt{2} + 2 = 8 + \sqrt{2};$
   $g(0,4) = 0 + 0 + 2 = 2,\ g(4,9) = 3(4)\sqrt{9} + 9\sqrt{4} + 2 = 56.$

7. $h(1,e) = \ln e - e \ln 1 = \ln e = 1;\ h(e,1) = e \ln 1 - \ln e = -1;$
   $h(e,e) = e \ln e - e \ln e = 0.$

9. $g(r,s,t) = re^{s/t};\ g(1,1,1) = e,\ g(1,0,1) = 1,\ g(-1,-1,-1) = -e^{-1/(-1)} = -e.$

11. The domain of f is the set of all ordered pairs (x, y) where x and y are real numbers.

13. All real values of u and v except those satisfying the equation $u = v$.

15. The domain of g is the set of all ordered pairs (r,s) satisfying $rs \geq 0$, that is the set of all ordered pairs where both $r \geq 0$ and $s \geq 0$, or in which both $r \leq 0$ and $s \leq 0$.

17. The domain of f is the set of all ordered pairs (x, y) such that $x + y > 5$.

19. The level curves of $z = f(x, y) = 2x + 3y$ for $z = -2, -1, 0, 1, 2$, are shown in the figure that follows.

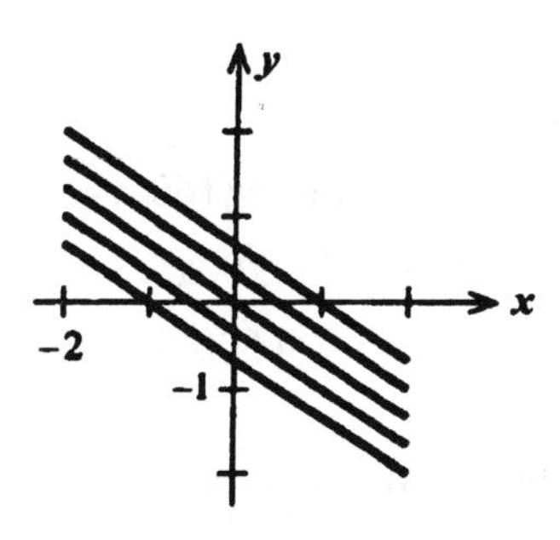

21. The level curves of $f(x,y) = 2x^2 + y$ for $z = -2, -1, 0, 1, 2,$ are shown below.

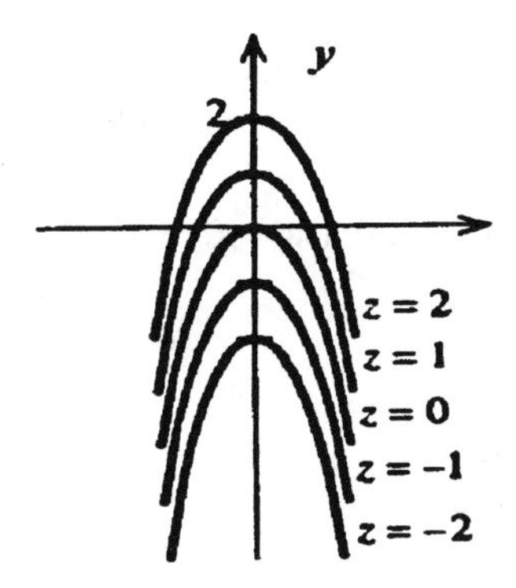

23. The level curves of $f(x,y) = \sqrt{16 - x^2 - y^2}$ for $z = 0, 1, 2, 3, 4$ are shown below.

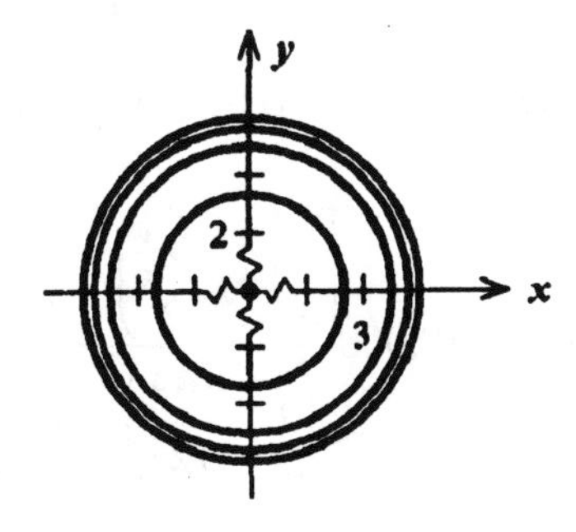

25. $V = f(1.5,4) = \pi(1.5)^2(4) = 9\pi$, or 9π cu ft

27. $R(4,0.1)=\dfrac{4k}{(0.1)^4}=40{,}000\,k$ dynes.

29. $R(x,y)=xp+yq=x(200-\frac{1}{5}x-\frac{1}{10}y)+y(160-\frac{1}{10}x-\frac{1}{4}y)$

$$=-\tfrac{1}{5}x^2-\tfrac{1}{4}y^2-\tfrac{1}{5}xh+200x+160y.$$

$$R(100,60)=-\tfrac{1}{5}(10{,}000)-\tfrac{1}{4}(3600)-\tfrac{1}{5}(6000)+200(100)+160(60)$$

$$=25{,}500,$$

and this says that the revenue from the sales of 100 units of the finished and 60 units of the unfinished furniture per week is \$25,500,

$$R(60,100)=-\tfrac{1}{5}(3600)-\tfrac{1}{4}(10{,}000)-\tfrac{1}{5}(6000)+200(60)+160(100)$$

$$=23{,}580,$$

and this says that the revenue from the sales of 60 units of the finished and 100 units of the unfinished furniture per week is \$23,580.

31. $R(300, 200)=-0.005(90000)-0.003(40000)-0.002(60000)+20(300)+15(200)$

$$=8310, \text{ or } \$8310.$$

$$R(200, 300)=-0.005(40000)-0.003(90000)-0.002(60000)+20(200)+15(300)$$

$$=7910, \text{ or } \$7910.$$

33. a. The domain of S is the set of all ordered pairs (W,H) such that W and H are nonnegative real numbers.

b. $S=0.007184(70)^{0.425}(178)^{0.725}\approx 1.87$ sq meters.

35. $A=f(10{,}000,\ 0.1,\ 3)=10{,}000e^{(0.1)(3)}=10{,}000e^{0.3}$, or \$13,498.59.

37. $$B=f(80{,}000,\ 0.09,\ 30,\ 60)=80{,}000\left[\frac{\left(1+\frac{0.09}{12}\right)^{60}-1}{\left(1+\frac{0.09}{12}\right)^{360}-1}\right]=3295.89.$$

Therefore, they owe 80,000 − 3295.89, or \$76,704,11.

$$B = f(80{,}000,\ 0{,}09,\ 30,\ 240) = 80{,}000\left[\frac{\left(1+\frac{0.09}{12}\right)^{240}-1}{\left(1+\frac{0.09}{12}\right)^{360}-1}\right] = 29{,}185.38.$$

Therefore, they owe 80,000 − 29,185.38, or \$50,814.62.

39. $f(20, 40, 5) = \sqrt{\frac{2(20(40)}{5}} = \sqrt{320} \approx 17.9$, or 18 bicycles.

EXERCISES 16.2, page 1100

1. $f_x = 2,\ f_y = 3$

3. $g_x = 4x,\ g_y = 4$

5. $f_x = -\frac{4y}{x^3};\ f_y = \frac{2}{x^2}.$

7. $g(u,v) = \frac{u-v}{u+v};\ \frac{\partial g}{\partial u} = \frac{(u+v)(1)-(u-v)(1)}{(u+v)^2} = \frac{2v}{(u+v)^2}.$
$\frac{\partial g}{\partial v} = \frac{(u+v)(-1)-(u-v)(1)}{(u+v)^2} = -\frac{2u}{(u+v)^2}.$

9. $f(s,t) = (s^2 - st + t^2)^3;\ f_s = 3(s^2 - st + t^2)^2(2s - t)$ and $f_t = 3(s^2 - st + t^2)^2(-s + 2t)$

11. $f(x,y) = (x^2 + y^2)^{2/3}; f_x = \frac{2}{3}(x^2 + y^2)^{-1/3}(2x) = \frac{4}{3}x(x^2 + y^2)^{-1/3}$. Similarly, $f_y = \frac{4}{3}y(x^2 + y^2)^{-1/3}$.

13. $f(x,y) = e^{xy+1};\ f_x = ye^{xy+1},\ f_y = xe^{xy+1}.$

15. $f(x,y) = x\ln y + y\ln x;\ f_x = \ln y + \frac{y}{x},\ f_y = \frac{x}{y} + \ln x.$

17. $g(u,v) = e^u \ln v.\ g_u = e^u \ln v,\ g_v = \frac{e^u}{v}.$

19. $f(x,y,z) = xyz + xy^2 + yz^2 + zx^2; f_x = yz + y^2 + 2xz,$

$f_y = xz + 2xy + z^2, f_z = xy + 2yz + x^2$.

21. $h(r,s,t) = e^{rst}$; $h_r = ste^{rst}$, $h_s = rte^{rst}$, $h_t = rse^{rst}$.

23. $f(x,y) = x^2y + xy^2$; $f_x(1,2) = 2xy + y^2\big|_{(1,2)} = 8$; $f_y(1,2) = x^2 + 2xy\big|_{(1,2)} = 5$.

25. $f(x,y) = x\sqrt{y} + y^2 = xy^{1/2} + y^2$; $f_x(2,1) = \sqrt{y}\ \big|_{(2,1)} = 1$,

$f_y(2,1) = \dfrac{x}{2\sqrt{y}} + 2y\ \big|_{(2,1)} = 3$.

27. $f(x,y) = \dfrac{x}{y}$; $f_x(1,2) = \left.\dfrac{1}{y}\right|_{(1,2)} = \dfrac{1}{2}$, $f_y(1,2) = -\left.\dfrac{x}{y^2}\right|_{(1,2)} = -\dfrac{1}{4}$.

29. $f(x,y) = e^{xy}$. $f_x(1,1) = ye^{xy}\big|_{(1,1)} = e$, $f_y(1,1) = xe^{xy}\big|_{(1,1)} = e$.

31. $f(x,y,z) = x^2yz^3$; $f_x(1,0,2) = 2xyz^3\big|_{(1,0,2)} = 0$; $f_y(1,0,2) = x^2z^3\big|_{(1,0,2)} = 8$.
$f_z(1,0,2) = 3x^2yz^2\big|_{(1,0,2)} = 0$.

33. $f(x,y) = x^2y + xy^3$; $f_x = 2xy + y^3$, $f_y = x^2 + 3xy^2$.
Therefore, $f_{xx} = 2y$, $f_{xy} = 2x + 3y^2 = f_{yx}$, $f_{yy} = 6xy$.

35. $f(x,y) = x^2 - 2xy + 2y^2 + x - 2y$; $f_x = 2x - 2y + 1$, $f_y = -2x + 4y - 2$; $f_{xx} = 2$,
$f_{xy} = -2$, $f_{yx} = -2$, $f_{yy} = 4$.

37. $f(x,y) = (x^2 + y^2)^{1/2}$; $f_x = \frac{1}{2}(x^2 + y^2)^{-1/2}(2x) = x(x^2 + y^2)^{-1/2}$; $f_y = y(x^2 + y^2)^{-1/2}$.

$f_{xx} = (x^2 + y^2)^{-1/2} + x(-\frac{1}{2})(x^2 + y^2)^{-3/2}(2x) = (x^2 + y^2)^{-1/2} - x^2(x^2 + y^2)^{-3/2}$

$= (x^2 + y^2)^{-3/2}(x^2 + y^2 - x^2) = \dfrac{y^2}{(x^2 + y^2)^{3/2}}$.

$f_{xy} = x(-\frac{1}{2})(x^2 + y^2)^{-3/2}(2y) = -\dfrac{xy}{(x^2 + y^2)^{3/2}} = f_{yx}$.

$f_{yy} = (x^2 + y^2)^{-1/2} + y(-\frac{1}{2})(x^2 + y^2)^{-3/2}(2y) = (x^2 + y^2)^{-1/2} - y^2(x^2 + y^2)^{-3/2}$

$$= (x^2 + y^2)^{-3/2}(x^2 + y^2 - y^2) = \frac{x^2}{(x^2 + y^2)^{3/2}}.$$

39. $f(x,y) = e^{-x/y}$; $f_x = -\dfrac{1}{y}e^{-x/y}$; $f_y = \dfrac{x}{y^2}e^{-x/y}$; $f_{xx} = \dfrac{1}{y^2}e^{-x/y}$;

$$f_{xy} = -\frac{x}{y^3}e^{-x/y} + \frac{1}{y^2}e^{-x/y} = \left(\frac{-x+y}{y^3}\right)e^{-x/y};$$

$$f_{yy} = -\frac{2x}{y^3}e^{-x/y} + \frac{x^2}{y^4}e^{-x/y} = \frac{x}{y^3}\left(\frac{x}{y} - 2\right)e^{-x/y}.$$

41. a. $f(x,y) = 20x^{3/4}y^{1/4}$. $f_x(256,16) = 15\left(\dfrac{y}{x}\right)^{1/4}\Bigg|_{(256,16)} = 15\left(\dfrac{16}{256}\right)^{1/4} = 15\left(\dfrac{2}{4}\right) = 7.5.$

$$f_y(256,16) = 5\left(\frac{x}{y}\right)^{3/4}\Bigg|_{(256,16)} = 5\left(\frac{256}{16}\right)^{3/4} = 5(80) = 40.$$

b. Yes.

43. $p(x,y) = 200 - 10(x - \tfrac{1}{2})^2 - 15(y-1)^2$. $\dfrac{\partial p}{\partial x}(0,1) = -20(x - \tfrac{1}{2})\big|_{(0,1)} = 10;$

At the location $(0,1)$ in the figure, the price of land is changing at the rate of \$10 per sq ft per mile change to the right.

$$\frac{\partial p}{\partial y}(0,1) = -30(y-1)\big|_{(0,1)} = 0;$$

At the location $(0,1)$ in the figure, the price of land is constant per mile change upwards.

45. $f(p,q) = 10{,}000 - 10p - e^{0.5q}$; $g(p,q) = 50{,}000 - 4000q - 10p$.

$$\frac{\partial f}{\partial q} = -0.5e^{0.5q} < 0 \text{ and } \frac{\partial g}{\partial p} = -10 < 0$$

and so the two commodities are complementary commodities.

47. $R(x,y) = -0.2x^2 - 0.25y^2 - 0.2xy + 200x + 160y.$

$$\frac{\partial R}{\partial x}(300,250) = -0.4x - 0.2y + 200\Big|_{(300,250)}$$

$$= -0.4(300) - 0.2(250) + 200 = 30$$

and this says that at a sales level of 300 finished and 250 unfinished units the revenue is increasing at the rate of \$30 per week per unit increase in the finished units.

$$\frac{\partial R}{\partial y}(300,250) = -0.5y - 0.2x + 160\Big|_{(300,250)}$$

$$= -0.5(250) - 0.2(300) + 160 = -25$$

and this says that at a level of 300 finished and 250 unfinished units the revenue is decreasing at the rate of \$25 per week per increase in the unfinished units.

49. $V = \dfrac{30.9T}{P}$. $\dfrac{\partial V}{\partial T} = \dfrac{30.9}{P}$ and $\dfrac{\partial V}{\partial P} = -\dfrac{30.9T}{P^2}$.

Therefore, $\left.\dfrac{\partial V}{\partial T}\right|_{T=300,\,P=800} = \dfrac{30.9}{800} = 0.039$, or 0.039 liters/degree.

$$\left.\frac{\partial V}{\partial P}\right|_{T=300,\,P=800} = -\frac{(30.9)(300)}{800^2} = -0.015$$

or -0.015 liters/mm of mercury.

USING TECHNOLOGY EXERCISES 16.2, page 1103

1. 1.3124; 0.4038 3. −1.8889; 0.7778 5. −0.3863; −0.8497

EXERCISES 16.3, page 1114

1. $f(x,y) = 1 - 2x^2 - 3y^2$. To find the critical point(s) of f, we solve the system

$$\begin{cases} f_x = -4x = 0 \\ f_y = -6y = 0 \end{cases}$$

obtaining (0,0) as the only critical point of f. Next,

$f_{xx} = -4$, $f_{xy} = 0$, and $f_{yy} = -6$.

In particular, $f_{xx}(0,0) = -4$, $f_{xy}(0,0) = 0$, and $f_{yy}(0,0) = -6$, giving

$D(0,0) = (-4)(-6) - 0^2 = 24 > 0.$

Since $f_{xx}(0,0) < 0$, the Second Derivative Test implies that (0,0) gives rise to a relative maximum of f. Finally, the relative maximum of f is $f(0,0) = 1$.

3. To find the critical points of f, we solve the system

$$\begin{cases} f_x = \quad 2x - 2 = 0 \\ f_y = -2y + 4 = 0 \end{cases}$$

obtaining $x = 1$ and $y = 2$ so that (1,2) is the only critical point.

$f_{xx} = 2, f_{xy} = 0$, and $f_{yy} = -2$.

So $D(x,y) = f_{xx}f_{yy} - f_{xy}^2 = -4$. In particular, $D(1,2) = -4 < 0$ and so (1,2) affords a saddle point of f and $f(1,2) = 4$.

5. $f(x,y) = x^2 + 2xy + 2y^2 - 4x + 8y - 1$. To find the critical point(s) of f, we solve the system

$$\begin{cases} f_x = 2x + 2y - 4 = 0 \\ f_y = 2x + 4y + 8 = 0 \end{cases}$$

obtaining (8,−6) as the critical point of f. Next, $f_{xx} = 2, f_{xy} = 2, f_{yy} = 4$. In particular, $f_{xx}(8,-6) = 2, f_{xy}(8,-6) = 2, f_{yy}(8,-6) = 4$, giving $D = 2(4) - 4 = 4 > 0$. Since $f_{xx}(8,-6) > 0$, (8,−6) gives rise to a relative minimum of f. Finally, the relative minimum value of f is $f(8,-6) = -41$.

7. $f(x,y) = 2x^3 + y^2 - 9x^2 - 4y + 12x - 2$.. To find the critical points of f, we solve the system

$$\begin{cases} f_x = 6x^2 - 18x + 12 = 0 \\ f_y = 2y - 4 = 0 \end{cases}$$

The first equation is equivalent to $x^2 - 3x + 2 = 0$, or $(x - 2)(x - 1) = 0$ which gives $x = 1$ or 2. The second equation of the system gives $y = 2$. Therefore, there are two critical points, (1,2) and (2,2). Next, we compute

$f_{xx} = 12x - 18 = 6(2x - 3), f_{xy} = 0, f_{yy} = 2$.

At the point (1,2):

$f_{xx}(1,2) = 6(2 - 3) = -6, f_{xy}(1,2) = 0$, and $f_{yy}(1,2) = 2$.

Therefore, $D = (-6)(2) - 0 = -12 < 0$ and we conclude that (1,2) gives rise to a saddle point of f. At the point (2,2):

$f_{xx}(2,2) = 6(4 - 3) = 6, f_{xy}(2,2) = 0$, and $f_{yy}(2,2) = 2$.

Therefore, $D = (6)(2) - 0 = 12 > 0$. Since $f_{xx}(2,2) > 0$, we see that (2,2) gives rise to a relative minimum with value $f(2,2) = -2$.

9. To find the critical points of f, we solve the system

$$\begin{cases} f_x = 3x^2 - 2y + 7 = 0 \\ f_y = 2y - 2x - 8 = 0 \end{cases}$$

Adding the two equations gives $3x^2 - 2x - 1 = 0$, or $(3x + 1)(x - 1) = 0$. Therefore, $x = -1/3$ or 1. Substituting each of these values of x into the second equation gives $y = 8/3$ and $y = 5$, respectively. Therefore, $(-\frac{1}{3}, \frac{11}{3})$ and $(1,5)$ are critical points of f. Next, $f_{xx} = 6x$, $f_{xy} = -2$, and $f_{yy} = 2$. So $D(x,y) = 12x - 4 = 4(3x - 1)$. Then

$$D(-\tfrac{1}{3}, \tfrac{11}{3}) = 4(-1-1) = -8 < 0$$

and so $(-\frac{1}{3}, \frac{11}{3})$ gives a saddle point. Next, $D(1,5) = 4(3 - 1) = 8 > 0$ and since $f_{xx}(1,5) = 6 > 0$, we see that $(1,5)$ gives rise to a relative minimum.

11. To find the critical points of f, we solve the system

$$\begin{cases} f_x = 3x^2 - 3y = 0 \\ f_y = -3x + 3y^2 = 0 \end{cases}$$

The first equation gives $y = x^2$ which when substituted into the second equation gives $-3x + 3x^4 = 3x(x^3 - 1) = 0$. Therefore, $x = 0$ or 1. Substituting these values of x into the first equation gives $y = 0$ and $y = 1$, respectively. Therefore, $(0,0)$ and $(1,1)$ are critical points of f. Next, we find $f_{xx} = 6x$, $f_{xy} = -3$, and $f_{yy} = 6y$. So $D = f_{xx}f_{yy} - f_{xy}^2 = 36xy - 9$. Since $D(0,0) = -9 < 0$, we see that $(0,0)$ gives a saddle point of f. Next, $D(1,1) = 36 - 9 = 27 > 0$ and since $f_{xx}(1,1) = 6 > 0$, we see that $f(1,1) = -3$ is a relative minimum value of f.

13. Solving the system of equations

$$\begin{cases} f_x = y - \frac{4}{x^2} = 0 \\ f_y = x - \frac{2}{y^2} = 0 \end{cases}$$

we obtain $y = \frac{4}{x^2}$. Therefore, $x - 2\left(\frac{x^4}{16}\right) = 0$ and $8x - x^4 = x(8 - x^3) = 0$, and $x = 0$, or $x = 2$. Since $x = 0$ is not in the domain of f, $(2,1)$ is the only critical point of f. Next, $f_{xx} = \frac{8}{y^3}$, $f_{xy} = 1$, and $f_{yy} = \frac{4}{y^3}$. Therefore,

$D(2,1) = \left.\frac{32}{x^3y^3} - 1\right|_{(2,1)} = 4 - 1 = 3 > 0$ and $f_{xx}(2,1) = 1 > 0$. Therefore, the relative minimum value of f is $f(2,1) = 2 + 4/2 + 2/1 = 6$.

15. Solving the system of equations $f_x = 2x = 0$ and $f_y = -2ye^{y^2} = 0$, we obtain $x = 0$ and $y = 0$. Therefore, (0,0) is the only critical point of f. Next,

$$f_{xx} = 2, f_{xy} = 0, f_{yy} = -2e^{y^2} - 4y^2e^{y^2}.$$

Therefore, $D(0,0) = -4e^{y^2}(1+2y^2)\Big|_{(0,0)} = -4(1) < 0$, and we conclude that (0,0) is a saddle point.

17. $f(x,y) = e^{x^2+y^2}$

Solving the system

$$\begin{cases} f_x = 2xe^{x^2+y^2} = 0 \\ f_y = 2ye^{x^2+y^2} = 0 \end{cases}$$

we see that $x = 0$ and $y = 0$ (recall that $e^{x^2+y^2} \neq 0$). Therefore, (0,0) is the only critical point of f. Next, we compute

$$f_{xx} = 2e^{x^2+y^2} + 2x(2x)e^{x^2+y^2} = 2(1+2x^2)e^{x^2+y^2}$$

$$f_{xy} = 2x(2y)e^{x^2+y^2} = 4xye^{x^2+y^2}$$

$$f_{yy} = 2(1+2y^2)e^{x^2+y^2}.$$

In particular, at the point (0,0), $f_{xx}(0,0) = 2$, $f_{xy}(0,0) = 0$, and $f_{yy}(0,0) = 2$. Therefore, $D = (2)(2) - 0 = 4 > 0$. Furthermore, since $f_{xx}(0,0) > 0$, we conclude that (0,0) gives rise to a relative minimum of f. The relative minimum value of f is $f(0,0) = 1$.

19. $f(x,y) = \ln(1+x^2+y^2)$. We solve the system of equations

$$f_x = \frac{2x}{1+x^2+y^2} = 0 \text{ and } f_y = \frac{2y}{1+x^2+y^2} = 0,$$

obtaining $x = 0$ and $y = 0$. Therefore, (0,0) is the only critical point of f. Next,

$$f_{xx} = \frac{(1+x^2+y^2)2-(2x)(2x)}{(1+x^2+y^2)^2} = \frac{2+2y^2-2x^2}{(1+x^2+y^2)^2}$$

$$f_{yy} = \frac{(1+x^2+y^2)2-(2y)(2y)}{(1+x^2+y^2)^2} = \frac{2+2x^2-2y^2}{(1+x^2+y^2)^2}$$

$$f_{xy} = -2x(1+x^2+y^2)^{-2}(2y) = -\frac{4xy}{(1+x^2+y^2)^2}.$$

Therefore, $D(x,y) = \dfrac{(2+2y^2-2x^2)(2+2x^2-2y^2)}{(1+x^2+y^2)^4} - \dfrac{16x^2y^2}{(1+x^2+y^2)^4}$.

Since $D(0,0) = \frac{4}{1} > 0$ and $f_{xx}(0,0) = 2 > 0$, $f(0,0) = 0$ is a relative minimum value.

21. $P(x) = -0.2x^2 - 0.25y^2 - 0.2xy + 200x + 160y - 100x - 70y - 4000$
$= -0.2x^2 - 0.25y^2 - 0.2xy + 100x + 90y - 4000.$

Then $\begin{cases} P_x = -0.4x - 0.2y + 100 = 0 \\ P_y = -0.5y - 0.2x + 90 = 0 \end{cases}$

implies that $\begin{cases} 4x + 2y = 1000 \\ 2x + 5y = \ \ 900 \end{cases}$. Solving, we find $x = 200$ and $y = 100$.

Next, $P_{xx} = -0.4, P_{yy} = -0.5,\ P_{xy} = -0.2$, and

$D(200,100) = (-0.4)(-0.5) - (-0.2)^2 > 0$. Since $P_{xx}(200, 100) < 0$, we conclude that $(200,100)$ is a relative maximum of P. Thus, the company should manufacture 200 finished and 100 unfinished units per week. The maximum profit is

$$P(200,100) = -0.2(200)^2 - 0.25(100)^2 - 0.2(100)200) + 100(200) + 90(100) - 4000$$
$$= 10{,}500, \text{ or } \$10{,}500 \text{ dollars.}$$

23. $p(x,y) = 200 - 10(x - \frac{1}{2})^2 - 15(y-1)^2$. Solving the system of equations

$$\begin{cases} p_x = -20(x - \frac{1}{2}) = 0 \\ p_y = -30(y-1) = 0 \end{cases}$$

we obtain $x = 1/2$, $y = 1$. We conclude that the only critical point of f is $(\frac{1}{2},1)$.

Next, $\quad p_{xx} = -20, p_{xy} = 0, p_{yy} = -30$

so $\quad D(\frac{1}{2},1) = (-20)(-30) = 600 > 0.$

Since $p_{xx} = -20 < 0$, we conclude that $f(\frac{1}{2},1)$ gives a relative maximum. So we conclude that the price of land is highest at $(\frac{1}{2},1)$.

25. We want to minimize
$f(x,y) = D^2 = (x-5)^2 + (y-2)^2 + (x+4)^2 + (y-4)^2 + (x+1)^2 + (y+3)^2.$

Next, $\begin{cases} f_x = 2(x-5) + 2(x+4) + 2(x+1) = 6x = 0, \\ f_y = 2(y-2) + 2(y-4) + 2(y+3) = 6y - 6 = 0 \end{cases}$

and we conclude that $x = 0$ and $y = 1$. Also,

$$f_{xx} = 6, f_{xy} = 0, f_{yy} = 6 \text{ and } D(x,y) = (6)(6) = 36 > 0.$$

Since $f_{xx} > 0$, we conclude that the function is minimized at $(0,1)$ and so $(0,1)$ gives the desired location.

27. The volume is given by $V = xyz = xz(108 - 2x - 2z) = 108xz - 2x^2z - 2xz^2$.
Solving the system of equations, $V_x = 108z - 4xz - 2z^2 = 0$ and
$V_z = 108x - 2x^2 - 4xz = 0$, we obtain

$(108 - 4x - 2z)z = 0$, or $108 - 4x - 2z = 0$
and $(108 - 4z - 2x)x = 0$, or $108 - 2x - 4z = 0$. Thus

$$\begin{cases} 108 - 4x - 2z = 0 \\ 216 - 4x - 8z = 0 \end{cases}$$

gives $-108 + 6z = 0$, or $z = 18$. So $x = \frac{1}{4}(108 - 36) = 18$ and

$y = 108 - 2x - 2z = 108 - 72 = 36$, and $(18,18)$ is the critical point of V. Next,

$$V_{xx} = -4z,\ V_{zz} = -4x,\ V_{xz} = 108 - 4x - 4z, \text{ and}$$
$$D(18,18) = -4(18)(-4)(18) - [108 - 4(18) - 4(18)]^2 > 0$$

and $V_{xx}(18,18) < 0$. We conclude that the dimensions yielding the maximum volume are 18" $\times$ 36" $\times$ 18".

29. Since $V = xyz$, $z = \dfrac{48}{xy}$. Then the amount of material used in the box is given by

$$S = xy + 2xz + 3yz = xy + \frac{48}{xy}(2x + 3y) = xy + \frac{96}{y} + \frac{144}{x}.$$ Solving the system of

equations $\begin{cases} S_x = y - \frac{144}{x^2} = 0 \\ S_y = x - \frac{96}{y^2} = 0 \end{cases}$, we have $y = \dfrac{144}{x^2}$. Therefore

$x - \dfrac{96x^4}{144^2} = 0$, $144^2 x - 96x^4 = 0$, $96x(216 - x^3) = 0$, and $x = 0$ or $x = 6$. Then

$y = \dfrac{144}{36} = 4$. Next, $S_{xx} = \dfrac{288}{x^3}$, $S_{yy} = \dfrac{192}{y^3}$, and $S_{xy} = 1$.

At the point (6,4):

$$D(x,y) = \frac{(288)(192)}{x^3 y^3} - 1\bigg|_{(6,4)} = \frac{288(192)}{216(64)} - 1 = 3 > 0,$$

and $S_{xx} > 0$. Therefore, we conclude that the function is minimized when its dimensions are 6" $\times$ 4" $\times$ 2".

EXERCISES 16.4, page 1127

1. We form the Lagrangian function $F(x,y,\lambda) = x^2 + 3y^2 + \lambda(x + y - 1)$. We solve the

system $\begin{cases} F_x = 2x + \lambda = 0 \\ F_y = 6y + \lambda = 0 \\ F_\lambda = x + y - 1 = 0. \end{cases}$

Solving the first and the second equations for x and y in terms of λ we obtain $x = -\frac{\lambda}{2}$ and $y = -\frac{\lambda}{6}$ which, upon substitution into the third equation, yields $-\frac{\lambda}{2} - \frac{\lambda}{6} - 1 = 0$ or $\lambda = -\frac{3}{2}$. Therefore, $x = \frac{3}{4}$ and $y = \frac{1}{4}$ which gives the point $(\frac{3}{4}, \frac{1}{4})$ as the sole critical point of F. Therefore, $(\frac{3}{4}, \frac{1}{4}) = \frac{3}{4}$ is a minimum of F.

3. We form the Lagrangian function $F(x,y,\lambda) = 2x + 3y - x^2 - y^2 + \lambda(x + 2y - 9)$. We

then solve the system

$$\begin{cases} F_x = 2 - 2x + \lambda = 0 \\ F_y = 3 - 2y + 2\lambda = 0. \\ F_\lambda = x + 2y - 9 = 0 \end{cases}$$

Solving the first equation λ, we obtain $\lambda = 2x - 2$. Substituting into the second equation, we have $3 - 2y + 4x - 4 = 0$, or $4x - 2y - 1 = 0$. Adding this equation to the third equation in the system, we have $5x - 10 = 0$, or $x = 2$. Therefore, $y = 7/2$ and $f(2, \frac{7}{2}) = -\frac{7}{4}$ is the maximum value of f.

5. Form the Lagrangian function $F(x,y,\lambda) = x^2 + 4y^2 + \lambda(xy - 1)$. We then solve the system $\begin{cases} F_x = 2x + \lambda y = 0 \\ F_y = 8y + \lambda x = 0 \\ F_\lambda = xy - 1 = 0 \end{cases}$. Multiplying the first and second equations by x and y, respectively, and subtracting the resulting equations, we obtain $2x^2 - 8y^2 = 0$, or $x = \pm 2y$. Substituting this into the third equation gives

$2y^2 - 1 = 0$ or $y = \pm\frac{\sqrt{2}}{2}$.

We conclude that $f(-\sqrt{2}, -\frac{\sqrt{2}}{2}) = f(\sqrt{2}, \frac{\sqrt{2}}{2}) = 4$ is the maximum value of f.

7. We form the Lagrangian function

$$F(x,y,\lambda) = x + 5y - 2xy - x^2 - 2y^2 + \lambda(2x + y - 4).$$

Next, we solve the system

$$\begin{cases} F_x = 1 - 2y - 2x + 2\lambda = 0 \\ F_y = 5 - 2x - 4y + \lambda = 0 \\ F_\lambda = 2x + y - 4 = 0 \end{cases}$$

Solving the last two equations for x and y in terms of λ, we obtain

$$y = \tfrac{1}{3}(1+\lambda) \text{ and } x = \tfrac{1}{6}(11-\lambda)$$

which, upon substitution into the first equation, yields

$$1 - \tfrac{2}{3}(1+\lambda) - \tfrac{1}{3}(11-\lambda) + 2\lambda = 0$$

$$\text{or } 1 - \tfrac{2}{3} - \tfrac{2}{3}\lambda - \tfrac{11}{3} + \tfrac{\lambda}{3} + 2\lambda = 0$$

or $\lambda = 2$. Therefore, $x = 3/2$ and $y = 1$. The maximum of f is

$$f(\tfrac{3}{2}, 1) = \tfrac{3}{2} + 5 - 2(\tfrac{3}{2}) - (\tfrac{3}{2})^2 - 2 = -\tfrac{3}{4}.$$

9. Form the Lagrangian $F(x,y,\lambda) = xy^2 + \lambda(9x^2 + y^2 - 9)$. We then solve

$$\begin{cases} F_x = y^2 + 18\lambda x = 0 \\ F_y = 2xy + 2\lambda y = 0 \\ F_\lambda = 9x^2 + y^2 - 9 = 0. \end{cases}$$

The first equation gives $\lambda = -\dfrac{y^2}{18x}$. Substituting into the second gives

$$2xy + 2y\left(-\frac{y^2}{18x}\right) = 0, \text{ or } 18x^2y - y^3 = y(18x^2 - y^2) = 0,$$

giving $y = 0$ or $y = \pm 3\sqrt{2}x$. If $y = 0$, then the third equation gives $9x^2 - 9 = 0$ or $x = \pm 1$. If $y = \pm 3\sqrt{3}/3$. Therefore, the points $(-1,0)$, $(-\sqrt{3}/3, -\sqrt{6})$, $(-\sqrt{3}/3, \sqrt{6})$, $(\sqrt{3}/3, -\sqrt{6})$ and $(\sqrt{3}/3, \sqrt{6})$ give rise to extreme values of f subject to the given constraint. Evaluating $f(x,y)$ at each of these points, we see that

$$f(\sqrt{3}/3, -\sqrt{6}) = (\sqrt{3}/3, \sqrt{6}) = 2\sqrt{3}$$

is the minimum value of f.

11. We form the Lagrangian function $F(x,y,\lambda) = xy + \lambda(x^2 + y^2 - 16)$. To find the critical points of F, we solve the system

$$\begin{cases} F_x = y + 2\lambda x = 0 \\ F_y = x + 2\lambda y = 0 \\ F_\lambda = x^2 + y^2 - 16 = 0 \end{cases}$$

Solving the first equation for λ and substituting this value into the second equation yields $x - 2\left(\dfrac{y}{2x}\right)y = 0$, or $x^2 = y^2$. Substituting the last equation into the third equation in the system, yields $x^2 + x^2 - 16 = 0$, or $x^2 = 8$, that is, $x = \pm 2\sqrt{2}$. The corresponding values of y are $y = \pm 2\sqrt{2}$. Therefore the critical points of F are $(-2\sqrt{2}, -2\sqrt{2})$, $(-2\sqrt{2}, 2\sqrt{2})$, $(2\sqrt{2}, -2\sqrt{2})(2\sqrt{2}, 2\sqrt{2})$. E valuating f at each of these values, we find that $f(-2\sqrt{2},2\sqrt{2})=-8$ and $f(2\sqrt{2},-2\sqrt{2})=-8$ are relative minimum values and $f(-2\sqrt{2},-2\sqrt{2})=8$ and $f(2\sqrt{2},2\sqrt{2})=8$, are relative maximum values.

13. We form the Lagrangian function $F(x,y,\lambda)=xy^2+\lambda(x^2+y^2-1)$. Next, we solve the system
$$\begin{cases} F_x = y^2+2x\lambda=0 \\ F_y=2xy+2y\lambda=0 \\ F_\lambda=x^2+y^2-1=0 \end{cases}.$$
We find that $x=\pm\sqrt{3}/3$ and $y=\pm\sqrt{6}/3$ and $x=\pm1$, $y=0$. Evaluating f at each of the critical points $(-\frac{\sqrt{3}}{3},-\frac{\sqrt{6}}{3}),(-\frac{\sqrt{3}}{3},\frac{\sqrt{6}}{3})(\frac{\sqrt{3}}{3},-\frac{\sqrt{6}}{3})(\frac{\sqrt{3}}{3},\frac{\sqrt{6}}{3})$,(-1,0), and (1,0), we find that $f(-\frac{\sqrt{3}}{3},-\frac{\sqrt{6}}{3})=-\frac{2\sqrt{3}}{9}$ and $f(-\frac{\sqrt{3}}{3},\frac{\sqrt{6}}{3})=-\frac{2\sqrt{3}}{9}$ are relative minimum values and $f(\frac{\sqrt{3}}{3},-\frac{\sqrt{6}}{3})=\frac{2\sqrt{3}}{9}$ and $f(\frac{\sqrt{3}}{3},\frac{\sqrt{6}}{3})=\frac{2\sqrt{3}}{9}$ are relative maximum values.

15. Form the Lagrangian function $F(x,y,z,\lambda)=x^2+y^2+z^2+\lambda(3x+2y+z-6)$. We solve the system
$$\begin{cases} F_x=2x+3\lambda=0 \\ F_y=2y+2\lambda=0 \\ F_z=2x+\lambda=0 \\ F_\lambda=3x+2y+z-6=0. \end{cases}$$
The third equation give $\lambda=-2z$. Substituting into the first two equations gives
$$\begin{cases} 2x-6z=0 \\ 2y-4z=0. \end{cases}$$
So $x=3z$ and $y=2z$. Substituting into the third equation yields $9z+4z+z-6=0$, or $z=3/7$. Therefore, $x=9/7$ and $y=6/7$. Therefore, $f(\frac{9}{7},\frac{6}{7},\frac{3}{7})=\frac{18}{7}$ is the minimum value of F.

17. We want to maximize P subject to the constraint $x+y=200$. The Lagrangian function is
$$F(x,y,\lambda)=-0.2x^3-0.25y^2-0.2xy+100x+90y-4000+\lambda(x+y-200).$$
Next, we solve
$$\begin{cases} F_x=-0.4x-0.2y+100+\lambda=0 \\ F_y=-0.5y-0.2x+90+\lambda=0 \\ F_\lambda=x+y-200=0. \end{cases}$$

Subtracting the fist equation from the second yields

$$0.2x - 0.3y - 10 = 0, \text{ or } 2x - 3y - 100 = 0.$$

Multiplying the third equation in the system by 2 and subtracting the resulting equation from the last equation, we find

$$-5y + 300 = 0 \text{ or } y = 60.$$

So $x = 140$ and the company should make 140 finished and 60 unfinished units.

19. Suppose each of the sides made of pine board is x feet long and those of steel are y feet long. Then $xy = 800$. The cost is $C = 12x + 3y$ and is to be minimized subject to the condition $xy = 800$. We form the Lagrangian function

$$F(x,y,\lambda) = 12x + 3y + \lambda(xy - 800).$$

We solve the system

$$\begin{cases} F_x = 12 + \lambda y = 0 \\ F_y = 3 + \lambda x = 0 \\ F_\lambda = xy - 800 = 0. \end{cases}$$

Multiplying the first equation by x and the second equation by y and subtracting the resulting equations, we obtain $12x - 3y = 0$, or $y = 4x$. Substituting this into the third equation of the system, we obtain $4x^2 - 800 = 0$, or $x = \pm\sqrt{200} = \pm 10\sqrt{2}$. Since x must be positive, we take $x = 10\sqrt{2}$. So $y = 40\sqrt{2}$. So the dimensions are approximately 14.14ft by 56.56 ft.

21. Refer to the following figure.

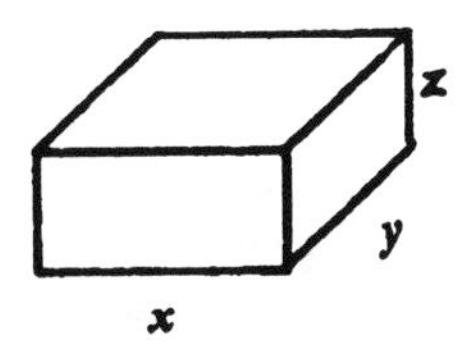

We form the Lagrangian function $F(x,y,\lambda) = xyz + \lambda(3xy + 2xz + 2yz - 36)$. Then, we solve the system

$$\begin{cases} F_x = yz + 3\lambda y + 2\lambda z = 0 \\ F_y = xz + 3\lambda x + 2\lambda z = 0 \\ F_z = xy + 2\lambda x + 2\lambda y = 0 \\ F_\lambda = 3xy + 2xz + 2yz - 36 = 0 \end{cases}$$

Multiplying the first, second, and third equation by x, y, and z, respectively, we obtain

$$\begin{cases} xyz + 3\lambda xy + 2\lambda xz = 0 \\ xyz + 3\lambda xy + 2\lambda yz = 0 \\ xyz + 2\lambda xz + 2\lambda yz = 0. \end{cases}$$

Subtracting the second equation from the first and the third equation from the second, yields

$$\begin{cases} 2\lambda(x-y)z = 0 \\ \lambda x(3y-2z) = 0. \end{cases}$$

Solving this system, we find that $x = y$ and $x = 3/2y$. Substituting these values into the third equation, we find that

$$3y^2 + 2y(\tfrac{3}{2})y + 2y(\tfrac{3}{2}) - 36 = 0$$

and $y = \pm 2$. We reject the negative root, and find that $x = 2$, $y = 2$, and $z = 3$ provides the desired relative maximum and the dimensions are $2' \times 2' \times 3'$.

23. We want to maximize $f(x,y) = 90x^{1/4}y^{3/4}$ subject to $x + y = 60{,}000$. We form the Lagrangian function $F(x,y,\lambda) = 90x^{1/4}y^{3/4} + \lambda(x + y - 60{,}000)$. Now set

$$\begin{cases} F_x = \frac{45}{2}x^{-3/4}y^{3/4} + \lambda = 0 \\ F_y = \frac{135}{2}x^{1/4}y^{-1/4} + \lambda = 0 \\ F_\lambda = x + y - 60{,}000 = 0. \end{cases}$$

Eliminating λ in the first two equations gives

$$\frac{45}{2}\left(\frac{y}{x}\right)^{3/4} - \frac{135}{2}\left(\frac{x}{y}\right)^{1/4} = 0$$

$$\frac{y}{x} - 3 = 0, \text{ or } y = 3x.$$

Substituting this value into the third equation in the system, we find $x + 3x = 60{,}000$ and $x = 15{,}000$ and $y = 45{,}000$. So the company should spend \$15,000 on newspaper advertisements and \$45,000 on television advertisements.

25. We want to minimize $C = 2xy + 8xz + 6yz$ subject to $xyz = 12000$. Form the Lagrangian function $F(x,y,z,\lambda) = 2xy + 8xz + 6yz + \lambda(xyz - 12000)$. Next, we solve the system

$$\begin{cases} F_x = 2y + 8z + \lambda yz = 0 \\ F_y = 2x + 6z + \lambda xz = 0 \\ F_z = 8x + 6y + 3xy = 0 \\ F_\lambda = xyz - 12{,}000 = 0. \end{cases}$$

Multiplying the first, second, and third equations by x, y, and z, we obtain

$$\begin{cases} 2xy+8yz+3xyz=0 \\ 2xy+6yz+\lambda xyz=0 \\ 8xz+6yz+\lambda xyz=0. \end{cases}$$

The first two equations imply that $z(8x-6y)=0$ or since $z \neq 0$, we have $x=(3/4)y$. The second and third equations imply that $x(8z-2y)=0$ or $x=(1/4)y$. Substituting these values into the third equation of the system, we have

$$(\tfrac{3}{4}y)(y)(\tfrac{1}{4}y)=12{,}000 \text{ or } y^3=64{,}000, \text{ or } y=40.$$

Therefore, $x=30$ and $z=10$. So the heating cost is

$$C=2(30)(40)+8(30)(10)+6(40)(10)=7200,$$

or \$7200 as obtained earlier.

EXERCISES 16.5, page 1141

1. $\int_1^2\int_0^1(y+2x)\,dy\,dx=\int_1^2 \tfrac{1}{2}y^2+2xy\Big|_{y=0}^{y=1}dx=\int_1^2(\tfrac{1}{2}+2x)\,dx=\tfrac{1}{2}x+x^2\Big|_1^2=5-\tfrac{3}{2}=\tfrac{7}{2}.$

3. $\int_{-1}^1\int_0^1 xy^2\,dy\,dx=\int_{-1}^1 \tfrac{1}{3}xy^3\Big|_0^1 dx=\int_{-1}^1\tfrac{1}{3}x\,dx=\tfrac{x^2}{6}\Big|_{-1}^1=\tfrac{1}{6}-(\tfrac{1}{6})=0.$

5. $\int_{-1}^2\int_1^{e^3}\dfrac{x}{y}\,dy\,dx=\int_{-1}^2 x\ln y\Big|_1^{e^3}dx=\int_{-1}^2 x\ln e^3\,dx=\int_{-1}^2 3x\,dx=\tfrac{3}{2}x^2\Big|_{-1}^2$

   $=\tfrac{3}{2}(4)-\tfrac{3}{2}(1)=\tfrac{9}{2}.$

7. $\int_{-2}^0\int_0^1 4xe^{2x^2+y}\,dx\,dy=\int_{-2}^0 e^{2x^2+y}\Big|_{x=0}^{x=1}dy=\int_{-2}^0(e^{2+y}-e^y)\,dy=(e^{2+y}-e^y)\Big|_{-2}^0$

   $=[(e^2-1)-(e^0-e^{-2})=e^2-2+e^{-2}=(e^2-1)(1-e^{-2}).$

9. $\int_0^1\int_1^e \ln y\,dy\,dx=\int_0^1 y\ln y-y\Big|_{y=1}^{y=e}dx=\int_0^1 dx=1.$

11. $\int_0^1\int_0^x(x+2y)\,dy\,dx=\int_0^1(xy+y^2)\Big|_{y=0}^{y=x}dx=\int_0^1 2x^2\,dx=\tfrac{2}{3}x^3\Big|_0^1=\tfrac{2}{3}.$

13. $\int_1^3\int_0^{x+1}(2x+4y)\,dy\,dx = \int_1^3 2xy+2y^2\Big|_{y=0}^{y=x+1} dx = \int_1^3[2x(x+1)+2(x+1)^2]dx$

$= \int_1^3(4x^2+6x+2)dx = (\frac{4}{3}x^3+3x^2+2x)\Big|_1^3$

$= (36+27+6)-(\frac{4}{3}+3+2) = \frac{188}{3}.$

15. $\int_0^4\int_0^{\sqrt{y}}(x+y)\,dx\,dy = \int_0^4 \frac{1}{2}x^2+xy\Big|_{x=0}^{x=\sqrt{y}} dy = \int_0^4(\frac{1}{2}y+y^{3/2})dy$

$=(\frac{1}{4}y^2+\frac{2}{5}y^{5/2})\Big|_0^4 = 4+\frac{64}{5} = \frac{84}{5}.$

17. $\int_0^2\int_0^{\sqrt{4-y^2}} y\,dx\,dy = \int_0^2 xy\Big|_0^{\sqrt{4-y^2}} dy = \int_0^2 y\sqrt{4-y^2}\,dy = -\frac{1}{2}(\frac{2}{3})(4-y^2)^{3/2}\Big|_0^2$

$=\frac{1}{3}(4^{3/2}) = \frac{8}{3}.$

19. $\int_0^1\int_0^x 2xe^y\,dy\,dx = \int_0^1 2xe^y\Big|_{y=0}^{y=x} dx = \int_0^1(2xe^x-2x)\,dx$

$= 2(x-1)e^x - x^2\Big|_0^1 = (-1)+2 = 1.$

21. $\int_0^1\int_{\sqrt{x}}^{x} ye^x\,dy\,dx = -\int_0^1\int_x^{\sqrt{x}} ye^x\,dy\,dx = \int_0^1 -\frac{1}{2}y^2e^x\Big|_{y=\sqrt{x}}^{y=x} dx = -\frac{1}{2}\int_0^1(x^2e^x - xe^x)dx$

$= -\frac{1}{2}[x^2e^x\Big|_0^1 - 2\int_0^1 xe^x\,dx - \int_0^1 xe^x\,dx] = -\frac{1}{2}[x^2e^x\Big|_0^1 - 3\int_0^1 xe^x\,dx]$

$= -\frac{1}{2}[x^2e^x - 3xe^x + 3e^x]\Big|_0^1 = -\frac{1}{2}[e-3e+3e-3] = \frac{1}{2}(3-e).$

23. $\int_0^1\int_{2x}^2 e^{y^2}\,dy\,dx = \int_0^2\int_0^{y/2} e^{y^2}\,dx\,dy = \int_0^2 xe^{y^2}\Big|_{x=0}^{x=y/2} dy = \int_0^2 \frac{1}{2}ye^{y^2}\,dy$

$= \frac{1}{4}e^{y^2}\Big|_0^2 = \frac{1}{4}(e^4-1).$

25. $\int_0^2\int_{y/2}^1 ye^{x^3}\,dx\,dy = \int_0^1\int_0^{2x} ye^{x^3}\,dy\,dx = \int_0^1 \frac{1}{2}y^2e^{x^3}\Big|_{y=0}^{y=2x} dx = \int_0^1 2x^2e^{x^3}\,dx$

$= \frac{2}{3}e^{x^3}\Big|_0^1 = \frac{2}{3}(e-1).$

27. $V = \int_0^4\int_0^3(6-x)\,dy\,dx = \int_0^4(6-x)y\Big|_{y=0}^{y=3} dx = 3\int_0^4(6-x)\,dx$

$= 3(6x-\frac{1}{2}x^2)\Big|_0^4 = 3(24-8) = 48$, or 48 cu units.

29. $V=\int_0^2\int_x^{3-(3/2)z}(6-2y-3z)dy\,dz=\int_0^2 6y-y^2-3yz\Big|_{y=0}^{y=3-(3/2)z}dz$

$=\int_0^2[(6(3-\frac{3}{2}z)-(3-\frac{3}{2}z)^2-3(3-\frac{3}{2}z)z]dz$

$=-2(3-\frac{3}{2}z)^2-\frac{2}{9}(3-\frac{3}{2}z)^3-\frac{9}{2}z^2+\frac{3}{2}z^3\Big|_0^2$

$=(-18+12)-(-18+6)=6$, or 6 cu units.

31. $V=\int_0^1\int_0^{-2x+2}(4-x^2-y^2)dy\,dx=\int_0^1(4y-x^2y-\frac{1}{3}y^3)\Big|_{y=0}^{y=2(1-x)}dx$

$=\int_0^1[8(1-x)-2x^2+2x^3-\frac{8}{3}(1-x)^3]dx$

$=[(8x-4x^2-\frac{2}{3}x^3+\frac{1}{2}x^4)+\frac{2}{3}(1-x)^4]\Big|_0^1$

$=(8-4-\frac{2}{3}+\frac{1}{2})-\frac{2}{3}=\frac{19}{6}$, or $\frac{19}{6}$ cu units.

33. $V=\int_0^2\int_0^2 5e^{-x-y}dx\,dy=\int_0^2-5e^{-x-y}\Big|_{x=0}^{x=2}dy=\int_0^2-5(e^{-2-y}-e^{-y})dy$

$=-5(-e^{-2-y}+e^{-y})\Big|_0^2=-5(-e^{-4}+e^{-2})+5(-e^{-2}+1)$

$=5(1-2e^{-2}+e^{-4})$ cu units.

35. $V=\int_0^2\int_0^{2x}(2x+y)dy\,dx=\int_0^2 2xy+\frac{1}{2}y^2\Big|_0^{2x}dx=\int_0^2(4x^2+2x^2)\,dx$

$=\int_0^2 6x^2\,dx=2x^3\Big|_0^2=16$, or 16 cu units.

37. $V=\int_0^1\int_0^{-x+1}e^{x+2y}\,dy\,dx=\int_0^1\frac{1}{2}e^{x+2y}\Big|_{y=0}^{y=-x+1}dx=\frac{1}{2}\int_0^1(e^{-x+2}-e^x)\,dx$

$=\frac{1}{2}(-e^{-x+2}-e^x)\Big|_0^1=\frac{1}{2}(-e-e+e^2+1)=\frac{1}{2}(e^2-2e+1)$

$=\frac{1}{2}(e-1)^2$ cu units.

39. $V=\int_0^4\int_0^{\sqrt{x}}\frac{2y}{1+x^2}dy\,dx=\int_0^4\frac{y^2}{1+x^2}\Big|_0^{\sqrt{x}}dx=\int_0^4\frac{x}{1+x^2}dx$

$=\frac{1}{2}\ln(1+x^2)\Big|_0^4=\frac{1}{2}(\ln 17-\ln 1)=\frac{1}{2}\ln 17$ cu units.

41. $V = \int_0^4 \int_0^{\sqrt{16-x^2}} x\,dy\,dx = \int_0^4 xy\Big|_{y=0}^{y=\sqrt{16-x^2}} dx = \int_0^4 x(16-x^2)^{1/2}\,dx$

$= (-\tfrac{1}{2})(\tfrac{2}{3})(16-x^2)^{3/2}\Big|_0^4 = \tfrac{1}{3}(16)^{3/2} = \tfrac{64}{3}.$

43.

$$A = \frac{1}{\frac{1}{2}} \int_0^1 \int_0^x (x+2y)\,dy\,dx = 2\int_0^1 xy + y^2\Big|_0^x dx = 2\int_0^1 (x^2+x^2)\,dx = 4\int_0^1 x^2\,dx = \tfrac{4x^3}{3}\Big|_0^1 = \tfrac{4}{3}.$$

45. The area of R is 1/2. The average value of f is

$$\frac{1}{1/2}\int_0^1 \int_0^x e^{-x^2}\,dy\,dx = 2\int_0^1 e^{-x^2} y\Big|_{y=0}^{y=x} dx = 2\int_0^1 xe^{-x^2}\,dx = -e^{-x^2}\Big|_0^1 = -e^{-1}+1 = 1-\frac{1}{e}.$$

47. The area of the region is, by elementary geometry, $[4+\tfrac{1}{2}(2)(4)]$, or 8 sq units. Therefore, the required average value is

$$A = \tfrac{1}{8}\int_1^3 \int_0^{2x} \ln x\,dy\,dx = \tfrac{1}{8}\int_1^3 (\ln x)y\Big|_0^{2x} dx = \tfrac{1}{4}\int_1^3 x\ln x\,dx$$

$$= \tfrac{1}{4}(\tfrac{x^2}{4})(2\ln x - 1)\Big|_1^3 \quad \text{(Integrating by parts)}$$

$$= \tfrac{9}{16}(2\ln 3 - 1) - \tfrac{1}{16}(-1) = \tfrac{1}{8}(9\ln 3 - 4).$$

49. The average population density inside R is $\dfrac{43{,}329}{20} = 2166.45$ people/sq mile.

51. The average price is

$$\tfrac{1}{2}\int_0^1 \int_0^2 [200 - 10(x-\tfrac{1}{2})^2 - 15(y-1)^2]\,dy\,dx$$

$$= \tfrac{1}{2}\int_0^1 [200y - 10(x-\tfrac{1}{2})^2 y - 5(y-1)^3]\Big|_0^2 dx$$

$$= \tfrac{1}{2}\int_0^1 [400 - 20(x-\tfrac{1}{2})^2 - 5 - 5]\,dx$$

$$= \tfrac{1}{2}\int_0^1 [390 - 20(x-\tfrac{1}{2})^2]\,dx = \tfrac{1}{2}[390x - \tfrac{20}{3}(x-\tfrac{1}{2})^3]\Big|_0^1$$

$= \tfrac{1}{2}[390 - \tfrac{20}{3}(\tfrac{1}{8}) - \tfrac{20}{3}(\tfrac{1}{8})] \approx 194.17$ or approximately \$194 per sq ft.

CHAPTER 16 REVIEW EXERCISES, page 1146

1. $f(0,1) = 0; f(1,0) = 0, f(1,1) = \dfrac{1}{1+1} = \dfrac{1}{2}.$

   $f(0,0)$ does not exist because the point (0,0) does not lie in the domain of f.

3. $h(1,1,0) = 1 + 1 = 2,\ h(-1,1,1) = -e - 1 = -(e + 1),\ \ h(1,-1,1) = -e - 1 = -(e + 1).$

5. $D = \{(x, y) | y \neq -x\}$

7. The domain of f is the set of all ordered triplets (x,y,z) of real numbers such that $z \geq 0$ and $x \neq 1,\ y \neq 1$, and $z \neq 1$.

9. $z = y - x^2$

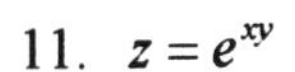

11. $z = e^{xy}$

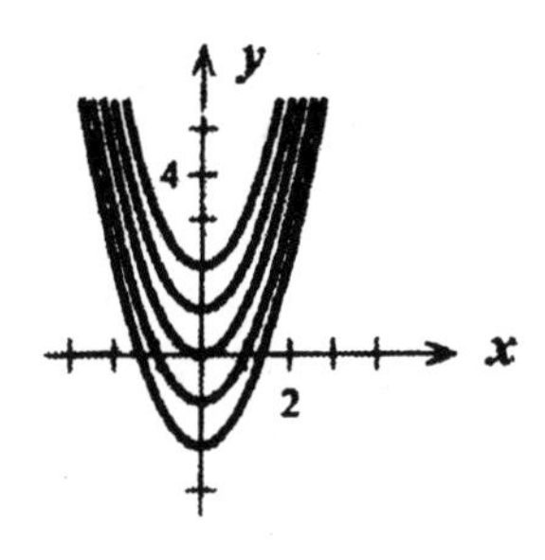

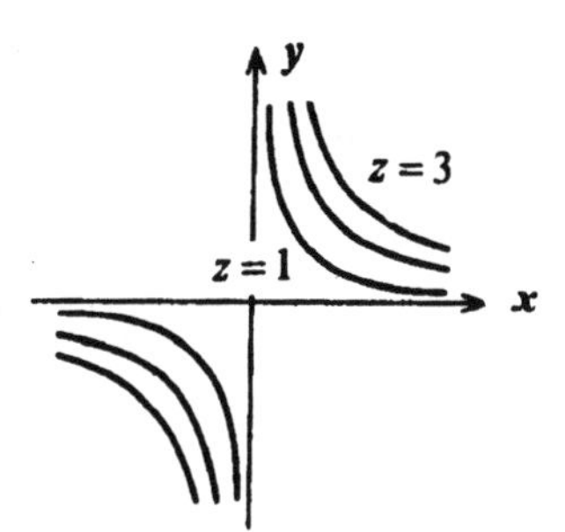

13. $f(x,y) = x\sqrt{y} + y\sqrt{x};\ f_x = \sqrt{y} + \dfrac{y}{2\sqrt{x}};\ f_y = \dfrac{x}{2\sqrt{y}} + \sqrt{x}$

15. $f(x,y) = \dfrac{x-y}{y+2x}.\ f_x = \dfrac{(y+2x)-(x-y)(2)}{(y+2x)^2} = \dfrac{3y}{(y+2x)^2}.$

    $f_y = \dfrac{(y+2x)(-1)-(x-y)}{(y+2x)^2} = \dfrac{-3x}{(y+2x)^2}.$

17. $h(x,y) = (2xy+3y^2)^5; h_x = 10y(2xy+3y^2)^4;\ h_y = 10(x+3y)(2xy+3y^2)^4.$

19. $f(x,y) = (x^2+y^2)e^{x^2+y^2};$

    $f_x = 2xe^{x^2+y^2} + (x^2+y^2)(2x)e^{x^2+y^2} = 2x(x^2+y^2+1)e^{x^2+y^2}.$

    $f_y = 2ye^{x^2+y^2} + (x^2+y^2)(2y)e^{x^2+y^2} = 2y(x^2+y^2+1)e^{x^2+y^2}.$

21. $f(x,y) = \ln\left(1 + \frac{x^2}{y^2}\right)$. $f_x = \frac{\frac{2x}{y^2}}{1 + \frac{x^2}{y^2}} = \frac{2x}{x^2 + y^2}$; $f_y = \frac{-\frac{2x^2}{y^3}}{1 + \frac{x^2}{y^2}} = -\frac{2x^2}{y(x^2 + y^2)}$.

23. $f(x,y) = x^4 + 2x^2y^2 - y^4; f_x = 4x^3 + 4xy^2, f_y = 4x^2y - 4y^3;$
$f_{xx} = 12x^2 + 4y^2,\ f_{xy} = 8xy = f_{yx},\ f_{yy} = 4x^2 - 12y^2.$

25. $g(x,y) = \frac{x}{x + y^2}$; $g_x = \frac{(x + y^2) - x}{(x + y^2)^2} = \frac{y^2}{(x + y^2)^2}$,
$g_y = \frac{-2xy}{(x + y^2)^2}$.

Therefore, $g_{xx} = -2y^2(x + y^2)^{-3} = -\frac{2y^2}{(x + y^2)^3}$,

$$g_{yy} = \frac{(x + y^2)^2(-2x) + 2xy(2)(x + y^2)2y}{(x + y^2)^4} = \frac{2x(x^2 + y^2)[-x - y^2 + 4y^2]}{(x + y^2)^4}$$
$$= \frac{2x(3y^2 - x)}{(x + y^2)^3}.$$

and
$$g_{xy} = \frac{(x + y^2)2y - y^2(2)(x + y^2)2y}{(x + y^2)^4} = \frac{2(x + y^2)[xy + y^3 - 2y^3]}{(x + y^2)^4}$$
$$= \frac{2y(x - y^2)}{(x + y^2)^3} = g_{yx}.$$

27. $h(s,t) = \ln\left(\frac{s}{t}\right)$. Write $h(s,t) = \ln s - \ln t$. Then

$h_s = \frac{1}{s},\ h_t = -\frac{1}{t}.$

Therefore, $h_{ss} = -\frac{1}{s^2},\ h_{st} = h_{ts} = 0,\ h_{tt} = \frac{1}{t^2}$.

29. $f(x,y) = 2x^2 + y^2 - 8x - 6y + 4$; To find the critical points of f, we solve the

system $\begin{cases} f_x = 4x - 8 = 0 \\ f_y = 2y - 6 = 0 \end{cases}$

obtaining $x = 2$ and $y = 3$. Therefore, the sole critical point of f is (2,3). Next,

$$f_{xx} = 4, f_{xy} = 0, f_{yy} = 2.$$

Therefore, $D = f_{xx}(2,3)f_{yy}(2,3) - f_{xy}(2,3)^2 = 8 > 0.$

Since $f_{xx}(2,3) > 0$, we see that $f(2,3) = -13$ is a relative minimum.

31. $f(x,y) = x^3 - 3xy + y^2$. We solve the system of equations $\begin{cases} f_x = 3x^2 - 3y = 0 \\ f_y = -3x + 2y = 0 \end{cases}$

obtaining $x^2 - y = 0$, or $y = x^2$. Then $-3x + 2x^2 = 0$, and $x(2x - 3) = 0$, and $x = 0$, or $x = 3/2$ and $y = 0$, or $y = 9/4$. Therefore, the critical points are (0,0) and $(\frac{3}{2}, \frac{9}{4})$.
Next, $f_{xx} = 6x, f_{xy} = -3$, and $f_{yy} = 2$ and $D(x,y) = 12x - 9 = 3(4x - 3)$. Therefore, $D(0,0) = -9$ so (0,0) is a saddle point.

$$D(\tfrac{3}{2}, \tfrac{9}{4}) = 3(6-3) = 9 > 0, \quad \text{and} \quad f_{xx}(\tfrac{3}{2}, \tfrac{9}{4}) > 0$$

and therefore,

$$f(\tfrac{3}{2}, \tfrac{9}{4}) = \tfrac{27}{8} - \tfrac{81}{8} + \tfrac{81}{16} = -\tfrac{27}{16}$$

is the relative minimum value.

33. $f(x,y) = f(x,y) = e^{2x^2+y^2}$. To find the critical points of f, we solve the system

$$\begin{cases} f_x = 4xe^{2x^2+y^2} = 0 \\ f_y = 2ye^{2x^2+y^2} = 0 \end{cases}$$

giving (0,0) as the only critical point of f. Next,

$$f_{xx} = 4(e^{2x^2+y^2} + 4x^2e^{2x^2+y^2}) = 4(1+4x^2)e^{2x^2+y^2}$$

$$f_{xy} = 8xye^{2x^2+y^2}$$

$$f_{yy} = 2(1+2y^2)e^{2x^2+y^2}.$$

Therefore,

$$D = f_{xx}(0,0)f_{yy}(0,0) - f_{xy}^2(0,0) = (4)(2) - 0 = 8 > 0$$

and so (0,0) gives a relative minimum of f since $f_{xx}(0,0) > 0$. The minimum value of f is $f(0,0) = e^0 = 1$.

35. We form the Lagrangian function $F(x,y,\lambda) = -3x^2 - y^2 + 2xy + \lambda(2x + y - 4)$.
Next, we solve the system

$$\begin{cases} F_x = 6x + 2y + 2\lambda = 0 \\ F_y = -2y + 2x + \lambda = 0. \\ F_\lambda = 2x + y - 4 = 0 \end{cases}$$

Multiplying the second equation by 2 and subtracting the resultant equation from the first equation yields $6y - 10x = 0$ so $y = 5x/3$. Substituting this value of y into the third equation of the system gives

$$2x + \tfrac{5}{3}x - 4 = 0.$$

So $x = \frac{12}{11}$ and consequently $y = \frac{20}{11}$. So $(\frac{12}{11}, \frac{20}{11})$ gives the maximum value for f subject to the given constraint.

37. The Lagrangian function is $F(x,y,\lambda) = 2x - 3y + 1 + \lambda(2x^2 + 3y^2 - 125)$. Next, we solve the system of equations

$$\begin{cases} F_x = 2 + 4\lambda x = 0 \\ F_y = -3 + 6\lambda y = 0 \\ F_\lambda = 2x^2 + 3y^2 - 125 = 0. \end{cases}$$

Solving the first equation for x gives $x = -1/2\lambda$. The second equation gives $y = 1/2\lambda$. substituting these values of x and y into the third equation gives

$$2\left(-\frac{1}{2\lambda}\right)^2 + 3\left(\frac{1}{2\lambda}\right)^2 - 125 = 0$$

$$\frac{1}{2\lambda^2} + \frac{3}{4\lambda^2} - 125 = 0$$

$$2 + 3 - 500\lambda^2 = 0, \text{ or } \lambda = \pm\frac{1}{10}.$$

Therefore, $x = \pm 5$ and $y = \pm 5$ and so the critical points of f are $(-5,5)$ and $(5,-5)$. Next, we compute

$$f(-5,5) = 2(-5) - 3(5) + 1 = -24.$$
$$f(5,-5) = 2(5) - 3(-5) + 1 = 26.$$

So f has a maximum value of 26 at $(5,-5)$ and a minimum value of -24 at $(-5,5)$.

39. $$\int_{-1}^{2}\int_{2}^{4}(3x - 2y)\,dx\,dy = \int_{-1}^{2} \tfrac{3}{2}x^2 - 2xy\Big|_{x=2}^{x=4} dy = \int_{-1}^{2}[(24 - 8y) - (6 - 4y)]\,dy$$
$$= \int_{-1}^{2}(18 - 4y)\,dy = (18y - 2y^2)\Big|_{-1}^{2} = (36 - 8) - (-18 - 2) = 48.$$

41. $$\int_{0}^{1}\int_{x^3}^{x^2} 2x^2 y\,dy\,dx = \int_{0}^{1} x^2 y^2\Big|_{y=x^3}^{y=x^2} dx = \int_{0}^{1} x^2(x^4 - x^6)\,dx$$
$$= \int_{0}^{1}(x^6 - x^8)\,dx = \tfrac{1}{7} - \tfrac{1}{9} = \tfrac{2}{63}.$$

43. $\int_0^2 \int_1^x (4x^2 + y^2)dy\,dx = \int_0^2 4x^2y + \frac{1}{3}y^3\Big|_{y=0}^{y=1} dx = \int_0^2 (4x^2 + \frac{1}{3})dx$

$$= (\tfrac{4}{3}x^3 + \tfrac{1}{3}x)\Big|_0^2 = \tfrac{32}{3} + \tfrac{2}{3} = \tfrac{34}{3}.$$

45. The area of R is

$$\int_0^2 \int_{x^2}^{2x} dy\,dx = \int_0^2 y\Big|_{y=x^2}^{y=2x} dx = \int_0^2 (2x - x^2)\,dx = (x^2 - \tfrac{1}{3}x^3)\Big|_0^2 = \tfrac{4}{3}.$$

Then

$$AV = \frac{1}{4/3}\int_0^2 \int_{x^2}^{2x} (xy + 1)dy\,dx = \frac{3}{4}\int_0^2 \frac{xy^2}{2} + y\Big|_{x^2}^{2x} dx$$

$$= \tfrac{3}{4}\int_0^2 (-\tfrac{1}{2}x^5 + 2x^3 - x^2 + 2x)\,dx = \tfrac{3}{4}(-\tfrac{1}{12}x^6 + \tfrac{1}{2}x^4 - \tfrac{1}{3}x^3 + x^2)\Big|_0^2$$

$$= \tfrac{3}{4}(-\tfrac{16}{3} + 8 - \tfrac{8}{3} + 4) = 3.$$

47. $f(p,q) = 900 - 9p - e^{0.4q}$; $g(p,q) = 20{,}000 - 3000q - 4p$.

We compute $\frac{\partial f}{\partial q} = -0.4e^{0.4q}$ and $\frac{\partial g}{\partial p} = -4$. Since $\frac{\partial f}{\partial q} < 0$ and $\frac{\partial g}{\partial p} < 0$ for all $p > 0$ and $q > 0$, we conclude that compact disc players and audio discs are complementary commodities.

49. Refer to the following diagram.

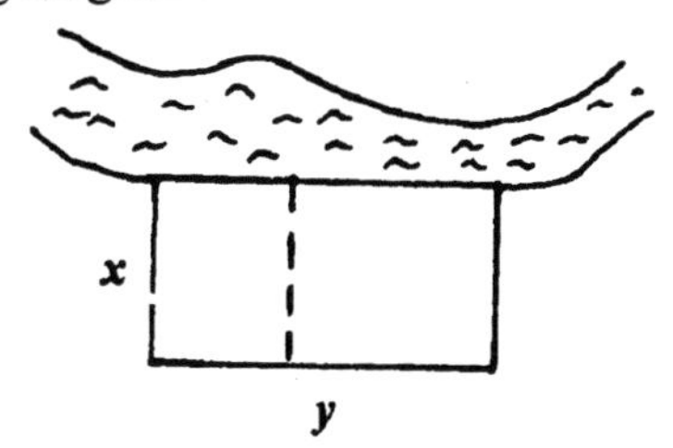

We want to minimize $C(x,y) = 3(2x) + 2(x) + 3y = 8x + 3y$ subject to $xy = 303{,}750$. The Lagrangian function is

$$F(x,y,\lambda) = 8x + 3y + \lambda(xy - 303{,}750).$$

Next, we solve the system

$$\begin{cases} F_x = 8 + \lambda y = 0 \\ F_y = 3 + \lambda x = 0 \\ F_\lambda = xy - 303{,}750 = 0 \end{cases}$$

Solving the first equation for y gives $y = -8/\lambda$. The second equation gives $x = -3/\lambda$.

Substituting this value into the third equation gives

$$\left(-\frac{3}{\lambda}\right)\left(-\frac{8}{\lambda}\right) = 303{,}750 \text{ or } \lambda^2 = \frac{24}{303{,}750} = \frac{4}{50{,}625},$$

or $\lambda = \pm\frac{2}{225}$. Therefore, $x = 337.5$ and $y = 900$ and so the required dimensions of the pasture are 337.5 yd by 900 yd.